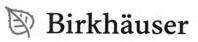

Geodynamics and Earth Tides Observations from Global to Micro Scale

Edited by
Carla Braitenberg
Giuliana Rossi
Geodynamics and Earth Tides Editor group

Janusz Bogusz
Luca Crescentini
David Crossley
Richard S. Gross
Kosuke Heki
Jacques Hinderer
Thomas Jahr
Bruno Meurers
Harald Schuh

Previously published in *Pure and Applied Geophysics* (PAGEOPH),
Volume 175, No. 5, 2018

 Birkhäuser

Editors
Carla Braitenberg
Mathematics and Geosciences
University of Trieste
Trieste, Italy

Giuliana Rossi
Centre for Seismological Research
Istituto Nazionale di Oceanografia
e di Geofisica Sperimentale –OGS
Trieste, Italy

Earth Tides Editor group

ISSN 2504-3625
Pageoph Topical Volumes
ISBN 978-3-319-96276-4

Library of Congress Control Number: 2018951585

This book is published under the imprint Birkhäuser, www.birkhauser-science.com by the registered company Springer Nature Switzerland AG.
The registered company address is: Gewerbestrasse 11, 6330 Cham, Switzerland

Contents

Booknote

Geodynamics and Earth Tides Editor group:

Janusz Bogusz, Military University of Technology, Faculty of Civil Engineering and Geodesy, Warsaw, Poland, janusz.bogusz@wat.edu.pl

Luca Crescentini, Department of Physics "E.R.Caianiello", University of Salerno, Salerno, Italy, luca.crescentini@sa.infn.it

David Crossley, Department of Earth and Atmospheric Sciences, Saint Louis University, Saint Louis, USA, crossley@eas.slu.edu

Richard S. Gross, Jet Propulsion Laboratory, California Institute of Technology, Pasadena, USA, richard.s.gross@jpl.nasa.gov

Kosuke Heki, Department of Natural History Sciences, Hokkaido University, Sapporo, Japan, heki@mail.sci.hokudai.ac.jp

Jacques Hinderer, University of Strasbourg, EOST/IPG, Strasbourg, France, jhinderer@unistra.fr

Thomas Jahr, Institute of Geosciences, Friedrich Schiller University Jena, Jena, Germany, thomas.jahr@uni-jena.de

Bruno Meurers, Department of Meteorology and Geophysics, University of Vienna, Vienna, Austria, bruno.meurers@univie.ac.at

Harald Schuh, GFZ German Research Centre for Geosciences, Potsdam, Germany, and Institute of Geodesy and Geoinformation Science, Technische Universität Berlin, Berlin, Germany, schuh@gfz-potsdam.de

---

The original version of this book was revised. The correction is available at https://doi.org/10.1007/978-3-319-96277-1_23

Pure Appl. Geophys. 175 (2018), 1595–1597
© 2018 Springer International Publishing AG, part of Springer Nature
https://doi.org/10.1007/s00024-018-1875-0

Geodynamics and Earth Tides Observations from Global to Micro Scale: Introduction

C. Braitenberg,[1] G. Rossi,[2] J. Bogusz,[3] L. Crescentini,[4] D. Crossley,[5] R. Gross,[6] K. Heki,[7] J. Hinderer,[8] T. Jahr,[9] B. Meurers,[10] and H. Schuh[11,12]

The volume collects papers submitted to Pure and Applied Geophysics following a call on the topic "Geodynamics and Earth Tides". Partly, the authors had participated in the 18th Geodynamics and Earth Tides Symposium held in Trieste, Italy, in June 2016. The Earth tides constitute the leading thread through the book, since instrumentation sensitive enough to observe them, also records a broad spectrum of signals generated by Earth dynamic processes.

The topics discussed belong to space geodesy, terrestrial geodesy, seismology, tectonophysics, hydrology, and geodynamics, demonstrating the interdisciplinarity needed for understanding the observations.

The deepest Earth is studied with the core resonance at diurnal periods, which alters expected amplitude and phase of the diurnal frequencies of the observed Earth tide. The phenomenon is described from observations at surface, allowing conclusions on the core to be taken (Agnew 2017; Bán et al. 2017). The Earth globe yields to the tidal forces through deformation, from which Earth rheology is retrieved (Varga et al. 2017), and which could be a cause of triggering earthquakes (Varga and Grafarend 2017).

The ocean tides produce a change of ocean height driven by the same frequencies of the Earth tides. The changing mass produces a variable load on the crust, which in turn responds isostatically by flexure. Observation of the oceanic load tides with sophisticated geodetic instruments (Ruotsalainen 2017; Virtanen and Raja-Halli 2017) allows to improve the ocean tide models (Amoruso et al. 2017). Even river estuaries can be subject to level variations that must be modeled to correct their effect on these precise instruments (Oreiro et al. 2017). The very sensitive instrumentation as tilt and strain meters and continuously recording gravity meters sense hydrologic flows. The modeling of the induced signals is a complex topic of its own (Weise and Jahr 2017), with applications in hydrology and induced seismicity (Grillo et al. 2018; Vinogradov et al. 2017). Continuous gravity has been made at the sea floor for monitoring a gas field, but many other signals must be taken into account (Rosat et al. 2017). The terrestrial measurements of the time-variable gravity field are analyzed to match the observed with the theoretical tidal gravity field (Yu et al. 2018), with surprising results for co-located instruments (Virtanen and Raja-Halli 2017). The cryogenic gravimeters have highest precision, and require particular attention for checking scale factors and instrumental drift (Crossley et al. 2018).

[1] Department of Mathematics and Geosciences, University of Trieste, Trieste, Italy. E-mail: berg@units.it
[2] Centro di Ricerche Sismologiche, Istituto Nazionale di Oceanografia e di Geofisica Sperimentale – OGS, Sgonico (Trieste), Italy. E-mail: grossi@inogs.it
[3] Faculty of Civil Engineering and Geodesy, Military University of Technology, Warsaw, Poland. E-mail: janusz.bogusz@wat.edu.pl
[4] Department of Physics "E.R.Caianiello", University of Salerno, Salerno, Italy. E-mail: luca.crescentini@sa.infn.it
[5] Department of Earth and Atmospheric Sciences, Saint Louis University, St. Louis, USA. E-mail: crossley@eas.slu.edu
[6] Jet Propulsion Laboratory, California Institute of Technology, Pasadena, USA. E-mail: richard.s.gross@jpl.nasa.gov
[7] Department of Natural History Sciences, Hokkaido University, Sapporo, Japan. E-mail: heki@mail.sci.hokudai.ac.jp
[8] University of Strasbourg, EOST/IPG, Strasbourg, France. E-mail: jhinderer@unistra.fr
[9] Institute of Geosciences, Friedrich Schiller University Jena, Jena, Germany. E-mail: thomas.jahr@uni-jena.de
[10] Department of Meteorology and Geophysics, University of Vienna, Vienna, Austria. E-mail: bruno.meurers@univie.ac.at
[11] GFZ German Research Centre for Geosciences, Potsdam, Germany. E-mail: harald.schuh@gfz-potsdam.de
[12] Institute of Geodesy and Geoinformation Science, Technische Universität Berlin, Berlin, Germany.

Recent improvements in the development of VLBI (Very Long Baseline Interferometry) and other space geodetic techniques like the global navigation satellite systems (GNSS) require very precise a priori information of short-period Earth rotation variations. Within the work of Karbon et al. (2018), a new model for the short-period ocean tidal variations in Earth rotation is developed with up to 251 partial constituents, based on modern ocean tides models and a reexamined theoretical description. An alternative to the conventional models is found; however, no significant improvement in the geodetic results can be reached.

Space geodetic observations such as GNSS have the drawback of being less sensitive to deformation at local scale, compared to the sophisticated high precision gravity and deformation measurements, but have the advantage of easier installation, reaching dense and global coverage. The studies are concerned with identifying hydrologic and temperature effects, developing modern spatio-temporal analysis methods (Gruszczynski et al. 2018; Gruszczynska et al. 2018; Klos et al. 2017), or identifying non-hydrologic common GNSS transient signals (Rossi et al. 2017).

The deformation at an active volcano (Elbruz, Caucasus) (Milyukov et al. 2017) and for a big earthquake (Gorkha, Nepal) (Morsut et al. 2017), demonstrates the importance of geodetic monitoring in hazard assessment.

In parallel to the Topical Volume in Pure and Applied Geophysics, a special volume in the open access journal Geodesy and Geodynamics was arranged, in which all abstracts of the meeting are published, and some selected manuscripts. A review of the meeting (Braitenberg 2018) includes a description of today existing geodetic stations worldwide. Instrumental and software aspects of continuous measurement of gravity with the superconducting (Meurers 2017) and automated Burris gravity meter (Jentzsch et al. 2018; Schulz 2017) are discussed. Use of Kalman filter in GNSS network monitoring is demonstrated by (Shults and Annenkov 2017). The detection of pore pressure changes induced by hydrologic pumping is recorded with tilt and strain observations at the geodetic station Moxa (Germany) (Jahr 2017). Hydrology in karstic areas is dominated by hydrologic flows in a macroscopic channel system, which in the classical Karst (Italy–Slovenia) gives rise to deformations during floods and an impressive river emerging at the foot of the Karst (Braitenberg et al. 2017). Geodynamic thermomechanical modeling of the subduction of the central Andes is presented by (Salomon 2018), while (Hazrati-Kashi et al. 2018) study inversion methods to define slow slip during the preparing phase of a large-scale earthquake at subduction zones.

This volume provides a representative cross-section on the recording, analysis and interpretation of the spectrum of signals generated by Earth dynamic processes. The material is of interest to scientists and students interested in the 4D Earth and keen to learn the latest achievements.

Acknowledgements

The 18th Geodynamics and Earth Tides Symposium held in Trieste, Italy, in June 2016, was scientifically supported by the IAG: Commission 3, the IAG Subcommission 3.1 and International Geodynamics and Earth Tide Service. The University of Trieste and the sponsors of the Symposium, namely the OGS (Istituto Nazionale di Oceanografia e di Geofisica Sperimentale), the Dipartimento di Fisica E. Caianiello, University of Salerno, the Department of Mathematics and Geosciences of the University of Trieste, Leica Geosystems S.P.A., International Association of Geodesy (3 IAG Travel Awards for young scientists), the European Geosciences Union (support to 8 young scientists), the Rector Maurizio Fermeglia of the University of Trieste and the President Maria Cristina Pedicchio of OGS, are gratefully acknowledged for supporting this event. We thank the organizational Secretariat "The Office" (http://www.theoffice.it/) for perfect assistance in all phases of the Symposium.

REFERENCES

Agnew, D. C. (2017). An improbable observation of the diurnal core resonance. *Pure Applied Geophysics*. https://doi.org/10.1007/s00024-017-1522-1.

Amoruso, A., Crescentini, L., Bayo, A., Fernández Royo, S., & Luongo, A. (2017). Two high-sensitivity laser strainmeters installed in the Canfranc Underground Laboratory (Spain): Instrument features from 100 to 0.001 mHz. *Pure and Applied Geophysics*. https://doi.org/10.1007/s00024-017-1553-7.

Bán, D., Mentes, G., Kis, M., & Koppán, A. (2017). Observation of the earth liquid core resonance by extensometers. *Pure and Applied Geophysics*. https://doi.org/10.1007/s00024-017-1724-6.

Braitenberg, C. (2018). The deforming and rotating Earth–a review of the 18th International Symposium on Geodynamics and Earth Tide, Trieste 2016. *Geodesy Geodyn*. https://doi.org/10.1016/j.geog.2018.03.003.

Braitenberg, C., Pivetta, T., Rossi, G., Ventura, P., & Betic, A. (2017). Karst caves and hydrology between geodesy and archeology: Field trip notes. *Geodesy Geodyn*. https://doi.org/10.1016/j.geog.2017.06.004.

Crossley, D., Calvo, M., Rosat, S., & Hinderer, J. (2018). More thoughts on AG-SG comparisons and SG scale factor determinations. *Pure and Applied Geophysics*. https://doi.org/10.1007/s00024-018-1834-9.

Grillo, B., Braitenberg, C., Nagy, I., Devoti, R., Zuliani, D., & Fabris, P. (2018). Cansiglio Karst Plateau: 10 years of geodetic–hydrological observations in Seismically Active Northeast Italy. *Pure and Applied Geophysics*. https://doi.org/10.1007/s00024-018-1860-7.

Gruszczynska, M., Rosat, S., Klos, A., Gruszczynski, M., & Bogusz, J. (2018). Multichannel singular spectrum analysis in the estimates of common environmental effects affecting GPS observations. *Pure and Applied Geophysics*. https://doi.org/10.1007/s00024-018-1814-0.

Gruszczynski, M., Klos, A., & Bogusz, J. (2018). A filtering of incomplete GNSS position time series with probabilistic Principal Component Analysis. *Pure and Applied Geophysics*. https://doi.org/10.1007/s00024-018-1856-3.

Hazrati-Kashi, M., Mirzaei, N., & Moshiri, B. (2018). An image segmentation based algorithm for imaging of slow slip earthquakes. *Geodesy and Geodynamics*. https://doi.org/10.1016/j.geog.2018.04.001.

Jahr, T. (2017). Non-tidal tilt and strain signals recorded at the Geodynamic Observatory Moxa, Thuringia/Germany. *Geodesy and Geodynamics*. https://doi.org/10.1016/j.geog.2017.03.015.

Jentzsch, G., Schulz, R., & Weise, A. (2018). Automated Burris gravity meter for single and continuous observation. *Geodesy and Geodynamics*. https://doi.org/10.1016/j.geog.2017.09.007.

Karbon, M., Balidakis, K., Belda, S., Nilsson, T., Hagedoorn, J., & Schuh, H. (2018). Long-term evaluation of ocean tidal variation models of polar motion and UT1. *Pure Appl Geophys*. https://doi.org/10.1007/s00024-018-1866-1.

Klos, A., Gruszczynska, M., Bos, M. S., Boy, J.-P., & Bogusz, J. (2017). Estimates of vertical velocity errors for IGS ITRF2014 stations by applying the improved singular spectrum analysis method and environmental loading models. *Pure and Applied Geophysics*. https://doi.org/10.1007/s00024-017-1494-1.

Meurers, B. (2017). Scintrex CG5 used for superconducting gravimeter calibration. *Geodesy Geodyn*. https://doi.org/10.1016/j.geog.2017.02.009.

Milyukov, V., Rogozhin, E., Gorbatikov, A., Mironov, A., Myasnikov, A., & Stepanova, M. (2017). Contemporary State of the Elbrus Volcanic Center (The Northern Caucasus). *Pure and Applied Geophysics*. https://doi.org/10.1007/s00024-017-1595-x.

Morsut, F., Pivetta, T., Braitenberg, C., & Poretti, G. (2017). Strain accumulation and release of the Gorkha, Nepal, earthquake (M w 7.8, 25 April 2015). *Pure and Applied Geophysics*. https://doi.org/10.1007/s00024-017-1639-2.

Oreiro, F. A., Wziontek, H., Fiore, M. E., D'Onofrio, E. E., & Brunini, C. (2017). Non-tidal ocean loading correction for the Argentinean-German geodetic observatory using an empirical model of storm surge for the Rio de la Plata. *Pure and Applied Geophysics*. https://doi.org/10.1007/s00024-017-1651-6.

Rosat, S., Escot, B., Hinderer, J., & Boy, J.-P. (2017). Analyses of a 426-day record of seafloor gravity and pressure time series in the North Sea. *Pure and Applied Geophysics*. https://doi.org/10.1007/s00024-017-1554-6.

Rossi, G., Fabris, P., & Zuliani, D. (2017). Overpressure and fluid diffusion causing non-hydrological transient GNSS displacements. *Pure and Applied Geophysics*. https://doi.org/10.1007/s00024-017-1712-x.

Ruotsalainen, H. (2017). Interferometric water level tilt meter development in Finland and comparison with combined earth tide and ocean loading models. *Pure and Applied Geophysics*. https://doi.org/10.1007/s00024-017-1562-6.

Salomon, C. (2018). Finite element modelling of the geodynamic processes of the Central Andes subduction zone: A reference model. *Geodesy and Geodynamics*. https://doi.org/10.1016/j.geog.2017.11.007.

Schulz, H. R. (2017). A new PC control software for ZLS-Burris gravity meters. *Geodesy and Geodynamics*. https://doi.org/10.1016/j.geog.2017.09.002.

Shults, R., & Annenkov, A. (2017). Investigation of the different weight models in Kalman filter: A case study of GNSS monitoring results. *Geodesy and Geodynamics*. https://doi.org/10.1016/j.geog.2017.09.003.

Varga, P., & Grafarend, E. (2017). Influence of tidal forces on the triggering of seismic events. *Pure and Applied Geophysics*. https://doi.org/10.1007/s00024-017-1563-5.

Varga, P., Grafarend, E., & Engels, J. (2017). Relation of different type Love-Shida numbers determined with the use of time-varying incremental gravitational potential. *Pure and Applied Geophysics*. https://doi.org/10.1007/s00024-017-1532-z.

Vinogradov, E., Gorbunova, E., Besedina, A., & Kabychenko, N. (2017). Earth tide analysis specifics in case of unstable aquifer regime. *Pure and Applied Geophysics*. https://doi.org/10.1007/s00024-017-1585-z.

Virtanen, H., & Raja-Halli, A. (2017). Parallel observations with three superconducting gravity sensors during 2014–2015 at Metsähovi Geodetic Research Station, Finland. *Pure and Applied Geophysics*. https://doi.org/10.1007/s00024-017-1719-3.

Weise, A., & Jahr, T. (2017). The improved hydrological gravity model for Moxa observatory, Germany. *Pure and Applied Geophysics*. https://doi.org/10.1007/s00024-017-1546-6.

Yu, H., Guo, J., Kong, Q., & Chen, X. (2018). Gravity tides extracted from relative gravimeter data by combining empirical mode decomposition and independent component analysis. *Pure Appl Geophys*. https://doi.org/10.1007/s00024-018-1864-3.

(Published online May 8, 2018)

Pure Appl. Geophys. 175 (2018), 1599–1609
© 2017 Springer International Publishing
https://doi.org/10.1007/s00024-017-1522-1

An Improbable Observation of the Diurnal Core Resonance

DUNCAN CARR AGNEW[1]

Abstract—The resonance associated with the ellipticity of the core-mantle boundary is usually measured with observations of either the Earth's nutations, or of tidal gravity, strain, or tilt. But, improbably, it can also be seen in a dataset collected and processed with older and simpler technologies: the harmonic constants for the ocean tides. One effect of the resonance is to decrease the ratio of the amplitude of the P_1 constituent to the amplitude of the K_1 constituent to 0.96 of the ratio in the equilibrium tidal potential. The compilation of ocean-tide harmonic constants prepared by the International Hydrographic Bureau between 1930 and 1980 shows considerable scatter in this ratio; however, if problematic stations and regions are removed, this dataset clearly shows a decreased ratio. While these data apply only a weak constraint to the frequency of the resonance, they also show that the effect could have been observed long before it actually was.

Key words: Nearly diurnal free wobble, ocean tides.

1. Introduction

Using nutations and tides to determine properties of the Earth goes back to the nineteenth century (Brush 1979; Kushner 1990), but this approach was overtaken by seismological methods early in the twentieth. Even the early researches identified a mode of oscillation in which the interior fluid, acting as a solid, precessed because of pressure forces on the ellipsoidal solid–fluid boundary. The first models of this with realistic properties for the core and mantle were those of Jeffreys and Vicente (1957a, b) and Molodensky (1961), both of whom showed that this mode of oscillation, now called either the free core nutation (FCN) or nearly diurnal free wobble (NDFW) would result in a resonance in the response

of the Earth to tidal forces that produce a net torque, as the diurnal tides do. The relevant theory was further developed using both normal-mode theory for the rotational modes (Wahr 1981a, b) and semi-analytic models (Sasao et al 1980; Dehant et al 1993; Mathews et al 1995; Mathews 2001; Dehant and Mathews 2015); these have included other kinds of core–mantle coupling as well as improved Earth models.

The first observation of this resonance was claimed by Melchior (1966), who used tilt data; subsequent measurements of it using Earth tides have focused more on gravity (Cummins and Wahr 1993; Sato et al 1994; Ducarme et al 2007; Rosat et al 2009) and strain (Polzer et al 1996; Amoruso et al 2012). The strongest observational constraints come from observations of the Earth's nutations using high-precision VLBI astrometry; these led Gwinn et al (1986) to the finding that the frequency of the resonance differed from that expected for a core–mantle boundary with the ellipticity for a hydrostatic Earth model. Subsequent measurements (Herring et al 2002; Koot et al 2008; Rosat and Lambert 2009; Chao and Hsieh 2015) have come in tandem with improvements in models: paradoxically, we learn about the core–mantle boundary by measurements of very distant quasars.

Wahr and Sasao (1981) showed that the core resonance should also be present in ocean-tide data, a more complicated case than other tidal measurements because the physics has to include not just the Earth's response to external forcing but also to the loading from the tide itself: the loading, just like the external forcing, is affected by the NDFW resonance. Wahr and Sasao (1981) used gravity-tide data to argue that this effect was present, but the only systematic search for it has been by Ooe and Tamura (1985), using tide-gauge data from Japan. Ray (2017) has recently

---

To the memory of John Wahr.

[1] Institute of Geophysics and Planetary Physics, Scripps Institution of Oceanography, University of California San Diego, La Jolla, CA, USA. E-mail: dagnew@ucsd.edu

pointed out that the resonance in ocean tides needs to be included in the "inference" of tidal constituents, notably of P_1 from K_1; as evidence for this, a high-quality set of open-ocean pressure data (Ray 2013) shows the resonance effect quite clearly.

Given this, we can ask if the resonance could also be seen in the tidal data collected along coastlines. Somewhat improbably, the answer is yes, although the precision is too low to contribute new information about the NDFW. It is startling to realize that this effect, usually sought using VLBI and supercon-ducting gravimeters, can be observed using a sensor no more complicated than a float, recording with a pencil writing on rolls of paper, and with data pro-cessing done mostly by human computers with (sometimes) mechanical calculators.

Section 2 of this paper reviews the theory for the resonance in ocean-tide data, while Sect. 3 describes the sources of ocean-tide data, and Sect. 4 describes the results, which rely on there being large amounts of data to average.

2. Effect of the NDFW on Ocean Tides

As Wahr and Sasao (1981) point out, the full equation for the ocean-tide resonance is complicated by the need to include loading. The resonance effect is confined to the spherical harmonic of degree two and order one, but this will perturb the total tide, couple into all spherical harmonics, and so needs to be computed from the tidal model itself, just as with the self-attraction and loading effects that have to be included in the tidal equations (Ray 1998). Wahr and Sasao (1981) give the full solution for an equilibrium ocean tide for an ocean-covered Earth; for a more realistic situation they extend this to the following approximate expression, giving the response $R(f)$ rel-ative to a reference frequency f_0:

$$R(f,f_0) = \frac{\gamma(f)}{\gamma(f_0)} \frac{1}{1 + K(\gamma'(f) - \gamma'(f_0))/\gamma'(f_0)} \quad (1)$$

where γ and γ' are the Love-number combinations relating the equilibrium tide to the tidal potential and to a tidal load, in both cases for a spherical harmonic of degree two and order one: $\gamma = 1 + k - h$ and

$\gamma' = 1 + k' - h'$, with h, k, h', and k' being the Love numbers and load Love numbers. The constant K is

$$K = \frac{3\rho t_{21}(f_0)}{5\rho_E g\phi(f_0)} \quad (2)$$

where ρ and ρ_E are the mean density of ocean water (1.035) and the solid Earth (5.51), respectively; $g\phi(f_0)$ is the height of the driving potential for the constituent at f_0; and $t_{21}(f_0)$ is the (2,1) spherical harmonic coefficient for a tidal model at frequency f_0, which is complex-valued. The approximation consists of assuming that t_{21} is the same for another tide at frequency f as it is for the tide at f_0.

The expression for the frequency-dependent γ can be derived from those for the Love numbers h and k in Pétit and Luzum (2010). These expressions include the resonances for the Chandler wobble and the Inner Core Nutation; ignoring the latter and treating the former as a constant term, we obtain

$$\gamma(f) = 0.7021 - \frac{8.98 \times 10^{-5}}{f - f_N} \quad (3)$$

where f_N is the frequency of the NDFW, and both it and f are expressed in cycles per solar day (cpd; hereafter "day" means solar day unless otherwise specified). The NDFW frequency in an earth-fixed frame is related to its period in a space-fixed frame, T_N, by $T_N = (f_N - 1.0027379)^{-1}$; for a hydrostatic earth $T_N = 460$ days, but both tidal and nutation data give values of about 430 days.

A similar expression applies to the combination for load Love numbers:

$$\gamma'(f) = 1.6877 - \frac{1.39 \times 10^{-4}}{f - f_N} \quad (4)$$

Spherical-harmonic expansions of modern tide mod-els give values of $gt_{21}(f)/\phi(f)$ which are all close to a complex number with amplitude 0.31 and phase 132°; putting this into Eq. (1) shows that the ampli-tude of K is 0.034. Using the K_1 frequency for f_0 and the P_1 for f, the total loading term in Eq. (1) is approximately 1.4×10^{-3}. We, therefore, ignore it and take $R(f,f_0) = \gamma(f)/\gamma(f_0)$. Figure 1 shows R as a function of f_N, to make the point that, over the range of plausible values for f_N, the variation in R is small; for observed values of T_N it is 0.964.

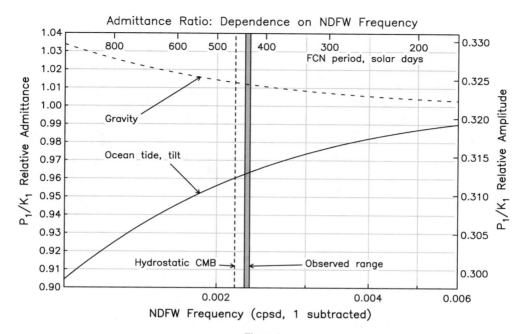

Figure 1

The *solid line* shows the relative size of P_1 and K_1 in the ocean tide (or tilt) as a function of the core resonance frequency f_C, in cycles per sidereal day (cpsd). The *dashed line* shows the same for tidal gravity. The *left axis* shows relative size as admittance (normalizing by the relative amplitude of the constituents in the tidal potential), and the *right axis* shows this as the ratio of measured constituent amplitudes

3. Ocean-Tide Data

My analysis uses the amplitudes of the P_1 and K_1 harmonic constituents for ocean tides, measured by coastal tide gauges. I first give some history to help readers understand how, why, and when this information became available—and, more recently, ceased to be so. A key point is that, interesting as the ocean tides are as a scientific problem, and important as sea-level change is, most tidal measurements have been (and are) collected and processed to predict future tides for the benefit of maritime trade, and to define the (legally important) boundary between land and sea. So there has been much more data collected than there would have been for scientific research. But the commercial value of predicted tides means that, like other economically (and militarily) valuable information, tidal parameters may not be freely available.

This is well illustrated by tide prediction in the early nineteenth century. Very simple methods (Cartwright 1999) were widely available, but better methods of tide prediction for active ports such as Liverpool and London were kept secret by those who profited by supplying them to almanac publishers (Rossiter 1972; Woodworth 2002). This changed (Hughes 2006; Reidy 2008) in the 1830s when Lubbock and Dessiou developed non-harmonic methods. This decade also saw the invention of self-recording tide gauges, which provided the first continuous record of the tides, as opposed to times and heights of high and low water. Such continuous records were in turn crucial to two new techniques: the harmonic method of tidal analysis developed by Thomson, Roberts, and Darwin between 1867 and 1883 (Darwin 1883) and the tide predicting machine invented by Thomson and Roberts in the 1870s (Anonymous 1926b; Cartwright 1999; Woodworth 2016). Together these could predict the tides more accurately for more diverse tidal regimes than the nonharmonic method, using much less data. The harmonic method also meant that the tidal behavior of any location could be described by a relatively small collection of numbers, namely the harmonic constants. Publishing these numbers made it relatively easy to predict the tides throughout the world, with great benefit to mariners (Hughes and Wall 2007).

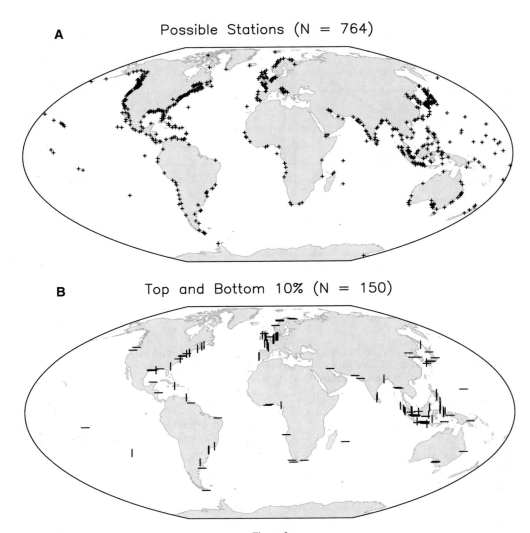

A Possible Stations (N = 764)

B Top and Bottom 10% (N = 150)

Figure 2

a (*Top*) shows distribution of "winnowed" stations from the IHO Tidal Constituent Data Bank. These have an analysis interval longer than 354 days, K_1 amplitude of at least 5 cm, diurnal nonlinear tides (MP or SO) no more than 20% of the P_1 tide, and a phase difference between the P_1 and K_1 constituents less than 15°. **b** (*Bottom*) shows the stations from this dataset for which the P_1/K_1 amplitude ratio is in the bottom or top 10% of the distribution of observed ratios (*vertical lines for top, horizontal for bottom*). Equal-area projection (McBryde–Thomas flat-polar quartic)

Thus, it is unsurprising that the International Hydrographic Bureau (IHB) founded in 1921 to promote standardization and the exchange of hydrographic information (Bermejo 1997) included tidal matters—perhaps also because one of its Directors, Pfaff, worked in this area (Pfaff 1926, 1927). In 1924 the IHB found that many hydrographic offices did not use harmonic analysis and prediction. To encourage such use, in 1926 the IHB published tables (Anonymous 1926a) for harmonic-constant calculation, including a list giving constants for over a thousand locations. The 1926 International Hydrographic Conference resolved that the IHB should collect and publish even more constants (Ritchie 1980), bringing them up to date annually. This was done by creating IHB Special Publication 26, in which groups of constants were published as separate fascicles, produced as different sources contributed them. The first fascicle was published in 1931, and by 1940 constants were available for 1464 locations, along with (usually) the time and duration of data analyzed, and information on the datums used.

Publication and distribution of harmonic constants in fascicles continued until 1968, when the IHB stopped the practice, though it continued to compile information, available as individual photocopies on request (Anonymous 1976). To make this information more available, the 1972 International Hydrographic Conference resolved that it should be put into machine-readable form. This was done by the Canadian Hydrographic Service, which created the International Hydrographic Organization (IHO) Tidal Data Bank. (The IHB had been renamed as the IHO in 1970 (Bermejo 1997), though the IHB designation remained for the central office in Monaco). This data bank was updated as additional harmonic constants were added, and a magnetic tape of it was distributed to each national office belonging to the IHO; this tape could be purchased, though at a high price.

Unfortunately, it soon became apparent that the data could be used in ways that were both unexpected and, for some, unwelcome. With the advent of personal computers, a market developed for tide-prediction software, the developers of which relied on the IHO Data Bank for the harmonic-constant values—though these were regarded by the IHO as belonging to the agencies contributing them. For a time the managers of the Data Bank separated values that could be made generally available from those for which this was not allowed without consulting the original contributor, but this became a significant burden, while at the same time new harmonic constants were rarely contributed. In 2000 the IHO, therefore, requested its members to vote on continuing to operate the Data Bank (Anonymous 2000); the outcome was that it was terminated, making the harmonic constants unavailable except to those who already had copies. In addition, legal threats by some of the contributing entities discouraged commercial use of this information by third parties.

The harmonic constants in the IHO Data Bank are almost all for coastal sites, especially those in locations of maritime activity. At many locations the tidal observations were made for a month or less, sometimes using continuously recording gauges and sometimes using tide poles. The number of tidal constituents varies from the four largest tides to up to

50 in areas with large nonlinear effects. And, given that many of the data come from sites not directly open to the ocean, such nonlinear behavior is not uncommon. The data were collected at different times from the 1850s to the 1980s; a wide range of methods (most using hand computation, and few using least squares) were used to estimate the harmonic constants from the actual sea-level measurements. All of this means that the quality of these constants is much more variable, and much less representative of the global tides, than those obtained by Ponchaut et al (2001), not to mention those estimated from open-ocean pressure data by Ray (2013, 2017).

Nevertheless, I shall show that the NDFW resonance can be clearly seen in this collection of values, as a ratio of P_1 to K_1 amplitudes that is much closer to that described in Sect. 2 than the ratio in the driving potential. This can only be done by combining some data selection with substantial averaging: it is not that many sites show the appropriate ratio, but that their average is close to it.

4. Results for Different Data Sets

To get the best results from the rather heterogeneous dataset just described, it is important to remove, or winnow out, locations for which the harmonic constants might be questionable. To some extent this can be done on an a priori basis by looking at other criteria than the actual ratio of P_1 to K_1—though an initial look at this ratio was useful in flagging possible gross errors in the database. Appendix A describes these relatively few problems in more detail.

It is not possible to determine the P_1 and K_1 tides reliably if the record length is less than six months. Looking at the distribution of durations in the database, it was apparent that there were not many stations with durations longer than 0.5 year and shorter than 0.96 year (355 days, a popular length for classical harmonic analysis). Of the 4090 stations available, 2974 had a duration that was unknown or less than 0.96 year, leaving 1116 to be considered further. Of these, 26 lacked P_1 or K_1 amplitudes, and 111 had K_1 amplitudes less than 0.05 m. This amplitude cutoff is the same used by Ray (2017), though it might be reasonable to make it larger in view of the greater noise level for many coastal locations.

As noted above, many stations in the IHO Data Bank are in locations more likely to be affected by nonlinearity. To ameliorate this problem, no station was included if the P_1 amplitude was less than five times that of the larger of two nonlinear tides in the diurnal band, MP_1 (frequency 0.9350 cpd) and SO_1 (frequency 1.0705 cpd): these are well enough separated from other constituents to be reliably measured. This criterion removed 133 stations.

Clearly, if there is a local resonance in the ocean with a frequency close to those of the P_1 and K_1 tides, the ratio between them could be more affected by this than by the difference in the driving potential. Such a resonance might also be expected to create a difference in phases; for this reason any station for which the P_1 and K_1 phases differed by more than $15°$ was eliminated. There were 80 such stations.

This winnowing left 766 stations; as described in Appendix A one was eliminated as having an outlying value for the P_1/K_1 ratio, and not agreeing with nearby ocean models derived from satellite ranging. Of the remaining 765, 30% had been put into the data bank before 1941, and another 30% between 1949 and 1968, suggesting that close to 60% were derived using hand computation. The oldest sea-level data used are from Cat Island, Mississippi, which were collected in 1848 and 1849; the oldest harmonic constants are probably for Hilbre Island, near Liverpool, computed by Roberts in the 1870s (Baird and Darwin 1885). The newest data and constants come from the late 1970s.

For each station, I computed the P_1/K_1 ratio and converted this to relative admittance by multiplying by 0.36883/0.12205, the ratio of K_1 to P_1 given in the tables of Cartwright and Edden (1973) for the mid-twentieth century, the epoch most appropriate to when most of these amplitudes were estimated.

The resulting 765 values of R have a mean value of 0.960 and a median of 0.948, with the extremes being 0.567 and 2.418. Figure 3 shows the distribution of ratios; this is presented as a cumulative distribution function. Such a presentation, by avoiding the binning needed for a histogram, can show features not otherwise visible: in this case, an excess of values with P_1/K_1 equal to one-third. The original values are in centimeters but only given to one decimal place; a Monte Carlo simulation using resampled values of K_1, a range of ratios, and rounding the P_1 amplitude to one decimal, shows the same effect at about the same level: so these are simply the effect of computation, rather than evidence for P_1 having been inferred from the potential values (which gives a ratio close to this).

Figure 2a shows the distribution of the winnowed station set. This shows, what is common in all sea-level measurements, a high density around Japan, North America, and northwest Europe—along with certain gaps caused by geopolitical issues, such as on the coasts of the former Soviet Union.

Figure 2b shows the locations of the stations with the lowest 10% and highest 10% of values for R, the question being whether there are particular regions that might be pruned from the dataset to reduce outliers. Two regions stand out for a relatively high proportion of extreme values for R and a plausible reason why this might be locally distorted. The first is along the Belgian, Dutch, and German coasts of the North Sea, where almost all the stations show high values of R, some extremely high (Antwerp, on the Scheldt, has $R = 2.3$). The other region is Indonesia, where complex bathymetry and coastlines create a great many short-wavelength variations in the tides. Also, many of harmonic constants in this region come from data collected in the 1890s—and, as reference to the original printed fascicles has shown, quite often with tide poles rather than tide gauges.

Figure 4 shows these two regions with the areas within which stations were pruned from the list: it is important to realize that all stations in these areas were removed, not just those with extreme values. This leaves 670 values of R, with a mean value of 0.943, and a median value of 0.946. The extremes are 0.618 (Ogusi, in Kyushu, Japan) and 1.295 (Bristol, Rhode Island, USA). It is perhaps worth noting that the first of these two uses data from 1915 and the second from 1890, while both are in regions somewhat separate from the open ocean: Omura and Narragansett Bays, respectively.

Figure 5 shows the cumulative distribution function for these pruned data, along with a probability plot. The latter shows that the data, while still slightly heavy-tailed, are close to Gaussian. The average value, mean or median, for R is robustly determined

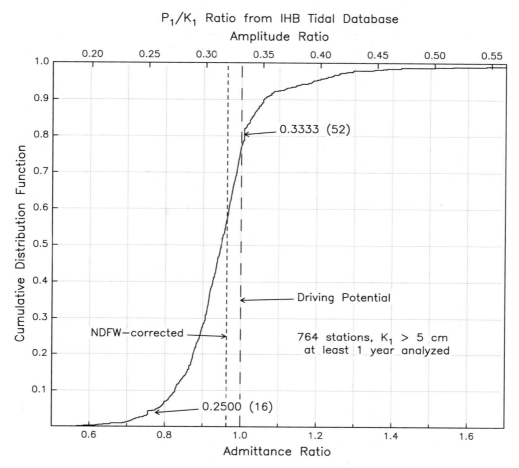

Figure 3
Empirical cumulative distribution for the P_1/K_1 admittance and amplitude ratios for all stations shown in Fig. 2a, along with the expected ratio for an earth with and without a core resonance

as 0.94. The nominal standard deviation if the data are further pruned by removing the tails beyond 2σ is 0.06, though it is not clear how meaningful this value is in terms of actual uncertainty. This value is about 2% below the value of R expected from theory and (much more accurate) observations (Fig. 1).

5. Conclusions

The main result of this paper is a simple one: ocean-tide data, even from such a heterogeneous source as the IHO Tidal Data Bank, clearly show the presence of the NDFW resonance in the ratio of the P_1 to the K_1 tides. This adds to the evidence educed by Ray (2017) that, when inferring the P_1 tide, the ratio should be the ratio of the constituents in the driving potential, modified by the NDFW admittance.

Two questions remain, one scientific and the other historical and hypothetical. The scientific question is what, if any, conclusions we can draw from the difference between the value of $R = 0.94$ found here, and the NDFW value of $R = 0.964$. Ray (2017) finds a value of $R = 0.961$ from open-ocean measurements, and his histogram of values obtained is peaked for P_1/K_1 amplitude ratios from 0.315 to 0.320, while the R found here gives a ratio of 0.311. One plausible explanation lies in the geographic distribution of stations shown in Fig. 2, which does not sample the coastline, much less the ocean, in anything approaching an even distribution. It may be that the discrepancy between coastal and open-ocean values

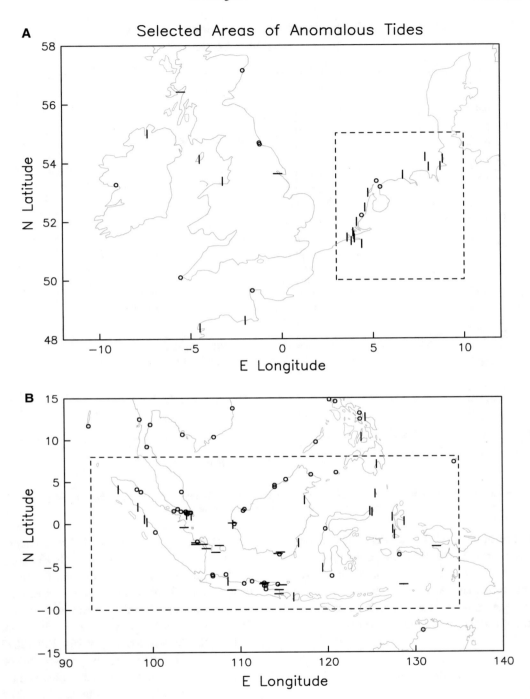

Figure 4
Detail maps of two regions from Fig. 2b that show a high concentration of extreme P_1/K_1 amplitude ratios. The *dashed lines* show the regions removed from the list

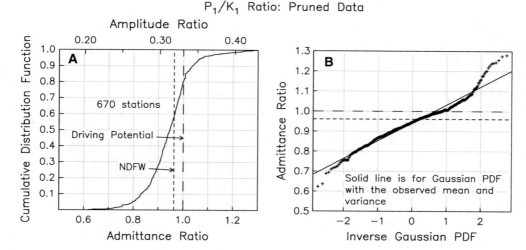

Figure 5
Probability distribution for the P_1/K_1 amplitude ratio for all stations shown in Fig. 2a, omitting those from the regions shown in Fig. 4. This distribution is shown on the *left* (**a**) as an empirical cumulative distribution and on the *right* (**b**) as a probability plot. The *sloping line* in panel (**b**) corresponds to a normal distribution with the same mean and variance as these data

of R could be explained by large-scale variations in the admittance for the ocean tide, perhaps in terms of the diurnal modes determined by Skiba et al (2013). But this is well beyond the scope of this paper, and probably beyond, as well, what can realistically be extracted from these data.

The historical and hypothetical question is, could the ocean-tide data, properly examined, have provided observational evidence for the NDFW prior to the tilt observations of Melchior (1966)? The answer is, yes, these data could have provided such evidence. Taking the "pruned" dataset described in Sect. 4 and limiting it to harmonic constants sent to the IHB before 1940 gives a median value of $R = 0.94$, based on 164 stations. If we extend the cutoff date to 1960, the same dataset gives a median value of $R = 0.93$ based on 327 stations. But anyone working in this field could be forgiven for not attempting such an estimate—it remains startling that this can be done, and could fairly have been judged to be improbable.

Acknowledgements

I thank Bernie Zetler for making NOAA's copy of the IHO Data Bank tape available to me in 1981, and Walter Zürn and Richard Ray for comments on an early draft of

this paper. Spherical-harmonic expansions of modern tide models are from Richard Ray at http://bowie.gsfc. nasa.gov/ggfc/tides/harmonics.html, and his recent paper on tidal inference stimulated me to write this one.

Appendix: Corrections to the IHO Data Bank

The following modifications were made to the values in the IHO Tidal Data Bank prior to the processing described here. The tidal constants were originally published as separate sheets bound into fascicles: sheets 1–1967 are each for individual stations, though only sheets 1–1180 were published. Groups of stations, usually with fewer constituents, were published on sheets 2000–2347 and 3000–3055.

Sheet 167 (Bass Harbour, Malaysia). The data bank value for P_1 is 0.5 cm; reference to the original published sheet shows that this should be 5.5 cm. (This location was not actually part of the winnowed data because of the ratio of P_1 to the nonlinear tide SO_1.)

Sheet 169 (Sydney, Australia). The data bank value for P_1 is 0.5 cm; reference to the original published sheet shows that this should be 4.7 cm.

Sheet 670 (Stockton, California). The data bank value for P_1 is 2.0 cm; the original published sheet shows a value of 1.999 cm, but this sheet also gives the amplitude in feet (the original units), and this amplitude corresponds to 19.99 cm, so in this case there is a typographical error on the sheet.

Sheet 1445 (Yeosu, Korea). The data bank gives two values for K_2 and none for K_1; looking at the phase of other diurnal tides it is clear that the first K_2 value should be assigned to K_1.

Sheet 1780 (Nagapatnam, India). The data bank value for K_1 is 0.5 cm and for P_1 is 22.3 cm. I have instead used the values given in Darwin (1888): 6.8 cm for K_1 and 2.2 cm for P_1.

Sheet 2313 (Santander, Spain). The data bank and published sheet both give P_1 an amplitude of 9.0 cm, larger than K_1 (6.4 cm). The values for K_1 match a number of global models (EOT11A, FES2004, TPXO7.2ATLAS, GOT4P7), which is to be expected since this is a harbor open to the ocean. But these models all give values around 2 to 3 cm for P_1. I have therefore rejected this station.

REFERENCES

Amoruso, A., Botta, V., & Crescentini, L. (2012). Free Core Resonance parameters from strain data: Sensitivity analysis and results from the Gran Sasso (Italy) extensometers. *Geophysical Journal International, 189*, 923–936. doi:10.1111/j.1365-246X.2012.05440.x.

Anonymous. (1926a). Tables for the calculation of tides by means of harmonic constants. *International Hydrographic Bureau Special Publication, 12a*, 1–136.

Anonymous. (1926b). Tide Predicting Machines. *International Hydrographic Bureau Special Publication, 13*, 110.

Anonymous. (1976). Tidal harmonic constants: index of stations. *International Hydrographic Bureau Special Publication, 26*.

Anonymous. (2000). *IHO Constituent Data Bank. IHO Circular Letter 19/2000*, International Hydrographic Office, Monaco, available at http://www.iho.int/mtg_docs/circular_letters/english/2000/Cl19e

Baird, A. W., & Darwin, G. H. (1885). Results of the harmonic analysis of tidal observations. *Proceedings of the Royal Society, 39*, 135–207. doi:10.1098/rspl.1885.0009.

Bermejo, F. (1997). *The History of the International Hydrographic Bureau*. International Hydrographic Bureau, Monaco, available at http://www.iho.int/iho_pubs/misc/HistoryIHBrevisedJan05

Brush, S. G. (1979). Nineteenth-century debates about the inside of the earth solid, liquid, or gas? *Annals of Science, 36*, 225–254.

Cartwright, D. E. (1999). *Tides: a Scientific History*. New York: Cambridge University Press.

Cartwright, D. E., & Edden, A. C. (1973). Corrected tables of tidal harmonics. *Geophysical Journal of the Royal Astronomical Society, 33*, 253–264.

Chao, B., & Hsieh, Y. (2015). The Earth's free core nutation: Formulation of dynamics and estimation of eigenperiod from the very-long-baseline interferometry data. *Earth and Planetary Science Letters, 432*, 483–492. doi:10.1016/j.epsl.2015.10.010.

Cummins, P. R., & Wahr, J. M. (1993). A study of the Earth's free core nutation using International Deployment of Accelerometers gravity data. *Journal of Geophysical Research, 98*, 2091–2103.

Darwin, G. H. (1883). Harmonic analysis of tidal observations. *British Association for the Advancement of Science: Annual Report, 53*, 49–117.

Darwin, G. H. (1888). Second series of results of the harmonic analysis of tidal observations. *Proceedings of the Royal Society, 45*, 556–611. doi:10.1098/rspl.1888.0127.

Dehant, V., & Mathews, P. M. (2015). *Precession, nutation and wobble of the earth*. Cambridge: Cambridge University Press.

Dehant, V., Hinderer, J., Legros, H., & Lefftz, M. (1993). Analytical approach to the computation of the earth, the outer core and the inner core rotational motions. *Physics of the Earth and Planetary Interiors, 76*, 25–28.

Ducarme, B., Sun, H. P., & Xu, J. Q. (2007). Determination of the free core nutation period from tidal gravity observations of the GGP superconducting gravimeter network. *Journal of Geodesy, 81*, 179–187. doi:10.1007/s00190-006-0098-9.

Gwinn, C. R., Herring, T. A., & Shapiro, I. I. (1986). Geodesy by radio interferometry: Studies of the forced nutations of the Earth: 2. Interpretation. *Journal of Geophysical Research, 91*, 4755–4765. doi:10.1029/JB091iB05p04755.

Herring, T. A., Mathews, P. M., & Buffett, B. A. (2002). Modeling of nutation-precession: Very long baseline interferometry results. *Journal of Geophysical Research, 107*(B4), 2069. doi:10.1029/2001JB000165.

Hughes, P. (2006). The revolution in tidal science. *Journal of Navigation, 59*, 445–459. doi:10.1017/S0373463306003870.

Hughes, P., & Wall, A. D. (2007). The ascent of extranational tide tables. *Mariner's Mirror, 93*, 51–64. doi:10.1080/00253359.2007.10657027.

Jeffreys, H., & Vicente, R. O. (1957a). The theory of nutation and the variation of latitude. *Monthly Notices of the Royal Astronomical Society, 117*, 142–161. doi:10.1093/mnras/117.2.142.

Jeffreys, H., & Vicente, R. O. (1957b). The theory of nutation and the variation of latitude: The Roche core model. *Monthly Notices of the Royal Astronomical Society, 117*, 162–173. doi:10.1093/mnras/117.2.162.

Koot, L., Rivoldini, A., de Viron, O., & Dehant, V. (2008). Estimation of Earth interior parameters from a Bayesian inversion of very long baseline interferometry nutation time series. *Journal of Geophysical Research, 113*. doi:10.1029/2007JB005409.

Kushner DS (1990) *The emergence of geophysics in nineteenth-century Britain*. Ph.D. thesis, Princeton University, Princeton.

Mathews, P. M. (2001). Love numbers and gravimetric factor for diurnal tides. *Journal of the Geodetic Society of Japan, 47*, 231–236.

Mathews, P. M., Buffett, B. A., & Shapiro, I. I. (1995). Love numbers for diurnal tides: Relation to wobble admittances and resonance expansion. *Journal of Geophysical Research, 100*, 9935–9948.

Melchior, P. (1966). Diurnal earth tides and the Earth's liquid core. *Geophysical Journal of the Royal Astronomical Society, 12*, 15–21. doi:10.1111/j.1365-246X.1966.tb03097.x.

Molodensky, M. S. (1961). The theory of nutations and diurnal earth tides. *Communications de l'Observatoire Royale de Belgique, Serie Geophysique, 58*, 25–56.

Ooe, M., & Tamura, Y. (1985). Fine structures of tidal admittance and the fluid core resonance effect in the ocean tide around Japan. *Manuscripta Geodaetica, 10*, 37–49.

Pétit, G., Luzum, B., & (2010) IERS Conventions,. (2010). *IERS Technical Note 36*. Frankfurt am Main: Verlag des Bundesamts für Kartographie und Geodäsie.

Pfaff, J. H. (1926). Investigation of harmonic constants, prediction of tide and current, and their description by means of these constants. *International Hydrographic Bureau Special Publication, 12*, 1–80.

Pfaff, J. H. (1927). A rapid method of the calculation of harmonic tidal constants by a system of cards and machines. *International Hydrographic Review, 4*, 25–32.

Polzer, G., Zürn, W., & Wenzel, H. G. (1996). NDFW analysis of gravity, strain and tilt data from BFO. *Bulletin d'Information de Marees Terrestres, 125*, 9514–9545.

Ponchaut, F., Lyard, F., & LeProvost, C. (2001). An analysis of the tidal signal in the WOCE sea level dataset. *Journal of Atmospheric and Oceanic Technology, 18*, 77–91.

Ray, R. D. (1998). Ocean self-attraction and loading in numerical tidal models. *Marine Geodesy, 21*, 181–192.

Ray, R. D. (2013). Precise comparisons of bottom-pressure and altimetric ocean tides. *Journal of Geophysical Research, 118*, 4570–4584. doi:10.1002/jgrc.20336.

Ray, R.D. (2017). On tidal inference in the diurnal band. *Journal of Atmospheric and Oceanic Technology, 35*, 437–446. doi:10.1175/JTECH-D-16-0142.1

Reidy, M. S. (2008). *Tides of history: Ocean science and her majesty's navy*. Chicago: University of Chicago Press.

Ritchie, G. S. (1980). Some aspects of the history of oceanography as seen through the publications of the International Hydrographic Bureau 1919–1939. In M. Sears & D. Merriman (Eds.), *Oceanography: The Past* (pp. 148–156). New York: Springer.

Rosat, S., & Lambert, S. B. (2009). Free core nutation resonance parameters from VLBI and superconducting gravimeter data. *Astronomy and Astrophysics, 503*, 287–291. doi:10.1051/0004-6361/200811489.

Rosat, S., Florsch, N., Hinderer, J., & Llubes, M. (2009). Estimation of the free core nutation parameters from SG data: Sensitivity study and comparative analysis using linearized least-squares and Bayesian methods. *Journal of Geodynamics, 48*, 331–339. doi:10.1016/j.jog.2009.09.027.

Rossiter, J. R. (1972). The history of tidal predictions in the United Kingdom before the twentieth century. *Proceedings of the Royal Society of Edinburgh. Section B, 73*, 13–23. doi:10.1017/S0080455X00002071.

Sasao, T., Okubo, S., Saito, M. (1980). A simple theory on the dynamical effects of a stratified fluid core upon nutational motion of the Earth. In: E.P. Fedorov, M.L. Smith, P.L. Bender (Eds.) Proc. of IAU Symp. (Number 78, pp 165–183). Norwalk, Mass: D. Reidel.

Sato, T., Tamura, Y., Higashi, T., Takemoto, S., Nakagawa, I., Morimoto, N., et al. (1994). Resonance parameters of the free core nutation measured from three superconducting gravimeters in Japan. *Journal of Geomagnetism and Geoelectricity, 46*, 571–586.

Skiba, A. W., Zeng, L., Arbic, B. K., Müller, M., & Godwin, W. J. (2013). On the resonance and shelf/open-ocean coupling of the global diurnal tides. *Journal of Physical Oceanography, 43*, 1301–1324. doi:10.1175/JPO-D-12-054.1.

Wahr, J. M. (1981a). A normal mode expansion for the forced response of a rotating Earth. *Geophysical Journal of the Royal Astronomical Society, 64*, 651–675.

Wahr, J. M. (1981b). Body tides on an elliptical, rotating, elastic and oceanless Earth. *Geophysical Journal of the Royal Astronomical Society, 64*, 677–703.

Wahr, J. M., & Sasao, T. (1981). A diurnal resonance in the ocean tide and in the Earth's load response due to the resonant free "core nutation". *Geophysical Journal of the Royal Astronomical Society, 64*, 747–765. doi:10.1111/j.1365-246X.1981.tb02693.x.

Woodworth, P. L. (2002). Three Georges and one Richard Holden: The Liverpool tide table makers. *Transactions of the Historic Society of Lancashire and Cheshire, 151*, 19–51.

Woodworth, P.L. (2016). *An Inventory of Tide-Predicting Machines*. Research and Consultancy Report 56, National Oceanography Centre, Liverpool, available at http://eprints.soton.ac.uk/394662/.

(Received October 17, 2016, revised January 25, 2017, accepted March 9, 2017, Published online March 23, 2017)

Pure Appl. Geophys. 175 (2018), 1611–1629
© 2018 Springer International Publishing AG, part of Springer Nature
https://doi.org/10.1007/s00024-018-1866-1

Pure and Applied Geophysics

CrossMark

Long-Term Evaluation of Ocean Tidal Variation Models of Polar Motion and UT1

Maria Karbon,[1] Kyriakos Balidakis,[2] Santiago Belda,[4] Tobias Nilsson,[3] Jan Hagedoorn,[5] and Harald Schuh[2,3]

Abstract—Recent improvements in the development of VLBI (very long baseline interferometry) and other space geodetic techniques such as the global navigation satellite systems (GNSS) require very precise a-priori information of short-period (daily and sub-daily) Earth rotation variations. One significant contribution to Earth rotation is caused by the diurnal and semi-diurnal ocean tides. Within this work, we developed a new model for the short-period ocean tidal variations in Earth rotation, where the ocean tidal angular momentum model and the Earth rotation variation have been setup jointly. Besides the model of the short-period variation of the Earth's rotation parameters (ERP), based on the empirical ocean tide model EOT11a, we developed also ERP models, that are based on the hydrodynamic ocean tide models FES2012 and HAMTIDE. Furthermore, we have assessed the effect of uncertainties in the elastic Earth model on the resulting ERP models. Our proposed alternative ERP model to the IERS 2010 conventional model considers the elastic model PREM and 260 partial tides. The choice of the ocean tide model and the determination of the tidal velocities have been identified as the main uncertainties. However, in the VLBI analysis all models perform on the same level of accuracy. From these findings, we conclude that the models presented here, which are based on a re-examined theoretical description and long-term satellite altimetry observation only, are an alternative for the IERS conventional model but do not improve the geodetic results.

Key words: Ocean tidal model, VLBI analysis, Euler–Liouville.

1. Introduction

Since the investigations of Brosche et al. (1989, 1991) different authors have studied the influence of diurnal and semi-diurnal ocean tides on Earth rotation. Due to its importance for the estimation processes of space geodetic techniques, a first version of a model for ocean tidal ERP variations was presented in the IERS Conventions 1996 (McCarthy 1996). This model considered eight major tides by ocean tidal angular momentum (OTAM), which have been derived from the satellite altimetry-based model TPXO.2 (Egbert et al. 1994). The recent conventional model (see Petit and Luzum 2010) is an improved version, which considers the influence of additional minor tides and in total 71 partial tides. The calculation of Earth rotation variation is based on a frequency domain approach for high-frequency variations as presented by Gross (1993) and Brzezinski (1994). In both approaches, the elastic behavior of the Earth is prescribed by tidal and load Love numbers according to Sasao and Wahr (1981) for a radial elastic Earth structure based on the model 1066A (see Gilbert and Dziewonski 1975). Several studies have compared the predicted short-period Earth rotation variation ERP observed by space geodetic techniques such as VLBI and GNSS and have shown a significant misfit between the theoretical models and the empirical results (e.g. Artz et al. 2011; Böhm et al. 2011). The development of models of the influence of high-frequency ocean tides on Earth rotation was done separately for ocean tide modeling and the corresponding Earth rotation variation.

Within this work, this development of an improved model for high-frequency ocean tidal influence on Earth rotation is carried out consistently and cooperatively by experts in the field of ocean tide and Earth rotation modeling. One of the original ideas of the project was the use of an alternative empirical ocean tide model instead of ocean tide models

---
[1] Universität Bonn, Bonn, Germany. E-mail: karbon@uni-bonn.de
[2] Technische Universität Berlin, Berlin, Germany.
[3] GFZ German Research Centre for Geosciences, Telegrafenberg A17, Potsdam, Germany.
[4] Department of Applied Mathematics, EPS, University of Alicante, Alicante, Spain.
[5] University of Potsdam, Potsdam, Germany.

derived by assimilation of altimetry observations into hydrodynamic ocean models. Therefore, the focus here was on the empirical model EOT11a (see Savcenko and Bosch 2012), and the results are compared to those based on the hydrodynamic models HAMTIDE (Taguchi et al. 2014) and FES2012 (Lyard et al. 2006; Carrere et al. 2012). All these models provide information for a set of individual partial tides with a fixed frequency, which motivated the development of an approach in the frequency domain. For this purpose, the ocean tidal angular momentum was integrated for each individual partial tide. The so-called mass term, considering the angular momentum variation due to mass redistribution characterized by the tidal heights, was directly integrated. In contrast, the relative angular momentum, the so-called motion term, was calculated from tidal velocities, determined by an inversion approach, first published by Ray (2001) and examined in Madzak et al. (2016). This approach was applied to nine major tides (Q_1, O_1, P_1, K_1, $2N_2$, N_2, M_2, S_2, K_2), which can also be resolved from satellite altimetry observations. Within this work, the influence of additional minor tides on Earth rotation was computed by admittance theory, i.e. linear or quadratic interpolation in the frequency domain, for ocean tidal angular momentum . Therefore, we reexamined the so-called high-frequency Euler–Liouville equation, which considers the Eigen frequencies of an elastic mantle, liquid outer core and solid inner core for diurnal and sub-diurnal periods. For radial distribution of elastic parameter according to elastic models from Alterman et al. (1961) (model Gutenberg and Bullen A 1961), Bullen and Haddon (1967), Gilbert and Dziewonski (1975) (model 1066a 1975), Wang (1972) and Dziewonski and Anderson (1981) (PREM 1981), the corresponding Love numbers were calculated and used to solve the Euler–Liouville equation in the frequency domain. The aim was the development of a model describing the impact of diurnal and sub-diurnal ocean tides on Earth rotation parameters, which is applicable to the analysis of space geodetic techniques such as VLBI. We have focused on the consistency of the theoretical description and the influence of the chosen ocean tide model on the resulting ERP model. In the time domain, we consider only prograde diurnal contribution to our ERP

model to exclude nutation from our investigation. We reexamine the formulation of the ocean tidal angular momentum given in Ray (1998) and followed in Madzak et al. (2016). Here we focus on the consistent determination of OTAM for minor tides by quadratic admittance theory, as they are commonly not considered in tidal models. In addition, we tried to detect minor tides in long time series of satellite altimetry and compared them with results of assimilating hydrodynamical models; concluding that a restricted set of minor tides can be determined empirically using satellite altimetry observations.

Finally, we tested various model realizations within the VLBI analysis.

2. Ocean Tide Models

We consider two hydrodynamic ocean tide models HAMTIDE11a (Zahel 1995; Taguchi et al. 2014) and FES2012 (Carrere et al. 2012; Lyard et al. 2006), which both assimilate altimetry observations and consider a set of hydrodynamic equations. As a third model, we chose the empirical altimetry-based tidal model EOT11a (Savcenko and Bosch 2012). Table 1 summarizes the basic characteristics. Here, only a small introduction shall be given, for more information on the models we refer the interested reader to the given references. An extensive comparison was made by Stammer et al. (2014).

Table 1

Basic features of the used tidal models

Model	Type	Grid	Partial tides	Σ
HAMTIDE11a	hyd.	$7.5'$	Q_1, O_1, P_1, K_1, $2N_2$, N_2, M_2, S_2, K_2	9
FES2012	hyd.	$3.75'$	Z_0, Ssa, Mm, MSf, Mf, Mtm, Q_1, O_1, P_1, S_1, K_1, J_1, E_2, $2N_2$, μ_2, N_2, ν_2, M_2, MKS_2, λ_2, L_2, T_2, S_2, R_2, K_2, M_3, N_4, MN_4, M_4, MS_4, S_4, M_6, M_8	33
EOT11a	emp.	7.5	Mm, Mf, Q_1, O_1, P_1, S_1, K_1, $2N_2$, N_2, M_2, S_2, K_2, M_4	13

emp. empirical modeling, *hyd.* hydrodynamic modeling with assimilation

2.1. EOT11a

A series of global tidal models (EOT08a, EOT10a, EOT11a; Savcenko and Bosch 2012) was developed by DAROTA and COTAGA, two consecutive projects of the DFG priority program SPP1257 (mass distribution and mass transports in the system Earth). EOT11a was derived purely empirically from the measurements of different altimeter missions. The empirically determined partial tides of EOT11a are based on the sea level topography measured by TOPEX, Jason-1, Jason-2, ERS-2 and ENVISAT over the sea-ice-free oceans in the period from October 1992 to March 2010. The estimation of the partial tides was performed by a least-squares solution of a residual harmonic analysis. The partial tides of EOT11a are available (like FES2004) on a global regular lattice of $7.5'$ spacing. For geographical latitudes higher than $81.5°$ EOT11a falls back to FES2004. The loading effect of the partial tides was calculated according to Cartwright et al. (1991).

2.2. HAMTIDE11a

The HAMTIDE model (Hamburg direct data assimilation methods for Tides) was developed at the University of Hamburg and is similar to the representer-method of Bennett (1992). However, the solution is obtained by a least-squares minimization of the residuals between model and data. The Laplace equations are linear elliptical equations driven by the second-degree tidal potential. According to Zahel et al. (2000), the target function is supplemented by spatial derivatives of dynamic residuals and secondary effects such as loading and self-attraction (Henderscott 1972; Zahel 1991). HAMTIDE12 was calculated on a global grid with $7.5'$ resolution.

2.3. FES2012

Finite-element solutions of the flat-water equation, formulated on an ever finer-mesh, with unstructured grating have been described by Le Provost et al. (1994) and led to a series of

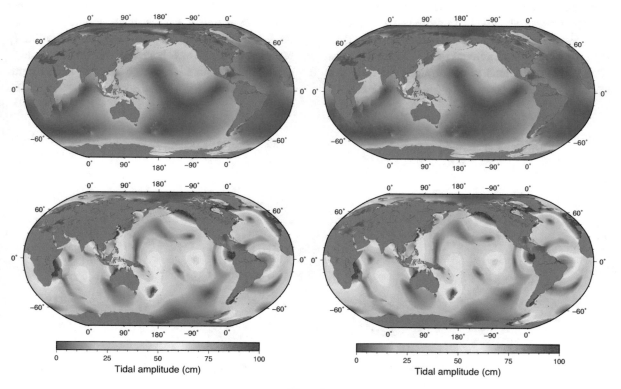

Figure 1
Tidal height amplitudes for K_1 (top) and M_2 (bottom) taken from ocean tide model FES2012 (left) and EOT11a (right)

hydrodynamic tidal models (FES1994, FES1999, FES2004). The model used in our study, FES2012 (Carrere et al. 2012), was created by Noveltis, Legos and CLS Space Oceanography Division and distributed by Aviso, with support from CNES. The integration grating has a spatial resolution of a few kilometers in coastal proximity up to 25 km mesh width in the open ocean. The tabulated partial tides are provided on a regular mesh with a distance of 3.75'. FES2012 is based on an improved global bathymetry and assimilates harmonious time constants, which consist of 20 years of altimetry (especially TOPEX, Jason-1 and Jason-2). In addition to the main tides the model offers a large number of smaller tides (see Table 1).

Figure 1 shows exemplarily the tidal height amplitudes for K_1 on the top and M_2 on the bottom panels as given by FES2012 (left) and EOT11a (right). The plots are very similar, only in high latitudes differences arise, e.g. in the Antarctic sea or the Ross sea.

While global ocean tide models have reached an impressive level of accuracy, there still remain big uncertainties, especially in shallow-water and high-latitudes (Stammer et al. 2014). It is also in these regions, where the individual models differ the most. Also, the currents tend to be larger by 10–20% than what in situ measurements deliver, whereas the

observed and modeled phases show a good agreement. Overall, the models agree within a 10% margin, however for certain tides, the differences can also reach more than 50%. Hence, it can already be ascertained, that the biggest uncertainty in our study stems from the choice of the ocean tide model.

3. ERP Models

A detailed description of the derivation of tidal velocities as well as the calculation of the oceanic tidal angular momentum (OTAM) from the ocean tidal models is given in Madzak et al. (2016) and shall not be repeated here. Thus, we will give only a short walk through the processing strategy, as shown in the flowchart in Fig. 2 and discuss the results of each step.

3.1. Tidal Velocities

Ray (2001) proposed a linear inversion method to derive tidal currents from tidal heights. Simplified and depth-averaged shallow water momentum equations, supplemented by a weighted continuity equation and no-flow boundary conditions from a set of linear equations, which can be solved iteratively, as shown in Madzak et al. (2016). The

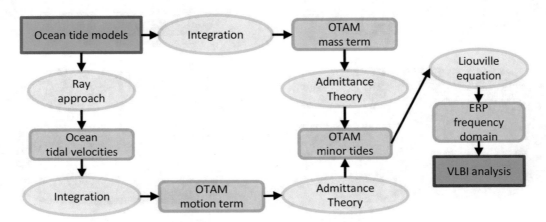

Figure 2
Flowchart of the determination of the ERP models from ocean tide models

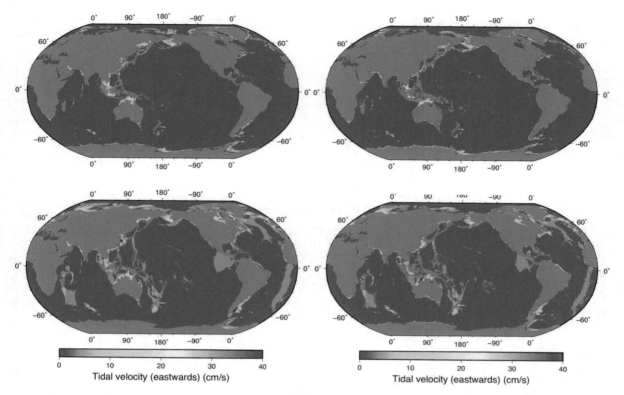

Figure 3
Tidal velocity (eastward component) for partial tide K_1 (top) and M_2 (bottom) given by FES2012 (left) and inverted from EOT11a, i.e. VEOT, (right) in cm/s

resulting ocean tide model, which consists of tidal heights of EOT11a and inverted tidal current velocities, is labeled VEOT. To correct equilibrium tidal forcing, and self-attraction and loading, we followed Pugh and Woodworth (2014) and Ray (1998), respectively.

At a first glance, the tidal velocities (eastward components) in Fig. 3 determined from FES2012 and VEOT seem to agree quite well; looking closer, however, significant discrepancies become obvious. First, the different grid sizes become clear, as the plot for VEOT looks much more pixelated than the one for FES2012. Generally, the main differences occur in small areas of large velocity and topography gradients, such as in the Weddell Sea, south-east of Argentina, or in shallow waters, like in the Indonesian archipelago. Overall, the derived volume transports from VEOT in these areas are smaller than volume transports from FES2012. For VEOT

several peaks arise around islands, for example, in the Pacific or around Madagascar. Very pronounced are also the peaks above the Chagos-Laccadive Ridge, south of India. For FES2012 such features are less discernable, as the overall picture is on one hand much smoother, on the other more detailed.

3.2. OTAM: Motion and Mass

The angular momentum approach is based on the balance of the angular momentum of the Earth and the angular momentum of the geophysical fluids (Wahr and Bergen 1986). Considering angular momentum conservation, any mass redistribution on the Earth excites Earth rotation variations (Chao and Ray 1997). Changes in polar motion and LOD are excited by two mechanisms: (1) mass redistribution, changing the inertia tensor ΔI, usually referred to as mass term; (2) motion relative to the rotating

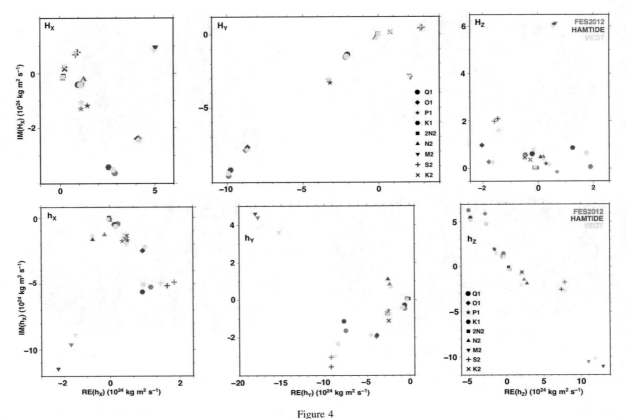

Figure 4
On top OTAM mass term, on the bottom the motion term for X, Y, and Z components for the nine major tides estimated from ocean tide models FES2012, HAMTIDE and VEOT, respectively. All values are given in units of 10^{24} kg m^2 s^{-1}

reference frame changing the relative angular momentum h, usually referred to as motion term (Gross 2007; Munk and MacDonald 1960). The mass term is represented by the tidal heights of the ocean surface, the motion term arises from oceanic currents. The OTAM for each frequency are computed by integration using the formulation given in Chao and Ray (1997). For more details see Madzak et al. (2016).

In Fig. 4 we plotted the mass (top) and motion terms (bottom) for the X, Y, and Z components for the various models. One can see that the components of the major tides group together; however, they deviate significantly depending on the individual partial tide. For instance, the models agree well for P_1, Q_1 and $2N_2$ for all components, for M_2 and S_2, however, they give significantly different results. Noticeable is also the fact, that the two hydrodynamic models FES2012

and HAMTIDE often disagree substantially, e.g. in case of the X component of N_2 where HAMTIDE and VEOT closely match, same for the Z component of O_1. For the mass term the agreement of the individual models is as expected much better, just in the Z component bigger differences appear, e.g. in case of K_1 or O_1.

3.3. Estimation of OTAM for Minor Tides

In the following, a *small partial tide* is understood as a tide that does not belong to the usually provided dominant tides of the diurnal and semi-diurnal species, but as tides which are not present in the set: $Sa, Ssa, Mf, Mm, K_1, P_1, Q_1, M_2, S_2, N_2, K_2, 2N_2$, and M_4. These small tides generally have amplitudes which are one order of magnitude less than the principal ones, and rarely exceed 5 cm. In sum, however, they can have significant effects. Of the many thousand partial tides, which are indicated by a development of the tidal potential (e.g. Hartmann and Wenzel 1995), we limit ourselves here to M_1, J_1 and OO_1 (diurnal), and $\mu2$, $v2$, $L2$ and $T2$ (semi-diurnal), because their (theoretical) amplitudes are at least 1

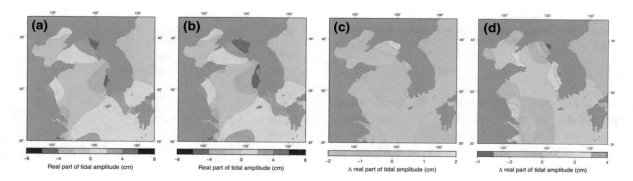

Figure 5
From left to right: **a** L_2 estimated through linear admittance, **b** L_2 estimated through quadratic admittance, **c** difference between the two methods, **d** difference w.r.t. FES2012 L_2 tide and the quadratic admittance theory

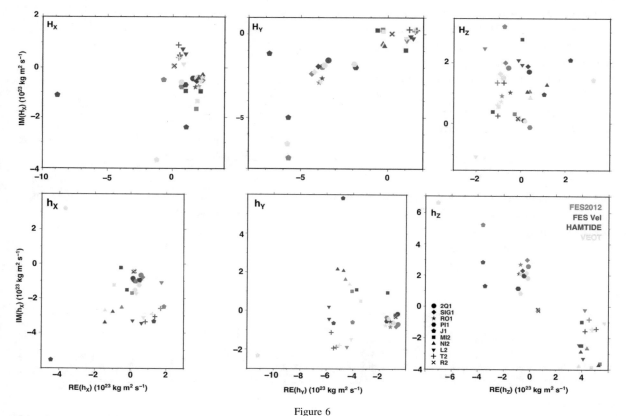

Figure 6
On top OTAM mass term, on the bottom the motion term for X, Y, and Z components for minor tides estimated from ocean tide models FES2012, HAMTIDE and VEOT, respectively. In addition, the minor tides delivered by FES2012 are shown for selected tides (FES Vel). All values are given in units of 10^{23} kg m^2 s^{-1}

Table 2

Load Love numbers of degree 2

Model	k_2^L	h_2^L	l_2^L
1066A	− 0.308118	− 1.008702	0.024567
Bullen	− 0.310369	− 1.007451	0.030426
PREM	− 0.305548	− 0.992544	0.023704
Wang	− 0.308190	− 1.009550	0.022050

Table 3

Pro- and retrograde ocean tidal terms of diurnal (O_1, K_1) and semi-diurnal (M_2, K_2) polar motion for HAMTIDE

Model	A_p	B_p	A_r	B_r
O_1	Prograde			
1066A	− 125.06	65.92		
Bullen	− 125.78	66.34		
PREM	− 126.52	66.76		
Wang	− 126.66	66.83		
K_1	Prograde			
1066A	131.61	− 89.47		
Bullen	132.29	− 90.19		
PREM	132.99	− 90.83		
Wang	133.15	− 90.95		
M_2	Prograde		Retrograde	
1066A	41.25	− 60.19	4.64	278.04
Bullen	41.26	− 60.09	3.79	276.61
PREM	41.50	− 60.20	4.58	277.52
Wang	41.56	− 60.23	3.78	277.43
K_2	Prograde		Retrograde	
1066A	− 0.90	− 6.42	− 18.99	30.03
Bullen	− 0.87	− 6.38	− 18.92	29.82
PREM	− 0.87	− 6.38	− 18.96	29.97
Wang	− 0.86	− 6.38	− 18.98	29.91

Units are (μas)

cm, in limited regions; however, they can also reach amplitudes up to 5 cm. Since a hydrodynamic modeling of some small tides has been performed with FES2012, we have a mean to compare our estimates.

The IERS2010 conventions suggest to apply a linear admittance to 71 constituents of the tidal potential in the diurnal and semi-diurnal band. Under the assumption that OTAM is a smooth function of frequency, we, however, consider a quadratic admittance applied to 29, 71 and 260 constituents up to a certain threshold value of the tidal potential from HW95 (Hartmann and Wenzel 1995), respectively. We follow hereby the quadratic admittance theory described by Munk and Cartwright (1966). The relation of the minor tidal height h_m to the amplitude of the tidal potential V_i is approximated by

$$h_m = \sum_i a_i^m \frac{V_m}{V_i} h_i. \tag{1}$$

The linear admittance theory considers two admittance factors based on the tidal frequencies. Here exactly two major tides with the frequencies σ_1 and σ_2 are used to determine the admittance factor a_1^m for the smaller tide with the frequency σ_m:

$$a_1^m = \frac{\sigma_m - \sigma_2}{\sigma_1 - \sigma_2}, \quad a_2^m = \frac{\sigma_m - \sigma_1}{\sigma_1 - \sigma_2}, \tag{2}$$

whereas, the quadratic admittance theory considers three admittance factors:

$$a_1^m = \frac{(\sigma_m - \sigma_2)(\sigma_m - \sigma_3)}{(\sigma_1 - \sigma_2)(\sigma_1 - \sigma_3)}, \quad a_2^m = \frac{(\sigma_m - \sigma_1)(\sigma_m - \sigma_3)}{(\sigma_2 - \sigma_1)(\sigma_2 - \sigma_3)},$$
$$a_3^m = \frac{(\sigma_m - \sigma_1)(\sigma_m - \sigma_2)}{(\sigma_3 - \sigma_1)(\sigma_3 - \sigma_2)}. \tag{3}$$

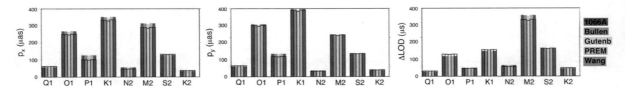

Figure 7

Impact of the various Earth models given in the different colors on the ERP components, the black outlines give the differences w.r.t. the IERS standard model. Red: 1066A, orange: Bullen, yellow: Gutenberg, green: PREM, blue: Wang

Table 4

Impact of the Earth models on selected tidal constituents, in terms of mean values, standard deviation and mean difference w.r.t. the standard model (all values in µas)

	Q_1	O_1	P_1	K_1	N_2	M_2	S_2	K_2
$\overline{p_x}$	58.34	266.76	127.20	350.83	56.50	313.10	129.59	34.48
$\sigma\,(p_x)$	0.89	2.25	1.05	2.87	0.47	2.55	1.03	0.27
$\Delta\,\overline{p_x}$	4.28	16.64	25.57	17.37	8.89	23.31	3.49	2.81
$\overline{p_y}$	107.42	536.53	235.02	691.77	62.60	429.85	237.28	65.10
$\sigma\,(p_y)$	0.93	4.62	1.99	5.85	0.50	3.43	1.91	0.52
$\Delta\overline{p_y}$	7.71	10.42	36.85	9.66	4.41	3.00	0.68	3.46
$\overline{\Delta LOD}$	24.04	103.57	42.77	129.45	56.69	312.76	141.82	39.28
$\sigma\,(\Delta LOD)$	0.19	0.81	0.33	1.01	0.44	2.45	1.11	0.31
$\overline{\Delta\Delta LOD}$	3.48	9.95	1.57	7.08	6.16	21.84	1.68	2.98

Hence, this theory is basically a linear, or in our case, a quadratic approximation of the minor tidal height, considering two or three major tidal heights, respectively. These heights are weighted by the admittance factors a_i^m, which are based on the related tidal frequencies σ. However, such an approximation is only applicable under the assumption of a smooth variation of the tidal potential as a function of frequency.

Exemplarily in Fig. 5 we want to show the minor tide L_2 in the Yellow Sea region, estimated by linear admittance (a) based on M_2 and S_2 and quadratic admittance (b) based on N_2, M_2 and K_2, in both cases the tides are based on FES2012 and HW95. The third plot (c) shows the differences, which rise up to $\pm$ 1.5 cm.

The last plot in Fig. 5 shows the difference between the quadratic admittance estimate and the L_2 amplitudes given directly by FES2012. As one can see, in the open water the differences are below 2 cm, within the bays the differences, however, can rise up to $\pm$ 4 cm. The reason are the large amplitudes along the Korean coastline in FES2012, whereas the quadratic interpolation seems to underestimate these large amplitudes. Comparisons with M_2 and N_2 deliver similar plots, whereas for T_2 and J_1 the differences barely exceed 3 cm. Also in basins without closed bays, the agreement is much better. On a global scale, the differences between the results from the quadratic admittance theory and FES2012 show an RMS of 0.268 cm.

In Fig. 6 we plotted the motion and mass terms of selected minor tides for the different ocean tide models and the three components. The tidal components given directly by FES2012 are plotted in blue. One can notice immediately, that the scatter between the individual models increases substantially compared to the plots in Fig 4. A drastic case is J_1, where all solutions differ substantially for all components for the motion, as well as for the mass terms. J_1 is the smallest lunar elliptic diurnal constituent with an amplitude of 0.03 mm [HW95 catalog by Pugh and Woodworth (2014)].

Concluding, it can be said that the empirical estimation of small astronomical tides is quite possible. The application of the admittance theory results in minor tides which differ from FES2012 within an RMS of 0.75 cm and below. Larger differences occur in particular near the coasts, which could indicate a violation of the admittance assumptions in these regions. Also, the choice of the ocean tidal model for the estimation of the minor tides, seems again to be the crucial point. Whereas for some tides the agreement is quite good, most of them disagree on a substantial level.

3.4. The Euler–Liouville Equations and the Earth Models

For the IERS 2010 conventional model the Euler–Liouville equations (Eq. 4) were solved following Gross (1993), as given in Eq. (4). It is based on the theoretical description of Sasao and Wahr (1981) and considers an elastic Earth model: 1066A of Gilbert and Dziewonski (1975). Let c^T be increment of the inertia tensor induced by a specific tide in the frequency domain, and Ωc^T the mass term of OTAM

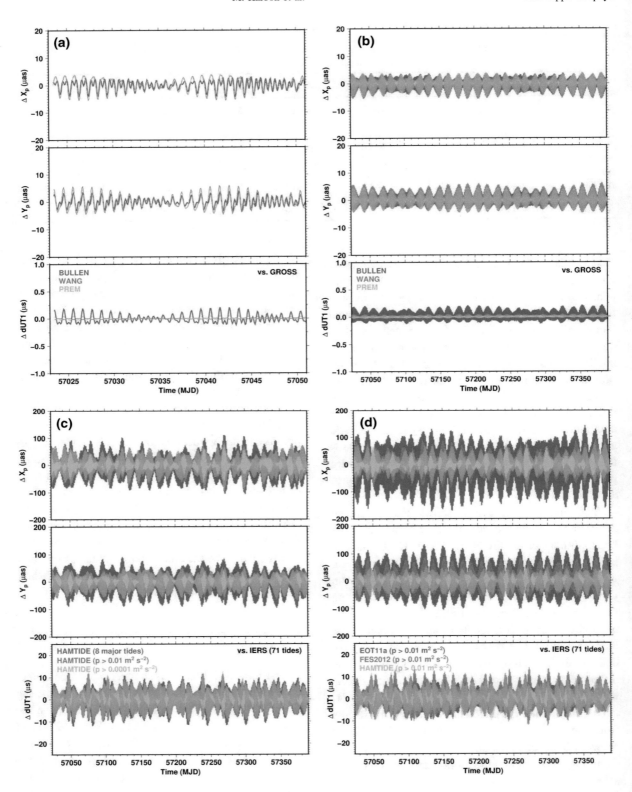

◄Figure 8
a,b Differences between results for three elastic Earth models and Gross (1993) formula considering ocean tides of HAMTIDE for January 2015 (**a**), and the entire year (**b**), respectively. **c** Differences between results for HAMTIDE / PREM combination, considering 8, 29, 307 tidal constituents (only mass terms for minor tides) and the IERS2010 conventional model for 2015. **d** Comparison of differences between different ocean tidal models and the IERS2010 conventional model for 2015

relative to the tidally-induced angular momentum h^T. Under the conservation of angular momentum in the solid Earth–ocean system, in presence of these perturbations, the linearized Liouville equations for the pole position changes $p\sigma$ are given as:

$$p(\sigma) = \left[2.686 \times 10^{-3} \frac{\Omega}{\sigma_{cw} - \sigma}\right.$$
$$+ 2.554 \times 10^{-4} \frac{\Omega}{\sigma_{ndfw} - \sigma}\right] \frac{\Omega c^T(\sigma)}{A\Omega\tau}$$
$$+ \left[1.124 \frac{\Omega}{\sigma_{cw} - \sigma} + 6.170 \times 10^{-4} \frac{\Omega}{\sigma_{ndfw} - \sigma}\right] \frac{h^T(\sigma)}{A\Omega},$$
$$(4)$$

where σ is the frequency of the considered tide, σ_{cw} the frequency of the Chandler wobble, σ_{ndfw} the frequency of the nearly-diurnal-free wobble, Ω the mean angular velocity of the Earth, A the mean moment of inertia of the entire Earth, $\tau = \Omega^2 R_E^5/(3GA)$, G being the universal gravitational constant, and R_E the mean Earth radius.

For our models we chose the Euler–Liouville equation for the so-called Molodensky model, i.e. elastic mantle and fluid core. Following Sasao et al. (1980):

$$p(\sigma) = \left[\frac{A}{A_m} \tau \left(1 + k^L\right) \frac{\Omega}{\sigma - \sigma_{cw}}\right.$$
$$- \frac{A}{A_m} \tau \left(h^d - k^d\right) \frac{\Omega}{\sigma - \sigma_{ndfw}}\right] \frac{\Omega c^T(\sigma)}{A\Omega\tau}$$
$$+ \left[\frac{A}{A_m} \tau \frac{\Omega}{\sigma - \sigma_{cw}}\right.$$
$$+ \frac{A}{A_m} \frac{A_c}{A_m} (\epsilon - \beta) \frac{\Omega}{\sigma - \sigma_{ndfw}}\right] \frac{h^T(\sigma)}{A\Omega}.$$
$$(5)$$

Here the additional variables are: A_m, A_f are the moments of inertia of the mantle and the fluid core, respectively, k^L is the load Love numbers, h^d and k^d

the dynamic Love number, ϵ describes the ellipticity of the core and

$$\beta = -\frac{1}{5GA_c} \left[r^4 \left[y_6^{(d)}(r) - 2y_5^{(d)}(r)/r\right]\right]\Big|_{r_{icb}}^{r_{cmb}}. \quad (6)$$

We tested five different Earth models: 1066A, Bullen, Gutenberg, PREM, Wang. For this task also the Love numbers had to be recalculated individually. Wang et al. (2012) show significant differences between the Load Love numbers for different Earth models, especially for degree larger than 20. For high degrees ($n > 100$) the differences can reach up to 40%. However, for our purpose only the degree 2 Love numbers have to be considered, which differ on the level of a few percent (see Table 2). If cosine- and sine-amplitudes are used instead of amplitudes and phases, the pro- and retrograde components can be calculated as well. The cosine- and sine-amplitudes are commonly used when tidal terms are fitted to time series, since they give linear equations in a least-squares adjustment. They are given as:

$$A_p = -\frac{1}{2}(px_c + py_s)$$
$$B_p = +\frac{1}{2}(px_s - py_c)$$
$$A_r = -\frac{1}{2}(px_c - py_s)$$
$$B_r = +\frac{1}{2}(px_s + py_c)$$
$$(7)$$

Table 5

Parametrization, models, and a-priori of the analysis

Parameters and estimation	
Models and frames	
Station coordinates	Session wise
Source coordinates	Session wise
Source velocities	Session wise
ERP	24 h
CPO	24 h
clock parameters	1 h
ZWD and gradients	1 and 6 h
CRF	ICRF2
ERP	USNO Finals
Celestial pole offset	IAU 2006/2000A
Atm. tidal loading	vanDam
Non-tidal atm. loading	uniSTR$_{ERAin}$
Ocean tidal loading	FES2004
Hyd. loading	GFZ$_{LSDM}$

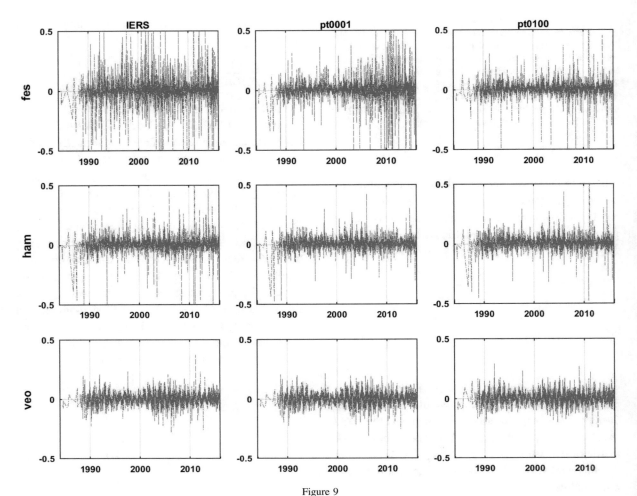

Figure 9

Differences in x-pol using the conventional IERS 2010 model and the other models, based on FES2012 (fes), HAMTIDE12 (ham), and VEOT (veo) using the same tidal constituents as IERS 2010, 251 minor tides (pt0001), or 20 minor tides (pt0100). The units are mas

For the diurnal contribution the retrograde amplitudes A_r and B_r are neglected to exclude nutation in the ERP model in the time domain. Table 3 lists the resulting ocean tidal terms of two diurnal and two semi-diurnal tides using different Earth models, exemplarily for HAMTIDE. The differences in percent are in a single digit range.

Figure 7 shows the resulting amplitudes of selected major tides based on the various Earth models. The black line gives the IERS standard model. Just by looking at the plots, one can see that the impact of the different Earth models is very small. To make the comparisons easier, Table 4 summarizes

the mean values of the models ($\overline{p}$), the standard deviation between them ($\sigma(p)$) and the mean difference w.r.t. the standard model ($\Delta\overline{p}$), all in µas. However, the agreement of the individual models varies, depending on the component, as well as the tide. For example, the overall agreement of all the models for K_2 is very good, for O_1; however, the agreement varies appreciably for each component. In case of M_2 the disagreement increases by a factor of 10 for all components. There is no clear pattern discernible, neither for individual tides, nor for the ERP. Still, where the differences between the ocean models have an impact of up to 10%, the Earth models only contribute less than 1% to the overall differences and thus are considered to be negligible. All considered Earth models are solely elastic models. For further effects such as anelasticity the underlying theoretical description has to be adopted (see e.g. Koot and de Viron 2011).

4. ERP in the Time Domain

Figure 8a shows the differences between the results using different Earth models, (a, b) considering the Gross-model, (c, d) w.r.t. the IERS standard model considering 71 tides. For this comparison we restricted the models to the eight major tides, as they reflect the dominant behavior. Plot (b) shows exactly the same, but instead of only covering January 2015, for the entire year 2015. One can see that Bullen deviates more from the other two, most significantly in dUT1. Nonetheless, the maximal differences are below 1% throughout the long time series. Hence, we decided to use PREM as Earth model in the further investigations.

Figure 8c shows the impact of tidal constituents, again in terms of the differences w.r.t. the IERS standard model, in this case exemplarily for HAM-TIDE. In blue only the eight major tides are considered, in yellow 29 constituents, and in orange 307 constituents. For the minor tides only the mass term is considered. The time spans the entire year 2015. We found that the maximal differences are in the order of 10% of the IERS model and are reduced significantly when extending the number of tidal constituents from eight to 29; a further expansion up to 307 constituents has only a marginal effect. This is due to the fact that for the minor tides we consider only the mass terms.

Figure 8d shows the differences between the standard model and the results when using different ocean tidal models for 29 tidal constituents. For polar motion we can see, that HAMTIDE shows the smallest deviation from the conventional model, but for dUT1 EOT11a is much closer to the standard model than the others. Concluding, it can be said that the contributions from minor tides' relative angular momentum (motion term) have an considerable impact and could further the improvement. However,

currently, the motion term of the minor tides is not included in our modeling.

5. VLBI Data Analysis

For this study we used 3441 VLBI experiments, also called sessions, observed by the International VLBI Service for Geodesy and Astrometry (IVS, Nothnagel et al. 2015) within the years 1979–2016. The sessions used encompass a polyhedron of more than 1 Mm^3 and contain more than three stations, to assure a global station network, and thus, a stable geometry. The single session analysis followed the IERS 2010 conventions (Petit and Luzum 2010), the involved a-prioris and the models are given in Table 5. The terrestrial datum was realized with no-net-rotation/no-net-translation (NNR/NNT) conditions on the stations contained in the VTRF2014 catalog (Bachmann et al. 2016), the celestial one with a NNR condition on the ICRF2 defining sources.

We analyzed all the sessions using the different ERP model with its different number of tidal constituents, adding up to a total of nine additional solutions. As reference we used the solution determined with the conventional ERP model given by the IERS 2010 conventions.

The naming convention for the individual solutions is as follows: tidalmodel_setofpartialtides, with fes for FES2012, ham for HAMTIDE11a, and veo for VEOT. The set of partial tides is in case of IERS given by nine major plus 62 minor tides as in the IERS conventional model, in case of pt0100 by 20 tidal constituents, with a threshold of 0.01 $m^2 \, s^{-2}$, and in case of pt0001 by 251 tidal constituents, with a threshold of 0.0001 $m^2 \, s^{-2}$.

Figure 9 shows exemplarily the differences of the resulting x-pol residuals between the conventional model and the nine alternative models. One can see, that the models based on FES2012 show the largest deviation and a very noisy behavior compared to the other two ocean tidal models. The extension of the model to include 20 minor tides (pt0100, right column) diminishes the differences considerably and the further extension to 251 minor tides only seems to aid in the years until 2010. After that the differences increase again. For the other two models,

29

HAMTIDE11a (ham) in the middle row, and VEOT (veo) at the bottom, no such large differences between the used number of constituents arise. A more detailed picture can be drawn from the resulting weighted RMS depicted in Fig. 10: the largest discrepancies result from the model FES2012 in blue, where the result can be improved by 50% including 20 minor tides; expanding the model leads again to a

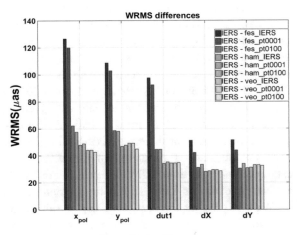

Figure 10
Weighted RMS of the differences of the EOP derived from the conventional model and the other models, based on FES2012 (fes), HAMTIDE12 (ham), and VEOT (veo) using the same tidal constituents as the conventional model (IERS), 251 minor tides (pt0001), or 20 minor tides (pt0100)

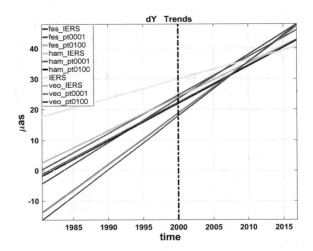

Figure 11
Trends of the EOP differences w.r.t. USNO Finals using different a-priori models in µas/year. Reference epoch is J2000.0

degradation. In case of HAMTIDE11a the model with 251 minor tides delivers the smallest WRMS, the result we would have intuitively expected. In case of VEOT (yellow), however, the model using 20 minor tides seems again to perform slightly better. The reason behind this could be again the missing motion term for the minor constituents. Overall it can be said, that the tidal models HAMTIDE11a and VEOT deliver overall results that are in a good agreement with the IERS conventional model, whereas the different realizations, especially the extension of the number of minor tides from 20 to 251 has an impact in the digit percentage range.

As an external validation we used the USNO Final EOP as a reference, which are also used as a-priori. Since the models applied are all periodic, we do not expect any long-term trends; however, trends do appear in the differences. The origin of these trends might be some long-periodic variations in the differences, for example, caused by errors in the modeling of tides being very close in frequency or of tides with periods very close to (but not exactly) 1 day. This period corresponds to the estimation interval of the EOP. Since the period of these differences (e.g. 18.6 years) are not short compared to the investigated period, these could cause to the estimated trends to be non-zero.

Looking at these trends resulting from the difference between the estimated EOP and the USNO Finals, all are in good agreement with each other. An exception can be found in dY (Fig. 11): for VEOT (green, light and dark brown) the trends for all realizations lie between 1.68 and 1.76 µas/year, for HAMTIDE11a (cyan, magenta, black) the trends are closer together ranging from 1.21 to 1.22 µas/year. For FES2012 the differences in trend are larger, ranging between 1.11 and 1.43 µas/year. However, the trend of the difference between the USNO Finals and the EOP time series estimates using the IERS conventions stands out. Here the trend results in 0.63 µas/year.

The trends and WRMS values components are summarized in Table 6, the smallest WRMS values per component are marked as bold values. As one can see, the values, trends and WRMS, vary for each model and number of tidal constituents by only a few

Table 6

Trends and WRMS of the resulting EOP w.r.t. USNO Finals

Model	x-pole		y-pole		dUT1		dX		dY	
	Trend	WRMS	Trend	WRMS	Trend	WRMS	Trend	WRMS	Trend	WRMS
IERS	2.48	**249.38**	4.40	287.47	− 6.86	239.22	− 2.68	**202.94**	0.63	194.67
fes_IERS	2.19	255.49	4.43	291.14	− 7.32	235.72	− 2.28	208.71	1.43	196.15
fes_pt0001	1.98	254.56	4.35	289.52	− 7.07	**233.69**	− 2.40	207.14	1.25	196.24
fes_pt0100	2.13	251.03	4.53	289.63	− 7.19	235.88	− 2.38	205.71	1.11	194.92
ham_IERS	2.19	250.30	4.31	290.38	− 7.24	237.25	− 2.22	206.15	1.21	195.22
ham_pt0001	2.24	249.72	4.30	288.25	− 7.26	240.05	− 2.22	206.00	1.21	**194.85**
ham_pt0100	2.26	249.79	4.29	288.36	− 7.27	239.80	− 2.24	206.18	1.22	195.12
veo_IERS	2.21	251.43	4.52	289.37	− 7.05	242.85	− 2.38	205.98	1.68	197.57
veo_pt0001	2.24	251.68	4.51	289.51	− 7.03	242.82	− 2.36	206.13	1.68	197.71
veo_pt0100	2.19	251.21	4.58	**288.16**	− 7.12	242.73	− 2.40	205.78	1.76	197.22

Reference epoch is J2000.0, WRMS is calculated after subtracting the trends in the difference. The smallest WRMS are marked as bold values. Units are [μas year^{-1}]/[μs year^{-1}], [μas]/[μs], respectively

percent. No model significantly diminishes the differences w.r.t. the USNO Finals.

We also assessed the impact of the different models on the station baseline repeatability. The number of improved baselines w.r.t. the IERS 2010 solution ranges between 44% (ham_IERS) and 51% (fes_IERS), where the average improvement of the improved baselines is 7% in case of fes_IERS, corresponding to 1 mm. From that we can conclude that the baseline lengths are only affected on the sub-millimeter level.

6. Conclusion

In this work, we developed alternative models for short-period ocean tidal Earth rotation variation and compared them to the IERS 2010 conventional model. The choice of the ocean tide model and the determination of the tidal velocities have been identified as the main uncertainties. Therefore, we focused on the implementation of the tidal velocity inversion, as proposed by Ray (2001), and the investigation of the influence of the prescribed parameters therein Madzak et al. (2016) . Besides the obvious effect of the chosen ocean tide model on the oceanic tidal angular momentum by the mass redistribution, which is characterized by the tidal heights, the resulting ERP model is significantly affected by

associated tidal velocities, which enter into the relative ocean tidal angular momentum (motion term). Moreover, the choice of a set of parameters for the inversion approach, such as internal friction coefficients and/or the weighting of the continuity equation in the set of simplified equation of motion, results in larger variations of the ERP model than the exchange of the commonly considered elastic Earth models. Therefore, we have focused on a further development of the proposed tool for the tidal velocity determination (Ray 2001).

To consider ocean tidal angular momentum of minor tides, which are neither considered in hydro-dynamical models (such as FES2012 or HAMTIDE), nor could be detected in long-term satellite altimetry, an interpolation scheme was developed and tested. Here, the interpolation of tidal heights according to admittance theory was transferred into a scheme of interpolated ocean tidal angular momentum in the frequency domain. This approach was applied on different ocean tide models (FES2012, HAMTIDE and EOT11a) and for different partial tides, from a set of the dominant 29, 71 (like in the IERS conventional model) or for given threshold values of the tide generating potential.

We tested the alternative models in VLBI analysis for the time span 1979 until 2016. The weighted root-mean squares (WRMS) of the delays with and without additional estimation of ERP show no significant differences between the IERS conventional model and the approach based on the tidal models FES2012, HAMTIDE11a and EOT11a. All models reduce significantly the WRMS in comparison with an analysis without any a-priori model for short-period variation of Earth rotation due to ocean tides. From these findings, we conclude that the models presented here are an alternative for the IERS conventional model, which are based on a re-examined theoretical description and long-term satellite altimetry observation only.

Acknowledgements

This work was supported by the FWF Austrian Science Fund, Project 24813. We also thank the IVS for providing the VLBI data. Thanks also go to Wolfgang Bosch for his input.

Appendix

In Table 7 we summarize the differences between the IERS model and the other models for the major tides. The columns headed by A_s and A_c show the corrections to the amplitudes of the sine and cosine terms for the relevant 71 tides in pole coordinates (x_p and y_p) and UT1. For the sake of concision, the values given in the table are obtained as the median and range of the correction differences between the conventional IERS model and the other models, based on FES2012, HAMTIDE12, and VEOT. Table also displays the range of amplitude corrections (i.e., the

Table 7

Median and range of the amplitude differences (i.e. the maximum difference between the maximum and minimum values) to the coefficients of sin(argument) and cos(argument) of semidiurnal variations caused by ocean tides in pole coordinates (x_p and y_p) and UT1 between the conventional IERS model and the other models, based on FES2012, HAMTIDE12, and VEOT

| | | | | | | | Median | | | | | | Range | | | | | |
| | | | | | | | x | | y | | UT1 | | x | | y | | UT1 | |
Tide	Υ	1	1'	F	D	Ω	A_s	A_c	A_s	A_c	A_s	A_c	A_s	A_c	A_s	A_c	A_s	A_c
	1	− 1	0	− 2	− 2	− 2	0.29	− 0.57	0.57	0.29	− 0.03	− 0.08	0.45	0.91	0.91	0.45	0.14	0.1
	1	− 2	0	− 2	0	− 1	0.02	− 0.2	0.2	0.02	0.01	− 0.03	0.15	0.26	0.26	0.15	0.06	0.03
$2Q_1$	1	− 2	0	− 2	0	− 2	0.1	− 1.06	1.06	0.1	0.04	− 0.15	0.82	1.39	1.39	0.82	0.31	0.18
	1	0	0	− 2	− 2	− 1	0	− 0.22	0.22	0	0.01	− 0.03	0.16	0.27	0.27	0.16	0.07	0.04
σ_1	1	0	0	− 2	− 2	− 2	0.02	− 1.13	1.13	0.01	0.06	− 0.16	0.89	1.41	1.41	0.89	0.35	0.19
	1	− 1	0	− 2	0	− 1	− 0.58	− 0.88	0.88	− 0.58	0.07	− 0.09	0.39	0.62	0.62	0.39	0.24	0.08
Q_1	1	− 1	0	− 2	0	− 2	− 3.17	− 4.63	4.63	− 3.18	0.4	− 0.44	2.02	3.27	3.27	2.02	1.28	0.41
	1	1	0	− 2	− 2	− 1	− 0.12	− 0.16	0.16	− 0.12	0.01	− 0.01	0.06	0.1	0.1	0.06	0.04	0.01
RO_1	1	1	0	− 2	− 2	− 2	− 0.65	− 0.84	0.84	− 0.65	0.07	− 0.07	0.31	0.54	0.54	0.31	0.22	0.06
	1	0	0	− 2	0	0	0.1	− 0.01	0.01	0.1	0	0	0.03	0.11	0.11	0.03	0.02	0
	1	0	0	− 2	0	− 1	− 3.31	0.39	− 0.39	− 3.31	0.08	− 0.07	0.83	3.88	3.88	0.83	0.6	0.15
O_1	1	0	0	− 2	0	− 2	− 17.64	2.24	− 2.23	− 17.64	0.41	− 0.35	4.4	20.53	20.53	4.4	3.15	0.78
	1	− 2	0	0	0	0	0.1	− 0.02	0.02	0.1	0	0	0.04	0.11	0.11	0.04	0.02	0.01
TO_1	1	0	0	0	− 2	0	0.1	− 0.12	0.12	0.1	0.01	0.01	0.24	0.12	0.12	0.24	0.04	0.02
	1	− 1	0	− 2	2	− 2	− 0.15	− 0.2	0.2	− 0.15	− 0.01	0.03	0.37	0.62	0.62	0.37	0.05	0.03
	1	1	0	− 2	0	− 1	− 0.11	− 0.14	0.14	− 0.11	− 0.01	0.02	0.27	0.45	0.45	0.27	0.04	0.02
	1	1	0	− 2	0	− 2	− 0.57	− 0.76	0.76	− 0.57	− 0.05	0.1	1.41	2.41	2.41	1.41	0.2	0.12
M_1	1	− 1	0	0	0	0	− 1.59	− 2.13	2.13	− 1.59	− 0.15	0.29	3.92	6.69	6.69	3.92	0.55	0.34
	1	− 1	0	0	0	− 1	− 0.32	− 0.43	0.43	− 0.32	− 0.03	0.06	0.78	1.34	1.34	0.78	0.11	0.07
χ_1	1	1	0	0	− 2	0	− 0.3	− 0.41	0.41	− 0.3	− 0.03	0.05	0.71	1.24	1.24	0.71	0.1	0.06
π_1	1	0	− 1	− 2	2	− 2	0.07	0.5	− 0.5	0.07	0.04	− 0.03	0.26	0.59	0.59	0.26	0.09	0.06
	1	0	0	− 2	2	− 1	0.03	− 0.08	0.08	0.03	− 0.01	0	0.03	0.05	0.05	0.03	0.01	0.01
P_1	1	0	0	− 2	2	− 2	− 2.35	7.47	− 7.47	− 2.35	0.52	− 0.21	2.93	4.48	4.48	2.93	1.15	0.79
	1	0	1	− 2	2	− 2	0.03	− 0.05	0.05	0.03	0	0	0.04	0.02	0.02	0.04	0.01	0
S_1	1	0	− 1	0	0	0	0.11	− 0.14	0.14	0.11	− 0.01	0	0.11	0.05	0.05	0.11	0.02	0.01
	1	0	0	0	0	1	− 0.34	0.32	− 0.32	− 0.34	0.03	0.02	0.45	0.49	0.49	0.45	0.04	0.02

Table 7

continued

Tide	Υ	l	l'	F	D	Ω	Median x A_s	A_c	y A_s	A_c	UT1 A_s	A_c	Range x A_s	A_c	y A_s	A_c	UT1 A_s	A_c
K_1	1	0	0	0	0	0	17.79	− 16.15	16.15	17.79	− 1.58	− 0.9	23.06	25.63	25.63	23.06	1.79	1.04
	1	0	0	0	0	− 1	2.43	− 2.15	2.15	2.43	− 0.22	− 0.13	3.19	3.63	3.63	3.19	0.24	0.14
	1	0	0	0	0	− 2	− 0.05	0.04	− 0.04	− 0.05	0.01	0	0.07	0.08	0.08	0.07	0.01	0
ψ_1	1	0	1	0	0	0	0.16	− 0.11	0.11	0.16	− 0.02	− 0.01	0.26	0.37	0.37	0.26	0.01	0
$\varphi 1$	1	0	0	2	− 2	2	0.39	− 0.12	0.12	0.39	− 0.03	− 0.04	0.64	1.01	1.01	0.64	0.04	0.01
TT_1	1	− 1	0	0	2	0	1.28	0.43	− 0.43	1.28	0.05	− 0.14	2.21	3.87	3.87	2.21	0.19	0.11
J_1	1	1	0	0	0	0	7.88	2.94	− 2.94	7.88	0.32	− 0.87	13.62	24	24	13.62	1.22	0.72
	1	1	0	0	0	− 1	1.56	0.59	− 0.59	1.56	0.06	− 0.17	2.71	4.78	4.78	2.71	˙0.24	0.14
S_{01}	1	0	0	0	2	0	2.89	1.32	− 1.32	2.89	0.15	− 0.31	4.97	8.97	8.97	4.97	0.48	0.29
	1	2	0	0	0	0	1.57	0.73	− 0.73	1.57	0.08	− 0.17	2.71	4.88	4.88	2.71	0.26	0.16
O_{01}	1	0	0	2	0	2	10.6	4.91	− 4.91	10.61	0.57	− 1.14	18.23	32.96	32.96	18.23	1.79	1.08
	1	0	0	2	0	1	6.81	3.16	− 3.16	6.81	0.37	− 0.73	11.71	21.18	21.18	11.71	1.15	0.69
	1	0	0	2	0	0	1.43	0.66	− 0.66	1.43	0.08	− 0.15	2.47	4.45	4.45	2.47	0.24	0.15
ν_1	1	1	0	2	0	2	3.59	1.74	− 1.74	3.59	0.23	− 0.37	6.16	11.28	11.28	6.16	0.62	0.38
	1	1	0	2	0	1	2.3	1.11	− 1.11	2.3	0.15	− 0.24	3.96	7.23	7.23	3.96	0.4	0.24
	2	− 3	0	− 2	0	− 2	0.17	− 0.07	0.23	0.08	− 0.01	0.04	1.99	0.39	0.33	1.68	0.12	0.17
	2	− 1	0	− 2	− 2	− 2	0.35	− 0.18	0.53	0.21	− 0.03	0.11	4.56	0.85	0.77	3.86	0.28	0.38
$2N_2$	2	− 2	0	− 2	0	− 2	− 0.76	0	0.77	0.63	− 0.15	0.24	5.24	1.18	1.6	4.26	0.32	0.42
μ_2	2	0	0	− 2	− 2	− 2	− 1.15	0.15	1.01	0.75	− 0.18	0.26	4.94	1.28	2.02	3.93	0.3	0.39
	2	0	1	− 2	− 2	− 2	− 0.08	0.02	0.07	0.05	− 0.01	0.02	0.28	0.08	0.14	0.22	0.02	0.02
	2	− 1	− 1	− 2	0	− 2	0.07	0	0.05	− 0.03	0	0	0.06	0.08	0.1	0.04	0.01	0.01
	2	− 1	0	− 2	0	− 1	0.3	0.02	− 0.22	− 0.14	0.01	− 0.02	0.3	0.34	0.41	0.27	0.04	0.04
N_2	2	− 1	0	− 2	0	− 2	− 8.15	− 0.27	6.2	3.75	− 0.24	0.53	8.17	9.17	11.1	7.13	0.95	0.95
	2	− 1	1	− 2	0	− 2	− 0.08	0	0.06	0.03	0	0.01	0.09	0.09	0.1	0.08	0.01	0.01
ν_2	2	1	0	− 2	− 2	− 2	− 1.47	− 0.1	1.06	0.67	− 0.04	0.11	2.02	1.71	1.93	1.78	0.19	0.18
	2	1	1	− 2	− 2	− 2	− 0.06	− 0.01	0.04	0.03	0	0.01	0.11	0.08	0.08	0.09	0.01	0.01
	2	− 2	0	− 2	2	− 2	− 0.06	− 0.03	0.03	0.02	0	0	0.17	0.07	0.05	0.15	0.01	0
	2	0	− 1	− 2	0	− 2	− 0.08	− 0.04	0.04	0.03	0	0	0.2	0.07	0.05	0.17	0.01	0
	2	0	0	− 2	0	− 1	− 0.99	− 0.49	0.41	0.38	− 0.02	− 0.04	2.16	0.79	0.53	1.92	0.1	0.04
M_2	2	0	0	− 2	0	− 2	26.89	12.77	− 11.15	− 10.25	0.57	0.98	57.96	21.08	14.24	51.34	2.61	1.05
	2	0	1	− 2	0	− 2	0.04	− 0.02	− 0.06	0	0	0	0.22	0. 1	0.07	0.17	0.01	0.01
$\lambda 2$	2	− 1	0	− 2	2	− 2	0.26	0.54	0.26	0.31	− 0.02	0	0.98	0.87	0.5	1.24	0.04	0.12
L_2	2	1	0	− 2	0	− 2	0.9	1.98	0.94	1.31	− 0.07	0	3.64	3.28	1.9	4.69	0.17	0.47
	2	− 1	0	0	0	0	− 0.22	− 0.49	− 0.24	− 0.33	0.02	0	0.9	0.81	0.47	1.17	0.04	0.12
	2	− 1	0	0	0	− 1	− 0.1	− 0.21	− 0.1	− 0.15	0.01	0	0.39	0.36	0.21	0.51	0.02	0.05
T_2	2	0	− 1	− 2	2	− 2	1.09	0	− 0.27	− 1.55	0.06	0.06	0.89	0.26	0.82	0.93	0.05	0.16
S_2	2	0	0	− 2	2	− 2	24.49	8.65	− 3.57	− 26.4	1.24	1.11	7.87	11.38	11.35	5.32	0.65	1.68
R_2	2	0	1	− 2	2	− 2	− 0.19	− 0.14	0.02	0.25	− 0.01	− 0.01	0.07	0.16	0.07	0.1	0.01	0.01
	2	0	0	0	0	1	− 0.1	− 0.09	− 0.01	0.11	0	− 0.01	0.05	0.1	0.03	0.09	0	0
K_2	2	0	0	0	0	0	8.08	6.48	0.8	− 8.36	0.27	0.42	4.22	8.23	2.63	6.45	0.29	0.18
	2	0	0	0	0	− 1	2.46	1.99	0.29	− 2.5	0.08	0.13	1.29	2.5	0.8	1.98	0.09	0.06
	2	0	0	0	0	− 2	0.27	0.22	0.03	− 0.27	0.01	0.01	0.14	0.28	0.09	0.22	0.01	0.01
	2	1	0	0	0	0	2.39	2.34	0.86	− 0.45	− 0.01	0.07	2.64	3.65	1.47	4.22	0.14	0.36
	2	1	0	0	0	− 1	1.05	1.02	0.38	− 0.2	− 0.01	0.03	1.16	1.6	0.65	1.85	0.06	0.16
	2	0	0	2	0	2	1.4	1.38	0.48	− 0.08	− 0.01	0.03	1.7	2.29	0.98	2.75	0.09	0.25

Units: as

maximum difference between the maximum and minimum values). The comparison of both parameters is useful to get more insight into the magnitude.

REFERENCES

Alterman, Z., Jarosch, H., & Pekeris, C. (1961). Propagation of rayleigh waves in the earth. *Geophysical Journal International, 4*, 219–241.

Artz, T., Tesmer, S., & Nothnagel, A. (2011). Assessment of periodic sub-diurnal Earth rotation variations at tidal frequencies through transformation of the VLBI normal equations. *Journal of Geodesy, 85*, 565–584.

Bachmann, S., Thaller, D., Roggenbuck, O., Lösler, M., & Messerschmitt, L. (2016). IVS contribution to ITRF2014. *Journal of Geodesy, 90*, 631–654. https://doi.org/10.1007/s00190-016-0899-4.

Bennett, A. F. (1992). *Inverse methods in physical oceanography* (p. 347). New York: Cambridge University Press.

Böhm, S., Brzezinski, A., & Schuh, H. (2011). Complex demodulation in VLBI estimation of high frequency Earth rotation components. *Journal of Geodesy, 62*, 56–68.

Brosche, P., Wünsch, J., Campbell, J., & Schuh, H. (1991). Ocean tide effects in universal time detected by VLBI. *Astronomy and Astrophysics, 245*, 676–682.

Brosche, P., Wünsch, J., Seiler, U., & Sndermann, J. (1989). Periodic changes in due to oceanic tides. *Astronomy and Astrophysics, 220*, 318–320.

Brzezinski, A. (1994). Polar motion excitation by variations of the effective angular momentum function, II: Extended-model. *Manuscripta Geodaetica, 19*, 157–171.

Bullen, K., & Haddon, R. (1967). Earth models based on compressible theory. *Physics of the Earth and Planetary Interiors, 1*, 1–13.

Carrere, L., Lyard, F., Cancet, M., Guillot, A., & Roblou, L. (2012). FES2012: A new global tidal model taking advantage of nearly 20 years of altimetry. In: *Proceedings of Meeting 20 Years of Altimetry*, Venice.

Cartwright, David E., & Ray, Richard D. (1991). Energetics of global ocean tides from Geosat altimetry. *Journal of Geophysical Research: Oceans, 96*, 16897–16912.

Chao, B. F., & Ray, R. D. (1997). Oceanic tidal angular momentum and Earths rotation variations. *Progress in Oceanography, 40*, 399–421. (Pergamon).

Dziewonski, A., & Anderson, D. (1981). Preliminary reference earth model. *Physics of the Earth and Planetary Interiors, 25*, 297–356.

Egbert, G., Bennett, A., & Foreman, M. (1994). TOPEX/POSEIDON tides estimated using a global inverse model. *Journal of Geophysical Research: Oceans, 99*(C12), 24821–24852.

Gilbert, F., & Dziewonski, A. (1975). An application of normal mode theory to the retrieval of structural parameters and source mechanism from seismic spectra. *Philosophical Transactions of the Royal Society A, 278*, 187–269.

Gross, R. S. (2007). Earth rotation variations—Long period. In T. A. Herring (Ed.), *Physical geodesy, Treatise on geophysics* (vol. 3, pp. 239–294). Amsterdam: Elsevier.

Gross, R. (1993). The effect of ocean tides on the Earths rotation as predicted by the results of an ocean tide model. *Geophysical Research Letters, 20*, 293–296.

Hartmann, T., & Wenzel, H.-G. (1995). The HW95 tidal potential catalogue. *Geophysical Research Letters, 22*, 3553–3556.

Henderscott, M. C. (1972). The effects of solid earth deformation on global ocean tides. *Geophysical Journal International, 29*, 389–402.

Koot, L., & de Viron, O. (2011). Atmospheric contributions to nutations and implications for the estimation of deep Earth's properties from nutation observations. *Geophysical Journal International, 185*, 1255–1265. https://doi.org/10.1111/j.1365-246X.2011.05026.x.

Le Provost, C., Genco, M. L., Lyard, F., Vincent, P., & Canceil, P. (1994). Spectroscopy of the world ocean tides from a finite element hydrodynamic model. *Journal of Geophyical Research, 99*(C12), 24777–24797. https://doi.org/10.1029/94JC01381.

Lyard, F., Lefevre, F., Letellier, T., & Francis, O. (2006). Modelling the global ocean tides: modern insight from FES2004. *Ocean Dynamics, 56*, 394–415.

Madzak, M., Schindelegger, M., Böhm, J., Bosch, W., & Hagedoorn, J. (2016). High-frequency Earth rotation variation deduced from altimetry-based ocean tides. *Journal of Geodesy, 90*, 1237–1253.

McCarthy, D. D. (ed.) (1996). IERS conventions. (*IERS Technical Note: 21*). Paris: Central Bureau of IERS-Observatoire de Paris, p. 97.

Munk, W. H., & Cartwright, D. E. (1966). Tidal spectroscopy and prediction. *Philosophical Transactions of the Royal Society, 263*, 1–50.

Munk, W. H., & MacDonald, G. J. F. (1960). *The rotation of the earth*. New York: Cambridge University Press.

Nothnagel, A., Alef, W., Amagai, J., Andersen, P. H., Andreeva, T., Artz, T., et al. (2015). The IVS data input to ITRF2014. International VLBI Service for Geodesy and Astrometry: IVS, GFZ Data Services.

Petit, G., & Luzum. B. (eds.) (2010). IERS conventions.

Pugh, D., & Woodworth, P. (2014). *Sea-level science: Understanding tides, surges, tsunamis and mean sea-level changes.* Cambridge: Cambridge University Press.

Ray, R. (1998). Ocean self-attraction and loading in numerical tidal models. *Marine Geodesy, 21*, 181–192. https://doi.org/10.1080/01490419809388134.

Ray, R. (2001). Inversion of oceanic tidal currents from measured elevations. *Journal of Marine Systems, 28*, 1–18.

Sasao, T., Okubo, S., & Sato, M. (1980). A simple theory on dynamical effects of stratified fluid core upon nutational motion of the Earth. In: R. L. Duncombe (Ed.), *Proceedings from IAU Symposium no. 78 held in Kiev, USSR 23–28 May, 1977.* International Astronomical Union. Symposium no. 78. Nutation and the Earth's Rotation (p. 165). Dordrecht: D. Reidel Pub. Co.

Sasao, T., & Wahr, J. (1981). An excitation mechanism for the free core nutation. *Geophysical Journal International, 64*, 729–746.

Savcenko, R., & Bosch, W. (2012). EOT11a—Empirical ocean tide model from multi-mission satellite altimetry. 89, Deutsches Geodätisches Forschungsinstitut (DGFI), München.

Stammer, D., Ray, R. D., Andersen, O. B., Arbic, B. K., Bosch, W., Carrre, L., et al. (2014). Accuracy assessment of global barotropic ocean tide models. *Reviews of Geophysics, 52*(3), 1944–9208.

Taguchi, E., Stammer, D., & Zahel, W. (2014). Inferring deep ocean tidal energy dissipation from the global high-resolution data-assimilative HAMTIDE model. *Journal of Geophysical Research: Oceans, 119,* 4573–4592.

Wahr, J. M., & Bergen, Z. (1986). The effects of mantle anelasticity on nutations, Earth tides, and tidal variations in rotation rate. *Geophysical Journal International, 87,* 633–688.

Wang, C. (1972). A simple earth model. *Journal of Geophysical Research, 77,* 4318–4329.

Wang, H., Xiang, L., Lulu, J., Jiang, L., Wang, Z., Hu, B., et al. (2012). Load Love Numbers and Green's functions for elastic Earth models PREM, Iasp91, Ak135, and modified models with refined crustal structure from Crust 2.0. *Computers & Geosciences, 49,* 190–199.

Zahel, W. (1991). Modeling ocean tides with and without assimilating data. *Journal of Geophysical Research: Solid Earth, 96*(B12), 20379–20391.

Zahel, W. (1995). Assimilating ocean tide determined data into global tidal models. *Journal of Marine Systems, 6,* 3–13.

Zahel, W., Gavino, J. H., & Seiler, U. (2000). Angular momentum and energy budget of a global ocean tide model with data assimilation [in Spanish]. *GEOS, 20*(4), 400–413.

(Received January 29, 2018, revised April 5, 2018, accepted April 7, 2018, Published online April 21, 2018)

Pure Appl. Geophys. 175 (2018), 1631–1642
© 2017 Springer International Publishing AG, part of Springer Nature
https://doi.org/10.1007/s00024-017-1724-6

Observation of the Earth Liquid Core Resonance by Extensometers

DÓRA BÁN,[1,2] GYULA MENTES,[1] MÁRTA KIS,[3] and ANDRÁS KOPPÁN[3]

Abstract—We performed Earth tidal measurements by quartz tube extensometers of the same type at several observatories (Budapest, Pécs, Sopronbánfalva in Hungary and Vyhne in Slovakia). In this paper, the first attempts to reveal the effect of the Free Core Nutation (FCN) from strain measurements are described. The effect of the FCN on the P1, K1, Ψ1 and Φ1 tidal waves were studied on the basis of tidal results obtained in four observatories. Effectiveness of the correction of tidal data for temperature, barometric pressure and ocean load was also investigated. The obtained K1/O1 ratios are close to the theoretical values with exception of the Pécs station. We found a discrepancy between the observed and theoretical P1/O1 values for all stations with exception of the Budapest station. It was found that the difference between the measured and theoretical Ψ1/O1 and Φ1/O1 ratios was very large independently of correction of the strain data. These discrepancies need further investigations. According to our results, fluid core resonance effects can also be detected by our quartz tube extensometers but correction of strain data for local effects is necessary.

Key words: Extensometer, Earth's tide, free core resonance.

1. Introduction

Earth core resonance is investigated by means of superconducting gravimeters (e.g. Florsch et al. 1994; Florsch and Hinderer 2000; Sato et al. 2004; Sun et al. 2004; Rosat et al. 2009), extensometers (e.g. Mukai et al. 2004; Ping et al. 2006; Amoruso et al. 2012) and in the last decades by space geodetic (VLBI) methods (e.g. Krásná et al. 2013).

Extensometers have lower accuracy in comparison with superconductive gravimeters, but the free core nutation induces about ten times higher resonance in the strain than in the gravity tide (Zürn 1977; Boyarsky et al. 2003; Amoruso et al. 2012).

In the Pannonian Basin, extensometers are mainly used for measurement of local tectonic movements and deformations which also depend on global, regional, local geodynamic processes, as well as on properties of the solid Earth crust and mantle (Mentes 2008, 2010; Eper-Pápai et al. 2014; Brimich et al. 2016). In addition to deformation measurements, extensometric data can also be used to study Earth's tides and related phenomena, e.g. the nearly diurnal resonance of the Earth liquid core, among others. This resonance with a frequency of 1.004915 cycles per solar day influences the amplitudes of tidal diurnal waves close to this frequency. To detect free core nutation (FCN) or in other words, fluid core resonance (FCR), we investigate how the FCN modifies the tidal waves P1, K1, Ψ1 and Φ1 which are close to the FCN frequency (Zürn 1977; Boyarsky et al. 2003). In addition to the detection of the resonance, we investigate how the detection depends on the different corrections (temperature, barometric pressure, ocean load) of extensometric data.

2. Measurement Sites and Instruments

Strain data recorded in four observatories were subjected to study the free core resonance. Figure 1 shows the location of the observatories. Three stations are located in Hungary and one in Slovakia. The Pécs station is a deep observatory (1991–1999); the other stations are surface observatories with overlying rock of different height and topography and they have been functioning since the 1980s (Budapest and Vyhne) and 1991 (Sopronbánfalva). The coordinates,

Session (Poster): Tides and non-tidal loading (Bruno Meurers, David Crossley).

[1] Geodetic and Geophysical Institute, Research Centre for Astronomy and Earth Sciences, Hungarian Academy of Sciences, Csatkai E. u. 6-8, Sopron 9400, Hungary. E-mail: mentes@ggki.hu
[2] Kitaibel Pál Doctoral School of Environmental Sciences, University of West Hungary, Bajcsy-Zsilinszky u. 4., Sopron 9400, Hungary.
[3] Geological and Geophysical Institute of Hungary, Stefánia u. 14, Budapest 1143, Hungary.

Figure 1
Observatories with extensometer in Hungary and in Slovakia

Table 1

Parameters of the extensometers in the Carpathian-Balkan Region

Extensometer	Coordinates of the stations			Azimuth of the instrument	Length of the instrument (m)	Length of data series used for the investigations
	Latitude	Longitude	Height a.s.l. (m)			
Budapest E2	47°33′11″	19°20′24″	240	38°	13.8	2010–2012
Pécs	46°6′52″	18°7′49″	− 694	19°	20.5	1998
Sopronbánfalva	47°40′55″	16°33′32″	280	116°	22	2009–2015
Vyhne	48°29′52″	18°49′48″	420	55°	20.5	2006/2007, 2013–2015

the height of the observatories, the azimuth and the length of the instruments as well as the data series used for our investigations are given in Table 1.

The Mátyáshegy Gravity and Geodynamic Observatory (MGGO) is situated in the karstic environment of Mátyáshegy (Mátyás Hill) in the northwestern suburban part of Budapest. Mátyáshegy is a part of the Buda Mountains, in north central Hungary, belonging to the regional unit 'Transdanubian Mountains'. The galleries were created by thermal water from the upper Triassic flinty limestone and the

discordantly bedding marine upper Eocene nummulitic limestone. The tunnels and galleries of the observatory were based on the natural cave system and were artificially formed. The galleries run about 30 m under the rock cover. The level of the karstic water is about a hundred metres deeper than the level of the station. The Danube is approx. 2 km away from the observatory.

The Pécs station was established in a uranium mine at a depth of 1040 m relative to the surface, far away from the working area of the mine. It was working from 1991 till 1999 when the mine was closed (Mentes and Berta 1997).

The Sopronbánfalva Geodynamic Observatory (SGO) is located on the Hungarian–Austrian border in the Sopron Mountains. The area belongs to the extensions of the Eastern Alps (Alpokalja region). The Sopron Mountains consist of metamorphic rocks of Palaeozoic age such as gneiss and different mica schists (Haas 2001). The observatory is an artificial gallery driven into an outcrop of muscovite gneiss. The rock cover above the observatory is about 60 m.

The Vyhne Tidal Station is located underground, in the gallery of St. Anthony of Padua in the Vyhne valley, Štiavnické vrchy Mts., Central Slovakia. The area in the vicinity of the St. Anthony of Padua gallery is built of Palaeozoic, Mesozoic, Paleogene and Neogene rocks. The gallery is situated at 420 m above sea level. It was driven in the Variscan-age granites, which were disturbed tectonically more times. The first motion caused the mylonitization of the granites. The others were accompanied by young intrusions and mineralizations of the tectonically modified medium (Brimich 2006). The relative depth of the location of the extensometer in the gallery is 50 m.

A detailed description of the Budapest, Sopronbánfalva, Pécs and Vyhne observatories are given by Mentes (2008), Eper-Pápai et al. (2014) and Brimich et al. (2016).

In each observatory, the instruments are the same type of quartz tube extensometer built in scientific co-operation with the Institute of Physics of the Earth of the former USSR Academy of Sciences, Moscow. The capacitive sensor of the instruments and a high precision in situ calibrator were developed in the Geodetic and Geophysical Research Institute of the Hungarian Academy of Sciences (now: Geodetic and Geophysical Institute, RCAES, HAS) in 1991. The resolution of the extensometers is better than 10^{-9} m. The construction and calibration of the extensometers are described by Mentes (1991, 2010) in detail.

3. Methods

3.1. Method for Investigation of the FCN

Several attempts have been made for correction of strain data for temperature, barometric pressure and

ocean load. Strain data were not corrected for cavity effects because they were expected to be small in all stations, since the measurements were carried out in galleries with a small cross section. The topographic effects were also expected to be small (Brimich 2006; Eper-Pápai et al. 2014; Brimich et al. 2016). The tidal processing was carried out by the programme package ETERNA 3.40 (Wenzel 1996). We used the well-known method described by many authors (e.g. Zürn 1977; Boyarsky et al. 2003, etc.) for revealing the liquid core resonance from our extensometric observations. The principle of this method is that tidal waves close to the FCN frequency are stronger influenced by the resonance than the farther. We calculated the ratio of the amplitude factors of the K1 (1.0027379 cpd), P1 (0.9972621 cpd), $\Psi1$ (1.0054757 cpd) and $\Phi1$ (1.0082137) waves—being close to the FCN frequency (1.004915 cpd)—to the amplitude factor of O1 (0.9295357 cpd) which is far from resonance. However, this method does not take into account that local effects affect different tidal waves differently, but measurement errors can be reduced. Therefore, both the amplitude factors and the ratios of the amplitude factors calculated from data series measured in different years in four observatories were compared with the theoretical values and with each other. The comparison was made by using amplitude factors and ratios obtained from uncorrected extensometric data series as well as corrected for temperature, barometric pressure and ocean load to investigate the effect of different corrections.

The theoretical amplitude factors of K1, P1, $\Psi1$, $\Phi1$ and amplitude factor ratios K1/O1, P1/O1, $\Psi1$/O1 and $\Phi1$/O1 ratios (Table 2) were calculated for the azimuth of the instruments (Boyarsky et al. 2003; Melchior 1978; Wahr 1981). The amplitude factors were calculated using the Wahr-Dehant Earth model (Wahr 1981; Dehant 1987) and the HW95 tidal potential catalogue (Hartmann and Wenzel 1995). For the ocean tide prediction the SPOTL routine was applied using the Gutenberg-Bullen Earth model "gr.gbocen.wef.p02.ce.gz", the ocean model "csr4tr" and the local ocean model "osu.mediterranean" (Agnew 2013).

3.2. Data Pre-processing and Correction

Several methods have been carried out to correct strain data for temperature and barometric effects.

Table 2

Theoretical amplitude factors of the tidal waves K1, P1, Ψ1, Φ1, O1 calculated in the azimuth of the extensometers and their relation to the theoretical amplitude factor of O1

Extensometer	Theoretical amplitude factors					Ratios of the amplitude factors			
	K1	P1	Ψ1	Φ1	O1	K1/O1	P1/O1	Ψ1/O1	Φ1/O1
Sopronbánfalva (SGO)	0.31	0.38	0.76	0.46	0.40	0.78	0.94	1.89	1.16
Budapest (MGGO) E2	0.24	0.31	0.70	0.40	0.33	0.72	0.93	2.11	1.20
Pécs	0.19	0.26	0.66	0.35	0.28	0.67	0.91	2.31	1.23
Vyhne	0.29	0.36	0.74	0.44	0.38	0.76	0.94	1.95	1.17

Table 3

Results of tidal evaluation of strain data recorded in the SGO in 2014

Frequency (cpd) from	to	Wave	Theoretical amplitude	Ampl. factor	Ampl. fact. stdv.	Phase lead	Phase lead stdv.
0.50137	0.91139	Q1	1.2624	0.377	0.182	7.0114	26.7189
0.911391	0.947991	O1	6.5933	0.439	0.055	− 8.0442	7.5072
0.947992	0.981854	M1	0.5183	0.260	0.394	16.9533	86.4715
0.981855	0.998631	P1	3.0674	0.324	0.083	− 179.945	15.3547
0.981855	1.023622	K1	9.2691	0.342	0.048	− 15.4841	9.2815
1.023623	1.057485	J1	0.5185	0.613	0.459	4.9403	42.719
1.057486	1.470243	OO1	0.2835	1.441	1.071	− 0.0403	42.5532

The one-year-long strain, temperature and barometric pressure data recorded with a sampling rate of 1 min were used for correction by the T-soft programme (Van Camp and Vauterin 2005). The long-term drift was approximated by fitting a polynomial of 9th order to the raw data series and it was subtracted from strain data. The data series were despiked and ungapped. The gaps with a length longer than 1 day were filled with the theoretical strain fitted to the measured strain. Short gaps were linearly interpolated. Theoretical tide was subtracted from the strain data and the remaining data were corrected for the temperature and barometric pressure by simple linear or multivariable regression methods and after the correction the theoretical tide was added back. During the correction procedure time lags between strain and temperature were also taken into consideration. The time lag between the strain and the temperature varied between 18 and 22 days, while it was practically negligible between the strain and barometric pressure (see also Mentes 2000; Mentes and Eper-Pápai 2009). The long-term temperature variation correlated with the long-term strain (the correlation coefficient was usually greater than 0.8). The correlation coefficient in the tidal band was

smaller than 0.1. Therefore, the correction of the temperature effect was only limited to the elimination of the seasonal strain variation. The seasonal effect of the barometric pressure was negligible (see, e.g. Mentes 2000; Mentes and Eper-Pápai 2009). Then the data were low-pass filtered and decimated to 1-h sampling and processed by ETERNA 3.40. Results of tidal evaluation of strain data measured in the SGO in 2014 are given in Table 3. Comparing the obtained O1, K1, P1 amplitude factors with the theoretical values in Table 2 we find a good agreement within the error limit. Figure 2 shows the Fourier transform of the residuals and the noise level N, which was calculated as the average of the amplitudes with exception of the largest amplitudes. The value obtained by this method is slightly higher than that calculated by ETERNA in the diurnal frequency range. It can be seen that the P1 wave is above the noise level in the residuals, while K1 is below it. This is the reason for the better detection of P1 than that of K1.

Since continuous strain, temperature and barometric pressure records (with some short gaps of maximum some days) were available in the SGO, the effect of the FCN on the Ψ1 and Φ1 waves was also

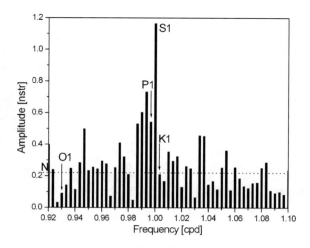

Figure 2
Fourier transform of the residuals of the strain data recorded in the SGO in year 2014. *N* noise level

investigated using data measured from 2009 to 2015. The yearly despiked, ungapped datasets were merged and the long-term drift (instrumental drift and tectonic movement) was approximated by a polynomial of 9th order which was subtracted from the data (see E in Fig. 3). The strain and temperature data were smoothed by a moving average filter using 201 adjacent data (FE and FT in Fig. 3). The smoothed strain (FE) was subtracted from the unsmoothed strain (E) to eliminate the seasonal variation of the strain data (SDCE). In Fig. 3, the correlation between the long-term variation of temperature and strain variation is obvious (correlation coefficient is greater than 0.8). The strain data cleansed from seasonal variations (SDCE in Fig. 3) were corrected for barometric pressure (DFPCE in Fig. 3). Due to

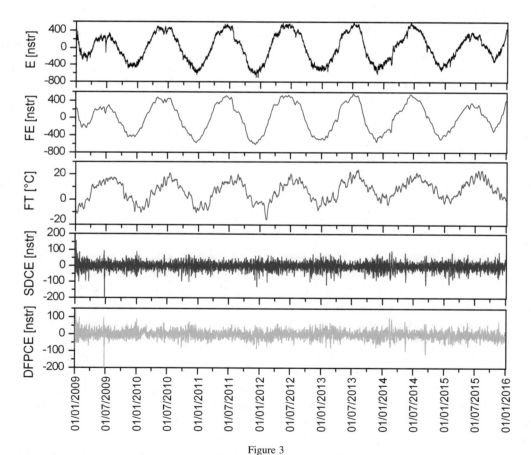

Figure 3
Correction of the strain data recorded in the SGO between 2009 and 2015 for temperature and barometric pressure. E, FE, FT denote the raw strain (long-term polynomial drift subtracted from it), the smoothed strain and the smoothed temperature data, respectively. *SDCE* seasonal drift-free strain [the seasonal variation of the strain data caused by temperature (FE) is subtracted from strain data (E)], *DFPCE* drift-free strain data corrected for barometric pressure

Table 4

Results of the tidal evaluation of strain data recorded in the SGO between 2009 and 2015

Frequency (cpd) from to		Wave	Theoretical amplitude	Ampl. factor	Ampl. fact. stdv.	Phase lead	Phase lead stdv.
Raw data							
0.50137	0.91139	Q1	1.2624	0.59999	0.06948	− 7.0102	6.6344
0.911391	0.947991	O1	6.5935	0.57803	0.01456	− 5.4504	1.4435
0.947992	0.981854	M1	0.5183	0.63496	0.16799	− 20.6682	15.1583
0.981855	0.998631	P1	3.0674	0.23737	0.03083	24.4849	7.4421
0.998632	1.001369	S1	0.0735	20.99471	1.82123	− 45.7234	4.9703
1.00137	1.004107	K1	9.2693	0.43621	0.01057	0.9136	1.3889
1.004108	1.006845	PSI1	0.0735	4.09174	1.29957	− 93.6278	18.1973
1.006846	1.023622	PHI1	0.1317	1.08956	0.71456	− 27.1402	37.5764
1.023623	1.057485	J1	0.5185	0.2473	0.18361	6.3389	42.5421
1.057486	1.470243	OO1	0.2836	0.86348	0.31176	− 11.0843	20.6861
Drift-free raw data							
0.50137	0.91139	Q1	1.2624	0.62603	0.06968	− 7.0993	6.3772
0.911391	0.947991	O1	6.5935	0.59307	0.01461	− 5.4423	1.411
0.947992	0.981854	M1	0.5183	0.62814	0.16848	− 20.5625	15.3681
0.981855	0.998631	P1	3.0674	0.22908	0.03092	24.5062	7.7342
0.998632	1.001369	S1	0.0735	20.24397	1.8266	− 45.7249	5.1699
1.00137	1.004107	K1	9.2693	0.4204	0.0106	0.9193	1.4453
1.004108	1.006845	PSI1	0.0735	3.93739	1.3034	− 93.7032	18.9665
1.006846	1.023622	PHI1	0.1317	1.05143	0.71667	− 27.233	39.0541
1.023623	1.057485	J1	0.5185	0.23886	0.18415	6.3879	44.174
1.057486	1.470243	OO1	0.2836	0.86479	0.31268	− 11.0864	20.7158
Drift-free raw data corrected for barometric pressure							
0.50137	0.91139	Q1	1.2624	0.62578	0.05602	− 6.9704	5.1286
0.911391	0.947991	O1	6.5935	0.59802	0.01174	− 4.9824	1.1249
0.947992	0.981854	M1	0.5183	0.58833	0.13544	− 22.5066	13.19
0.981855	0.998631	P1	3.0674	0.20678	0.02486	7.3488	6.8879
0.998632	1.001369	S1	0.0735	19.97928	1.46836	− 25.5594	4.211
1.00137	1.004107	K1	9.2693	0.41827	0.00852	− 1.3251	1.1678
1.004108	1.006845	PSI1	0.0735	3.8671	1.04778	− 92.3485	15.5238
1.006846	1.023622	PHI1	0.1317	1.32689	0.57612	− 22.4571	24.8773
1.023623	1.057485	J1	0.5185	0.28898	0.14804	− 4.0337	29.352
1.057486	1.470243	OO1	0.2836	0.80784	0.25136	− 10.2809	17.827

random weather fronts the barometric pressure does not show a clear seasonal variation, so we used it directly for correction. Since drift-free strain data do not correlate with either raw temperature data or temperature data cleaned from seasonal effects (the correlation coefficient is less than 0.2 in both cases), the strain data have not been corrected for temperature. The results of the tidal evaluation of the raw data, drift-free raw data and drift-free raw data corrected for barometric pressure, obtained by ETERNA 3.40 are listed in Table 4. The tidal factors obtained from the three datasets do not differ significantly while there is a discrepancy between them and the theoretical values in Table 2. Figure 4 shows the Fourier transform of the residuals: (a) raw

data (without any corrections), (b) drift-free raw data (long-term and temperature caused seasonal drift is subtracted from the raw data), (c) drift-free raw data corrected for barometric pressure. There is no significant difference between the residuals, which is consistent with the tidal results.

A two-year-long dataset recorded in 2013 and 2014 was similarly corrected for barometric pressure. The obtained tidal parameters and the Fourier transform of the residuals are shown in Table 5 and Fig. 5, respectively. The Fourier transform in Fig. 5 differs significantly from those shown in Figs. 2 and 4. Our experience was that the short-term temperature changes have no significant influence on the tidal strain (see also Mentes 2000; Mukai et al. 2004). We

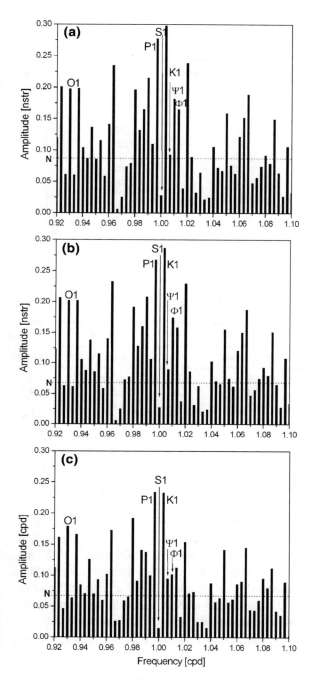

Figure 4

Fourier transform of the residuals of the strain data recorded in the SGO in years 2009–2015. **a** Raw data, **b** drift (long-term and temperature caused seasonal) subtracted from the raw data, **c** drift-free raw data corrected for barometric pressure. *N* noise level

found no large difference between the different methods of strain correction for temperature and barometric pressure. In the most cases the corrections

made by the above described methods did not give better results than the corrections made by linear regression methods before tidal processing either made the corrections by the programme package ETERNA 3.40 or using uncorrected strain data. It can be seen in Table 4 and Fig. 4 that the correction of strain data for temperature and barometric pressure variations considerably contribute to reducing the noise level (see also Amoruso et al. 2012; Botta 2012). In Table 6 we summarised the amplitude factors of the tidal waves O1, P1, K1, Ψ1, Φ1 and their standard deviations obtained in the SGO in years 2014, 2013–2014 and 2009–2015. It is clear from Table 6 that the standard deviations are smaller for the longer time series. This may be due to the fact that tidal components can be more accurately estimated from long data series than from short series of data.

4. Results and Discussion

In the SGO, the K1/O1, P1/O1, Ψ1/O1 and Φ1/O1 ratios were determined from the data series recorded from 2009 to 2015. The data were separately evaluated in years 2013, 2014, 2013–2014 and 2009–2015. Figure 6 shows the K1/O1 and P1/O1 amplitude ratios. The K1/O1 ratios are close to the theoretical values and show less variance than the P1/O1 values. The ratios calculated from the seven-year data series show the highest deviations. For this reason we have not found an explanation so far. The Ψ1/O1 and Φ1/O1 ratios were calculated from the two-year-long data series 2013–2014 and from the seven-year long data series 2009–2015 (Table 7). The deviations of the ratios from the theoretical values and the errors are very large independently from the data correction methods.

Figure 7 shows the K1/O1, P1/O1 amplitude factor ratios obtained by different correction of the strain data recorded in the SGO in 2014. THEOR denotes the theoretical amplitude factor ratios, UNC, TC, PC, PTC, OC, OTC, OPC, OPTC denote amplitude factor ratios obtained using uncorrected strain data, corrected for temperature, for barometric pressure, for temperature and barometric pressure, for ocean load, for ocean load and temperature, for ocean

Table 5

Results of tidal evaluation of strain data recorded in the SGO in years 2013 and 2014

Frequency (cpd)		Wave	Theoretical amplitude	Ampl. factor	Ampl. fact. stdv.	Phase lead	Phase lead stdv.
From	To						
0.50137	0.91139	Q1	1.2624	0.601	0.123	− 9.4744	11.7364
0.911391	0.947991	O1	6.5934	0.558	0.025	− 1.9864	2.5985
0.947992	0.981854	M1	0.5183	0.270	0.322	30.5848	68.3949
0.981855	0.998631	P1	3.0674	0.211	0.048	− 15.2387	13.0295
0.998632	1.001369	S1	0.0735	14.312	2.802	21.9104	11.2171
1.00137	1.004107	K1	9.2691	0.423	0.018	− 3.2489	2.388
1.004108	1.006845	PSI1	0.0735	5.815	2.052	− 113.555	20.2183
1.006846	1.023622	PHI1	0.1317	0.169	1.128	111.3932	381.8954
1.023623	1.057485	J1	0.5185	0.770	0.326	28.649	24.2903
1.057486	1.470243	OO1	0.2836	0.992	0.795	11.1606	45.9096

Table 6

Amplitude factors of the tidal waves O1, P1, K1, Ψ1, Φ1 and their standard deviations obtained in the SGO in years 2014, 2013–2014 and 2009–2015

Waves	Theoretical ampl. factors	2014		2013–2014		2009–2015	
		Ampl. fact.	± stdv	Ampl. fact.	± stdv	Ampl. fact.	± stdv
O1	0.40	0.439	0.055	0.558	0.025	0.598	0.012
P1	0.38	0.324	0.083	0.211	0.048	0.207	0.025
K1	0.31	0.342	0.048	0.423	0.018	0.418	0.009
PSI1	0.76	–	–	5.815	2.052	3.867	1.048
PHI1	0.46	–	–	0.169	1.128	1.327	0.576

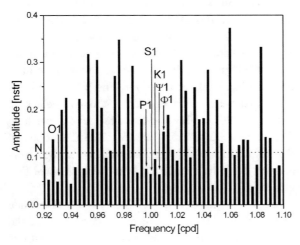

Figure 5
Fourier transform of the residuals of the strain data recorded in the SGO in years 2013–2014. *N* noise level

load and barometric pressure, and for ocean load, temperature and barometric pressure, respectively. The correction for barometric pressure (PC, OPC)

yields the best results for the K1/O1 ratios which also prove that the SGO is sensitive to barometric pressure variations (Gebauer et al. 2010; Mentes 2000). Involving temperature into corrections (TC, PTC, OTC, OTPC), the ratios are not better than without temperature correction. The P1/O1 ratios are far from the theoretical value in every case. The ocean load correction has no marked improvements in the ratios.

In the Vyhne station, the data series are much more disturbed than in the SGO. Therefore, there are sometimes long data gaps here. In spite of this fact, the obtained tidal parameters here are better than in the SGO. We chose the years from the middle of 2006 to the middle of 2007, 2013, 2014 and 2015 for the investigation. In this station, the barometric pressure was not recorded. So, we only made the correction of the raw extensometric data (UNC) for temperature (TC), ocean load (OC) and for temperature and ocean load (TOC). Strain data recorded in 2006/2007 were not corrected for temperature due to

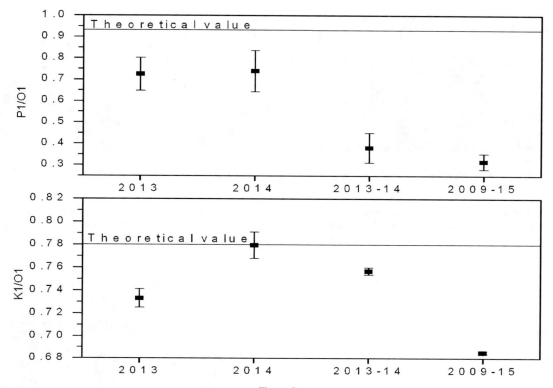

Figure 6
The K1/O1 and P1/O1 ratios and their errors obtained in the SGO

Table 7

Ψ1/O1 and Φ1/O1 ratios and their errors obtained in the SGO and in the MGGO

Observatory	Data series	Ψ1/O1 ± error	Φ1/O1 ± error
SGO	2013–2014	10.42 ± 3.71	0.30 ± 2.02
	2009–2015	6.47 ± 1.76	2.22 ± 0.96
MGGO	2010–2012	4.09 ± 1.14	1.83 ± 0.62

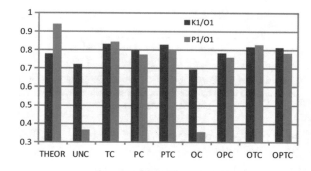

Figure 7
Comparison of the effect of different data corrections on the K1/O1 and P1/O1 amplitude factor ratios obtained from extensometric data recorded in the SGO in 2014. THEOR denotes the theoretical amplitude factor ratios, UNC, TC, PC, PTC, OC, OTC, OPC, OPTC denote amplitude factor ratios obtained using uncorrected strain data, corrected for temperature, for barometric pressure, for temperature and barometric pressure, for ocean load, for ocean load and temperature, for ocean load and barometric pressure, and for ocean load, temperature and barometric pressure, respectively

the very bad temperature data set containing many gaps. The resulting amplitude ratios, in particular the P1/O1 ratios, are higher than the theoretical values (Fig. 8). The nearest values to the theoretical are obtained for the uncorrected strain data. The amplitude factor ratios obtained for the years 2013 and 2015 are particularly high, especially the P1/O1 ratios. We have not found an explanation of the reason for these high values so far.

The Pécs station in the uranium mine (1040 m under surface, constant 49 °C at the instrument) was not disturbed by the outer temperature variations;

therefore, the extensometric data were corrected only for the barometric pressure, ocean load and for both. The data recorded in 1998 were chosen because in the

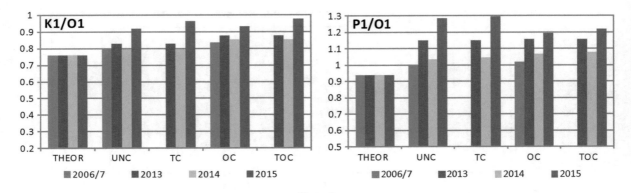

Figure 8
Comparison of the effect of different data corrections on the K1/O1 and P1/O1 amplitude factor ratios obtained from extensometric data recorded in the Vyhne station in the years 2006/7, 2013, 2014, 2015. THEOR, UNC, TC, OC, TOC denote theoretical, uncorrected extensometric data, corrected for temperature, for ocean load, for ocean load and temperature, respectively

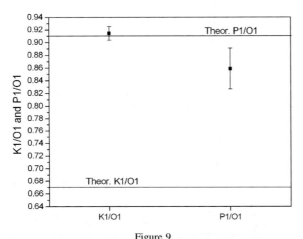

Figure 9
Comparison of the K1/O1 and P1/O1 amplitude factor ratios with the theoretical ratios obtained from extensometric data recorded in the Pécs station in 1998

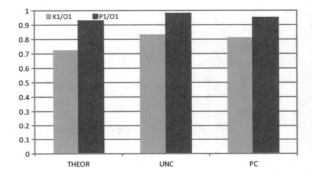

Figure 10
Comparison of the K1/O1 and P1/O1 amplitude factor ratios with the theoretical (THEOR) ratios obtained from extensometric data recorded in the MGGO in 2012. UNC and PC denote uncorrected extensometric data and corrected for barometric pressure, respectively

other years a lot of gaps occurred due to the handling difficulties of the instrument in the mine. Figure 9 shows that there is a large difference between the measured and the theoretical ratios. In this case the P1/O1 ratio is nearer to the theoretical value than the K1/O1 ratio.

In spite of the fact that the long-term tidal records are disturbed by the karstic water, the obtained tidal parameters in the MGGO are good (Eper-Pápai et al. 2014). For investigation of the free core resonance the data measured from 2010 to 2012 were chosen, since this time there were no long data gaps in the extensometric record. The data were analysed as a long data series. Unfortunately, the temperature

record with large gaps could not be used for the correction; therefore, the extensometric record was corrected only for the barometric pressure. In Fig. 10 the K1/O1 and P1/O1 ratios obtained from the uncorrected data and from data corrected for barometric pressure are presented. The Ψ1/O1 and Φ1/O1 ratios and their errors are in Table 7.

Comparison of the amplitude factor ratios obtained in the four observatories shows that the K1/O1 ratio closest to the theoretical ratio was obtained in the SGO, MGGO and in the Vyhne station. The deviation of the P1/O1 ratios from the theoretical value was the highest in SGO, and the lowest in the MGGO (see Fig. 10). With exception of the MGGO the dispersion of the P1/O1 ratios is greater than the dispersion of the K1/O1 ratios. Ψ1/O1 and Φ1/O1 ratios and their errors are similar in the SGO and in the MGGO.

5. *Conclusions and Perspectives*

According to our observations, fluid core resonance can also be detected by extensometric measurements. With the exception of the Pécs station, the K1/O1 ratios are close to the theoretical values. Except for MGGO in Budapest, there is a large difference between the obtained and theoretical P1/O1 ratios in each observatory. The Ψ1/O1 and Φ1/O1 ratios differ significantly from the theoretical values and have large errors. These discrepancies need further investigations. Our results have shown that the correction of temperature effects does not affect the errors of the obtained amplitude factors, in contrast to the correction of the atmospheric pressure, which slightly reduces them. Probably the correction of the extensometer data for local effects would contribute to reducing errors of the amplitude factors.

Besides the improvement of the recording and correction of extensometric data for local and environmental effects as a next step we also would like to estimate the FCN parameters using different inversion techniques (linearized least-squares, Bayesian or non-linear least-squares) and compare them with the results from other strainmeters.

Acknowledgements

The authors thank the two anonymous reviewers for their valuable comments and suggestions which helped to improve the paper. This work was funded by the Hungarian National Research Fund (OTKA) under Project K 109060. Special thanks to Ildikó Eperné-Pápai for her help in data pre-processing and Tibor Molnár for his careful maintenance of the instruments.

REFERENCES

Agnew, D. C. (2013). SPOTL: Some Programs for Ocean-Tide Loading. Institute of Geophysics and Planetary Physics, Scripps Institution for Oceanography, University of California. Technical Report.

Amoruso, A., Botta, V., & Crescentini, L. (2012). Free Core Resonance parameters from strain data: Sensitivity analysis and results from the Sasso (Italy) extensometers. *Geophysical Journal International, 189,* 923–936.

Botta, V. (2012). High sensitive strain measurements from underground interferometric stations: Geodynamic phenoma at Gran Sasso and first records from Canfranc. *Ph. D Thesis* (p. 242). Università degli Studi di Salerno. Dipartimento di Matematica.

Boyarsky, E. A., Ducarme, B., Latynina, L. A., & Vandercoilden, L. (2003). An attempt to observe the Earth liquid core resonance with extensometers at Protvino Observatory. *Bulletin d'Information Marées Terrestres, 138,* 10987–11009.

Brimich, L. (2006). Strain measurements at the Vyhne tidal station. *Contributions to Geophysics and Geodesy, 36*(4), 361–371.

Brimich, L., Bednarik, M., Vajda, P., Bán, D., Eper-Pápai, I., & Mentes, Gy. (2016). Extensometric observation of Earth tides and local tectonic processes at the Vyhne station, Slovakia. *Contribution to Geophysics and Geodesy, 46*(2), 75–90.

Dehant, V. (1987). Tidal parameters for an unelastic Earth. *Physics of the Earth and Planetary Interiors, 49,* 97–116.

Eper-Pápai, I., Mentes, G., Kis, M., & Koppán, A. (2014). Comparison of two extensometric stations in Hungary. *Journal of Geodynamics, 80,* 3–11.

Florsch, N., Chambat, F., Hinderer, J., & Legros, H. (1994). A simple method to retrieve the complex eigenfrequency of the Earth's nearly diurnal-free wobble; application to the Strasbourg superconducting gravimeter data. *Geophysical Journal International, 116,* 53–63.

Florsch, N., & Hinderer, J. (2000). Bayesian estimation of the free core nutation parameters from the analysis of precise tidal gravity data. *Physics of the Earth and Planetary Interiors, 117*(1), 21–35.

Gebauer, A., Steffen, H., Kroner, C., & Jahr, T. (2010). Finite element modelling of atmosphere loading effects on strain, tilt and displacement at multi-sensor stations. *Geophysical Journal International, 181,* 1593–1612.

Haas, J. (Ed.). (2001). *Geology of Hungary.* Budapest: Eötvös University Press.

Hartmann, T., & Wenzel, H. G. (1995). The HW95 tidal potential catalogue. *Geophysical Research Letters, 22*(24), 3553–3556.

Krásná, H., Böhm, J., & Schuh, H. (2013). Free core nutation observed by VLBI. *Astronomy and Astrophysics, 555*(A29), 1–5.

Melchior, P. (1978). *The tides of the planet Earth.* Oxford: Pergamon Press.

Mentes, G. (1991). Installation of a quartz tube extensometer at the Sopron Observatory. *Bulletin d'Information Marées Terrestres, 110,* 7936–7939.

Mentes, G. (2000). Influence of temperature and barometric pressure variations on extensometric deformation measurements at the Sopron Station. *Acta Geodaetica et Geophysica Hungarica, 35*(3), 277–282.

Mentes, G. (2008). Observation of recent tectonic movements by extensometers in the Pannonian Basin. *Journal of Geodynamics, 45*(4), 169–177.

Mentes, G. (2010). Quartz tube extensometer for observation of Earth tides and local tectonic deformations at the Sopronbánfalva Geodynamic Observatory. *Hungary. Review of Scientific Instruments, 81*(7), 074501.

Mentes, G., & Berta, Z. (1997). First results of the extensometric measurements in South Hungary. *Bulletin d'Information Marées Terrestres, 127,* 9744–9749.

Mentes, G., & Eper-Pápai, I. (2009). Relations between microbarograph and strain data. *Journal of Geodynamics, 48,* 110–114.

Mukai, A., Takemoto, S., & Yamamoto, T. (2004). Fluid core resonance revealed from a laser extensometer at the Rokko-Takao station, Kobe, Japan. *Geophysical Journal International, 156,* 22–28.

Ping, J., Tsubokawa, T., Tamura, Y., Heki, K., Matsumoto, K., & Sato, T. (2006). Observing long-term FCR variation using Esashi extensometers. *Journal of Geodynamics, 41*(1–3), 155–163.

Rosat, S., Florsch, N., Hinderer, J., & Llubes, M. (2009). Estimation of the Fee Core Nutation parameters from SG data: Sensitivity study and comparative analysis using linearized Least-Squares and Bayesian methods. *Journal of Geodynamics, 48*(3–5), 331–339.

Sato, T., Tamura, Y., Matsumoto, K., Imanishi, Y., & McQueen, H. (2004). Parameters of the fluid core resonance inferred from superconducting gravimeter data. *Journal of Geodynamics, 38,* 375–389.

Sun, H.-P., Jentzsch, G., Xu, J.-Q., Hsu, H.-Z., Chen, X.-D., & Zhou, J.-C. (2004). Earth's free core nutation determined using C032 superconducting gravimeter at station Wuhan/China. *Journal of Geodynamics, 38,* 451–460.

Van Camp, M., & Vauterin, P. (2005). Tsoft: Graphical and interactive software for the analysis of time series and Earth tides. *Computers and Geosciences, 31*(5), 631–640.

Wahr, J. M. (1981). Body tides on an elliptical, rotating, elastic and oceanless Earth. *Geophysical Journal of the Royal Astronomical Society, 64,* 677–703.

Wenzel, H. G. (1996). The nanogal software: Earth tide data processing package ETERNA 3.30. *Bulletin d'Information Marées Terrestres, 124,* 9425–9439.

Zürn, W. (1977). The nearly-diurnal free woble resonance. In H. Wilhelm, W. Zürn, & H.-G. Wenzel (Eds.), *Tidal Phenomena* (Vol. 66, pp. 95–109)., Lecture Notes in Earth Sciences Berlin: Springer.

(Received October 29, 2016, revised November 13, 2017, accepted November 16, 2017, Published online November 20, 2017)

Pure Appl. Geophys. 175 (2018), 1643–1648
© 2017 Springer International Publishing
https://doi.org/10.1007/s00024-017-1532-z

Relation of Different Type Love–Shida Numbers Determined with the Use of Time-Varying Incremental Gravitational Potential

PETER VARGA,[1] ERIK GRAFAREND,[2] and JOHANNES ENGELS[2]

Abstract—There are different equations to describe relations between different classes of Love–Shida numbers. In this study with the use of the time-varying gravitational potential an integral relation was obtained which connects tidal Love–Shida numbers (h, l, k), load numbers (h', l', k'), potential free Love–Shida numbers generated by normal (h'', l'', k'') and horizontal (h''', l''', k''') stresses. The equations obtained in frame of present study is the only one which

- holds for every type of Love–Shida numbers,
- describes a relationship not between different, but the same type of Love–Shida numbers,
- does not follow from the sixth-order differential equation system of motion usually applied to calculate the Love–Shida numbers.

Key words: Love–Shida numbers, load numbers, potential free Love–Shida numbers, inhomogeneous motion equation systems.

1. Introduction

The improvement of accuracy of Earth tidal observations and increase the length of the monitoring period, especially in the case of gravitational tidal observations, recently allows high resolution separation of near-frequency waves and precise determination of their amplitudes (Ducarme 2012; Calvo et al. 2014; Van Camp et al. 2016). At the same time interpretation of observation results requires a calibration accuracy at the 1‰ level (Meurers et al. 2016). This is important for the interpretation of Earth tide data, because of relatively poorly known amplitude ratios. The amplitude ratios obtained from observations carried out with gravimeters are $\delta_2 = 1 + h_2 - 3/2k_2$, with tiltmeters $\gamma_2 = 1 + k_2 - h_2$, with strainmeters $S_{\phi\phi2} = h_2 - 4l_2$ and $S_{\lambda\lambda} = h_2 - 2l_2$ and with dilatometers $f_2 = 2h_2 - 6l_2[1 - \Lambda/(\Lambda + 2\mu)]$, where Λ and μ are Lamé parameters and ϕ and λ the geographical latitude and longitude. On the basis of theoretical model calculations it can be assumed that $k_2 \sim 0.5h_2$ and $l_2 \sim 0.15h_2$ and at the surface of the Earth ($r = a$) $\Lambda(a) \sim \mu(a)$, $\delta_2 = 1 + 1/4h_2$, $\gamma_2 = 1 - k_2$, $S_{\phi\phi2} = 1/3h_2$, $S_{\lambda\lambda2} = 2/3h_2$, $f_2 = 3/4h_2$. This means that deviation from the absolutely rigid, inelastic Earth in case of gravity Earth tides is only 15%. In case of tilt and strain observations along latitude and longitude this deviations are 30, 20 and 40%, respectively, while for dilatational observations it is 45%. The free core nutation due to the dynamical effect of the liquid core is more pronounced in case of strain and dilatation measurements than for gravimetric and tilts observations:

$$\delta_{O1} - \delta_{K1} = 0.021(1.8\%)$$
$$\gamma_{O1} - \gamma_{K1} = -0.041(5.8\%)$$
$$S_{\phi\phi O1} - S_{\phi\phi K1} = 0.098(33.7\%)$$
$$S_{\lambda\lambda O1} - S_{\lambda\lambda K1} = 0.192(37.5\%)$$
$$f_{O1} - f_{K1} = 0.117(23.7\%)$$

The amplitude ratios derived from observations provide linear combinations of Love and Shida numbers but do not provide the numbers themselves. At the same time for study of many different geodetic and geophysical problems the numbers and not their combinations are needed. It may be, therefore, important to determine relationships between the Love–Shida numbers. A similarly important task is to

---

[1]   Seismological Observatory, Institute of Geodesy and Geophysics, Budapest, Hungary. E-mail: varga@seismology.hu
[2]   Department of Geodesy and Geoinformatics, Stuttgart University, Stuttgart, Germany.

derive relations between the tidal and other types of Love–Shida numbers, the load and potential free load numbers.

For the study of aforementioned problems the time-varying gravitational potential was determined in case of deformation generated by tide, tidal, load and potential free load with the use of the Love and Shida numbers in case of a symmetric non-rotating isotropic elastic model Earth. The Love–Shida numbers are determined in this case by the radial profiles of elastic Lamé parameters and density. Due to the mass conservation in case of elastic deformation the incremental mass density was inferred from the divergence of the product of initial mass density and the vertical displacement field.

2. Equation of Melchior (1950) for the Ratio of Love Numbers h_n and k_n

For ratio of Love numbers k_n and h_n Melchior (1950) obtained

$$\frac{k_n}{h_n} = \frac{n+3}{(2n+1)a} \cdot \frac{\int_0^a \rho(r)r^{n+2}dr}{\int_0^a \rho(r)r^2dr} \quad (1)$$

These ratios for $n = 2$–10 and $n = 20$ are shown on Fig. 1.

In case of $n = 2$

$$\frac{k_2}{h_2} = \frac{1}{a} \cdot \frac{\int_0^a \rho(r)r^4dr}{\int_0^a \rho(r)r^2dr} \quad (2)$$

The r.h.s. of this equation is a ratio of two Stokes constants: the mass (M) and polar moment of inertia (C) of the Earth. Consequently the ratio of the two Love numbers can be determined without any hypothesis concerning the inner structure of the Earth. According to Jeffreys (1959)

$$\int_0^a \rho(r)r^4dr = \frac{3}{2}Ma^2\left(1 - \frac{2}{5}\sqrt{1 + \eta(a)}\right) \quad (3)$$

where $\eta(r)$ is the Radau function and for $r = a$ it is $\eta(a) = \frac{5}{2\alpha} \cdot \frac{\omega^2 a}{g(a)} = 0.572$ (Moritz 1990). Consequently $\frac{k_2}{h_2} = 0.493$.

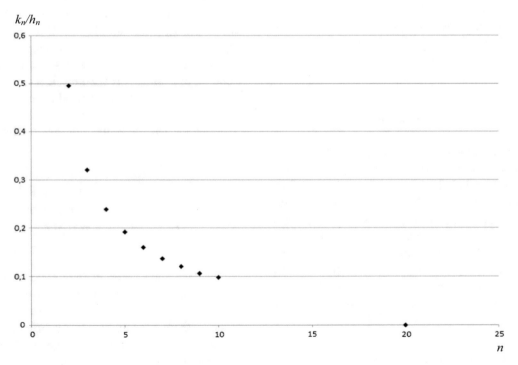

Figure 1
Ratios k_n/h_n for $n = 2$–10 and 20

3. Relationships Between Different Types of Love–Shida Numbers

The Love–Shida numbers can be obtained with the use of the inhomogeneous motion equation system given by Takeuchi (1953), Molodensky (1953), Alterman et al. (1959). The solution of this system must satisfy three boundary conditions on the surface of the

Earth which concern the normal stress ($N(a)$), the tangential stress ($M(a)$) and the derivative of the mass potential ($L(a)$) (Saito 1978):

$$N(a) = -\frac{2n+1}{3} \cdot \alpha_N \tag{3a}$$

$$M(a) = \frac{2n+1}{3 \cdot n \cdot (n+1)} \cdot \alpha_M \tag{3b}$$

$$L(a) = (2n+1) \cdot \alpha_L \tag{3c}$$

where $\alpha_N, \alpha_M, \alpha_L$ can take the value 0 or 1 and n is the order of spherical harmonic describing the deformation.

In case of Earth tides boundary conditions $\alpha_N = \alpha_M = 0$ and $\alpha_L = 1$ determine the Love–Shida numbers h_n, k_n and l_n. In case of a normal load acting on the surface of the Earth the triplet of parameters $\alpha_N = 1, \alpha_M = 0$ and $\alpha_L = 1$ allows to determine the load Love–Shida numbers h'_n, k'_n and l'_n. If a potential free stress acting on the Earth surface ($\alpha_L = 0$) potential free Love–Shida numbers can be introduced (Molodensky 1977) for normal $\alpha_N = 1, \alpha_M = 0, \alpha_L = 0$ and tangential $\alpha_N = 0, \alpha_M = 1, \alpha_L = 0$ stresses. The corresponding triplets are h'', l'', k'' and h''', k''', l'''. The three different triplets are connected due to their physical meaning. The load numbers are sums of deformation due to normal stress and gravitational effect of loading masses:

$$h'_n = h_n - \frac{2n+1}{3} \cdot h''_n \tag{4a}$$

$$k'_n = k_n - \frac{2n+1}{3} \cdot (1 + k''_n) \tag{4b}$$

$$l'_n = l_n - \frac{2n+1}{3} \cdot l''_n \tag{4c}$$

Furthermore, according Molodensky (1977) the following equations can be introduced

Table 1

Love numbers h_n, k_n and load number k'_n calculated with the use of differential equation system of Molodensky (1953) on the basis of PREM for the degrees $n = 2$–4

n	k_n	h_n	$k'_n = k_n - h_n$	k'_n
2	0.3035	0.6176	−0.3141	−0.3145
3	0.0944	0.2957	−0.2013	−0.2012
4	0.0425	0.1781	−0.1356	−0.1354

$$h'''_n = n(n+1) \cdot l''_n \tag{5a}$$

$$h_n = \frac{2n+1}{3} \cdot (1 + k''_n) \tag{5b}$$

$$l_n = \frac{2n+1}{3n(n+1)} \cdot (1 + k'''_n) \tag{5c}$$

From (5b) it follows

$$k'_n = k_n - h_n \tag{6}$$

The validity of (6) can be checked with computation of Love numbers h_n, k_n and load number k'_n (Table 1). A comparison of k'_n and $k_n - h_n$ where the Love–Shida numbers were obtained with the use of boundary conditions $\alpha_N = \alpha_M = 0; \alpha_L = 1$ and the load number k'_n with $\alpha_N = 1, \alpha_M = 0$ and $\alpha_L = 1$ allows to conclude that the calculations carried out with the differential equation system of Molodensky (1953) have deviation $\leq \pm 2 \times 10^{-4}$.

4. The Gravitational Potential of a Deformable Body Generated by Tidal, Load and Potential Free Potentials

The aim of the next part of present study is to derive a mathematical expression which is composed by all three components of triplets and which is equally applicable for Love–Shida, for load and for potential free Love numbers. For this purpose an approach was developed which is not based on an inhomogeneous motion equation system given by Takeuchi (1953), Molodensky (1953), Alterman et al. (1959).

The Newtonian equation for the time dependent deformation potential is (Grafarend et al. 1997)

$$\Delta\omega(x_a, t) = G \iiint\limits_{r\lambda\phi} \frac{\mathrm{div}\rho(x)\mathrm{d}(x,t)}{\|x_a - x\|} \mathrm{d}r\mathrm{d}\lambda\mathrm{d}\phi \quad (7)$$

where x is a source point within the Earth, x_a is the field point on the surface $(r = a)$, $\rho(x)$ and $d(x,t)$ characterize the initial mass density and the time dependent displacement vector of the deformable Earth. In (7) the continuity equation has been used

$$\Delta\varrho(x, t) = -\mathrm{div}[\rho(x)\mathrm{d}(x, t)]$$

what means that there is no loss of mass during the deformation. Furthermore,

$$\Delta\varrho(x, t) = -\mathrm{div}[\rho(x)\mathrm{d}(x, t)]$$
$$= -\varrho(x)\mathrm{div}\,\mathrm{d}(x, t) + \frac{\mathrm{grad}\varrho(x)}{\mathrm{d}(x, t)}$$

The denominator in (7) could be expanded into the scalar surface spherical surface functions

$$\frac{1}{\|x_a - x\|} = \frac{1}{a}\sum_{n=0}^{\infty}\sum_{m=-n}^{n}\frac{1}{2n+1}\left(\frac{r}{a}\right)^n Y_{n,m}(\lambda_a, \phi_a)Y_{n,m}(\lambda, \phi)$$
$$(8)$$

In the most general case the displacement vector expressed by orthonormal vector surface spherical functions is

$$\mathrm{d}(x, t) = \sum_{n=0}^{\infty}\sum_{m=-n}^{n}[r^{nm}(r, t)R_{nm}(\lambda, \phi) + s^{nm}(r, t)S_{nm}$$
$$(\lambda, \phi) + t^{nm}(r, t)T_{nm}((\lambda, \phi)]$$
$$(9)$$

On the r.h.s. of (9) the first two terms represent the spheroidal displacements. The third term stands for the toroidal displacement fields connected to phenomena not discussed in this study. Furthermore, with unit vectors e_r, e_λ, e_ϕ one can write

$$R_{nm} = e_r Y_{n,m}(\lambda, \phi) \quad (10a)$$

$$S_{nm} = \frac{1}{\sqrt{n(n+1)}}\left(\frac{e_\lambda}{\cos\phi}\frac{\partial}{\partial\lambda}Y_{n,m}(\lambda, \phi) + e_\phi\frac{\partial}{\partial\phi}Y_{n,m}(\lambda, \phi)\right)$$
$$(10b)$$

With the use of Eqs. (7)–(10a, 10b) the time dependent deformation potential is

$$\Delta\omega(x_a, t) =$$
$$-\frac{G}{a}\iiint\limits_{r\lambda\phi}\sum_{n=0}^{\infty}\sum_{m=-n}^{n}\frac{1}{2n+1}\left(\frac{r}{a}\right)^n Y_{nm}(\lambda_a, \phi_a)\cdot Y_{nm}(\lambda, \phi)\cdot$$
$$\sum_{i=0}^{\infty}\sum_{j=-i}^{i}\left\{\left[\frac{2}{r}r^{ij} + \frac{\mathrm{d}}{\mathrm{d}r}r^{ij} - \frac{\sqrt{i(i+1)}}{r}s^{ij}\right]\right.$$
$$\cdot\rho(x)Y_{ij}(\lambda, \phi) + \frac{1}{r\cos\phi}\cdot\frac{\partial\rho}{\partial\lambda}$$
$$\cdot\left[\frac{s^{ij}}{\sqrt{i(i+1)}}\frac{1}{\cos\phi}\frac{\partial}{\partial\lambda}Y_{ij}(\lambda, \phi)\right]$$
$$\left.+\frac{1}{r}\frac{\partial\varrho}{\partial\phi}\frac{s^{ij}}{\sqrt{i(i+1)}}\frac{\partial}{\partial\phi}Y_{ij}(\lambda, \phi) + \frac{\partial\varrho}{\partial r}\left[r^{ij}Y_{ij}(\lambda, \phi)\right]\right\}\mathrm{d}r\mathrm{d}\lambda\mathrm{d}\phi$$
$$(11)$$

The tidal displacement components could be represented as

$$d_\lambda = l(r)(\gamma r\cos\phi)^{-1}\frac{\partial V_T}{\partial\lambda}$$

$$d_\phi = l(r)(\gamma r)^{-1}\frac{\partial V_T}{\partial\phi}$$

$$d_r = h(r)\gamma^{-1}V_T$$

where γ is the mean gravity acceleration and the tidal potential expressed with spherical surface function is

$$V_T = \sum_{n=0}^{\infty}\sum_{m=-l}^{l}V_{nm}Y_{nm}(\lambda_a, \phi_a)$$

Consequently, $r^{nm}(r) = h_n(r)\gamma^{-1}V_{nm}$ and $s^{nm}(r) = l_n(r)\gamma^{-1}\sqrt{n(n+1)}V_{nm}$. Due to the radial symmetry of the Earth model used in this paper $\partial\varrho/\partial\lambda = \partial\varrho/\partial\phi = 0, \partial h/\partial\lambda = \partial h/\partial\phi = 0$ and $\partial l/\partial\lambda = \partial l/\partial\phi = 0$. Consequently, (11) can be given in a simpler form:

$$\Delta\omega_{nm}(a, t) =$$
$$-G\iiint\limits_{r\lambda\phi}\left\{\left[\rho(r)\left(\frac{1}{a}\right)\cdot\frac{1}{2n+1}\left(\frac{r}{a}\right)^n Y_{nm}(\lambda_a, \phi_a)Y_{nm}(\lambda, \phi)\right]\right.$$
$$\cdot\left\{\left[\frac{2}{\gamma r}h_n(r) + \left(\frac{1}{\gamma}\right)\frac{\partial h_n(r)}{\partial r} - \frac{i(i+1)}{\gamma r}l_n(r)\right]V_{ij}Y_{ij}(\lambda, \phi)\right.$$
$$\left.\left.+\left[\left(\frac{1}{\gamma}\right)\frac{\partial\varrho(r)}{\partial r}h_n(r)\right]^i V_{ij}Y_{ij}(\lambda, \phi)\right\}\right\}$$
$$(12)$$

With the use of orthonormality relations of the surface spherical harmonics (12) can be given in form:

$$
\begin{aligned}
&\Delta\omega_{nm}(a,t) \\
&= -4\pi G \int_0^a \left\{ \frac{\varrho(r)}{a} \left[\frac{1}{2n+1} Y_{nm}(\lambda_a, \phi_a) \right] \right. \\
&\quad \cdot \left[\frac{2}{\gamma r} h_n(r) + \frac{1}{r} \frac{dh_n(r)}{dr} - \frac{n(n+1)}{\gamma r} l_n(r) \right] V_{nm} \\
&\quad \left. + \left[\frac{d\varrho(r)}{dr} \cdot \frac{1}{\gamma} h_n V_{nm} \right] \right\} \cdot \left(\frac{r}{a}\right)^n r^2 dr \\
&= -\frac{4\pi G Y_{nm}(\lambda_a, \phi_a) V_{nm}}{\gamma a (2n+1)} \\
&\quad \int_0^a \left\{ \rho(r) \left[2\frac{h_n(r)}{r} + \frac{dh_n(r)}{dr} - \frac{n(n+1)l_n(r)}{r} \right] \right. \\
&\quad \left. + h_n(r) \frac{d\varrho(r)}{dr} \right\} \left(\frac{r}{a}\right)^n r^2 dr
\end{aligned}
\tag{13}
$$

Using some simplifications (13) can be written as

$$
\begin{aligned}
&\Delta\omega_{nm}(a,t) = -4\pi G / \left[\gamma(2n+1) a^{n+1} \right] V_{nm} Y_{nm}(\lambda_a, \phi_a) \\
&\quad \cdot \int_0^a \left\{ \rho(r) \left[\frac{dh_n(r)}{dr} r^{n+2} + 2h_n(r) r^{n+1} - n(n+1) l_n(r) r^{n+1} \right] \right. \\
&\quad \left. + \frac{d\varrho(r)}{dr} h_n(r) \cdot r^{n+2} \right\} dr
\end{aligned}
\tag{14}
$$

The partial integration of the last term of (14) gives

$$
\begin{aligned}
&\int_0^a \frac{d\varrho(r)}{dr} h_n(r) \cdot r^{n+2} = \\
&\quad - \int_0^a \varrho(r) \left[\frac{dh_n(r)}{dr} r^{n+2} + h_n(r)(n+2) r^{n+1} \right] dr
\end{aligned}
$$

Therefore, (14) may be simplified:

$$
\begin{aligned}
&\Delta\omega_{nm}(a,t) = 4\pi G / \left[\gamma(2n+1) a^{n+1} \right] V_{nm} Y_{nm}(\lambda_a, \phi_a) \\
&\quad \cdot \int_0^a \rho(r) \left[n(n+1) l_n(r) r^{n+1} + n h_n(r) r^{n+1} \right] dr
\end{aligned}
\tag{15}
$$

The r.h.s. of (15) gives the potential variation generated by deformation of the Earth which can be written as

$$
\Delta\omega_{nm}(a,t) = k_n(a) V_{nm} Y_{nm}(\lambda_a, \phi_a)
\tag{16}
$$

where $k_n(a)$ is the Love number to describe the potential variations due to deformations. Connection of (15) and (16) leads to a relation between the Love–Shida numbers:

$$
\begin{aligned}
k_n(a) &= 4\pi G n / \left[\gamma(2n+1) a^{n+1} \right] \\
&\quad \cdot \int_0^a -\rho(r) r^{n+1} [n(n+1) l_n(r) + h_n(r)] dr
\end{aligned}
\tag{17}
$$

With the use of relative system of units: the unit of distance is a, the unit of the gravity the mean gravity acceleration γ and the unit of the density is the mean density of the Earth ($\rho_{\text{mean}} = 5514 \text{ kg/m}^3$). In this system, $G = 3/4\pi$, and (17) will be

$$
k_n(a) = \frac{3}{2n+1} \int_0^a \varrho(r) \cdot r^{n+1} [(n+1) l_n(r) + h_n(r)]
\tag{18}
$$

And consequently for $n = 2, 3, 4$

$$
k_2(a) = \frac{3}{5} \int_0^a \varrho(r) \cdot r^3 [3 l_n(r) + h_n(r)]
\tag{19}
$$

$$
k_3(a) = \frac{3}{7} \int_0^a \varrho(r) \cdot r^4 [4 l_n(r) + h_n(r)]
\tag{20}
$$

$$
k_4(a) = \frac{3}{9} \int_0^a \varrho(r) \cdot r^5 [5 l_n(r) + h_n(r)]
\tag{21}
$$

5. Conclusions

The importance of Eqs. (18)–(21) is that they are only relations which do not follow from the differential equation of motion usually used to calculate the Love–Shida numbers and consist the complete triplet h_n, k_n and l_n of the same type. They were obtained without any considerations concerning the boundary conditions (3) and, therefore, they are valid both for tidal, load and potential free Love–Shida numbers.

With the use of (19) $k_2(a) = 0.3031$ which is close to value in Table 1 (0.3035).

On the r.h.s. of Eqs. (18)–(21) the first term describes the horizontal and the second the vertical displacements. Their contribution to the value of Love number k is almost equal to 0.148 and 0.146, respectively, in the case of $n = 2$. In case of the load numbers this ratio is different and significantly decreases with increasing n: 0 589 in case of $n = 2$ and 0.097 for $n = 10$.

Acknowledgements

We thank the Guest Editor David Crossley and an anonymous reviewer colleague for their helpful comments. The research described in this paper was completed during research stay of P. Varga (01.03.2016–31.05.2016) supported by the Alexander Humboldt Foundation at the Department of Geodesy and Geoinformatics, Stuttgart University. P. Varga thanks Professor Nico Sneeuw for the excellent research conditions provided by him. Financial support from the Hungarian Scientific Research Found OTKA (Project K125008) is acknowledged.

REFERENCES

Alterman, Z., Jarosch, H., & Pekeris, C. L. (1959). Oscillations of the Earth. *Proceedings of the Royal Society London A, 252,* 80–95.

Calvo, M., Hinderer, J., Rosat, S., Legros, H., Boy, J.-P., Ducarme, B., et al. (2014). Time stability of spring and superconducting gravimeters, through the analysis of very long gravity records. *Journal of Geodynamics, 80*(2014), 20–33.

Ducarme, B. (2012). Determination of the main lunar waves generated by the third degree tidal potential and validity of the corresponding body tides models. *Journal of Geodesy, 86*(1), 65–75.

Grafarend, E., Engels, J., & Varga, P. (1997). The spacetime gravitational field of a deformable body. *Journal of Geodesy, 72,* 11–30.

Jeffreys, H. (1959). *The Earth its origin, history and physical constitution.* Cambridge: University Press.

Melchior, P. J. (1950). Sur l'influence de la loi de répartition des densités á l'intérieur de la Terre dans les variations Luni-Solaires de lagravité en un point. *Geophysica Pura et Applicata, 16*(3–4), 105–112.

Meurers, B., Van Camp, M., Francis, O., & Pálinkáš, V. (2016). Temporal variation of tidal parameters in superconducting gravimeter time-series. *Geophysical Journal International, 205*(1), 284–300.

Molodensky, M. S. (1953). Elastic tides, free nutations and some questions concerning the inner structure of the Earth. *Trudi Geofizitseskogo Instituta Akademii Nauk of the USSR, 19*(146), 3–42.

Molodensky, S. M. (1977). On the relation between the Love numbers and the load coefficients. *Fizika Zemli, 3,* 3–7.

Moritz, H. (1990). *The figure of the Earth: theoretical geodesy and the Earth's interior.* Karlsruhe: Wichmann.

Saito, M. (1978). Relationship between tidal and load numbers. *Journal of Physics of the Earth, 26,* 13–16.

Takeuchi, H. (1953). On the Earth tide of the compressible Earth of variable density and elasticity. *Transactions American Geophysical Union, 31*(5), 651–689.

Van Camp, M., Meurers, B., de Viron, O., & Forbriger, Th. (2016). Optimized strategy for the calibration of superconducting gravimeters at the one per mille level. *Journal of Geodesy, 90*(1), 91–99.

(Received December 3, 2016, revised March 13, 2017, accepted March 15, 2017, Published online March 22, 2017)

Pure Appl. Geophys. 175 (2018), 1649–1657
© 2017 Springer International Publishing
https://doi.org/10.1007/s00024-017-1563-5

Pure and Applied Geophysics

Influence of Tidal Forces on the Triggering of Seismic Events

PÉTER VARGA[1] and ERIK GRAFAREND[2]

Abstract—Tidal stresses are generated in any three-dimensional body influenced by an external inhomogeneous gravity field of rotating planets or moons. In this paper, as a special case, stresses caused within the solid Earth by the body tides are discussed from viewpoint of their influence on seismic activity. The earthquake triggering effects of the Moon and Sun are usually investigated by statistical comparison of tidal variations and temporal distribution of earthquake activity, or with the use of mathematical or experimental modelling of physical processes in earthquake prone structures. In this study, the magnitude of the lunisolar stress tensor in terms of its components along the latitude of the spherical surface of the Earth as well as inside the Earth (up to the core-mantle boundary) were calculated for the PREM (Dziewonski and Anderson in Phys Earth Planet Inter 25(4):297–356, 1981). Results of calculations prove that stress increases as a function of depth reaching a value around some kPa at the depth of 900–1500 km, well below the zone of deep earthquakes. At the depth of the overwhelming part of seismic energy accumulation (around 50 km) the stresses of lunisolar origin are only $(0.0–1.0) \cdot 10^3$ Pa. Despite the fact that these values are much smaller than the earthquake stress drops (1–30 MPa) (Kanamori in Annu Rev Earth Planet Sci 22:207–237, 1994) this does not exclude the possibility of an impact of tidal forces on outbreak of seismic events. Since the tidal potential and its derivatives are coordinate dependent and the zonal, tesseral and sectorial tides have different distributions from the surface down to the CMB, the lunisolar stress cannot influence the break-out of every seismological event in the same degree. The influencing lunisolar effect of the solid earth tides on earthquake occurrences is connected first of all with stress components acting parallel to the surface of the Earth. The influence of load tides is limited to the loaded area and its immediate vicinity.

Key words: Spherical tidal stress tensor, zonal, tesseral and sectorial tides, oceanic tidal load.

[1] Research Centre for Astronomy and Earth Sciences, Hungarian Academy of Sciences, Geodetic and Geophysical Institute, Kövesligethy Seismological Observatory, Meredek 18, 1015 Budapest, Hungary. E-mail: varga@seismology.hu

[2] Department of Geodesy and Geoinformatics, Stuttgart University, Geschwister-Scholl-Str. 24D, 70174 Stuttgart, Germany.

1. Historical Overview and Current Research Results

One of the founders of modern seismology, professor of mathematics at University of Dijon Perrey realized first that the study of seismology needs international cooperation. Thanks to his efforts between 1844 and 1871 a significant (for that time) earthquake data set was collected by him in form of "Annual lists of Earthquakes". The total number of earthquakes described and catalogued by him is over 21,000. Perrey completed also regional catalogues. This way he initiated the "seismological geography". Although examination of the temporal distribution of earthquakes has been started already in the eighteenth century (Baldivi in 1703 and Toaldo in 1797 see Davison 1927) Perrey was the first who recognised regularities in the temporal distribution of seismic events. Based on his far from complete catalogue Perrey (1875) concluded, that the series of earthquake occurrences contains periodicities. According to Davison (1927) he has identified three regularities: earthquakes are more frequent at the syzygies than at the quadratures, they are more frequent at perigee than at apogee and the seismic events are more frequent when the moon is near to the meridian.

Since Perrey's time the problem of tidal triggering is one of the "evergreen" problems of earthquake research. The committee appointed by the French Academy of Sciences (Élie de Beaumont, Liouville and Lamé) reported favourably on Perrey's conclusions, while the other great French seismologist F. de Montessus de Ballore (1911) strongly criticised them. Schuster (1897) was the first who applied to the study of temporal variation of seismic events the tools of statistics and on this basis he has arrived at an optimistic conclusion: "The reality of the period (i.e. tidal periodicity of earthquake occurrence) would be thereby established beyond reasonable doubt" (Schuster 1911). In his study Cotton (1922) suggested tidal stresses in the Earth as a secondary cause of earthquakes and according to him it might be possible to predict earthquake occurrence "with sufficient accuracy" by consideration of the position of the Sun and the Moon, what allows "to provide timely warnings of disastrous shocks". From the early twentieth century, many researchers have dealt with the relationship between tides and earthquakes, in the hope that this way a tool can be found for a more accurate estimation of earthquake hazard. In addition, study of potential effects of tides may be important for seismology because it modifies the spatial and temporal distribution of earthquakes, distorts the image of the nature of seismo-tectonics, what makes more difficult understanding of the nature of earthquakes. A comprehensive summary of research dealing with tidal triggering of earthquakes and volcanic events during the first nine decades of the twentieth century was compiled by Emter (1997).

A significant number of recent studies found relationship between the lunisolar effect and temporal distribution of seismicity. Most of them show this relationship on a regional scale. A number of such previous studies were mentioned in Métivier et al. (2009). Nowadays many papers using significant database and advanced statistical methods discuss relationship between different seismological and tidal phenomena. Vergos et al. (2015) detected evidence for tidal triggering of earthquakes of the Hellenic arc, Arabelos et al. (2016) describe a lunisolar influence on seismic activity in central Greece. Chen et al. (2012a) suggest that in case of earthquakes which occurred in Taiwan between 1973 and 2008 "the lunar tidal force is likely a factor in the triggering of earthquakes". According to Cochran et al. (2004) the earth tides can trigger shallow thrust fault seismic events. Wilcock (2009) found a weak correlation between tidal influence and earthquake occurrences in continental areas which can be increased in settings influenced by loading effect of oceanic tides. Stroup et al. (2007) conclude that the lunisolar periodic stress changes are dominated by oceanic or by solid earth tide in dependence of the position of seismic source zone relative to the distribution of area

loaded by oceanic tidal load. At the same time Métivier et al. (2009) on basis of a study of global earthquake activity come to the conclusion that "it is highly probable that the observed triggering is caused by the solid Earth tide rather than by loading from the ocean or atmospheric tides". Tanaka (2010, 2012) and Ide et al. (2016) found significant lunisolar triggering in case of foreshocks prior to the Sumatra megathrust earthquakes of 26 December 2004 (M_w 9.0), 28 March 2005 (M_w 8.6), 12 September 2007 (M_w 8.5), and the Tohoku-Oki earthquake (2011, $M_w = 9.1$). Li and Xu (2013) have found that seismicities of some provinces of China show higher correlation with the lunisolar effect during several years preceding a large (destructive) earthquake than during other times. Chen et al. (2012b) in the case of Christchurch earthquake (2011, $M_w = 6.3$) found a correlation between the aftershock sequence and diurnal tide. At the same time these authors found also a global correlation in case of seismic events $M_S \geq 7$ since 1900 and the semidiurnal tidal wave M_2. Examinations carried out by different authors show slight dependence of tidal triggering effect and seismological activity on focal mechanisms. Earlier Heaton (1975) concluded that first of all shallow oblique-slip and dip-slip earthquakes $M \geq 5$ correlate with tidally generated shear stresses (Heaton 1982 retracted this statement after using an extended earthquake catalogue). Tsuruoka et al. (1995) have found a slightly higher probability of earthquake occurrence for normal fault earthquakes. Métivier et al. (2009) found that more tidal triggering is exhibited in case of normal strike-slip faulting than in other cases. It was concluded in a recent paper by Ide et al. (2016), that probability of occurrence of great seismic events is higher "during periods of high tidal stresses" (i.e. during spring tides). Houston (2015) combined seismic data and calculated tidal stress and found rising sensitivity to lunisolar stresses during six large slow-slip events of tremors in Cascadia.

Summarising the results of the studies detecting significant tidal triggering it can be concluded.

- the conclusions are almost exclusively based on statistical investigations, only sometimes supported by schematic seismic source models.

- correlation was found in case of shallow-dipping thrust events in case of earthquakes shallower than 50 km depth.
- the tides can act only in case of faults that are already close to rupture.
- the normal and shear stress are essential factors of tidal triggering.

On the other hand, many researchers could not detect any significant interdependence between time distribution of tidal force and seismic activity. Young and Zürn (1979), Vidale et al. (1998), Stein (2004), Tanaka et al. (2006) could not find statistically significant relationship between tidal force and earthquake activity. Based on their investigations Tormann et al. (2015) concluded that large earthquakes "may not have a characteristic location, size or recurrence interval, and might therefore occur more randomly distributed in time".

To contribute to the interpretation of the differences of research results the study of equations of lunisolar tidal elastic stress tensor components can be probably helpful because it could explain these differences between the research results. In the present investigation for calculations carried out the mathematical tools of author's former study (Varga and Grafarend 1996) have been used. Our previously used mathematical procedure and—as will be seen in the following sections—the interpretation of the calculation results we have developed in the following ways:

- magnitudes of stresses caused by the lunisolar effect and by surface load are expressed in SI units instead of the relative units used in our previous paper
- a new interpretation was given for the new numerical results obtained both in case of tidal and load models.

2. The Spherical Tidal Stress Tensor

With the use of spherical harmonics $Y_{2i}(\Phi, \lambda)$ (where Φ and λ denotes spherical latitude and longitude) in cases of zonal ($i = 0$), tesseral ($i = 1$) and sectorial ($i = 2$) tides the spherical tidal stress tensor components (Fig. 1) on the basis of Varga and Grafarend (1996) can be written in the following forms:

$$\sigma_{rr_i}(r, \lambda, \Phi) = \frac{(\Lambda + 2\mu)\partial H}{\partial r} + \Lambda\left[\frac{2}{r}H - \frac{n(n+1)}{r^2}T\right]$$
$$\cdot Y_{2i}(\lambda, \Phi)$$
$$= N(r) \cdot Y_{2i}(\lambda, \Phi)$$
$$(1)$$

$$\sigma_{\Phi\Phi_i}(r, \lambda, \Phi) = \left\{\Lambda\left[\frac{\partial H}{\partial r} + \frac{2H}{r} - \frac{n(n+1)}{r^2}T\right] + 2\mu\frac{H}{r}\right\}$$
$$\cdot Y_{2i}(\lambda, \Phi) + 2\mu\frac{T}{r^2} \cdot \frac{\partial^2 Y_{2i}(\lambda, \Phi)}{\partial\Phi^2}$$
$$(2)$$

$$\sigma_{\lambda\lambda_i}(r, \lambda, \Phi) = \left\{\Lambda\left[\frac{\partial H}{\partial r} + \frac{2H}{r} - \frac{n(n+1)}{r^2}T\right] + 2\mu\frac{H}{r}\right\}$$
$$\cdot Y_{2i}(\lambda, \Phi) + 2\mu\frac{T}{r^2}$$
$$\times \left[\frac{\partial^2 Y_{2i}(\lambda, \Phi)}{\partial\lambda^2}(\cos^{-2}\Phi) + \tan\Phi\frac{\partial Y_{2i}(\lambda, \Phi)}{\partial\Phi}\right]$$
$$(3)$$

$$\sigma_{r\Phi_i}(r, \lambda, \Phi) = \frac{\mu}{r}\left[\frac{\partial T}{\partial r} - \frac{2T}{r} + H\right]\frac{\partial Y_{2i}(\lambda, \Phi)}{\partial\Phi} \quad (4)$$

$$\sigma_{r\lambda_i}(r, \lambda, \Phi) = \frac{\mu}{r}\left[\frac{\partial d_{ri}(r, \lambda, \Phi)}{\partial\lambda}\frac{1}{\cos\Phi} + r\frac{\partial d_{\lambda i}(r, \lambda, \Phi)}{\partial r}\right.$$
$$\left. - d_{\lambda i}(r, \lambda, \Phi)\right]$$
$$= \frac{\mu}{r\cos\Phi}\left[\frac{\partial T}{\partial r} - \frac{2T}{r} + H\right]\frac{\partial Y_{2i}(\lambda, \Phi)}{\partial\lambda}$$
$$(5)$$

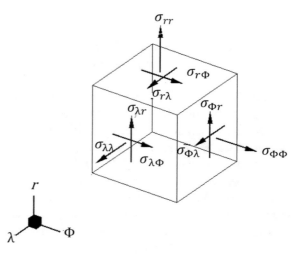

Figure 1
Stress tensor components. Normal stresses: $\sigma_{rr}, \sigma_{\Phi\Phi}, \sigma_{\lambda\lambda}$ and shear stresses: $\sigma_{r\lambda} = \sigma_{\lambda r}$, $\sigma_{r\Phi} = \sigma_{\Phi r}$, $\sigma_{\Phi\lambda} = \sigma_{\lambda\Phi}$

$$\sigma_{\Phi\lambda_i}(r, \lambda, \Phi) = 2\frac{\mu T}{r^2 \cos\Phi}\left[\frac{\partial^2(Y_{2i}(\lambda, \Phi)}{\partial\Phi\partial\lambda} - \tan\Phi\frac{\partial Y_{2i}(\lambda, \Phi)}{\partial\lambda}\right] \tag{6}$$

(of course $\sigma_{r\Phi_i} = \sigma_{\Phi r_i}$; $\sigma_{r\lambda_i} = \sigma_{\lambda r_i}$; $\sigma_{\Phi\lambda_i} = \sigma_{\lambda\Phi_i}$).

In Eqs. (1)–(6) $H(r)$ and $T(r)$ are functions of the distance r from the centre of the Earth and describe radial and horizontal elastic deformations within the Earth and at the surface ($r = a$) they are the Love–Shida numbers $H(a) = h$ and $T(a) = l \cdot \mu(r)$, $\Lambda(r)$ are the Lamé parameters and $\rho(r)$ is density.

Since the right hand side of the Eq. (1) for the radial normal stress is proportional to the equation of boundary condition at $r = a$ $\frac{(\Lambda+2\mu)\partial H}{\partial r} + \Lambda\left[\frac{2}{a}H - \frac{n(n+1)}{a^2}T\right] = 0$ implies, that $\sigma_{rr_i}(a, \lambda, \Phi) = 0$. Similarly at the surface of the Earth $\sigma_{r\Phi_i}(a, \lambda, \Phi) = \sigma_{r\lambda_i}(a, \lambda, \Phi) = 0$ whereas the terms on the right hand side of the equation are proportional to the boundary condition $\mu(\frac{\partial T}{\partial r} - \frac{2}{a}T + H) = 0$ at $r = a$.

Therefore, it can be concluded that three out of six independent components of the stress tensor are zero at $r = a$ and they cannot be involved in tidal triggering. It can be seen from the results of calculations (see next section) that $\sigma_{\Phi\lambda_i}(a, \lambda, \Phi) = 0$. That is why the trigger effects at the surface of the Earth ($r = a$) can be associated only with two normal tensor components $\sigma_{\Phi\Phi_i}(a, \lambda, \Phi)$ and $\sigma_{\lambda\lambda_i}(a, \lambda, \Phi)$ acting parallel to the surface of the Earth.

3. Calculation of Spherical Tidal Stress Tensor Components

As a first step similarly to the procedure described in Varga and Grafarend (1996), the numerical values of $H(r)$ and $T(r)$ functions were determined for the elastic mantle of the Earth from the surface ($r = a$) to the core-mantle boundary ($r = b$). For this purpose the equation of motion introduced by Takeuchi (1953), Molodensky (1953), Alterman et al. (1959) should be used. For the solution Molodensky used auxiliary functions for normal ($N(r)$), tangential ($M(r)$) stresses and gravity potential (Poisson equation) ($L(r)$):

$$N = (\Lambda + 2\mu)\partial H/\partial r + \Lambda[2/rH - n(n+1)/r^2 T] \tag{7}$$

$$M = \mu\left(\frac{\partial T}{\partial r} - \frac{2}{r}T + H\right) \tag{8}$$

$$L = r^2\left(\frac{\partial R}{\partial r} - 4\pi G\rho H\right) \tag{9}$$

In Eq. (9) $R(a) - 1 = k$ is the second Love number and G the gravitational constant. In case of calculation of functions $N(r)$, $M(r)$ along the radius at the surface of the Earth ($r = a$) the boundary conditions are equal to zero: $N(a) = M(a) = L_n(a) = 0$. The boundary conditions at the core-mantle boundary ($r = b$) are $N(b) = -P$, $M(b) = L(b) = 0$, where P is the hydrostatic pressure.

With the use of numerical values of $H(r)$, $T(r)$, $N(r)$, and $M(r)$ in Eqs. (1)–(6) the stress tensor components were calculated in cases of zonal ($i = 0$), tesseral ($i = 1$) and sectorial ($i = 2$) tides, for the latitudes $0°$, $20°$, $40°$ and $60°$ from the surface of the Earth till the core-mantle boundary. The numerical values of Lamé parameters $\mu(r)$, $\Lambda(r)$, $\rho(r)$ for the elastic mantle ($a \geq r \geq b$) were taken from the PREM (Dziewonski and Anderson 1981).

The results of computations (Table 1) show that

- $\sigma_{r\lambda_{i=0}}(r, \lambda, \Phi)$ and $\sigma_{\Phi\lambda_{i=0}}(r, \lambda, \Phi)$ are equal to zero from the surface to the core-mantle boundary in case of zonal tides for all latitudes (therefore they are missing from the table)
- at the equator ($\Phi = 0°$) the zonal $\sigma_{r\Phi_{i=0}}(r, \lambda, 0°)$, the tesseral $\sigma_{rr_{i=1}}(r, \lambda, 0°)$, $\sigma_{\Phi\Phi_{i=1}}(r, \lambda, 0°)$ and $\sigma_{\lambda\lambda_{i=1}}(r, \lambda, 0°)$, and the sectorial $\sigma_{\Phi\lambda_{i=2}}(r, \lambda, 0°)$, $\sigma_{\lambda_{i=2}}(r, \lambda, 0°)$ tensor components have zero value through the mantle
- as it was already mentioned, at the surface of the Earth normal stresses $\sigma_{rr_i}(a, \lambda, \Phi) = \sigma_{r\Phi_i}(a, \lambda, \Phi) = \sigma_{r\lambda_i}(a, \lambda, \Phi) = 0$ ($i = 0, 1, 2$). The same is valid for shear stress: $\sigma_{\Phi\lambda_{i=0}}(a, \lambda, \Phi) = 0$. These stress tensor components are negligibly small to depths of approximately 200–300 km
- only the horizontal components of the normal stress, $\sigma_{\Phi\Phi}$ and $\sigma_{\lambda\lambda}$, have non zero values at $r = a$, which are ≤ 2 kPa.

Table 1

Lunisolar zonal, tesseral and sectorial stress tensor components (in kPa) within the mantle (a ≥ r ≥ 0.55—the approximate radius of the core) at latitudes 0°, 20°, 40° and 60°

	0°	20°	40°	60°
Zonal tides				
σ_{rr} r/a				
1.00	0	0	0	0
0.90	−0.38	−0.24	0.03	0.35
0.80	−0.52	−0.35	0.07	0.76
0.70	−0.41	−0.28	0.03	0.59
0.60	−0.28	−0.14	0.17	0.31
0.55	−0.69	−0.35	0.41	0.93
$\sigma_{r\varphi}$ r/a				
1.00	0.00	0.00	0.00	0.00
0.90	0.00	0.38	0.55	0.76
0.80	0.00	0.86	1.10	1.38
0.70	0.00	1.21	1.66	2.00
0.60	0.00	0.86	1.14	1.62
0.55	0.00	0.00	0.00	0.00
$\sigma_{\varphi\varphi}$ r/a				
1.00	−1.83	−0.41	0.07	0.69
0.90	−0.52	0.10	0.69	1.97
0.80	0.69	0.79	1.38	2.52
0.70	2.83	2.76	2.66	2.35
0.60	9.32	7.76	3.90	0.41
0.55	14.84	8.45	5.35	0.17
$\sigma_{\lambda\lambda}$ r/a				
1.00	0.00	0.00	0.00	1.38
0.90	−2.66	−0.52	1.28	4.55
0.80	−3.86	−2.07	2.83	12.42
0.70	−4.83	−2.42	5.00	16.91
0.60	−4.83	−1.38	7.07	17.77
0.55	−4.14	−1.04	7.59	18.11
Tesseral tides				
σ_{rr} r/a				
1.00	0.00	0.00	0.00	0.00
0.90	0.00	0.10	0.14	0.10
0.80	0.00	0.31	0.55	0.38
0.70	0.00	0.21	0.48	0.41
0.60	0.00	0.21	0.28	0.24
0.55	0.00	0.31	0.55	0.41
$\sigma_{r\varphi}$ r/a				
1.00	0.00	0.00	0.00	0.00
0.90	0.48	0.41	0.17	−0.28
0.80	1.04	0.76	0.28	−0.66
0.70	1.41	1.17	0.35	−0.83
0.60	1.10	0.76	0.10	−0.59
0.55	0.00	0.00	0.00	0.00
$\sigma_{\varphi\varphi}$ r/a				
1.00	0.00	0.28	0.41	0.35
0.90	0.00	0.59	0.83	0.59
0.80	0.00	0.41	0.66	0.48
0.70	0.00	0.07	0.10	0.14
0.60	0.00	−1.28	−1.55	1.38
0.55	0.00	−3.28	−3.28	−5.35

Table 1 *continued*

	0°	20°	40°	60°
$\sigma_{\varphi\lambda}$ r/a				
1.00	0.00	0.00	0.00	0.00
0.90	0.59	0.38	−0.10	−1.38
0.80	1.41	0.97	−0.24	−3.69
0.70	2.76	1.62	−0.35	−6.04
0.60	4.04	2.76	−1.38	−7.49
0.55	5.21	3.28	−1.62	−8.18
$\sigma_{\lambda\lambda}$ r/a				
1.00	0.00	0.10	0.24	0.08
0.90	0.00	1.38	1.73	0.10
0.80	0.00	2.42	2.76	−0.35
0.70	0.00	3.11	3.11	−2.76
0.60	0.00	3.45	1.73	−5.87
0.55	0.00	3.11	1.21	−6.73
$\sigma_{r\lambda}$ r/a				
1.00	0.00	0.00	0.00	0.00
0.90	0.00	0.17	0.28	0.41
0.80	0.00	0.38	0.62	0.93
0.70	0.00	0.59	1.00	1.31
0.60	0.00	0.35	0.72	1.04
0.55	0.00	0.00	0.00	0.00
Sectorial tides				
σ_{rr} r/a				
1.00	0.00	0.00	0.00	0.00
0.90	0.38	0.38	2.42	1.24
0.80	0.55	0.48	0.35	1.41
0.70	0.38	0.31	0.24	1.04
0.60	0.35	0.28	0.21	0.10
0.55	0.62	0.52	0.38	1.31
$\sigma_{r\varphi}$ r/a				
1.00	0.00	0.00	0.00	0.00
0.90	0.00	−0.10	−0.17	−0.14
0.80	0.00	−0.28	−0.41	−0.41
0.70	0.00	−0.59	−0.69	−0.55
0.60	0.00	−0.31	−0.59	−0.45
0.55	0.00	0.00	0.00	0.00
$\sigma_{\varphi\varphi}$ r/a				
1.00	0.69	0.59	0.52	−0.10
0.90	1.97	1.83	1.38	0.93
0.80	2.31	2.14	2.00	1.73
0.70	2.17	2.17	2.31	2.45
0.60	1.10	1.66	2.42	2.90
0.55	0.00	0.00	0.00	0.00
$\sigma_{\varphi\lambda}$ r/a				
1.00	0.00	0.00	0.00	0.00
0.90	0.00	−0.59	−1.10	−1.73
0.80	0.00	−1.66	−2.66	−0.41
0.70	0.00	−2.66	−4.07	−6.62
0.60	0.00	0.00	0.00	0.00
0.55	0.00	0.00	0.00	0.00
$\sigma_{\lambda\lambda}$ r/a				
1.00	0.93	0.62	0.55	0.24
0.90	1.38	1.14	0.41	−0.17
0.80	1.21	0.69	−0.17	−0.86
0.70	−0.52	−0.41	−1.21	−1.93
0.60	−3.62	−3.69	−3.76	−3.80
0.55	−5.87	−6.56	−7.59	−8.97

Table 1 *continued*

$\sigma_{r\lambda}$ r/a	0°	20°	40°	60°
1.00	0.00	0.00	0.00	0.00
0.90	0.48	0.45	0.38	0.24
0.80	0.97	0.90	0.79	0.55
0.70	1.52	1.45	1.14	0.76
0.60	1.10	1.04	0.93	0.62
0.55	0.00	0.00	0.00	0.00

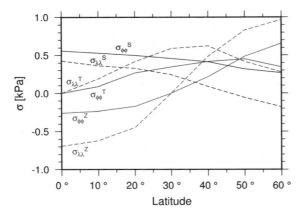

Figure 2

Latitude dependence of normal horizontal stress tensor components along the surface of the Earth (Z zonal, T tesseral and S sectorial tides)

Therefore it can be concluded that the influence of aforementioned conclusions that the influence of the earth tides on earthquake activity is a complex phenomenon. Their impact depends on the geographic location, on the depth and on the direction of stress tensor components. The strongest influence have the zonal (long-periodic) and sectorial (semi-diurnal) tides on and $\lambda\lambda$ (Fig. 2; Table 2). Their largest magnitude (~ 1 kPa) can be observed in the equatorial region and at high latitudes. In contrast the tesseral (diurnal) tides have probably less triggering effect as the two aforementioned ones.

4. Spherical Stresses Produced by Oceanic Load

The calculation of the functions $H'(r)$ and $T'(r)$ describing the response of the Earth to external

Table 2

Latitude dependence of the horizontal shear stress tensor components $\sigma\lambda\lambda$ and $\sigma_{\lambda\lambda}$ at the Earth's surface in kPa

	$\sigma_{\varphi\varphi}$	$\sigma_{\lambda\lambda}$
Zonal		
0°	−1.83	0.00
20°	−0.41	0.00
40°	0.07	0.00
60°	0.66	1.38
Tesseral		
0°	0.00	0.00
20°	0.27	0.17
40°	0.41	0.35
60°	0.35	0.07
Sectorial		
0°	0.69	0.93
20°	0.55	0.62
40°	0.52	0.06
60°	−1.04	0.24

load (relative to the solid Earth). Load is similar to the case of the lunisolar effect, but the surface boundary conditions are modified: two of them remain the same as in case of earth tides: $M'(a) = L'(a) = 0$, but the third one expresses that the loading mass is on the surface of the Earth: $N'(a) = p = \rho^* gh$, where p is the surface pressure, ρ^* is the density of sea water (1025 kg/m³) and h is tidal height in m. Because $N'(a) \neq 0$ consequently $\sigma_{rr}(a) / = 0$ (see Eq. 1). The gravitational potential generated by unit load acting on a square shaped ($\Delta\varphi = \Delta\lambda$) spheroidal layer on the surface ($r = a$) and characterised by a centri angle $\alpha = [(\Delta\phi \times \Delta\lambda)/\pi]^{1/2}$, at a spherical distance θ from the centre of the loaded area, can be given in the following form (Pertzev 1976):

$$W(\Theta) = \sum_{n=1}^{\infty} W_n = \frac{2\pi Ga}{g} \sum_{n=1}^{\infty} \frac{1}{2n+1}$$
$$[P_{n-1}(\cos\alpha) - P_{n+1}(\cos\alpha)]P_n(\cos\Theta)\left(\frac{r}{a}\right)^n \tag{10}$$

Similarly to the earth tidal stress tensor, the load stress tensor can be obtained, considering that the surface load is axially symmetric, and therefore the solution does not depend on the azimuth (Varga and Grafarend 1996):

$$\sigma_{rr} = \sum_{n=1}^{\infty} N_n' W_n(\Theta) \left(\frac{r}{a}\right)^n \tag{11}$$

$$\sigma_{\Theta\Theta} = \sum_{n=1}^{\infty} \left\{ \left[\Lambda\left(\frac{\partial H_n'}{\partial r} + \frac{2H_n'}{r} - \frac{n(n+1)}{r^2} T_n'\right) \right. \right.$$
$$\left. \left. + 2\mu \frac{H_n'}{r} \right] W_n(\Theta) + 2\mu \frac{T_n'}{r^2} \frac{\partial W_n}{\partial \Theta} \right\} \left(\frac{r}{a}\right)^n \tag{12}$$

$$\sigma_{\lambda\lambda} = \sum_{n=1}^{\infty} \left\{ \left[\Lambda\left(\frac{\partial H_n'}{\partial r} + \frac{2H_n'}{r} - \frac{n(n+1)}{r^2} T_n'\right) \right. \right.$$
$$\left. \left. + 2\mu \frac{H_n'}{r} \right] W_n(\Theta) + 2\mu T_n' \tan\Theta \frac{\partial W_n}{\partial \Theta} \right\} \left(\frac{r}{a}\right)^n \tag{13}$$

$$\sigma_{r\Theta} = \sum_{n=1}^{\infty} \frac{M_n'}{r} \frac{\partial W_n(\Theta)}{\partial \Theta} \left(\frac{r}{a}\right)^n \tag{14}$$

$$\sigma_{r\lambda} = 0 \tag{15}$$

$$\sigma_{\Theta\lambda} = 0 \tag{16}$$

Equations (11)–(16) show that the load generated stresses are mainly the normal ones (σ_{rr}, $\sigma_{\Phi\Phi}$ and $\sigma_{\lambda\lambda}$). The shear stress $\sigma_{r\Theta} = 0$, while $M'(a) = 0$. The magnitude of σ_{rr} is significant only locally, in the middle of the loaded area it may reach 100 kPa in case of a tidal height of 1 m on an area 10° by 10° (approx. 10^6 km^2), while the other two components $\sigma_{\Phi\Phi}$ and $\sigma_{\lambda\lambda}$ are 30 kPa (Fig. 3). If the area of the tidal load is reduced the amplitude of the load stress also decreases. According to model calculations carried out: the amplitudes of the load stresses in case of spherical segments 5° by 5° ($\sim 2.5 \times 10^5$ km^2),

and 2° by 2° ($\sim 4 \times 10^4$ km^2) are 25 and 7% of the amplitude obtained for spherical segment 10° by 10°. In the immediate vicinity of the loaded area the magnitude of load stress is (30–40)% of the maximum obtained for the middle of the loaded area. In contrast to the tidal stress, which inside the Earth moving away from the surface significantly increases its value (Table 3), the magnitude of the load stress decreases with depth rapidly.

5. What is the Impact of the Earth and Oceanic Tides on Earthquake Activity?

The results of the calculations show that in case of all tide types (zonal, tesseral, and sectorial) the stress components significantly increase with depth, but not approaching their maximum values in the depth range of interest from the viewpoint of earthquakes. The deepest earthquake focal depth data reliably determined is 68,410 km (South of Fiji, 17 June 1977). The radial and shear stress components show a monotonous increase at least until the middle of the mantle between 1000 and 2000 km depth zone (0.85–0.70) r/a. Moreover, some stress types reach their maximum value in this depth range. Results of model calculations show that distribution of tidal stress tensor components up to 1000 km depth does not depend to a substantial degree on the inner structure of the Earth. The distribution was compared in case of PREM and a model of the Earth with homogeneous mantle where the density depends only on the hydrostatic pressure. The biggest observed difference did not exceed 20%.

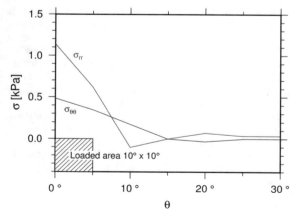

Figure 3
Radial normal and lateral stress tensor components ($\sigma_{rr}, \sigma_{\Theta\Theta}$) caused by 10^2 Pa normal load on a spherical segment $10^0 \times 10^0$ in function of spherical distance θ

Table 3

Distribution of vertical (σ_{rr}) and horizontal ($\sigma_{\theta\theta}$) load stress components (in kPa) at the surface in function of angular distance from the centre of the loaded area 10° × 10°. 1 m high water column load is considered

θ	σ_{rr}	$\sigma_{\theta\theta}$
0°	1.14	0.48
5°	0.62	0.35
10°	−0.10	0.17
15°	0.00	0.00
20°	0.07	−0.03
25°	0.03	0.00
30°	0.03	0.00

The depth distribution of stresses has a minor importance for the study possible tidal triggering, since 95% of the seismic energy is released within the depth interval (0–50) km (Varga et al. 2017). As it was shown in Sect. 2, the role of lunisolar effect can be significant in case of horizontal shear stresses $\sigma_{\phi\phi}$ and $\sigma_{\lambda\lambda}$. The likelihood of triggering effect in case of shallow seismicity in all probability can be increased by the fact, that the majority of the forces generating earthquakes are horizontal (or nearly horizontal) (e.g. normal, reverse or strike slips). If we accept that these two stress types may be the most likely earth tidal stresses, we must also take into account that the geographical distributions of $\sigma_{\phi\phi}$ and $\sigma_{\lambda\lambda}$ are not the same for different tides. In the equatorial region it is unlikely that tesseral tides, trigger while sectorial and zonal tides are big. At mid-latitudes the zonal tide generated stresses are small, while in polar regions only the zonal stress tensor component are significant.

Stresses produced by surface load are generated directly on the loaded area or in its immediate vicinity. Despite of the fact that the tidally generated load can be sometimes very significant, it has a local effect only (Fig. 3). If a huge area of about one million square kilometres ($\sim 10° \times 10°$) is loaded by 1 cm high tide (100 Pa) the generated vertical (σ_{rr}) and horizontal ($\sigma_{\theta\Theta}$) stresses are only 1.13 and 0.48 kPa (Table 3; Fig. 3). Of course in case of 1 m high tide the generated stresses are of the order of 10^{-1} MPa, that is still very small in comparison with the accumulated energy in earthquake foci of significant seismic events (10^5–10^7 Pa; 1–10 bar). Of course the area of $10° \times 10°$ is very large. In case of $1° \times 1°$ the 100 Pa load generates stresses about ten times smaller. As it was mentioned already earthquakes are shear fracture processes and consequently they are controlled predominantly by maximum stresses. Therefore—as in the case of tidal stress—for triggering the horizontal load stress ($\theta\theta$) is of importance.

6. Summary

Reviewing the results of the calculations it can be concluded that the magnitude of triggering effect of earth tides is different in case of zonal, tesseral, and sectorial tides and also significantly depends on the latitude.

The results of calculations carried out using theoretical models show that only the horizontal shear stresses $\sigma_{\phi\phi}$ and $\sigma_{\lambda\lambda}$ produced by earth tides are most likely to influence the outbreak of an earthquake.

The load caused by oceanic tides due to their local influence on solid Earth can have only a limited role in earthquake triggering by tides.

Acknowledgements

We thank our reviewers (Walter Zürn and an anonymous colleague) for their helpful comments. The research described in this paper was completed during research stay of P. Varga (01.03.2016–31.05.2016) supported by the Alexander Humboldt Foundation at the Department of Geodesy and Geoinformatics, Stuttgart University. P. Varga thanks Professor Nico Sneeuw for the excellent research conditions provided by him. Financial support from the Hungarian Scientific Research Found OTKA (Project K12508) is acknowledged.

References

Alterman, Z., Jarosch, H., & Pekeris, C. L. (1959). Oscillations of the Earth. *Proceedings of the Royal Society London A, 252,* 80–95.

Arabelos, D. N., Contadakis, M. E., Vergos, G., & Spatalas, S. (2016). Variation of the Earth tide-seismicity compliance parameter during the recent seismic activity in Fthiotida, central Greece. *Annals of Geophysics, 59*(1), 102. doi:10.4401/ag-6795.

Chen, H.-J., Chen, C.-Y., Tseng, J.-H., & Wang, J.-H. (2012a). Effect of tidal triggering on seismicity in Taiwan revealed by the empirical mode decomposition method. *Natural Hazards and Earth System Sciences, 12,* 2193–2202.

Chen, L., Chen, J. G., & Xu, Q. H. (2012b). Correlation between solid tides and worldwide earthquakes M $\geq$ 7 since 1900. *Natural Hazards and Earth System Sciences, 12,* 587–590.

Cochran, E. S., Vidale, J. E., & Tanaka, S. (2004). Earth tide can trigger shallow thrust fault earthquakes. *Science, 306,* 1164–1166.

Cotton, L. A. (1922). Earthquake frequency with special reference to tidal stresses in the lithosphere. *Bulletin of the Seismological Society of America, 12,* 47–198.

Davison, C. (1927). *Founders of seismology.* Cambridge: Cambridge University Press.

Dziewonski, A. M., & Anderson, D. L. (1981). Preliminary reference Earth model. *Physics of the Earth and Planetary Interiors, 25*(4), 297–356.

Emter, D. (1997). Tidal triggering of earthquakes and volcanic events. In H. Wilhelm, W. Zürn, & H.-G. Wenzel (Eds.), *Tidal phenomena* (Vol. 66, pp. 293–309). Lecture Notes in Earth Sciences Heidelberg: Springer.

Grafarend, E. (1986). Three-dimensional deformation analysis: Global vector spherical harmonic and local finite element representation. *Tectonophysics, 130*(1–4), 337–359.

Heaton, T. H. (1975). Tidal triggering of earthquakes. *Geophysical Journal of the Royal Astronomical Society, 43*, 307–326.

Heaton, T. H. (1982). Tidal triggering of earthquakes. *Bulletin of the Seismological Society of America, 72*(6), 2181–2200.

Houston, H. (2015). Low friction and fault weakening revealed by rising sensitivity of tremor to tidal stress. *Nature Geoscience, 8*, 409–415.

Ide, S., Yabe, S., & Tanaka, Y. (2016). Earthquake potential revealed by tidal influence on earthquake size–frequency statistics. *Nature Geoscience*. doi:10.1038/ngeo2796.

Kanamori, H. (1994). Mechanics of Earthquakes. *Annual Reviews of Earth and Planetary Sciences, 22*, 207–237.

Li, Q., & Xu, G.-M. (2013). Precursory pattern of tidal triggering of eartquakes. *Natural Hazards and Earth System Sciences, 13*, 2605–2618.

Métivier, L., de Viron, O., Conrad, C. P., Renault, S., Diament, M., & Patau, G. (2009). Evidence of earthquake triggering by the solid earth tides. *Earth and Planetary Science Letters, 278*, 370–375.

Molodensky, M. S. (1953). Elastic tides, free nutations and some questions concerning the inner structure of the Earth. *Trudi Geofizitseskogo Instituta Akademii Nauk of the USSR, 19*(146), 3–42.

Montessus de Ballore, F. (1911). *La sismologie moderne: les tremblements de terre*. Colin: Libraire A.

Perrey, A. (1875). Sur la fréquences des tremblements de terre relativement a l'age de la lune. *Comptes Rendus hebdomadaires des séances de la Académie des Sciences, 81*, 690–692.

Pertzev, B. P. (1976). Influence of the oceanic tides. *Physics of the Solid Earth, 1*, 13–27.

Schuster, A. (1897). On lunar and solar periodicities of earthquakes. *Proceedings of the Royal Society of London, 61*, 455–465.

Schuster, A. (1911). Some problems of seismology. *Bulletin of the Seismological Society of America, 1*, 97–100.

Stein, R. S. (2004). Tidal triggering caught in the act. *Science, 305*(5688), 1248–1249. doi:10.1126/sciencee.1100726.

Stroup, D. F., Bohnenstiehl, D. R., Tolstoy, M., Waldhauser, F., & Weekly, R. T. (2007). Pulse of the seafloor: Tidal triggering of microearthquakes at 9 500 N East Pacific Rise. *Geophysical Research Letters, 34*, L15301. doi:10.1029/2007GL030088.

Takeuchi, H. (1953). On the earth tide of the compressible earth of variable density and elasticity. *Transactions American Geophysical Union, 31*(5), 651–689.

Tanaka, S. (2010). Tidal triggering of earthquakes precursory to the recent Sumatra megathrust earthquakes of 26 December 2004 (Mw 9.0), 28 March 2005 (Mw 8.6), and 12 September 2007 (Mw 8.5). *Geophysical Research Letters, 37*, 2. doi:10.1029/2009GL041581.

Tanaka, S. (2012). Tidal triggering of earthquake prior to the 2011 Tohoku-Oki eartquake(M_W = 9.1). *Geophysical Research Letters, 39*, 7. doi:10.1029/2012GL051179.

Tanaka, S., Sato, H., Matsumura, S., & Ohtake, M. (2006). Tidal triggering of earthquakes in the subducting Philippine sea plate beneath the locked zone of the plate interface in the Tokai region. *Tectonophysics, 417*, 69–80.

Tormann, T., Enescu, B., Woessner, J., & Wiemer, S. (2015). Randomness of megathrust earthquakes implied by rapid stress recovery after the Japan earthquake. *Nature Geoscience, 8*(2), 152–158.

Tsuruoka, H., Ohtake, M., & Sato, H. (1995). Statistical test of the tidal triggering of earthquakes: Contribution of the ocean tide loading effect. *Geophysical Journal International, 122*(1), 183–194.

Varga, P., & Grafarend, E. (1996). Distribution of the lunisolar tidal elastic stress tensor components within the Earth's mantle. *Physics of the Earth and Planetary Interiors, 96*, 285–297.

Varga, P., Rogozhin, E. A., Süle, B., & Andreeva, N. V. (2017). A study of energy released by great (M7) deep focus seismic events having regard to the May 24, 2013 Mw 8.3 earthquake the Sea of Okhotsk, Russia. *Izvestiya, Physics of the Solid Earth, 53*(5), 385–409.

Vergos, G. S., Arabelos, D., & Contadakis, M. E. (2015). Evidence for tidal triggering on the earthquakes of the Hellenic arc, Greece. *Physics and Chemistry of the Earth Parts A/B/C, 85–86*, 210–215.

Vidale, J. E., Agnew, D. C., Johnston, M. J. S., & Oppenheimer, D. H. (1998). Absence of earthquake correlation with Earth tides: An indication of high preseismic fault stress rate. *Journal of Geophysical Research-Solid Earth, 103*, 24567–24572.

Wilcock, W. S. D. (2009). Tidal triggering of earthquakes in the Northeast Pacific Ocean. *Geophysical Journal International, 179*(2), 1055–1070.

Young, D., & Zürn, W. (1979). Tidal triggering of earthquakes in the Swabian Jura? *Journal of Geophysics, 45*, 171–182.

(Received October 13, 2016, revised April 23, 2017, accepted April 25, 2017, Published online May 10, 2017)

Reprinted from the journal

Pure Appl. Geophys. 175 (2018), 1659–1667
© 2017 The Author(s)
This article is an open access publication
https://doi.org/10.1007/s00024-017-1562-6

❙ Pure and Applied Geophysics

Interferometric Water Level Tilt Meter Development in Finland and Comparison with Combined Earth Tide and Ocean Loading Models

HANNU RUOTSALAINEN[1]

Abstract—A modern third-generation interferometric water level tilt meter was developed at the Finnish Geodetic Institute in 2000. The tilt meter has absolute scale and can do high-precision tilt measurements on earth tides, ocean tide loading and atmospheric loading. Additionally, it can be applied in various kinds of geodynamic and geophysical research. The principles and results of the historical 100-year-old Michelson–Gale tilt meter, as well as the development of interferometric water tube tilt meters of the Finnish Geodetic Institute, Finland, are reviewed. Modern Earth tide model tilt combined with Schwiderski ocean tide loading model explains the uncertainty in historical tilt observations by Michelson and Gale. Earth tide tilt observations in Lohja2 geodynamic station, southern Finland, are compared with the combined model earth tide and four ocean tide loading models. The observed diurnal and semidiurnal harmonic constituents do not fit well with combined models. The reason could be a result of the improper harmonic modelling of the Baltic Sea tides in those models.

Key words: Interferometric tilt meter, earth tides, ocean tide loading.

1. Introduction

Discussions on the rigidity of the earth were initiated already 150 years ago by Kelvin 1863 (Michelson 1914). The Earth was recognised not only as an elastic body, but also as a plastic yielding "modulus of relaxation", termed by Maxwell. Plastic yielding is realised by the lag of the distortion relative to the forces producing it (Michelson 1914).

Michelson (1914), Gale (1914) and staff at Yerkes Observatory, Williams Bay, Wisconsin, USA carried out preliminary studies on the earth's rigidity using east–west and north–south-oriented long water level tilt meters in autumn 1914. The water level tilt meters were installed at a 1.8-m-deep underground at the Yerkes Observatory. The tubes were 150 m long and half filled with water. Detailed descriptions are given in Michelson (1914) and in Gale (1914).

The amplitude ratio of measured tilt vs. calculated model tilt of an absolutely rigid earth gives the rate of deformation. The plastic yielding of the earth is observed from the retardation (lag) of the observed tilt phase to the tidal model tilt of absolutely rigid earth. The observed retardation of the earth tide signal must always be negative, because positive lag is meaningless (Michelson 1914). The mean amplitude ratio between observed east–west (EW) tilt to theoretical one was 0.710 and for north–south (NS) 0.523. The phase lag of total earth tide tilt for EW was −0.059 h and for NS was +0.007 h in the 1914 tilt observations of Michelson (1914) and Gale (1914).[1]

A similar difference between amplitude ratios in EW and NS directions was also observed earlier by Hecker, and he interpreted the reason to be the difference in earth rigidity (Michelson 1914).

Love and Schweydar (Michelson 1914, p. 124) had the opinion that the difference is attributed to the effect of ocean tides, and it causes differences in ratios of observed amplitudes and phases to theoretical.

Michelson and Gale (M–G) continued studies after 1914 by experimenting further in 1916–1917, using the water level tilt meters presented above with an interferometric recording system developed by Michelson in 1910 (Michelson 1914). The recording

---

[1]   Finnish Geospatial Research Institute, FGI, National Land Survey of Finland, Geodeetinrinne 2, 02430 Masala, Finland. E-mail: hannu.ruotsalainen@nls.fi

[1] A sign deviation exists in the observed phase of the north–south and east–west tilt meter results between pages 111 and 122 in Michelson's original paper 1914.

The original version of this chapter was revised. The correction to this chapter is available at https://doi.org/10.1007/978-3-319-96277-1_23

interferometers had direct internal absolute calibration. Figure 1 shows the principle of the recording setup (M–G 1919).

The refraction coefficient μ for water was 1.3408 with a wavelength of 435.8 nm. The number of fringes N caused by displaced water was calculated according to the formula

$$N = \frac{2(\mu - 1)d}{\lambda},$$

where λ was the wavelength of a mercury lamp light source with special arrangement and d was the displaced water level. One fringe corresponded to 1/1564 mm–639.4 nm/fringe. The tilt was estimated with 1/10 of fringe, and according to the formula above, the tilt rate is 0.173 ms-of-arc (mas), which means 0.839 nanoradian (nrad) resolution for a 152.4-m-long instrument. Using the conversion formula above for the tilt rate/fringe, it is possible to estimate tilts and compare them, e.g. with the combined earth tide model and ocean tide loading (OTL) model tilts at Yerkes observatory. In Sect. 2, a comparison of M–G observations with tilt predictions is given.

Kääriäinen (1979) constructed a water level tilt meter at the Finnish Geodetic Institute (FGI), which follows in principle the tube-pot technique developed by M–G (1919). He presented dimensions and properties of the instrument, hydrodynamical condition of the water in the tube-pot system, the instrument's thermal expansion modelling on environmental temperature change, and orientation of the instrument at the station. The level interferometer (diagram in Fig. 2) was a typical off-axis Fizeau interferometric setup, and interference fringe recording was carried out by film camera. The shape of varying interference fringes on the film in this construction was different, because the interferometric setup by M–G was completely different. A 177-m-long (EWWT) and 62-m-long (NSWT) water level tilt meter were built and installed in the Tytyri mine tunnel (geodynamic station Lohja2 of the FGI), in the vicinity of the city Lohja in southern Finland (Kääriäinen 1979; Kääriäinen and Ruotsalainen 1989). The location of the recording site is shown in Fig. 3. The reanalysed EWWT and NSWT results with OTL comparison are presented in Sect. 3.

The next step was a modern, redesigned, computer-controlled version of the laser interferometric water level tilt meter, installed at the same place as the NS-oriented instrument of the FGI (Ruotsalainen 2001). Construction details and earth tide analysis results with comparison to OTL models are described in Sect. 4.

2. Predicting Tilt Observations for Michelson–Gale Experiments Using Combined Earth Tide and Ocean Tide Loading Model Tilt

Using Agnew's (1997, 2012) ocean tide loading program, NLOADF, it is possible to determine harmonic ocean tide loading (OTL) amplitude and phase values for Yerkes Observatory (42°34.2′N, 88°33.4′W), e.g. using Schwiderski's ocean tide model. Figures 4, 5, 6 and 7 show the Schwiderski model's OTL vectors combined with the Wahr–Dehant–Zschau earth model tilt vectors (Schüller 2016) to predict harmonic model tilt observation in EW and NS directions. The green vectors are Wahr–Dehant–Zschau model earth tide tilt (nrad) with diminishing factor $\gamma_2 = 1 + k - h = 0.6948$, using Love numbers $h = 0.6032$, $k = 0.2980$ (PREM, Agnew 2009) and 0.0° phase, because Zschau (1978) argued that the observed earth tide phase lag is delayed only by 0.01°–0.001° to the theoretical model earth tide. Blue OTL vectors are subtracted from green earth tide model tilts, and red residual vectors are the prediction for tilt observation. They can be compared with M–G observations, e.g. by converting tilt values (nrad) to fringe values by the conversion formula above. In the following figures, all the amplitudes are nrad and phases in degrees, phase lags are negative and local. Terminologies A cos (alpha) and A sin(alpha) in the figures follow the convention by Melchior (1983, p. 332) for indirect effects.

Figure 4 shows that in the NS direction, the diurnal band harmonic amplitudes are quite small. The O1 and K1 wave groups have less than a 3.2 nrad tilt. The major energy NS direction is located in the semidiurnal wave band. The NS diurnal tilt harmonics have negative phase lags, but semidiurnal positive lags according to Schwiderski's OTL model. These explain the difficulties of amplitude and phase

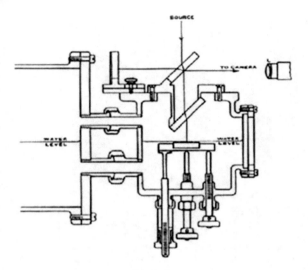

Figure 1
Principle of M–G interferometric water level tilt meter. Picture from (Michelson and Gale 1919), ©AAS. Reproduced with permission

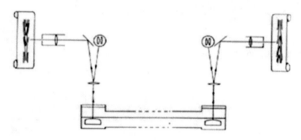

Figure 2
Operating principle of the old FGI water level tilt meter. Diagram from Kääriäinen (1979)

determinations 100 years ago. Love and Schweydar were right in their interpretation.

The only positive phase lag of predicted tilt in the EW direction exists in wave group N2. All others have negative lags.

In the EW direction, tilt phase lag for the K1 harmonic wave from Fig. 6 is as follows. The phase angle for K1 is

$$\alpha = \arctan\left(\frac{A(\sin(alpha))}{A(\cos(alpha))}\right) \approx -0.86°.$$

The EW/K1 predicted phase lag is −0.057 h in the time domain, and it is comparable to value −0.059 h, observed by Michelson (1914) and Gale (1914) as total EW tilt phase lag. The phase lag of the predicted NS/M2 vector in Fig. 5 is +0.015 h, and the value Michelson and Gale got for the total NS tilt phase lag

was +0.007 h. Michelson and Gale did not necessarily make an error in their calculation relating to earth tide tilt in 1914, because the positive phase lag in the NS direction complicates the comparison of the tilt observation and earth tide model tilt. Of course, all harmonic terms must be taken into account when determining the total diurnal or semidiurnal phase lags in each direction. By the least squares method, amplitude and phase values for diurnal and semidiurnal bands were determined again for interferometric setups by M–G (1919). The common diminishing of amplitude ratio in weighted mean is 0.690 and phase lag is 2°41′ for NS and 4°34′ for EW (the sign convention for lag is opposite than above) there.

3. Reanalysis of the Earth Tide Tilt of the FGI Tilt Meters 1977–1993

The resolution of tilt/fringe was determined according to the formula (Kääriäinen 1979),

$$S = \frac{\lambda \times \rho \times 10^3}{2 \times n \times L}, \ (\text{mas/fringe}),$$

where λ is the wavelength of light source, $\rho = 206{,}265$ is the conversion factor from radians to arc-seconds, n is the refraction coefficient of fluid and L is the length of the tilt meter. Half of the length of the water level inside tube indicates the tilt rate and, therefore, the length, L, in the formula above must be $L/2$. For sodium (Na), the light-based fluid level interferometer tilt value is then 0.515 mas/fringe and for helium (He) light, 0.514 mas/fringe.

The EWWT- and NSWT-tilt meter data were reanalysed by ETERNA 3.4 Earth tide analysis program (Wenzel 1996) and the newest version ET34-ANA-V52, developed by Schüller (2016). The OTL values based on Schwiderski (1980), TPXO7.0 (Egbert and Erofeeva 2002) and CSR4.0 (Eanes 1994) ocean tide models were determined using the NLOADF program by Agnew (1997). FES2004 (Lyard et al. 2006) OTL values were obtained using the OTL provided by Bos and Scherneck (2014) (http://holt.oso.chalmers.se/loading/). The phase lags in the OTL provider is relative to Greenwich meridian and lags positive. They must be converted from Greenwich meridian to local with sign convention using the formula by Agnew (2009).

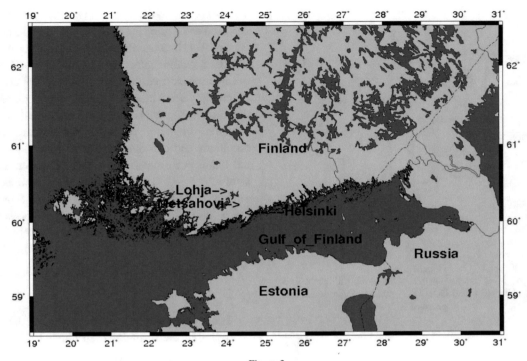

Figure 3
Location observation site Lohja in southern Finland

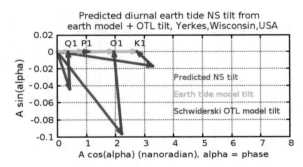

Figure 4
Predicted (*red*) earth tide tilt from OTL (*blue*) and earth tide model (*green*) in diurnal band in NS direction

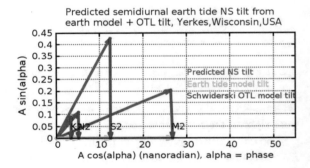

Figure 5
Predicted (*red*) earth tide tilt from OTL (*blue*) and earth tide model (*green*) in semidiurnal band in NS direction

The EW scale of diurnal CSR4.0 OTL model is three times larger than the theoretically predicted earth tide. In the diurnal band, Schwiderski's OTL amplitudes and phases are too small. The main reason for the wave group K1 phase deviation is the well-known core–mantle resonance (Fig. 8).

In the semidiurnal band in the EW direction, CSR4.0 amplitudes and phases excluding the N2 wave group fit better than Schwiderski and TPXO7.0.

The FES2004 model has the most deviating phases and amplitudes there (Fig. 9).

In the NS orientation, all OTL models deviate from observations, and the reason can be partly the improperly modelled Baltic Sea loading and partly the Norwegian Sea/Arctic Sea OTL modelling.

The OTL values in the diurnal frequency band in Q1, P1 and O1 wave groups have amplitude values in fraction of nanoradian in the Schwiderski, CSR4.0

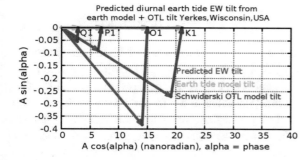

Figure 6
Predicted (*red*) earth tide tilt from OTL (*blue*) and earth tide model
(*green*) in diurnal band in EW direction

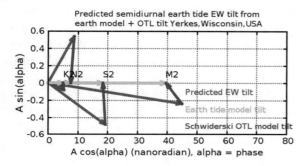

Figure 7
Predicted earth tide tilt from OTL and earth tide model in
semidiurnal band EW direction

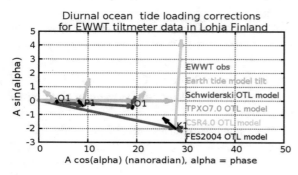

Figure 8
Observed earth tide tilt of EWWT with OTL models in the diurnal
band in the EW direction

and TPXO7.0 models. In the case of FES2004, the values are too big compared to others in the same wave group. The K1 wave group amplitudes are larger, but only Schwiderski and CSR4.0 show the phase angle in the right direction [the K1 vector in the CSR4.0 model for NSWT in Ruotsalainen et al. (2015) contained a combined TPXO7.0 model for northern latitudes; therefore, it is larger there]. The

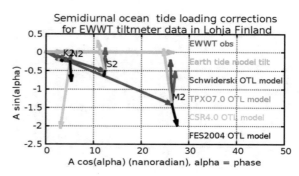

Figure 9
Observed earth tide tilt of EWWT with OTL models in semidiurnal
band in the EW direction

larger phase deviation from the theoretical earth tide model in the case of K1 is caused again by core–mantle resonance (Fig. 10).

In the semidiurnal band NS direction, nearly all models are deviating from the preferable phase. The Schwiderski model fits in the case of N2 and S2 (Fig. 11).

4. Modernisation of the FGI Water Level Tilt Meter

Mechanics, automation and a higher tilt resolution were the reasons for modernisation of fluid level sensing of the interferometric water level tilt meter of the FGI. The HeNe laser, digital camera and automated interference phase interpretation were used for modernisation (Ruotsalainen 2001). Some details were taken into account from innovations of the former tilt meter design of the FGI. In the new design, special stainless steel is used in the tube and pot constructions to avoid corrosion in a hostile mine

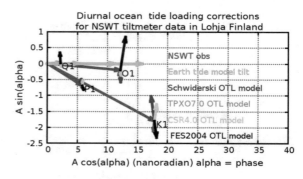

Figure 10
Observed earth tide tilt of NSWT with OTL models in diurnal band
in the NS direction

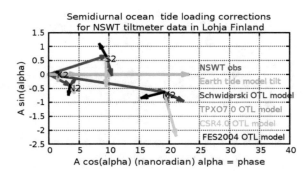

Figure 11
Observed earth tide tilt of NSWT with OTL models in semidiurnal band in the NS direction

environment. Fizeau–Kukkamäki interferometer principle (see Fig. 12) is used for a level sensing laser interferometer together with fibre optics.

Thorlabs HGR020 HeNe laser ($\lambda_{vacuum} = 543.0$ nm) is used as a light source for interferometer. Collimation of the beam is carried out by a telescope-type collimator connected to an optical fibre, as shown in Fig. 13. Basler A602f CMOS cameras are used for the recording of interference fringes with a sampling rate of 15 Hz. In Fig. 13, the Basler A602f camera system is located to the left of the end pot system, sealed against humidity inside a plastic box (Ruotsalainen et al. 2015).

5. Recordings and Analysis of the Earth Tide

The tilt resolution of the modern laser interferometer level sensing water level tilt meter (NSiWT) in Lohja2 is

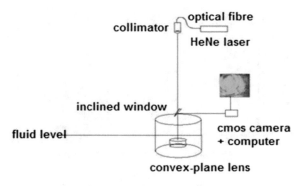

Figure 12
Principle of Fizeau–Kukkamäki fluid level interferometer

$$S = \frac{\lambda_{air}}{n \times L}, \ (nrad/fringe),$$

where λ_{air} is the wavelength of laser light (nanometres) and n is the refraction coefficient of water in physical conditions at the observation site. The wavelength value in the formula for laser light in the air is $\lambda_{air} = 542.8$ nm in the nominal physical condition of the station, when variations of the local air pressure, temperature and humidity are not yet corrected. These local variations cause a less than 10 pm variation in level sensing. The refraction coefficient of water is $n = 1.333$, determined by optical refraction observations. The length of the tube, $L = 50.40$ m, is measured with steel tape. The tilt resolution is then 8.0794 nrad/fringe and, for 1/100 of fringe (2.03 nm level sensing), 0.077 nrad (0.016 mas).

The example tilt recording of the NSiWT is given in Fig. 14. The red curve is the tilt recording, and the green curve is the theoretical tidal model tilt with amplitude factor 0.6948 (PREM, Agnew 2009) and zero phase (Heikkinen 1978). The observed tilt deviation from theoretical earth tide tilt is mainly caused by ocean tide loading, the Baltic Sea loading and atmospheric loading (Ruotsalainen et al. 2015, p. 160).

The NSiWT tilt meter data were also analysed by the ETERNA 3.4 Earth tide analysis program (Wenzel 1996) and its version ET34-ANA-V52, developed by Schüller (2016). Figures 15 and 16 show the analysis results for the main tidal harmonic wave groups.

Very small differences exist in earth tide analysis results between the old NSWT and new NSiWT water level tilt meters. The largest deviation between amplitude factors is 0.0315 in the O1 wave group, and other deviations are considerably smaller. In the tidal phase, the largest deviation is 6.10° in wave group Q1. In other wave group phases, they are within ±2.15°.

The deviation in phase of the wave group Q1 between instruments can be explained by the loading effect of the seiche oscillation phenomenon of the Baltic Sea. The oscillation of 26.2 h in the Gulf of Finland was determined by Lisitzin (1959), and this non-tidal period is harmfully located inside wave

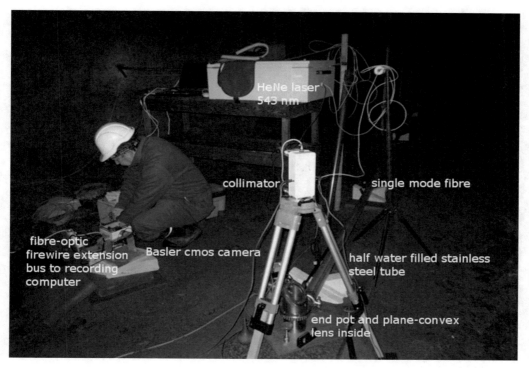

Figure 13
End pot-tube system, collimator connected with fibre to HeNe laser and Basler A602F CMOS camera system on the floor of the Tytyri mine at Lohja2 geodynamics station. Photo: M. Portin

group Q1 in the tidal frequency band. The phases of seiche oscillations frequencies are mainly wind generated; therefore, they strongly disturb both the earth tide tilt and the Baltic Sea tidal wave signals (Witting 1911) and their loading tilt at Lohja (Ruotsalainen et al. 2015, p. 160).

In the semidiurnal band, both in NSWT and NSiWT, the M2 amplitude factor diminishing to 0.56 can be recognised and none of the OTL models can correct the tilt to fit the earth tide model tilt. Amplitudes are of a preferable size, but the phases are not fitting? The Baltic Sea and atmospheric tidal loading harmonic presentations need to be taken into more careful consideration and combined for modelling.

The broad band of other geophysical phenomena (Ruotsalainen 2012) has been recorded since 2008, when the 50.4-m-long NSiWT instrument was set up as operational in the Lohja2 geodynamic station. These include Baltic Sea non-tidal loading and atmospheric loading (Ruotsalainen et al. 2015), free oscillations of the earth after great earthquakes (Ruotsalainen 2012), microseism and secular tilt recordings.

6. Conclusions

The semidiurnal earth tide tilt predictions in the NS direction using combined earth model tilt and Schwiderski OTL model tilt show positive lags and predicted diurnal amplitudes with negative lags smaller than 3 nrad in the NS direction for Yerkes observatory. The semidiurnal band in the NS tilt recording has a leading role, instead of diurnal, and this explains the uncertainty in the interpretation of the earth tide analysis of the Yerkes tilt observations 100 year ago.

The earth tide analysis of the tilt recordings between the old NSWT and new NSiWT tilt meters of the FGI has no significant differences. However, there are differences in ocean tide loading models compared to the tilt observations in the Lohja2

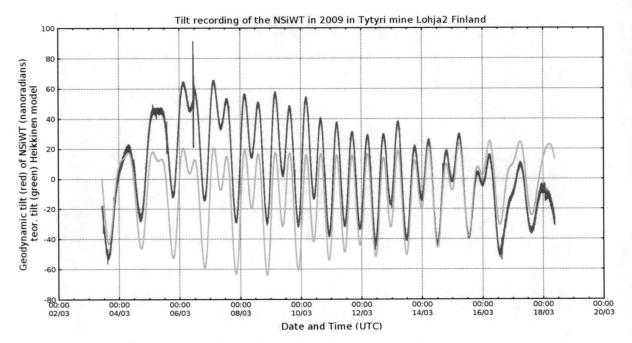

Figure 14
Tilt recording (*red*) of the new interferometric tilt meter of the FGI and theoretical tilt (nrad), (Heikkinen 1978)

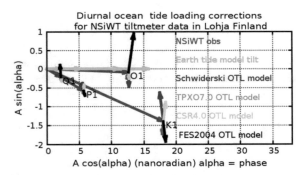

Figure 15
Observed earth tide tilt of NSiWT with OTL models in diurnal band in the NS direction

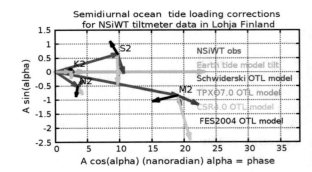

Figure 16
Observed earth tide tilt of NSiWT with OTL models in semidiurnal band in the NS direction

station. The best OTLs fit the earth tide model tilt in the semidiurnal EWWT tilt observation. The core–mantle resonances exist clearly in all three observation data sets.

OTL models do not explain the amplitude diminishing (0.6948 to >0.56) of M2 wave groups in NSWT and NSiWT tilt meter data in the NS direction. Baltic Sea and atmospheric loading harmonic modellings are the next steps in giving information on the deviating features of OTL models.

A modern NSiWT fluid level tilt meter is suitable for geodynamic and geophysical studies with an absolute scale.

Acknowledgements

Agnew's SPOTL program was used for computing ocean tidal loading. The ocean tide loading provider, developed by Bos and Scherneck, was also used for determination of OTL tilt. The modern version of the program ET34-ANA-V52 of the original ETERNA 3.4 program by Wenzel, developed further by Schüller, was used for the earth tide analysis. All of the above programs and data obtained are kindly acknowledged. Thanks go to two anonymous referees on their critical comments of the manuscript. The permission to reproduce material from The Astrophysical Journal, volumes 39 and 50, on behalf of the AAS by IOP Publishing, is kindly acknowledged.

REFERENCES

Agnew, D. C. (1997). NLOADF: A program for computing ocean-tide loading. *Journal of Geophysical Research, 102,* 5109–5110.

Agnew, D. C. (2009). Earth tides. In G. Schubert & T. Herring (Eds.), *Geodesy.* Oxford: Elsevier.

Agnew D. C. (2012). SPOTL: Some programs for ocean-tide loading, SIO Techn. Rep., Scripps Institution of Oceanography. http://escholarship.org/uc/item/954322pg.

Bos, M. S., Scherneck, H. G. (2014). Free ocean tide loading provider. Onsala Space Observatory, Chalmers University of Technology, Gothenburg, Sweden. http://holt.oso.chalmers.se/loading/.

Eanes, R. J. (1994). Diurnal and semidiurnal tides from TOPEX/POSEIDON altimetry. *Eos Transactions American Geophysical Union, 75*(16), 108.

Egbert, G. D., & Erofeeva, L. (2002). Efficient inverse modeling of barotropic ocean tides. *Journal of Atmospheric and Oceanic Technology, 19,* 183–204.

Gale, H. (1914). On an experimental determination of the earth's elastic properties. *Science New Series, 39*(1017), 927–933.

Heikkinen, M. (1978). On the tide generating forces. Publication of the Finnish Geodetic Institute, No. 85, Helsinki.

Kääriäinen, J. (1979). Observing the earth tides with a long water tube tilt meter. *Annales Academiae Scientiarum Fennicae A VI Physica, 424.*

Kääriäinen, J., Ruotsalainen, H. (1989). Tilt measurements in the underground laboratory Lohja 2, Finland in 1977–1988. Publication of the Finnish Geodetic Institute, No. 110, Helsinki.

Lisitzin, E. (1959). Uninodal seiches in the oscillation system Baltic proper, Gulf of Finland. *Tellus, 4,* 459–466.

Lyard, L., Lefevre, L., Letellier, T., & Francis, O. (2006). Modelling the global ocean tides: Insights from FES2004. *Ocean Dynamics, 56,* 394–415.

Melchior, P. (1983). *The tides of the planet earth.* Oxford: Pergamon Press.

Michelson, A. A. (1914). Preliminary results of measurement of the rigidity of the earth. *Astrophysical Journal, 39,* 105–128.

Michelson, A. A., & Gale, H. (1919). The rigidity of the earth. *Astrophysical Journal, 50,* 330–345.

Ruotsalainen, H. (2001). Modernizing the Finnish long water-tube tilt meter. *Journal of the Geodetic Society of Japan, 47*(1), 28–33.

Ruotsalainen, H. (2012). Broad band of geophysical signals recorded with an interferometrical tilt meter in Lohja, Finland. Geophysical Research Abstracts Vol. 14, EGU2012-9827, 2012 EGU General Assembly 2012.

Ruotsalainen, H., Nordman, M., Virtanen, J., & Virtanen, H. (2015). Ocean tide, Baltic Sea and atmospheric loading model tilt comparisons with interferometric geodynamic tilt observation—case study at Lohja2 geodynamic station, southern Finland. *Journal of Geodetic Science, 5*(1), 2081–9943. doi:10.1515/jogs-2015-0015. **(ISSN (online))**.

Schüller, K. (2016). User's guide ET34-ANA-V5.2, Installation Guide ETERNA34-ANA-V5.2, Surin.

Schwiderski, E. W. (1980). On charting global ocean tides. *Reviews of Geophysics and Space Physics, 18*(1), 243–268.

Wenzel, H. G. (1996). The nanogal software: Earth tide data processing package Eterna 3.30. *Bulletin d'Information des Marées Terrestres, 124,* 9425–9439.

Witting, R. J. (1911). Tidvattnen i Östersjön och Finska viken. In *Fennia 29* (p. 84). Helsinki: Simelius **(in Swedish)**.

Zschau, J. (1978). Tidal friction in the solid earth: Loading tides versus body tides. In P. Brosche & J. Sündermann (Eds.), *Tidal friction and the earth's rotation.* New York: Springer.

(Received December 16, 2016, revised April 11, 2017, accepted April 25, 2017, Published online May 3, 2017)

Pure Appl. Geophys. 175 (2018), 1669–1681
© 2017 Springer International Publishing AG, part of Springer Nature
https://doi.org/10.1007/s00024-017-1719-3

Pure and Applied Geophysics

Parallel Observations with Three Superconducting Gravity Sensors During 2014–2015 at Metsähovi Geodetic Research Station, Finland

HEIKKI VIRTANEN[1] and ARTTU RAJA-HALLI[1]

Abstract—The new dual-sphere superconducting gravimeter (SG) OSG-073 was installed at Metsähovi Geodetic Fundamental Station in Southern Finland in February 2014. Its two gravity sensors (N6 and N7) are side by side, not one on top of the other as in other earlier dual-sensor installations. The old SG T020 has been recording continuously since 1994–2016. This instrument is situated in the same room at a distance of 3 m from the dual-sphere SG. T020 observed simultaneously for 1 year with N6 and for 15 months with N7. The gravity signals observed by N6 and N7 are very similar, except for the initial exponential drift. We have calculated the power spectral density to compare the noise level of these instruments with other low noise SGs. In this paper we present the observed differences in the gravity time series of T020 and OSG-073, induced by local hydrology. We have observed a clear 10–20 nms^{-2} difference in the seasonal gravity variations of OSG-073 and T020. We have found clear gravity differences due to transient effect of heavy precipitation. In addition, we compare the remote effect on gravity due to variations in the Baltic Sea level and total water storage in Finland to the observed gravity signal. We also present modeling results of gravity variations due to local hydrology.

Key words: Superconducting gravimeter, gravity gradiometry, hydrology.

1. Introduction

1.1. Metsähovi Geodetic Fundamental Station

Finnish Geospatial Research Institute, FGI (Formerly Finnish Geodetic Institute) operates a specially designed gravity laboratory at Metsähovi Geodetic Fundamental station (ME). The building and the gravimeter piers stand on a knoll of Precambrian granite giving it a solid foundation. The station is located in a rural area. There are no industrial plants or transport arteries in the near vicinity of the station that could cause ground vibrations. Hence, ME belongs to the group of low background noise stations among other GGP stations (Rosat and Hinderer 2011). ME is a multi-technique geodetic research station and is a part of the GGOS's (Global Geodetic Observing System) core sites, including absolute gravity (AG), permanent GNSS, Satellite Laser Ranging (SLR), DORIS beacon and geodetic VLBI (Very long Baseline Interferometry). All techniques are influenced by the same environmental loading effects as the SG. Because of its high sensitivity, the SG is an excellent tool for testing and validating the pertinent correction models. Within a distance of 100 m from the gravity laboratory there are multiple automated hydrological sensors for hydrological studies: 3 deep boreholes in the bedrock, 11 groundwater observation tubes in the sediments, 12 arrays of soil-moisture sensors and a pluviometer (Mäkinen et al. 2014; Hokkanen et al. 2006). In addition, the water equivalent of snow is measured with Campbell CS725 as well as manually. In 2015, the station was equipped with a new Vaisala weather station AWS310. Besides the fundamental sensors (pressure, temperature, humidity, wind, precipitation), two pyranometers, one for global radiation and one for ground radiation, and an ultrasonic snow depth sensor are included in the weather station. These sensors are necessary for hydrological modeling. The rain gauge is heated (to measure also precipitation as snow) and is equipped with a Tretyakov windshield. The map of local area is presented in Fig. 1. Together with the gravity data it is possible to model the gravitational effect due to changes in atmospheric mass distributions, hydrological conditions and in the Baltic Sea level.

[1] Finnish Geospatial Research Institute-FGI, National Land Survey, Geodeetinrinne 2, 02430 Masala, Finland. E-mail: heikki.virtanen@nls.fi

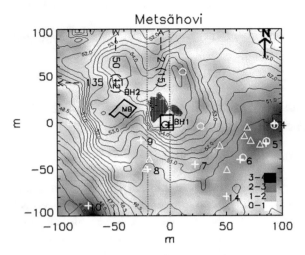

Metsähovi

Figure 1

Map of the Metsähovi area (200 × 200 m) included in our local modeling. The isolines at 0.5-m interval show surface elevations and the gray shades show soil thickness, dark areas North of the gravity lab are bare bedrock above gravity sensors. Maximum soil thickness is 3.7 m, mean value is 0.8 m. The SG T020 is in the center, marked by a cross inside the gravity laboratory and OSG-022 marked by a circle 3 m to the West. The SG sensor is at 55.6 m elevation. MB denotes main building. Numbered crosses mark the places of the groundwater tubes in soil. Three tubes are outside the plotted area. Tube 8 used in Fig. 7 is at about (− 20, − 50 m) in local coordinates. Triangles denote arrays of soil-moisture sensors. BH1 (just E from the gravity laboratory) and BH2 are borehole wells in bedrock. A third borehole well for water use is inside the main building (NW from the gravity laboratory). Small circles denote dry access tubes. They are not used in this paper. The dotted lines are height profiles shown in Fig. 12

1.2. ME Superconducting Gravimeters

The SG of the FGI, GWR T020, has operated continuously at ME since August 1994 (Virtanen and Kääriäinen 1995, 1997). The new dual-sphere

(sensor) gravimeter OSG-073 was installed in Metsähovi in February 2014 at a distance of 3 meters from SG T020 (Fig. 2). One sensor (N7) was the standard iGrav™, with a lightweight sphere (0.005 kg), which has a low drift rate. The second sensor (N6) uses a heavy 0.02 kg sphere which gives a very low noise with a much higher quality factor Q. Its novel design was unique: two gravity sensors are separate and side by side (15 cm), not one on top of the other as in most of the earlier dual-sensor installations (Goodkind 1999). Advantage of dual sensor is, e.g., correction of offsets and other instrumental errors (Hinderer et al. 2007).

There were four dual-sphere SGs operating around the world with normal lightweight sensors (0.005 kg sphere). At Black Forest Observatory, Germany (BFO), the lower sensor is a heavy (0.02 kg sphere) sensor (OSG-056L).

The noise level of these instruments and a comparison to NLNM (Peterson 1993) was estimated with a power spectral density (PSD) calculation (Fig. 3). For the PSD calculation we have used 5 days of data from days without microseism or other environmental or instrumental d levels of N7 and T020 are quite similar when compared. The sensor N6 has a very low noise, comparable to the best known gravimetric instruments in the world, such as at BFO.

We have two sets of simultaneous absolute gravity measurements with FG5X-221 at Metsähovi. In spite of the length of the SG time series (9 and 7 days), the accuracies were not very good. During the calibration measurements there were strong

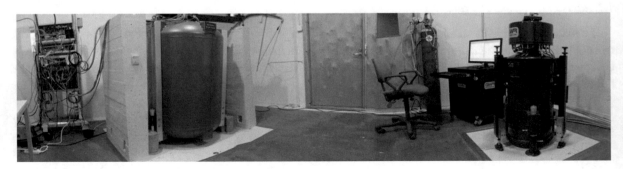

Figure 2

Superconducting gravimeters at Metsähovi on November 2014. The old T020 is on the left and the new OSG-073 on the right. The distance between gravity sensors is 3.0 m. The sensor height is about 20 cm higher in T020

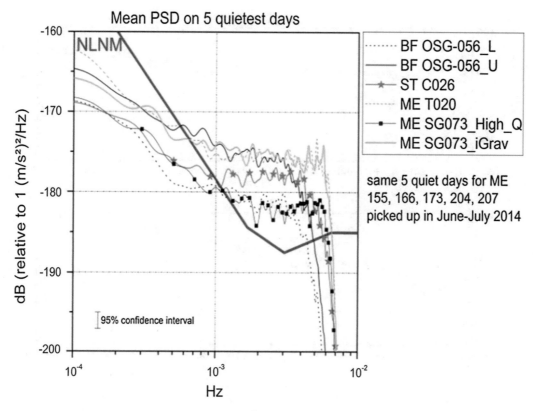

Figure 3
Noise levels for different gravity sensors (Rosat et al. 2016, Private communication). BF OSG-056_L means Black Forest Observatory (Schiltach) lower high-Q sensor, ST C026 refers to the high_Q sensor in Strasbourg, J9 station, ME SG073 High-Q is N6 and ME SG073_iGrav is N7. NLNM model is the red curve (Peterson 1993)

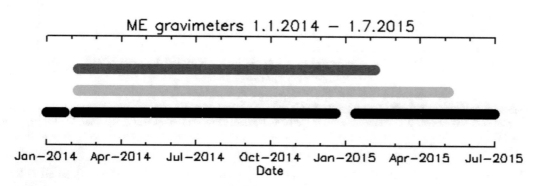

Figure 4
Parallel observations with the three gravimeters between 2014 and 2015. N6 is shown in red, N7 in green and T020 in black

microseisms, data gaps and technical problems with N6. The calculated values for the calibration factor were: N6 − 447.11 ± 0.37 and N7 − 932.47 ± 0.75 nms^2/V for N6 and N7, respectively. However, we have a very good calibration for SG T020 (Virtanen et al. 2014). We used 24 different datasets between 2003 and 2012, extending over 2–7 days with parallel FG5 measurements. The resulting calibration factor

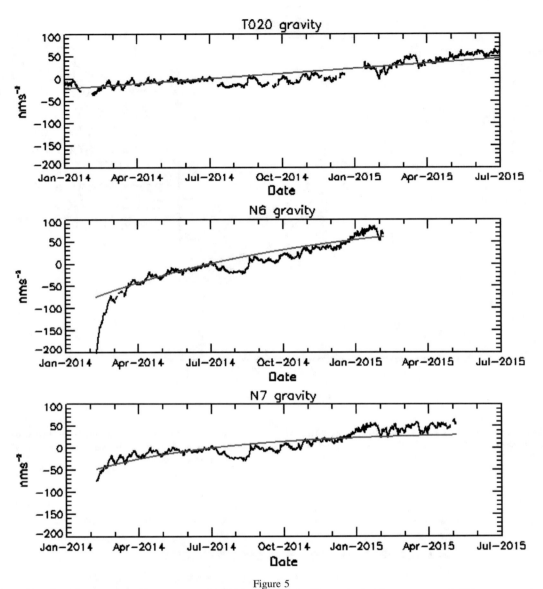

Figure 5
Gravity residuals of three sensors 1 Jan 2014–1 Jul 7 2015. Tides, atmospheric effect and pole tide were corrected. Offsets were corrected and traces of big earthquakes removed. Drift correction models for T020, N6 and N7 sensors are plotted in red lines. We have applied a linear trend for T020 sensor. Due to large non-linear drift after initialization, we have applied an exponential drift model to N6 and N7

for T020 is -1104.3 ± 1.2 nms²/V. Hence, using T020 to calibrate sensors N6 and N7 we got calibration factors -447.53 ± 0.04 (N6) and -933.34 ± 0.08 (N7) nms²/V. The usable length of the time series was about 164 days. We have taken into account different drift behavior of the instruments and have rejected disturbed data. Transfer functions for N6 and N7 were determined by step pulses.

The operation of T020 was finished in September 2016. The instrument has the second longest gravity data series in IGETS database, as shown in its documentation (Voigt et al. 2016). The OSG-073 was sent back to the manufacturer for improvements (GWR) in May 2015. Therefore, T020 observed simultaneously with N6 for one year and with N7 for 15 months. Observation periods for the three instruments are given in Fig. 4.

2. Data Processing

We have usable common data with T020 and N7 from Feb 2014 to 4th of May 2015, i.e., a total of 15 months. The common data set with N6 stopped on 4th of Feb 2015 due to technical problems. Some parts of T020 data were unusable due to breakage of its cooling system (Fig. 5). These problems caused an unmodeled drift to data and were handled as gaps. Original 1-s data were decimated to 1 min for cleaning and preprocessing. Cleaning process was standard (Hinderer et al. 2007, Virtanen 2006), consisting of, e.g., removal of spikes, offsets, traces of earthquakes and other disturbances. A few offsets of N6 and N7 were due to lightning detected by the Finnish Meteorological Institute. T020 had several offsets due to cooling problems, and were corrected using the OSG-073 observations. We have exploited TSOFT software and data analysis tools (Van Camp and Vauterin 2005). After cleaning the time series from disturbances we have applied an observed local tidal model (Virtanen 2006), air pressure (AP) correction with a single admittance ($-$ 3.10 nms^{-2} hPa^{-1}) and local pole tide correction, using IERS pole coordinates (https://www.iers.org).

Measured time lags for the sensors N7, T020 and N6 are correspondingly 9.5, 9.7 and 20.2 s. Time lag between N7 and T020 is very small and does not cause a significant error in the gravity signal. Instead the phase lag of N6 can cause errors in gravity up to 1.8 nms^{-2}. We have used tidal correction for N6 with a 10-s phase lag compared to T020. We have used for air pressure admittance a generic value based on several tidal analyses (Virtanen 2006).

In time domain the mean standard deviations (STD) of 1-min residual data were 0.20, 0.40, 0.50 nms^{-2}, respectively, for N6, N7 and T020.

For drift corrections and analyses shown in Table 1 we have used hourly values. Next step was the determination of drift models for the time series. We have used a linear model for the old SG T020. Due to a large non-linear drift after initialization of OSG-073, we have applied an exponential drift model by fitting the function

$$f(x_i) = c_0 \, e \, c_1 \, (x_i) + c_2,$$

where x_i are hourly gravity values, and c_0, c_1 and c_2 are coefficients to be determined by least-squares fitting.

Fits were applied to data after 1st of Mar 2014. For longer time series this approach is not valid, as the drift becomes nearly linear. Exponential modeling is evidently better for SGs, which are just initialized. Drift models are shown in Fig. 5. Approximate linear drifts (around Feb 2015) are for T020, N6 and N7: 4.5, 130 and 75 nms^{-2}/year. The drifts included both instrumental and geophysical parts. Geophysical part (7 nms^{-2} year^{-1}) is due to the post-glacial land uplift, which is about 2 mm/year at Metsähovi (Virtanen et al. 2014).

3. Results of Gravity Comparison Between the Three Sensors

3.1. Remote Effects on Gravity

We consider remote effects so that the gravity effect is uniform at ME, i.e., place of the gravity

Table 1

SG instrument, STD standard deviations of gravity residuals (1. col.) and regression coefficients (last col.) fitted to the remote water storage (TW) and Baltic Sea level (HSL) data

SG	TW			Baltic sea		
	STD	REG TW		REG HSL	COR	STD
T020	9.8	0.18		23.1	0.80	5.9
N6	10.8	0.23		18.2	0.86	5.5
N7	11.7	0.27		15.3	0.90	5.1

REG TW = regression coefficients for TW (nms^{-2} mm^{-1}), REG HSL = regression coefficient for HSL (nm/s^{-2} m^{-1}). Calculations are from 1st Apr–19th Dec 2014. The reduction of variance of N7 sensor is 81%

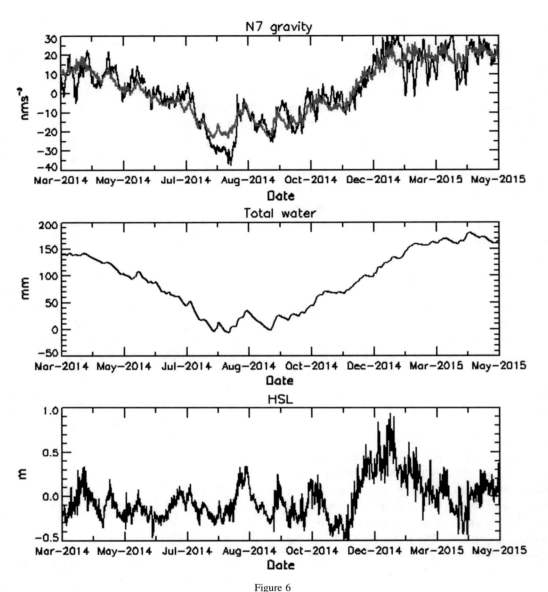

Figure 6
Top: drift-corrected N7 gravity residual (black) and fit to total water in Finland (WSFS) and HSL presented (red). Middle: total water (WSFS). Bottom: HSL (Baltic Sea level in Helsinki tide gauge) (1 Mar 2014–1 Mar 2015)

sensor at Metsähovi has no role. The drift-corrected gravity time series of N7 (1st of Mar 2014–1st of Mar 2015) is presented in Fig. 6 (top panel, black line). For comparison we have calculated STD of residuals for all three sensors (Table 1). We also calculated simultaneous linear regressions with remote water storage (TW, Fig. 6, middle panel) and Baltic Sea level changes at the Helsinki tide gauge (HSL, Fig. 6, bottom panel). TW estimates the total water storage

change of Finland based on WSFS (Watershed simulation and forecast System) provided by the Finnish Environmental Institute (Vehviläinen 2007). Gravity acts as independent data and remote effects are dependent variables. Results are shown in Table 1. There are remarkable differences between the coefficients. These can be due to imperfect initial drift estimates (N6, N7) and offset corrections of T020. We got the best results for the sensor N7, for

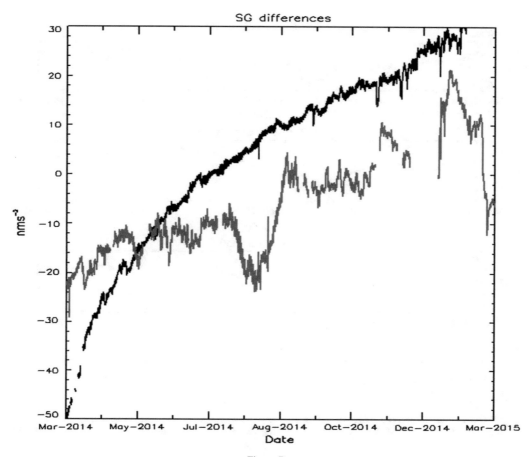

Figure 7
Difference of the gravity data series observed by N6 and N7 (black) and N7-T020 (red) from 1 Mar 2014 to 1 Mar 2015. Traces of big earthquakes and offsets were corrected, other corrections were not applied

which the variance is reduced by 81% (red color) when compared to the original time series. The fitted gravity response to TW and HSL (not residual) is shown in Fig. 6 (upper panel, red line). We have selected these environmental observations due to their well-known effect to gravity at ME (Virtanen et al. 2014).

3.2. Local Effect on Gravity

Local effects are mainly related due to mass changes at about 100 m distance around the gravity laboratory, where there is installed a variety of hydrological sensors (Fig. 1). The differences in the gravity observations of the three sensors are shown in Fig. 7. Gravitational effects of air pressure, tide and

polar motion, TW and HSL are presumably similar to all three sensors. Hence, differences are due to instrumental drift and local mass variations, i.e., local hydrology. Discrepancies between N6 and N7 sensors are small and mainly due to the different noise levels and drift of the sensors. Larger differences between OSG-073 sensors can arise due to very close masses, e.g. visitors inside the gravity laboratory. A human body near the gravimeter can cause a different signal at sensors which are 15 cm apart. We can clearly see a long-term difference between OSG–073 and T020 sensors in Fig. 7. In Fig. 8 we present the drift-corrected gravity difference (N7-T020) together with local hydrological observations using data from the boreholes in the bedrock, soil-moisture sensors and water tubes in the soil. The locations of the

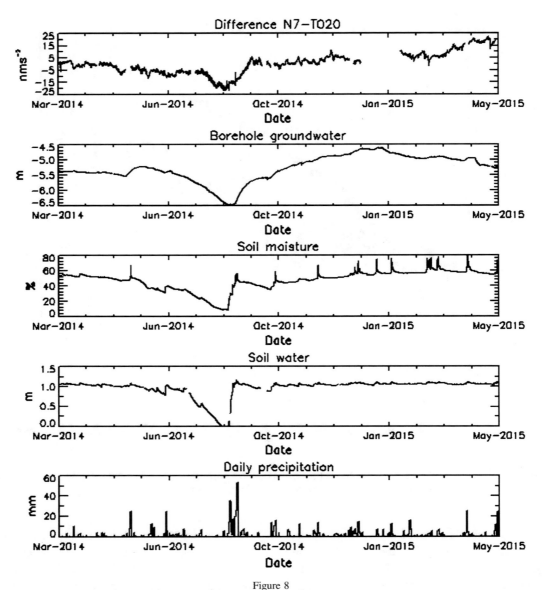

Figure 8
Top panel: gravity difference of drift-corrected N7 and T020 sensors (nms^{-2}). Second panel: groundwater in bedrock tube BH2 (m) below surface. Third panel: soil-moisture (%), sensor location is shown in ME map (Fig. 1). Fourth panel: water level in a soil ground access tube 8 m shown in ME map. Bottom panel shows daily precipitation (mm). Series are from 1 Mar 2014 to 1 May 2015

hydrological instruments are shown in Fig. 2. In addition, we get daily precipitation amount from a rain gauge, located near the gravity laboratory. All observations show similar long-term features.

A strong rain event can produce a different response of the gravity sensors. In Fig. 9 we show an example of heavy precipitation on the 20th of Aug 2014 at ME for sensors N6, N7 and T020. Daily

precipitation was approximately 50 mm and the maximum intensity was 30 mm within an half an hour period. That happens very seldom at ME. Peaks caused by this event can be seen in Figs. 7 and 8.

We only found 15 days, when the gravity data of all three sensors were usable for studying the gravitational effect of precipitation. In many cases during heavy rain we had to reject data due to, e.g.,

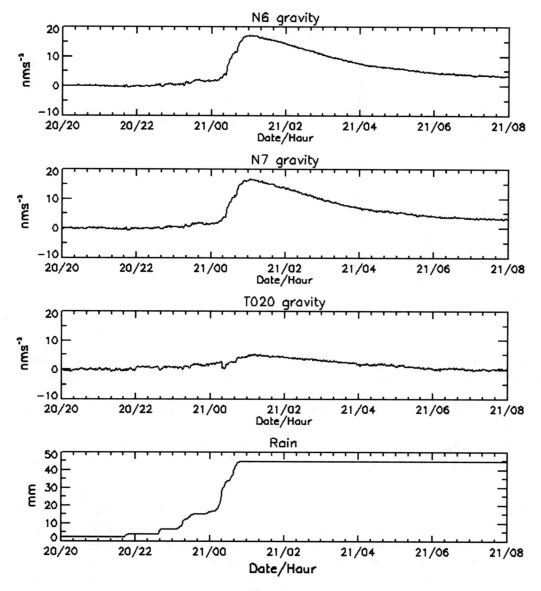

Figure 9
Gravity variations of three sensors during heavy rainfall (30 mm/0.5 h) on 20 August 2014

lightning, data gaps, saturation of the feedback of N6 or failure of the cooling system of T020. All studied rain events are presented in Table 2. The calculated gravity variation due to precipitation events are compiled in Table 2 and shown in Fig. 10. For sensor N7, we get the result 0.41 nm/s^2/mm, which is close to the Bouguer plate approximation. Time-varying gravity due to local hydrology is depending on precipitation and evaporation. Mathematical models are presented by Meurers (2007) and (Deville et al. 2013). To calculate the gravity effect of water in the soil, we have constructed a model, which takes into account both topography and the umbrella effect caused by the laboratory building. The model is constructed from rectangular 3D blocks (voxels) with a horizontal size of 1×1 m and a vertical size of 0.1 m. The model extends over a horizontal area of 201×201 m and to 15 m in depth, and hence

Table 2

Studied rainfall cases 2014

Date	Tot	Event	Int	T020	N7	N6	GD	Dur	CN
14-06-12	24.5	24.1	12.2	7.08	4.73	4.95	− 2.35	1011	1
14-07-16	6.6	6.4	39.6	1.0	1.31	1.16	0.31	49	2
14-08-07	5.5	5.0	64.8	0.03	0.63	0.34	0.60	6	3
14-08-13	35.5	9.0	72.0	2.92	2.62	2.61	− 0.30	15	4
14-08-13	35.5	8.6	100.8	2.61	2.73	2.59	0.12	17	5
14-08-14	5	4.2	11.7	2.02	2.37	2.58	0.35	97	6
14-08-16	16.9	8.8	35.6	1.42	1.21	1.39	− 0.21	40	7
14-08-16	16.9	6.4	64.8	1.34	0.99	0.88	− 0.35	13	8
14-08-18	19.6	4.0	57.6	0.13	0.46	0.32	0.33	9	9
14-08-19	26.2	8.9	79.2	1.37	1.35	1.13	− 0.02	30	10
14-08-20	53	28.1	126	1.99	13.74	13.92	11.75	48	11
14-09-22	13.6	13.1	6.0	2.74	1.97	2.16	0.77	526	12
14-10-19	12.4	11.4	11.9	4.45	6.07	6.09	1.62	628	13
14-11-02	6.2	2.8	6.0	1.32	1.13	0.96	− 0.19	47	14
14-11-06	13.5	14.3	6.0	7.32	7.54	7.77	0.22	944	15

Precipitation data are recorded every second, but for analyses decimated to 1 min. The mean square difference between N6 and N7 sensors computed from 15 cases was 0.21 nms^{-2}

Date event day, *Tot* total amount of precipitation per day (mm), *Event* precipitation of event (mm), *Int* maximum intensity of rain (mmh^{-1}), *T020, N7, N6* gravity effect of events (nm/s^{-2}), *GD* gravity difference (N7-N6) (nms-2), *Duration* event duration (min), *CN* case number

comprising a total of 6,060,150 voxels. The horizontal extension of the model is shown in Fig. 1. We calculated the gravity effect at the sensor position exactly for each rectangular voxel, using different densities (Nagy 1966). We used 2600 kg m^{-3} for bedrock and 1400 kg m^{-3} for soil. We did not consider the voxels directly below the gravity laboratory. The soil depth was determined by gravimetric methods (Elo 2001, 2006). Gravity measurements were carried out in a grid with a distance of about 5 m (within 50 m from the gravity laboratory) and 10 m further away. Topographic heights were determined using RTK GPS. Maximum thickness was 3.7 m and the mean was 0.8 m.

Water in the soil was simulated using higher densities for the respective voxels. We exploited rain gauges, soil-moisture sensors, the tubes measuring the water level in the sediments and weather data for estimating temporary water content in the soil. In addition, we can add to the model snow with variate densities above soil. In Fig. 11 we present the topography of the area shown in Fig. 1. Height profiles for surface and bedrock along the dotted lines shown in Fig. 1 are presented in Fig. 12. In Table 3

we provide results of some model calculations, for different soil water content, using two sensor locations separated by 3 m in WE direction (Fig. 1) representing T020 and N7, respectively. First we have calculated the gravity effects without extra water. Then we have added 100 mm water in the top of the soil layer, then 100 mm water on top of the bedrock and finally fully saturated soil. The total recorded precipitation for the study period was 990 mm. From Table 3 we can see that as a part of bedrock rises above of N6/N7, it causes a negative effect. Moderate added extra mass did not increase the difference between the two sensor sites. Only a large amount of extra mass produced an increasing difference. We have looked into other possibilities for "hidden" water.

Effects of the fracture water of bedrock on superconducting gravimeter data were studied by Hokkanen et al. (2007) using a ground penetration Radar (GPR) around the laboratory (22 × 22 m). Maximum effect could be several nms^{-2}. However, we do not have information of water exchange and fractures below the gravity laboratory.

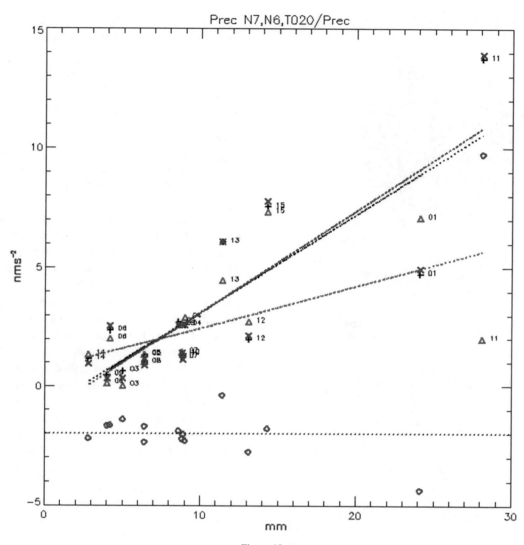

Figure 10

Observed gravity effect (nms^{-2}) due to rain events (mm) presented in Table 2. N7 shown with black + sign, N6 with blue x and T020 with red triangles. Dotted lines are regressions between gravity and amount of rain. Difference between sensors N7-T020 is presented by blue diamonds. Regression coefficient for N7 is 0.41 nms^{-2} mm^{-1} ($\pm$ 0.08), for N6 it is 0.43 nms^{-2} mm^{-1} ($\pm$ 0.08) and for T020 0.18 nms^{-2} mm^{-1} ($\pm$ 0.07)

4. Conclusions

The difference in annual variations between the two OSG-073 and T020 gravimeters is remarkable and reaches up to 20 nms^{-2}. By comparing two sensors separated by only 3 m, we have illustrated a horizontal gravitational effect due to local hydrology at Metsahovi station. The difference in the gravity signal is biggest in July–August, when the soil is dry.

Abundant precipitation in August and September reduce the difference. T020 is on the middle of a bedrock hill and N7 was situated closer to the soil. Model calculations show that 100 mm of water on soil areas of ME produces a gravity effect of 8 nms^{-2} (Mäkinen et al. 2014) (Table 3). It could mean that evaporation or runoff of about 250 mm water from the soil around the gravity laboratory corresponds to about 20 nms^{-2} in gravity. Annual variations in

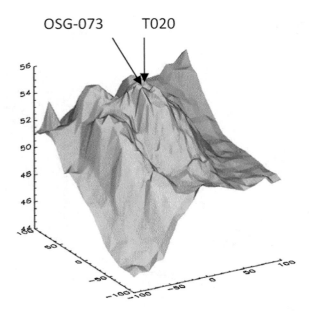

Figure 11
The map in Fig. 1 presented as shaded surfaces with position of gravimeters. Dimensions are in m

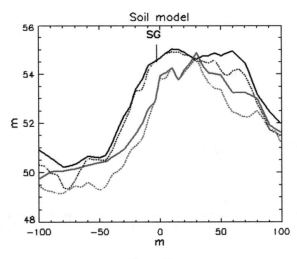

Figure 12
Height profiles presented in Fig. 1. Profile passing SG (T020) is shown in black and profile 20 m to west in red. Solid lines mean height of terrain and dotted lines are height of bedrock. The soil including water is modeled between these layers.

Table 3

Gravity response (nms^{-2}) of different gravimeters to water added to the model

Water	N6/N7	T020	Difference
No add water	− 6.9	0.0	− 6.9
Soil upper 100	+ 1.6	+ 8.5	− 6.9
Soil lower 100	+ 3.1	+ 10.1	− 7.0
Soil whole 100	+ 43.1	+ 53.1	− 10.0

Soil upper (100) means 100 mm water in upper part of soil; soil lower means 100 mm water above bedrock. Soil 100 means 100 mm water in the whole soil layer

the winter 2014–2015, snow cover was smaller than usual and there were no significant amount of snow on the laboratory roof. During rainfall, the gravity difference between the SG sensors seems (Table 2) to have a weak connection with the duration of the rain event (Table 2, cases 1 and 11). Rain events 1 and 11 (Table 2) have about the same amount of rain but different duration. In the special case 11, lot of water was quickly accumulated near the sensors N6 and N7. Water accumulation was probably due to the location of downspout, which leads the water from the roof to a pass between gravity laboratory and main building.

Annual precipitation can be 800 mm, thus moderate precipitation (100 mm) in our model cannot give observable gravity differences between different sensor locations. It is evident that we will need more detailed model near the sensors and observations from instantaneous runoff areas.

To get more information, we will continue parallel observation with two SGs, one is iOSG-022 (N6) at the same place as OSG-073 and iGrav-013 (N7) located on the same pier as T020 before.

Acknowledgements

Special thanks to Richard Warburton and Jyri Näränen for installation work of OSG-073.

References

Deville, S., Jacob, T., Chéry, J., & Champollion, C. (2013). On the impact of topography and building mask on time varying gravity due to local hydrology. *Geophysical Journal International, 192,* 82–93.

groundwater in bedrock and water in fractures have a lesser effect on T020. Uneven snow cover can cause gravity effects up to 20 nms^{-2}, especially if snow is on the roof of the laboratory (Virtanen 2000). During

Elo, S. (2001) Irtomaan paksuuden arviointi painovoimamittausten avulla (Working report in Finnish). Geological survey of Finland

Elo, S. (2006) 3D modelling of overburden thickness at Metsähovi. Abstract. Geological survey of Finland

Goodkind, J. M. (1999). The superconducting gravimeter. *Review of Scientific Instruments, 70*(11), 4131–4152.

Hinderer, Crossley, Warburton (2007) Treatise on geophysics, Vol 3, Superconducting Gravimetry, Elsevier.

Hokkanen, T., Korhonen, K., & Virtanen, H. (2006). Hydrogeological effects on superconducting gravimeter measurements at Metsähovi, Finland. *Journal of Environmental and Engineering Geophysics, 11*(4), 261–267.

Hokkanen, T., Korhonen, K., Virtanen, H., & Laine, E.-L. (2007). Effects of the fracture water of bedrock on superconducting gravimeter data. *Near Surface Geophysics, 5*(2), 133–139. **(82007)**.

Mäkinen, J., Hokkanen, T., Virtanen, H., Raja-Halli, A., Mäkinen R.P. (2014) Local hydrological effects on gravity at Metsähovi, Finland: implications for comparing observations by the superconducting gravimeter with global hydrological models and with GRACE. In: Proceedings of the International Symposium on Gravity, Geoid and Height Systems GGHS 2012, October 9–12, 2012, Venice, IAG Symposia 141

Meurers, B. (2007). Correcting superconducting gravity time-series using rainfall modeling at the Vienna and Membach stations and application to Earth tide analysis. *Journal of Geodesy, 81,* 703–712. https://doi.org/10.1007/s00190-007-0137-1.

Nagy, D. (1966). The gravitational attraction of right rectangular prism. *Geophysics, 31,* 361–371.

Peterson, J. (1993) Observations and modelling of seismic background noise, US Geol. Surv. Open-File Rept. 93–332, Albuquerque, New Mexico

Rosat, S., & Hinderer, J. (2011). Noise levels of superconducting gravimeters: updated comparison and time stability. *Bulletin of the Seismological Society of America, 101*(3), 1233–1241.

Rosat, S., Hinderer, J., Boy, J.P., Littel, F., Boyer, D., Bernard, J.D., Rogister, Y., Mémin, A., Gaffet, S. (2016). First analyses of the iOSG-type superconducting gravimeter at the low noise underground laboratory (LSBB URL) of Rustrel, France, E3S Web of Conf., 12, 06003. doi:https://doi.org/10.1051/e3sconf/20161206003

Van Camp, M., & Vauterin, P. (2005). Tsoft: graphical and interactive software for the analysis of time series and Earth tides. *Computers and Geosciences, 31*(5), 631–640. https://doi.org/10.1016/j.cageo.2004.11.015.

Vehviläinen, B. (2007). Hydrological forecasting and real-time monitoring: the watershed simulation and forecasting system (WSFS). https://doi.org/10.1002/9780470511121.ch2.

Virtanen, H. (2006). Studies of earth dynamics with the superconducting gravimeter, academic dissertation in geophysics, Helsinki. Also published as No. 133 in the series of: publications of the Finnish Geodetic Institute. http://hdl.handle.net/10138/23166. Accessed 14 Nov 2017.

Virtanen, H., Bilker-Koivula, M., Mäkinen, J., Näränen, J., & Ruotsalainen, H. (2014). Comparison between measurements with the superconducting gravimeter T020 and the absolute gravimeter FG5-221 at Metsähovi, Finland in 2003–2012. *Bulletin d'Information des Marées Terrestres, 148,* 11923–11928.

Virtanen, H., & Kääriäinen, J. (1995). The installation of and first results from the superconducting gravimeter GWR20 at the Metsähovi station. *Reports of the Finnish Geodetic Institute, 95,* 1.

Virtanen, H., & Kääriäinen, J. (1997). The GWR T020 Superconducting gravimeter 1994–1996 at the Metsähovi station, Finland. *Reports of the Finnish Geodetic Institute, 97,* 4.

Voigt, C., Förste, C., Wziontek, H., Crossley, D., Meurers, B., Pálinkáš, V., et al. (2016). *Report on the data base of the international geodynamics and earth tide service (IGETS), (scientific technical report STR—Data; 16/08)*. Potsdam: GFZ German Re-search Centre for Geosciences. https://doi.org/10.2312/GFZ.b103-16087.

(Received December 16, 2016, revised November 1, 2017, accepted November 5, 2017, Published online November 17, 2017)

Pure Appl. Geophys. 175 (2018), 1683–1697
© 2018 Springer International Publishing AG, part of Springer Nature
https://doi.org/10.1007/s00024-018-1864-3

Pure and Applied Geophysics

Gravity Tides Extracted from Relative Gravimeter Data by Combining Empirical Mode Decomposition and Independent Component Analysis

HONGJUAN YU,[1,2] JINYUN GUO,[1,3] QIAOLI KONG,[1,3] and XIAODONG CHEN[4]

Abstract—The static observation data from a relative gravimeter contain noise and signals such as gravity tides. This paper focuses on the extraction of the gravity tides from the static relative gravimeter data for the first time applying the combined method of empirical mode decomposition (EMD) and independent component analysis (ICA), called the EMD-ICA method. The experimental results from the CG-5 gravimeter (SCINTREX Limited Ontario Canada) data show that the gravity tides time series derived by EMD-ICA are consistent with the theoretical reference (Longman formula) and the RMS of their differences only reaches 4.4 μGal. The time series of the gravity tides derived by EMD-ICA have a strong correlation with the theoretical time series and the correlation coefficient is greater than 0.997. The accuracy of the gravity tides estimated by EMD-ICA is comparable to the theoretical model and is slightly higher than that of independent component analysis (ICA). EMD-ICA could overcome the limitation of ICA having to process multiple observations and slightly improve the extraction accuracy and reliability of gravity tides from relative gravimeter data compared to that estimated with ICA.

Key words: Gravity tides, relative gravimeter data, CG-5 gravimeter, empirical mode decomposition, independent component analysis, EMD-ICA.

1. Introduction

It is of great significance to study the static gravity observation, structure of the Earth's interior, and geodynamics. The change of gravity tides can also be used in research of seismic precursor monitoring. In addition, the analysis of continuous gravity recordings and the other deformation components is one of the most important factors to understand physical processes of earthquakes, slow deformation of the Earth, and the determination of geodynamic parameters (Crescentini et al. 1999; Kasahara 2002; Métivier et al. 2009).

The acquisition and research of gravity tides has been an important content of classical geodesy for decades. In 1997, a variety of space observation techniques were applied in the research project of "gravity tides in space geodesy technology" (Haas 2001), such as Global Navigation Satellite System (GNSS), satellite altimetry, Satellite Laser Ranging (SLR), Doppler Orbitography and Radio-positioning Integrated by Satellite (DORIS), Lunar Laser Ranging (LLR), and Very Long Baseline Interferometry (VLBI). The relevant information about the tidal effect was obtained, and the inconsistency between the results of the different spatial observations was analyzed. With the continuous efforts of many scholars, softwares and methods for harmonic analysis and calculation of gravity tides have been continuously improved. The program packages ETERNA developed by Wenzel (1997) and BAYTAP by Tamura et al. (1991) are considered two of high standard harmonic analysis softwares of gravity tides in the world. BAYTAP is based on the Bayesian principle for harmonic analysis and ETERNA 3.4 is currently the only earth tide data processing package

---
[1] College of Geodesy and Geomatics, Shandong University of Science and Technology, Qingdao 266590, People's Republic of China. E-mail: jinyunguo1@126.com
[2] College of Surveying and Geo-informatics, Tongji University, Shanghai 200092, People's Republic of China.
[3] State Key Laboratory of Mining Disaster Prevention and Control Co-founded by Shandong Province and Ministry of Science & Technology, Shandong University of Science and Technology, Qingdao 266590, People's Republic of China.
[4] State Key Laboratory of Geodesy and Earth's Dynamics, Institute of Geodesy and Geophysics, Chinese Academy of Sciences, Wuhan 430077, People's Republic of China.

with a model accuracy better than 1 nGal (Wenzel 1997). A lot of theoretical studies on the gravity tides in gravity measurements have also been done by many scientists and the gravity tides model has achieved high accuracy (Sun et al. 1999; Venedikov et al. 2003; Zhou et al. 2009), which corresponds to the tidal potential developments achieved using precise analysis methods (Cartwright and Tayler 1971; Cartwright and Edden 1973; Tamura 1987; Xi 1989; Hartmann and Wenzel 1995). With the development of ground observation technique and the improvement of the accuracy of gravimeters, the relative gravity measurement has been one of the main observation approaches of the gravity tides (Ducarme and Sun 2001; Sun et al. 2001; Timofeev et al. 2017). The superconducting gravimeter produced by GWR company is currently recognized by the international counterparts as the relative gravimeter with the highest accuracy, continuity, and sensitivity, whose sensitivity can achieve up to $0.001 \times 10^{-8} m/s^2$ in the frequency domain. The CG5 relative gravimeter is a new type of improved automatic electronic reading gravimeter designed and produced by Scintrex Company in Canada. This gravimeter adopts a microprocessor device to realize automatic measurement. The sensor is designed with a static fused quartz spring, so the gravimeter's accuracy can reach $5 \times 10^{-8} m/s^2$, and the reading resolution can be up to $1 \times 10^{-8} m/s^2$, both in the time and frequency domain. However, the recordings of gravimeters are usually influenced by many factors, especially nontidal changes such as hydrological effects, air mass, and loading changes, which consequently are superimposed by the real change of the gravity tides. Based on the gravity observation data, how to obtain the accurate time series of the gravity tides has become the main emphasis of this study.

Many scientists have applied different methods to process different signals via their essential characteristics. Independent component analysis (ICA) (Forootan and Kusche 2012, 2013), wavelet analysis in combination with independent component analysis (WICA) (Lin and Zhang 2005), and empirical mode decomposition (EMD) (Mijović et al. 2010a, b; Cai and Chen 2016) are widely used for signal processing. ICA is a kind of blind source separation method, which can be used to extract dominating signals from

the observations even if no apriori knowledge of the component is existing (Hyvärinen and Oja 1997; Cheung and Lei 2001; Stone 2002; Davies and James 2007; Guo et al. 2014). ICA is based on the assumption that the components belong to non-Gaussian distribution or not more than one meets Gaussian distribution. Besides, ICA assumes that the components are statistically independent from each other (Bell and Sejnowski 1995; Amari et al. 1997; Hyvärinen and Oja 2000; Zarzoso and Comon 2010). ICA can separate the independent signals, here the tides, from the superimposed signals, but it needs to deal with the multiple series of observations (Mijović et al. 2010a, b). WICA first uses the wavelet to decompose the signal into its sub-bands to expand a one-dimensional signal to two dimensions, and then ICA is applied to extract the source signals. However, some limitations remain in analysis for processing a practical signal (Lin and Zhang 2005; Shah et al. 2010): (1) there is no uniform criterion for the selection of an appropriate mother wavelet to extract signals, and we make decisions only by the intuition and experience of the analysts; (2) determining appropriate wavelet parameters without a prior knowledge of the signal is very difficult. EMD is a decomposition method for a non-stationary signal (Mijović et al. 2010a, b; Cai and Chen 2016), which can decompose one time series into a set of spectrally independent oscillatory modes called intrinsic mode functions (IMFs). However, EMD usually leads to the problem of model aliasing.

Since the static relative gravimeter data contain noise and signals, besides other signals, such as gravity tides and ocean tides, we can effectively combine the advantage of the empirical mode decomposition (EMD) and the independent component analysis (ICA) to determine the tides in the gravity observations recorded at a station. Therefore, the main focus of this paper is to use EMD-ICA to extract the gravity tides from the gravimeter time series to study the reliability and accuracy of this method, which could provide a new idea to obtain the gravity tides for the relative gravimetry correction.

The rest of this paper is organized as follows: Sect. 2, respectively, gives an introduction to the theory of EMD-ICA and ICA, and especially elaborates the boundary effect processing method. In

Sect. 3, the case study and analysis of the results are made to verify the reliability and applicability of EMD-ICA method by comparing with ICA and the theoretical model. Section 4 then presents the conclusions of the study.

2. Methodology

2.1. Empirical Mode Decomposition

Empirical mode decomposition (EMD) is a signal decomposition method proposed by Huang et al. (1998), which mainly aims at analyzing non-stationary and non-linear signals (Wu and Huang 2004; Mijović et al. 2010a, b; Jiang et al. 2015; Mariyappa et al. 2015; Cai and Chen 2016). Since the decomposition is based on the local characteristic time scale of the data, complicated signals can be adaptively decomposed into a finite set of intrinsic mode functions (IMFs) whose instantaneous frequency is generated from the high frequency to the low frequency.

Assuming that the gravity observation vector at a certain location is $x(t)$, EMD is applied to the process. The specific procedure can be formulated in detail as follows (Mijović et al. 2010b; Jiang et al. 2015; Cai and Chen 2016).

1. Identify all local minimum points and maximum points of the gravity data, and select all local minimums and maximums to one data set.
2. Using the cubic spline interpolation method, the local maxima and minima are connected by two special lines which are called the upper envelopes $e_{\max}(t)$ and the lower envelopes $e_{\min}(t)$. The original gravity record $x(t)$ lies between the upper and lower envelopes.
3. The mean of these two envelopes at any time is designated as $m_1(t) = [e_{\max}(t) + e_{\min}(t)]/2$, i.e., the instantaneous mean of both envelopes. The local mean function of the original gravity signal $x(t)$ is denoted by $m_1(t)$, and the difference between $x(t)$ and $m_1(t)$ is $h_1(t)$, that is $h_1(t) = x(t) - m_1(t)$.
4. Replace $x(t)$ with $h_1(t)$ and repeat the above steps until h_{1k} becomes a function as IMF, i.e., $imf_1(t) = h_{1k}$. Then, subtract the first $imf_1(t)$ from

$x(t)$ to obtain a new time series $r_1(t) = x(t) - imf_1(t)$ and make $x(t) = r_1(t)$. To determine whether $h_1(t)$ is IMF, Huang et al. (1998) proposed the standard stopping iterative criterion $\mathrm{SD} = \sum_{t=0}^{T} \left| h_{1(k-1)}(t) - h_{1k}(t) \right|^2 / \sum_{t=0}^{T} h_{1(k-1)}^2(t)$, where SD is generally between 0.2 and 0.3.

5. Repeat the process as described in step (4) until $r_n(t)$ or $imf_n(t)$ is smaller than the predetermined value, or the residue $r_n(t)$ becomes a monotonic function or a constant. $r_n(t)$ is a residual component after the complete decomposition and it is a trend term which represents the average trend of the signal. Then, the decomposition process of the gravity record stops.

According to the above theory of EMD, the gravimeter observation time series is decomposed and a certain amount of IMFs as well as one residual component can be generated (Mijović et al. 2010b; Jiang et al. 2015; Cai and Chen 2016), which can be expressed as follows:

$$x(t) = \sum_{i=1}^{n} \mathrm{imf}_i(t) + r_n(t) \tag{1}$$

where $x(t)$ is the static gravity observation series, $\mathrm{imf}_i(t)$ are the components of IMF, and $r_n(t)$ is a monotonic residue.

2.2. Independent Component Analysis

Independent component analysis (ICA) is a blind source separation technique, which can separate the dominating signals from the observations without any information about the 'real' signal. The observations of ICA must be non-Gaussian distribution, or only one of them is Gaussian distribution (Bell and Sejnowski 1995; Amari et al. 1997; Hyvärinen and Oja 2000; Zarzoso and Comon 2010). The original signals are statistically independent and the observed signals G are expressed as a linear combination of the S as follows:

$$G = MS, \tag{2}$$

where $G = (g_1, g_2, \ldots, g_n)^{\mathrm{T}}$ and $S = (s_1, s_2, \ldots, s_n)^{\mathrm{T}}$. In Eq. (2), S denotes the independent components to be extracted and M is the unknown mixing matrix with a full rank. Each component s_i of S is statistically independent and can be linearly expressed by the observation vector as follows:

$$S = M^{-1}G = WG, \tag{3}$$

where W is a separation matrix. Therefore, the original signal S can be obtained using the gravity observation G multiplied by matrix W.

Robust ICA can efficiently process the sub-Gaussian and super-Gaussian observations and has a higher convergence and robustness (Zarzoso et al. 2006; Bermejo 2007; Zarzoso and Comon 2010). Here, we use the popular robust ICA and the study is conducted in time domain. It maximizes the non-Gaussianity of observations and uses the kurtosis as the contrast function (Zarzoso and Comon 2010; Guo et al. 2014) to estimate matrix W. Every component s_i in vector S can be expressed by observation vector G from Eq. (3) as follows:

$$s = w^T G, \tag{4}$$

where w is a column vector, which can be solved by maximizing the kurtosis of signals in robust ICA approach. According to Eq. (4), the kurtosis, which is defined as the normalized fourth-order marginal cumulant, can be expressed as follows:

$$k(w) = \frac{E\{s^4\} - 3E^2\{s^2\}}{E^2\{s^2\}}, \tag{5}$$

where $E\{\}$ denotes the mathematical expectation and k denotes the kurtosis. Robust ICA can estimate w by the iteration to satisfy the following:

$$\mu_{opt} = \max_{\mu}|k(w+\mu g)|, \tag{6}$$

where μ is the initial iteration value depending on the statistical information of observations and g is the gradient of $k(w)$ as follows:

$$g = \nabla k(w) = \frac{4}{E^2\{s^2\}}\{E\{s^3 G\} - E\{sG\}E\{s^2\}$$
$$-\frac{E\{s^4\} - E^2\{s^2\} - E\{sG\}}{E\{s^2\}}\}, \tag{7}$$

which determines the iteration direction. Thus, every component s can be separated by Eq. (4) and corresponding w is also determined. Finally, we can obtain the separation matrix W.

2.3. Emd-Ica

The original gravity observation can be adaptively decomposed to several IMFs and a residue, but there usually exists a mode aliasing phenomenon. ICA can extract the dominating signals such as gravity tides signal from other non-tidal signals. However, ICA needs to process multiple observations. Otherwise, the underdetermined problem of ICA will not be solved. Therefore, taking good use of the advantages of the two signal processing methods, a new method of combining EMD and ICA is gradually developed for signal denoising and signal extraction (Hyvärinen and Oja 2000; Mijović et al. 2010b; Forootan and Kusche 2012).

Figure 1 shows the signal extraction process with EMD-ICA from the relative gravimeter data and the steps are detailed as follows:

1. The gravity record $x(t)$ is decomposed with EMD and the matrix imf_i is obtained.

The instantaneous frequency of IMFs decomposed by EMD is characterized from high to low. The high-frequency noise mainly concentrates in the first IMF component. Therefore, the first IMF should be first excluded and the remaining IMFs will be studied in the next steps.

2. The correlation coefficients between each component of IMF and the original gravity observation signal are calculated and analyzed.
3. Several $IMFs$ are selected which have a relatively strong independence from $x(t)$ to construct a virtual noise series, which is expressed as

$$ref\_EMD = [imf_1, imf_2, \ldots, imf_m], \tag{8}$$

where $1, 2, \cdots, m$ are the subscripts of the IMF, which correspond to the $IMFs$ with a relatively weak correlation with $x(t)$. Then, ref_EMD and $x(t)$ are used together as the input matrix of ICA, which can solve the underdetermined problem of ICA. Then, use ICA to realize effectively the separation of signals and noise to extract the tidal signal.

We take the relative gravimeter data in this paper as an example to introduce the extraction steps of the gravity tides. Assuming that the gravity observation

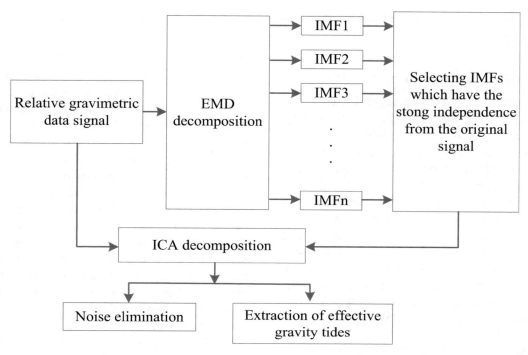

Figure 1
Sketch of the signal extraction process of EMD-ICA

series is a n × 1 column vector, i.e., G and G can be decomposed with EMD and m IMF components are obtained as follows:

$$G\_EMD = [imf_1, imf_2, \ldots, imf_m]. \qquad (9)$$

Then, the correlation coefficients between imf_i and G can be calculated, and according to Fig. 1, the input matrix G_ICA is calculated as follows:

$$G\_ICA = [ref\_EMD, G] = [imf_i, imf_j, \ldots, imf_k, G], \qquad (10)$$

where $i, j, \cdots, k$ are all from 1 to m, which denotes the IMF with a relatively weak correlation with G.

For this relative gravimeter data as listed in Table 1, if the correlation coefficient is smaller than 0.01, we define that the IMF or r has a relatively weak correlation with G. The matrix (10) can be expressed as follows:

$$G\_ICA = [imf_2, imf_3, imf_4, imf_5, imf_{10}, imf_{12}, imf_{13}, imf_{14}, G].$$

$$(11)$$

ICA is used to realize the extraction of the gravity tides from the matrix G_ICA and to achieve the separation of the gravity tides and noise, other signals

Table 1

Correlation coefficients between each IMF and the original signal G

	IMF_2	IMF_3	IMF_4	IMF_5	IMF_6	IMF_7	IMF_8
Correlation coefficient	0.0021	0.0048	0.0047	0.0041	0.4161	0.7877	0.4288
	IMF_9	IMF_{10}	IMF_{11}	IMF_{12}	IMF_{13}	IMF_{14}	r
Correlation coefficient	0.0210	0.0076	0.0174	0.0027	0.0049	0.0004	0.0083

like hydrological, air pressure effects, or other mass changes.

2.4. Boundary effect processing for EMD

When dealing with a discrete gravity record, i.e., $t \in [t(1), t(2), \ldots, t(n)] = [t_1, t_2, \ldots, t_n]$, $X(t) \in [x(t_1), x(t_2), \ldots, x(t_n)] = [x_1, x_2, \ldots, x_n]$, with the number of samples $T = t_n - t_1 + 1$, the traditional EMD performs the cubic spline interpolation at all the upper and lower extreme points of the time series. Then, the mean of the envelope $m(t)$ and the difference $h_1(t)$ can be obtained according to step (3) in Sect. 2.1. However, in the EMD algorithm of Huang et al. (1998), there is no signal on both ends of gravity observation and the cubic spline interpolation can create a boundary effect, which results in distortion near two endpoints and contamination for the gravity observation. To suppress the boundary effect, many researchers (Wang et al. 2007; Ye 2013; Jaber et al. 2014; An et al. 2015) take the whole period in the form of symmetry to extend the signal on the boundary in electronics and we take the half of the series, which achieved the desired results. The specific steps are detailed as follows.

1. Carry out the periodic symmetric extension processing on the boundaries of the gravity record, in other words, make $x(t)$ to mirror half of the series at both boundaries. Connect with $x(t)$ to create a new signal $x_1(t)$ with the series length of 2T, which is:

$$t \in [t(0.5n), \ldots, t(1), t(1), \ldots, t(n), t(n), \ldots, t(0.5n)]$$
$$= [t_{0.5n}, \ldots, t_1, t_1, \ldots, t_n, t_n, \ldots, t_{0.5n}]$$
$$x_1(t) \in [x(t_{0.5n}), \ldots, x(t_1), x(t_1) \ldots x(t_n), x(t_n) \ldots, x(t_{0.5n})]$$
$$= [x_{0.5n}, \ldots x_1, x_1, \ldots, x_n, x_n, \ldots, x_{0.5n}].$$

$$(12)$$

2. Perform EMD sifting process (Ref. Sect. 2.1) for $x_1(t)$, and then, $m(t)$ and $h_1(t)$ will be obtained. Intercept the corresponding section of the original gravity record and discard the parts with the length 0.5T on the left and right boundaries, respectively. Then, the corresponding section of the original gravity record after EMD processing is obtained.

3. Case Study and Analysis

3.1. Data

The static relative gravimeter data used in this experiment were observed inside the laboratory of Shandong University of Science and Technology, about 10 km far from the coast. The 30-day observations were obtained from April 5 to May 5 2016 with the CG5 gravimeter (serial number 140541221). In this period, the observations are used as a statistical variable. Because CG5 relative gravimeter itself has the function of seismic filtering, the low-frequency noise can be filtered and the high-frequency noise that is six times higher than the standard deviation can be discarded. In addition, adopting a 6 Hz sample frequency (a time resolution of 1 s), the readings of 1 min are averaged to get a final reading to which has been applied the tilt correction and temperature compensation (SCINTREX LIMITED 2009). Therefore, 1440×30 gravimeter readings are obtained in 30 days with the sample rate of 1 min.

Except the influence of the gravity tides, it is necessary to apply other corrections. The influence of temperature and ground vibration can be compensated by the gravimeter's own function. The drift characteristic of the CG-5 gravimeter is nearly linear, and hence, its effects can be corrected by the least square fit method. Since the ocean loading has the extremely similar frequencies with the gravity tides, we cannot separate them from each other. Therefore, the ocean tides (Lei et al. 2017) must be subtracted from the observational data. We compute the ocean tide loading values on the website http://holt.oso.chalmers.se/loading/ provided by Bos and Scherneck (2011). The website provides our first 11 tidal component parameters (M2, S2, N2, K2, K1, O1, P1, Q1, Mf, Mm, and Ssa from the ocean tide model NAO.99b) just when we select an ocean tide model, tick the required type of loading, and fill in the coordinates of the stations. Then, these parameters are used to generate the ocean loading with the software Tsoft (Van Camp and Vauterin 2005). The preprocessing is carried out to weaken the influence of gross errors, and a time series mainly containing the gravity tidal signal has been obtained. In this paper, the main focus is to use EMD-ICA for the

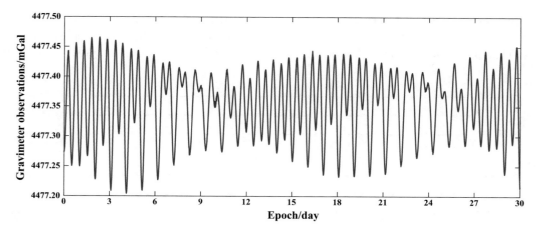

Figure 2
Time series of the gravimeter observations

extraction of the gravity tides from the time series to study the reliability and accuracy of this method.

3.2. Results and Analysis

Although the CG5 relative gravimeter data were observed at a fixed station, except for the tidal effects, there are some non-tidal effects such as air pressure or air mass variations, hydrological changes (ground water and soil moisture), or other effects due to mass changes. After the corrections mentioned in Sect. 3.1, the gravity observation series mainly present the periodic shape of the gravity tides changes, as shown in Fig. 2. In addition, the amplitude spectrum of this series mainly presents 11 obvious periodic terms, that are the semidiurnal and the diurnal wave groups, as shown in Fig. 3 and listed in Table 2, which all coincide with the wave groups of gravity tides. Therefore, the mentioned above illustrates that CG5 relative gravimeter data should contain the gravity tides. Then, the method of EMD-ICA can be used to estimate the gravity tides from thirty-day observations from April 5 to May 5 2016 by signal decomposition with EMD and signal reconstruction with ICA.

In the process of EMD decomposition, the boundary effect should be first taken into account and the results of boundary effect processing will be illustrated. According to the theory and the processing procedure in Sect. 2.4, the IMFs before and after

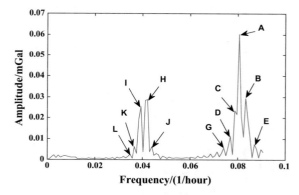

Figure 3
Frequency spectrum of the gravity observation series of 30 days

Table 2

Tidal wave groups of the frequency spectrum

	Wave	Frequency/(1/h)	Amplitude/mGal	Period/h
E	K2	0.086975	0.007368	11.49754386
B	S2	0.083313	0.029442	12.00293040
A	M2	0.080566	0.060012	12.41212121
C	N2	0.078735	0.023474	12.70077519
D	2N2	0.076904	0.011270	13.00317460
G	MNS2	0.074158	0.005721	13.48477366
J	K1	0.043945	0.006758	22.75555556
H	S1	0.042114	0.030144	23.74492754
I	M1	0.039368	0.025889	25.40155039
K	Q1	0.036621	0.006470	27.30666667
L	2Q1	0.034790	0.002801	28.74385965

the boundary effect processing were obtained, as shown in Figs. 4 and 5, respectively. In Fig. 4, there is a large distortion at the boundary caused by the boundary effect. In addition, there are only 12 IMFs decomposed by EMD, where the boundary effect is not eliminated. However, in Fig. 5, there are 15 IMFs

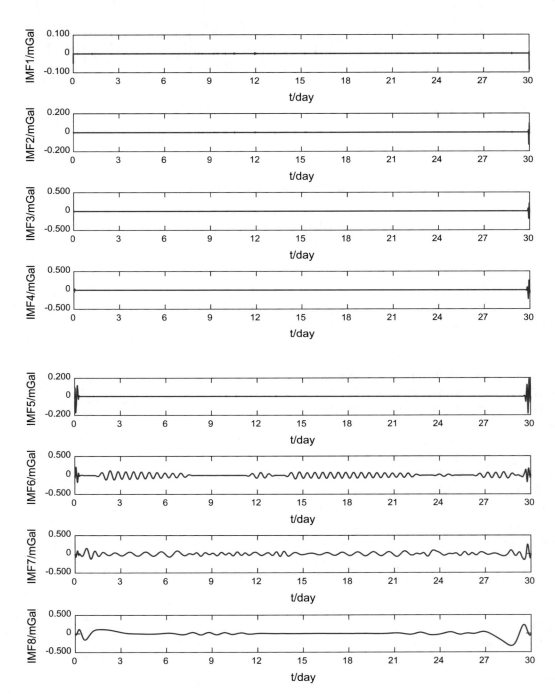

Figure 4
Components of **IMF** without processing the boundary effects

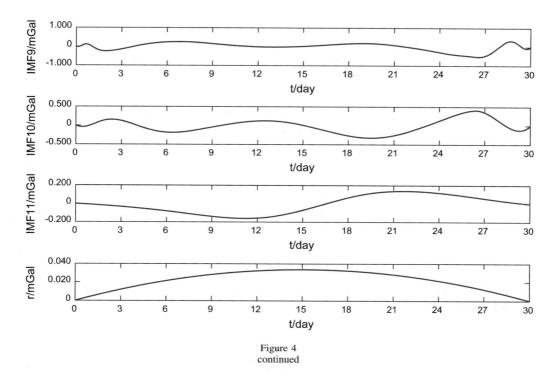

Figure 4
continued

when processing the boundary effect. Figure 5 also shows that the boundary distortion has been effectively suppressed.

Based on the theory in Sect. 2.3 and the process in Fig. 1, the gravity data are decomposed with EMD to obtain m IMFs, as shown in Eq. (9). Table 1 illustrates the correlation coefficients between each IMF and the original signal. As expressed in Eq. (11), IMFs are selected to construct the virtual noise series. Then, the gravity tides from April 5 to May 5 are estimated with EMD-ICA. For comparison, we also calculate the extraction results of the gravity tides using ICA alone. Here, GT is an abbreviation of the theoretical time series (Longman 1959) of the gravity tides. Figure 6 compares time series of the gravity tides derived from ICA, EMD-ICA, and GT from April 5 to May 5. Table 3 lists the statistical results of the gravity tides derived from ICA, EMD-ICA, and GT. Figure 7 shows the histograms of I-G, which denotes the difference GT subtracted from ICA.

As can be seen from Fig. 6, the results estimated by ICA and EMD-ICA are basically consistent with those of GT. In Table 3, the results of the ICA and GT are virtually identical in terms of the root mean squares (RMS). In addition, RMS of I-G is 5.4 μGal, which results the slightly systematic deviation of the ICA results from GT as the 'residuals' distributed in Fig. 7. The maximum value reaches 33 μGal, but the number of larger differences is quite low and most of the values focus between -10 μGal and +10 μGal by analyzing the histogram (Fig. 7).

The results estimated by EMD-ICA fit slightly better with GT results than those of ICA results (compare Fig. 6). In addition, Table 3 also shows that the EMD-ICA results are slightly closer to the GT results than that of ICA. The differences of the three statistics (EMD-ICA, GT, and ICA) are in the μGal order of magnitude and the RMS of E–G is less than 5 μGal. Comparing to ICA, the RMS of E–G is smaller than I-G, which indicates that the gravity tides obtained by EMD-ICA are more consistent with that of GT rather than that of ICA. From Fig. 6, we can also see that the three methods are very consistent, especially for EMD-ICA and GT with

97

slightly smaller difference. Table 4 illustrates that the results derived from ICA, EMD-ICA, and GT display strong correlations with correlation coefficients greater than 0.99. The correlations between EMD-ICA and GT are closer to 1.0 than between ICA and GT. The accuracy of the gravity tides estimated by

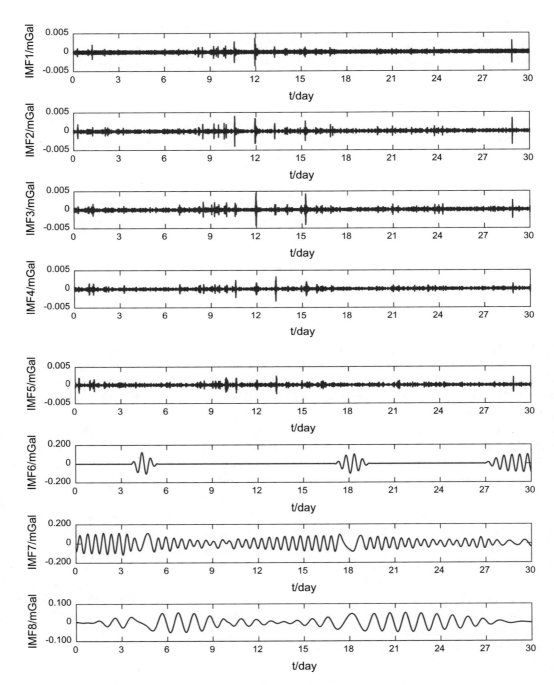

Figure 5
Components of **IMF** after processing the boundary effects

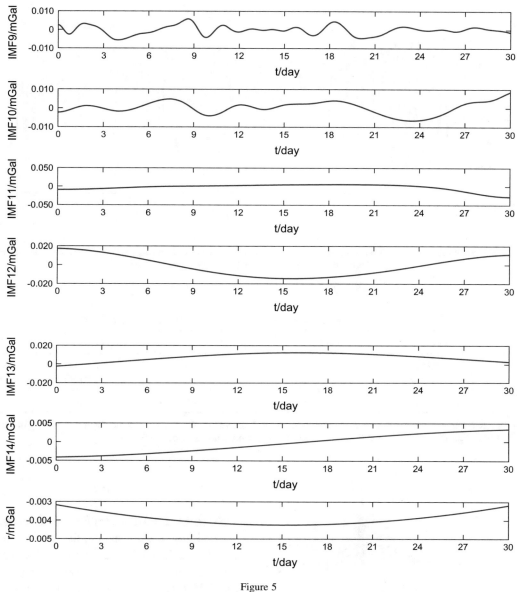

Figure 5
continued

EMD-ICA is comparable to that of GT and is slightly higher than that of ICA.

The 30-day residual time series of gravimeter readings can be obtained after the gravity tides corrections, as shown in Fig. 8. The residuals still present some slight perturbations of up to 4 microGal, which may contain some non-tidal gravity changes like hydrological, air pressure effects, or other mass changes that are not corrected. The mean change remains within ± 2 μGal (mainly range from 4477.354 μGal to 4477.357 μGal).

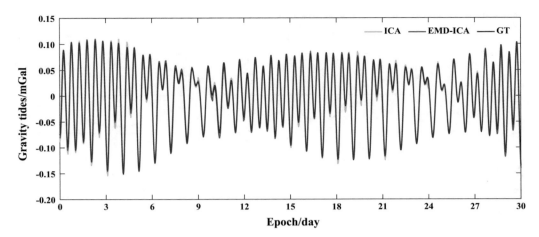

Figure 6
Time series of the gravity tides derived from ICA, EMD-ICA, and GT. *ICA* ICA-derived results, *EMD-ICA* EMD-ICA-derived results, *GT* theoretical time series (Longman 1959) of the gravity tides

Table 3

Statistical results of the gravity tides derived from ICA, EMD-ICA, and GT (mGal)

Statistics	ICA	EMD-ICA	GT	I-G	E-G
MAX	0.1092	0.1101	0.1070	0.0173	0.0127
MIN	− 0.1540	− 0.1512	− 0.1490	− 0.0337	− 0.0157
MEAN	0.0000	0.0000	0.0015	− 0.0015	− 0.0015
RMS	0.0572	0.0573	0.0577	0.0054	0.0044

ICA denotes the ICA-derived results, EMD-ICA denotes the EMD-ICA-derived results, GT denotes the theoretical gravity tides, I-G is the difference by subtracting GT from ICA, and E–G is the difference by subtracting GT from EMD-ICA

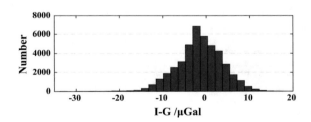

Figure 7
Histogram of I-G, which denotes the difference GT subtracted from ICA

4. Conclusions

Since CG5 relative gravimeter observations are obtained from a fixed station, the gravimeter data set is mainly affected by the gravity tides. This paper primarily focuses on the estimation of gravity tides with EMD-ICA from the relative gravimeter data. The original gravity observations can be adaptively decomposed to several IMFs and a residue with

Table 4

Correlation coefficients among ICA, EMD-ICA, and GT

Method	ICA/GT	EMD-ICA/GT	EMD-ICA/ICA
Correlation Coefficient	0.9964	0.9974	0.9968

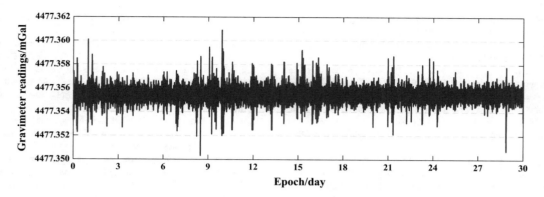

Figure 8
30-day gravimetric observations time series after gravity tides correction

EMD, but there usually exists mode aliasing phenomenon. ICA can separate the components from other non-tidal signals, but ICA needs to process multiple signals. Therefore, taking the advantage of the two signal processing methods into consideration, the method of combining EMD and ICA is proposed for the first application in the static relative gravimetric data processing, which can obtain the gravity tides for the relative gravimetry correction. Based on the characteristics of CG5 relative gravity data, EMD-ICA is used to decompose the gravity record into different components and then reconstruct them to obtain the gravity tides.

The experimental results indicate that the gravity tides derived from ICA, EMD-ICA, and GT, respectively, are of rather high agreement. The results from ICA, EMD-ICA, and GT display strong correlations with correlation coefficients greater than 0.99, but the correlation between EMD-ICA and GT is slightly closer to 1.0 than that between ICA and GT. Obviously, EMD-ICA is more suitable for the extraction of the gravity tides from the relative gravimeter data with a accuracy comparable to that of the theoretical model and slightly higher than that of ICA. Based on the above calculated results, the relative gravimeter observations after the gravity tides correction still present some small non-tidal changes, and the changes remain within ± 2 μGal in the mean, but within 4 or 5 microGal in fact.

Acknowledgements

The authors are grateful to the editors and anonymous reviewers for their helpful comments, which led a significant improvement in this paper. This study is partially supported by the National Natural Science Foundation of China (Grant No. 41774001, 41374009 & 41574072), the Special Project of Basic Science and Technology of China (Grant No. 2015FY310200), the Shandong Natural Science Foundation of China (Grant No. ZR2013DM009), and the SDUST Research Fund (Grant No. 2014TDJH101).

REFERENCES

Amari, S. I., Chen, T. P., & Cichocki, A. (1997). Stability analysis of learning algorithms for blind source separation. *Neural Networks, 10*(8), 1345–1351. https://doi.org/10.1016/S0893-6080(97)00039-7.

An, F. P., Lin, D. C., Zhou, X. W., & Sun, Z. H. (2015). Enhancing image denoising performance of bidimensional empirical mode decomposition by improving the edge effect. *International Journal of Antennas and Propagation, 2015*(2015), 1–12. https://doi.org/10.1155/2015/769478.

Bell, A. J., & Sejnowski, T. J. (1995). An information-maximization approach to blind separation and blind deconvolution. *Neural Computation, 7*(6), 1129–1159. https://doi.org/10.1162/neco.1995.7.6.1129.

Bermejo, S. (2007). Finite sample effects of the fast ICA algorithm. *Neurocomputing, 71*(1), 392–399. https://doi.org/10.1016/j.neucom.2006.09.015.

Bos, M.S., & Scherneck, H. G. (2011). http://holt.oso.chalmers.se/loading/.

Cai, J., & Chen, Q. (2016). De-noising for NMR oil well logging signals based on empirical mode decomposition and independent component analysis. *Arabian Journal of Geosciences, 9*(1), 1–11. https://doi.org/10.1007/s12517-015-2175-y.

Cartwright, D. E., & Edden, A. C. (1973). Corrected tables of tidal harmonics. *Geophysical Journal International, 33*(3), 253–264. https://doi.org/10.1111/j.1365-246X.1973.tb03420.x.

Cartwright, D. E., & Tayler, R. J. (1971). New computations of the tide-generating potential. *Geophysical Journal International, 23*(1), 45–73. https://doi.org/10.1111/j.1365-246X.1971.tb01803.x.

Cheung, Y. M., & Lei, X. (2001). Independent component ordering in ICA time series analysis. *Neurocomputing, 41*(1), 145–152.

Crescentini, L., Amoruso, A., & Scarpa, R. (1999). Constraints on slow earthquake dynamics from a swarm in central Italy. *Science, 286*(5447), 2132–2134. https://doi.org/10.1126/science.286.5447.2132.

Davies, M. E., & James, C. J. (2007). Source separation using single channel ICA. *Signal Process, 87*(8), 1819–1832. https://doi.org/10.1016/j.sigpro.2007.01.011.

Ducarme, B., & Sun, H. P. (2001). Tidal gravity results from GGP network in connection with tidal loading and earth response. *Journal of the Geodetic Society of Japan, 47*(1), 308–315. https://doi.org/10.11366/sokuchi1954.47.308.

Forootan, E., & Kusche, J. (2012). Separation of global time-variable gravity signals into maximally independent components. *Journal of Geodesy, 86*(7), 477–497. https://doi.org/10.1007/s00190-011-0532-5.

Forootan, E., & Kusche, J. (2013). Separation of deterministic signals, using independent component analysis (ICA). *Studia Geophysica et Geodaetica, 57*(1), 17–26. https://doi.org/10.1007/s11200-012-0718-1.

Guo, J., Mu, D., Liu, X., Yan, H. M., & Dai, H. L. (2014). Equivalent water height extracted from GRACE gravity field model with robust independent component analysis. *Acta Geophysica, 62*(4), 953–972. https://doi.org/10.2478/s11600-014-0210-0.

Haas, R. (2001). Tidal effects and space geodetic techniques. *Journal of the Geodetic Society of Japan, 47*(1), 161–168. https://doi.org/10.11366/sokuchi1954.47.161.

Hartmann, T., & Wenzel, H. G. (1995). The HW95 tidal potential catalogue. *Geophysical Research Letters, 22*(24), 3553–3556. https://doi.org/10.1029/95GL03324.

Huang, N. E., Shen, Z., Long, S. R., Wu, M., Shih, H. H., Zheng, Q., et al. (1998). The empirical mode decomposition and the Hilbert spectrum for nonlinear and non-stationary time series analysis. *Proceedings of the Royal Society of London A, 454*(1971), 903–995. https://doi.org/10.1098/rspa.1998.0193.

Hyvärinen, A., & Oja, E. (1997). A fast fixed-point algorithm for independent component analysis. *Neural Computation, 9*(7), 1483–1492. https://doi.org/10.1162/neco.1997.9.7.1483.

Hyvärinen, A., & Oja, E. (2000). Independent component analysis: algorithms and applications. *Neural Networks, 13*(4), 411–430. https://doi.org/10.1016/S0893-6080(00)00026-5.

Jaber, A. M., Ismail, M. T., & Altaher, A. M. (2014). Empirical mode decomposition combined with local linear quantile regression for automatic boundary correction. *Abstract and Applied Analysis, 90*(135), 1–8. https://doi.org/10.1155/2014/731827.

Jiang, X. P., Wu, F. H., Yu, H. W., & Wu, F. (2015). Mixed pixel decomposition of mineral spectrum based on EMD-ICA method.

Geometrical and Applied Optics, 119(5), 893–898. https://doi.org/10.1134/S0030400X15110260.

Kasahara, J. (2002). Tides, earthquakes, and volcanoes. *Science, 297*(5580), 348–349. https://doi.org/10.1126/science.1074601.

Lei, M., Wang, Q., Liu, X., Xu, B., & Zhang, H. (2017). Influence of ocean tidal loading on InSAR offshore areas deformation monitoring. *Geodesy and Geodynamics, 8*(1), 70–76. https://doi.org/10.1016/j.geog.2016.09.004.

Lin, J., & Zhang, A. (2005). Fault feature separation using wavelet-ICA filter. *NDT&E International, 38*(6), 421–427. https://doi.org/10.1016/j.ndteint.2004.11.005.

Longman, I. M. (1959). Formulas for computing the tidal accelerations due to the moon and sun. *Journal of Geophysical Research, 64*(12), 2351–2355. https://doi.org/10.1029/jz064i012p02351.

Mariyappa, N., Sengottuvel, S., Patel, R., Parasakthi, C., Gireesan, K., Janawadkar, M. P., et al. (2015). Denoising of multichannel MCG data by the combination of EEMD and ICA and its effect on the pseudo current density maps. *Biomedical Signal Processing and Control, 18*, 204–213. https://doi.org/10.1016/j.bspc.2014.12.012.

Métivier, L., de-Viron, O., Conrad, C. P., Renault, S., & Diament, M. (2009). Evidence of earthquake triggering by the solid earth tides. Earth and Planetary Science Letters, 278(3): 370–375. https://doi.org/10.1016/j.epsl.2008.12.024.

Mijović, B., De-Vos, M., Gligorijević, I., & Huffel, S. V. (2010a). Combining EMD with ICA for extracting independent sources from single channel and two-channel data. *Annual International Conference of the IEEE Engineering in Medicine and Biology, 2010*, 5387–5390. https://doi.org/10.1109/IEMBS.2010.5626482.

Mijović, B., De-Vos, M., Gligorijević, I., Taelman, J., & Huffel, S. V. (2010b). Source separation from single-channel recordings by combining empirical-mode decomposition and independent component analysis. *IEEE Transactions on Biomedical Engineering, 57*(9), 2188–2196. https://doi.org/10.1109/TBME.2010.2051440.

SCINTREX LIMITED. (2009). CG-5 Scintrex autograv system operation manual V5.0, 99–172.

Shah, V. P., Younan, N. H., Durbha, S. S., & King, R. L. (2010). Feature identification via a combined ICA-Wavelet method for image information mining. *IEEE Geoscience and Remote Sensing Letters, 7*(1), 18–22. https://doi.org/10.1109/LGRS.2009.2020519.

Stone, J. V. (2002). Independent component analysis: an introduction. *Trends in Cognitive Sciences, 6*(2), 59–64. https://doi.org/10.1016/S1364-6613(00)01813-1.

Sun, H. P., Takemoto, S., Hsu, H. T., Higashi, T., & Mukai, A. (2001). Precise tidal gravity recorded with superconducting gravimeters at stations Wuhan/China and Kyoto/Japan. *Journal of Geodesy, 74*(10), 720–729. https://doi.org/10.1007/s001900000139.

Sun, H. P., Xu, H. Z., Ducarme, B., & Hinderer, J. (1999). Comprehensive comparison and analysis of the tidal gravity observations obtained with superconducting gravimeters at stations in China. *Belgium and France. Chinese Science Bulletin, 44*(8), 750–755. https://doi.org/10.1007/BF02909719.

Tamura, Y. (1987). A harmonic development of the tide-generating potential. *Bulletin d'Informations Marées Terrestres, 99*, 6813–6855.

Tamura, Y., Sato, T., Ooe, M., & Ishiguro, M. (1991). A procedure for tidal analysis with a Bayesian information criterion. *Geophysical Journal International, 104*(3), 507–516. https://doi.org/10.1111/j.1365-246X.1991.tb05697.x.

Timofeev, V. Y., Kalish, E. N., Ardyukov, D. G., Valitov, M. G., Timofeev, A. V., Stus, Y. F., et al. (2017). Gravity observation at continental borderlands (Russia, Primorie, Cape Shults). *Geodesy and Geodynamics, 8*(3), 193–200. https://doi.org/10.1016/j.geog.2017.03.011.

Van Camp, M., & Vauterin, P. (2005). Tsoft: graphical and interactive software for the analysis of time series and Earth tides. *Computers & Geosciences, 31*(5), 631–640. https://doi.org/10.1016/j.cageo.2004.11.015.

Venedikov, A. P., Arnoso, J., & Vieira, R. (2003). VAV: a program for tidal data processing. *Computers & Geosciences, 29*(4), 487–502. https://doi.org/10.1016/S0098-3004(03)00019-0.

Wang, J., Peng, Y., & Peng, X. (2007). Similarity searching based boundary effect processing method for empirical mode decomposition. *Electronics Letters, 43*(1), 58–59. https://doi.org/10.1049/el:20072762.

Wenzel, H. G. (1997). The nanogal software: earth tide data processing package ETERNA 3.30. *Bull. Inf. Marées Terrestres, 124*, 9425–9439.

Wu, Z., & Huang, N. E. (2004). A study of the characteristics of white noises using the empirical mode decomposition method. *Proceedings of the Royal Society of London A: Mathematical, Physical and Engineering Sciences, 460*(2046), 1597–1611. https://doi.org/10.1098/rspa.2003.1221.

Xi, Q. W. (1989). The precision of the development of the tidal generating potential and some explanatory notes. *Bulletin d'Informations Marées Terrestres, 105*, 7396–7404.

Ye, Y. (2013). Adaptive boundary effect processing for empirical mode decomposition using template matching. *Applied Mathematics & Information Sciences, 7*(1L), 61–66. https://doi.org/10.12785/amis/071L10.

Zarzoso, V., & Comon, P. (2010). Robust independent component analysis by iterative maximization of the kurtosis contrast with algebraic optimal step size. *IEEE Transaction on Neural Networks, 21*(2), 248–261. https://doi.org/10.1109/TNN.2009.2035920.

Zarzoso, V., Comon, P., & Kallel, M. (2006). How fast is Fast ICA? Signal Processing Conference, 2006 14th European. *IEEE, 2006*, 1–5.

Zhou, J. C., Xu, J. Q., & Sun, H. P. (2009). Accurate correction models for tidal gravity in Chinese continent. *Chinese Journal of Geophysics, 52*(6), 1474–1482. https://doi.org/10.3969/j.issn.00015733.2009.06.008.

(Received April 20, 2017, revised April 4, 2018, accepted April 7, 2018, Published online April 16, 2018)

Pure Appl. Geophys. 175 (2018), 1699–1725
© 2018 Springer International Publishing AG, part of Springer Nature
https://doi.org/10.1007/s00024-018-1834-9

❙Pure and Applied Geophysics

More Thoughts on AG–SG Comparisons and SG Scale Factor Determinations

David Crossley,[1] Marta Calvo,[2] Severine Rosat,[3] and Jacques Hinderer[3]

Abstract—We revisit a number of details that arise when doing joint AG–SG (absolute gravimeter–superconducting gravimeter) calibrations, focusing on the scale factor determination and the AG mean value that derives from the offset. When fitting SG data to AG data, the choice of which time span to use for the SG data can make a difference, as well as the inclusion of a trend that might be present in the fitting. The SG time delay has only a small effect. We review a number of options discussed recently in the literature on whether drops or sets provide the most accurate scale factor, and how to reject drops and sets to get the most consistent result. Two effects are clearly indicated by our tests, one being to smooth the raw SG 1 s (or similar sampling interval) data for times that coincide with AG drops, the other being a second pass in processing to reject residual outliers after the initial fit. Although drops can usefully provide smaller SG calibration errors compared to using set data, set values are more robust to data problems but one has to use the standard error to avoid large uncertainties. When combining scale factor determinations for the same SG at the same station, the expected gradual reduction of the error with each new experiment is consistent with the method of conflation. This is valid even when the SG data acquisition system is changed, or different AG's are used. We also find a relationship between the AG mean values obtained from SG to AG fits with the traditional short-term AG ('site') measurements usually done with shorter datasets. This involves different zero levels and corrections in the AG versus SG processing. Without using the Micro-g FG5 software it is possible to use the SG-derived corrections for tides, barometric pressure, and polar motion to convert an AG–SG calibration experiment into a site measurement (and vice versa). Finally, we provide a simple method for AG users who do not have the FG5-software to find an internal FG5 parameter that allows us to convert AG values between different transfer heights when there is a change in gradient.

Key words: Superconducting gravimeters, absolute gravimeters, scale factor, calibration.

1. Introduction

1.1. The SG at Apache

The original work in this paper was done on the Apache Point (AP, New Mexico, USA) SG–AG data from 2011 to 2015, and later applied to data from the J9 installation in Strasbourg, France. It is necessary to briefly discuss the site situation at AP, which is a first class astronomical observatory that hosts one of the best lunar laser ranging (LLR) facilities in the world. In 2009 an SG was installed at the site to assist in constraining the displacement of the ground during the LLR experiments which use a 3.5 m optical dish attached to a solid pier. Due to logistic and financial considerations it was not possible to place the SG in its own isolated building as is common at other geodetic sites, and as a compromise the SG was located in the cone room, a small access room directly under the telescope housing. There are both advantages and disadvantages to this location, but the subsequent difficulty of providing a suitable environment for the AG instrument to calibrate the SG was not considered.

It was quickly discovered during the first calibration experiment in 2011 that the AG was subject to excessive disturbance during the nighttime operations, when the telescope was in constant use, and these disturbances severely compromised the quality of the AG data. There are two effects associated with the telescope motion, one being small self-correcting offsets in the SG data (at the level of 0.5 μGal or less) due to mass changes associated with the telescope position above the gravimeter. These can be removed by constructing a model using additional data from the telescope slews, but this is a time-consuming operation that has not been done systematically for all the AP data, and not for the SG data used in the

[1]   Earth and Atmospheric Sciences, Saint Louis University, St. Louis, MO, USA. E-mail: crossley@slu.edu
[2]   Observatorio Geofísico Central, IGN, Madrid, Spain.
[3]   Université de Strasbourg, CNRS, EOST, IPGS, UMR 7516, 67000 Strasbourg, France.

calibrations. For the AG, the cooling system beneath the telescope blows air directly into the cone room and onto the AG instrument itself which cause data disturbances that are not damped by the F5 super-spring. Unfortunately, there is no possibility to avoid this problem by moving the SG/AG to another location in the observatory complex, and the only remedy with the AG is to reject all the disturbed data.

Thus we are obliged to use the AG data as recorded, and cannot easily improve the situation. Coupled with this is the limitation on residence time for the AG. Site requirements permit the AG instrument to remain in the cone room for 5 days or less, except for the first experiment in 2011 where it was allowed to run over a weekend (and gave by far the best data), and this severely limits the amount of good data we can collect. Our site is, therefore, one of the noisiest and most challenging for an AG–SG calibration experiment. Ground accelerations induced by the movements of a nearby VLBI antenna have also been also detected in the SG recordings at Ishigakijima, Japan (Imanishi et al. 2018), therefore calibration experiments at such a site might encounter similar problems.

1.2. Motivation

One of the motivations for this paper is to share our experience with the calibration experiments at AP that were initially done without the collaboration with the Strasbourg group that later became available, and thus represents the situation that might face a less experienced team of SG–AG operators. For example, an initial assumption at AP was that both the SG series and the AG series must be compared without any AG corrections (i.e., for tides, ocean-tide loading, local pressure or polar motion), and so all such corrections were turned off in the FG5 setup. Later it became clear that, as is done routinely at many observatories, the experiment can be done with the standard AG corrections, and the FG5 settings can be changed to remove the corrections for the calibration and produce uncorrected files. At SG installations where there is no in-house or dedicated AG, one may need external assistance for the FG5 instrument, which for the case of AP is the National Geospatial-Intelligence Agency (NGA). Frequently, such FG5's

are in heavy demand which limits their availability. Further, as mentioned, site constraints at AP dictate not only a very small space for the SG and AG, in a room in the middle of a very complex building, but visitations by the FG5 are disruptive of local operations so only one or two AG measurements per year are preferred. This would be similar to an SG being located remotely (e.g., Syowa, Antarctica), or in a special underground environment for hydrological purposes. It is obvious that whenever we did an SG calibration we needed also to produce an AG site measurement, which is the normal 1–2 day occupation with all corrections turned on, unlike a calibration experiment which normally takes at least 5 days.

Thus, a major goal of the paper is to process the SG–AG calibration data using only the FG5 text files, without access to the software that is supplied with the Micro-g software (denoted 'g-software') http://www.microglacoste.com/pdf/g9Help.pdf. The g-software gives complete user control over the processing of the AG fringe data, and produces a set of internal binary files and a set of 3 ASCII files—the drop data file, the set data file, and the project file that summarize the results of the processing. For the AP station the binary files were available from NGA, as was some limited re-processing, but were not useful to us at AP; the situation in Strasbourg is of course entirely different. This limitation was perfectly anticipated in Sect. 2 of the paper by Palinkas et al. (2012) who noted that some users of the gravity data have no access to the g-software (see the "Appendix" for further information). We acknowledge that the g-software is available independently of the instrument from Micro-g, but perhaps the suggestions in our paper may help some users to avoid that necessity.

Accurate calibration of a superconducting gravimeter (SG) is fundamental in many geophysical and physical applications, for instance for the search for time-variability in the Earth's response to tides induced by internal process inside the Earth or by surface loading (Calvo et al. 2014), or the search for anisotropy in the Newtonian gravitational constant G (Warburton and Goodkind 1976). Many papers cite ocean tide loading as a prominent requirement for accurate SG scale factors (along with accurate phase calibration) e.g., Boy et al. (2003) and Baker and Bos (2003). Although much of the initial work at AP could have been avoided using the g-software, we hope some of the results are still of interest to those who contract out AG measurements, or perform only occasional calibration or drift checks on their SG.

There are numerous papers on the use of an AG to calibrate an SG, summarized in Hinderer et al. (2015). Here we investigate some small issues that arise in this type of comparison. In order of presentation, these are: (a) a discussion on the merits of various ways to use drop or set SG data, (b) the effect of adding the data acquisition time delay and a local trend to the solution, (c) combining multiple determinations of the scale factor for a particular station, and (d) comparing the AG offset from a calibration experiment to regular determinations of the AG site gravity. We use data from the Apache Point (AP) station in New Mexico, USA, and from J9 station in Strasbourg (ST), France to demonstrate the various points. As mentioned, AP is a site with especially high nighttime site noise, perhaps the extreme end of stations that have high cultural noise during some part of the day. This has been encountered at some older SG installations, e.g., Wuhan, China or Vienna (Meurers 2012) or a recent one at Ishigaki (Imanishi et al. 2018), whereas ST is typical of a station with quiet and fairly constant site noise. Van Camp et al. (2016) make the useful suggestion that higher drop rates (e.g., every 5 s) should be used, and this would be beneficial in future for AP measurements during the undisturbed daytime recording.

1.3. Basic Equations

To begin, we assume a simultaneous measurement of AG and SG gravity over a time period $T \sim 5$ days, to be assured of reaching a reasonable convergence in the scale factor (e.g., Francis 1997; Meurers 2012). All the SG data come from either the raw 1 s data, or the filtered 1 min files available at GGP/IGETS (Crossley and Hinderer 2010; Voigt et al. 2016). Very little of our SG data at AP required corrections for SG-specific disturbances such as He refills, disturbances, or offsets but simple pre-processing was done where necessary. Likewise there was no problematic data (such as a large earthquake)

that would have affected both instruments (in different ways) and therefore, to be avoided. The common time period T was chosen to span a period during the largest diurnal tides at the station, which occur fortnightly. Pre-processing of some SG data from ST was done to avoid data disturbances, as described in Rosat et al. (2009).

The FG5 data at both stations, denoted by $y(t)$, was collected drop-by-drop every 10 s, and accumulated each 20 min as a set mean of 100 drops. In our original processing, the SG data x(t) was 1 min smoothed from 1 s raw data by applying a low-pass filter, which avoids the problem of aliasing of the 5–10 s microseismic noise (Van Camp et al. 2016). This is still present even for station AP in the middle of the N. American continent, but less than ST in Central Europe. The SG data is normally cubic-splined to the AG drop or set times which are given at a sampling time t (Rosat et al. 2009). Later we also used 1 s data for comparison.

The AG data is composed of a constant mean value y_0 (over the time period T of the experiment) plus a time-varying part $y_1(t)$; similarly the SG data is composed of a constant part x_0 plus a time-varying part $x_1(t)$. We perform a least-squares (LSQ) fit of $y(t)$ (μGal, 10^{-8} m/s^2) to the SG data $x(t)$ (volt) using

$$y(t) = \alpha + \beta \times x(t) + \gamma \times t + \varepsilon \qquad (1)$$

where ε is assumed to be Gaussian random noise, and the sum of ε^2 is minimized. The parameters determined from the fit are α, the offset between the mean zero levels of the AG and SG data, β the scale factor SF (or calibration constant) of the SG (μGal or nm s^{-2}/volt), and γ is a trend to account for possible differential instrument drifts (see e.g., Imanishi et al. 2002). If no mean values are subtracted, then

$$x(t) = x_0 + x_1(t)$$
$$y(t) = y_0 + y_1(t) \qquad (2)$$

and after the LSQ fit (we use *lfit* from Numerical Recipes, but any similar code will do) for (α, β, γ) we can equate, within the errors, the constant and time-variable parts

$$y_0 = \alpha + \beta \times x_0$$

$$y_1(t) = \beta \times x_1(t) + \gamma \times (t - t_0) \qquad (3)$$

where t_0 has been added to indicate the time of the first AG drop. We refer to the quantity y_0 as the *AG mean value*, which depends on the fitted offset α, the SG mean x_0 (which can be computed separately prior to the fit), and β—the SG scale factor. With Gaussian errors, if the standard deviations for (α, β) are σ_α and σ_β, then the variance of y_0 is

$$\sigma_0^2 = \sigma_\alpha^2 + x_0^2 \times \sigma_\beta^2 + 2x_0 \times \sigma_{\alpha\beta} \qquad (4)$$

where $\sigma_{\alpha\beta}$ is the covariance of (α, β).

A few points need to be mentioned about these equations. First, many authors do not consider the offset α, nor the mean values x_0 or y_0, as being of sufficient interest to mention, and others ignore the trend γ, thus leaving β as the only parameter of interest. This is understandable if one chooses to get a regular AG site value at the site by reprocessing the AG data using the g-software. Later we show another method to do this based on y_0. As for the offset α, Imanishi et al. (2002) discussed in some detail its variations for a month-long series of AG-SG measurements, and ascribed the cause to possible AG instrumental drift during the experiments. The possibility of such an effect is one reason we include the term $\gamma \times t$ in (1). Unfortunately we could not repeat the experiment of Imanishi et al. because we could only record at AP over a few days, but any linear trend in α will appear in the $\gamma \times t$ term in (1). Short-term effects over the time T of the calibration are distinct from the classic long-term SG drift, but it is reasonable that the latter be removed first from the SG data, though it is unnecessary here.

Wziontek et al. (2006) add explicitly a drift function to the SG component, and an offset to the AG data, but no AG trend. Although arbitrarily adding a drift parameter to (1) without being able to identify the reason might seem unjustified, Meurers (2012) clearly showed that a linear trend can perturb the amplitude ratio between the AG and SG data (which is the goal of the calibration), and so there is a good reason to include it. Many other papers also advocate a drift parameter, for instance Hinderer

et al. (1991) included drift when using an earlier JILA-5 instrument with twin laser drift problems, and a drift is explicitly included by Tamura et al. (2005) and Van Camp et al. (2016). Meurers (2002) explored the effect of unmodeled drift on the calibration factor using synthetic and real datasets. We also received a suggestion that He gas from the SG might leak into the room and affect the AG; this could have a preferential effect on one instrument and not the other (B. Meurers, editorial comment). This phenomenon has also been reported by Mäkinen et al. (2015), but it is not usually a problem for closed cycle SG's such as the current observatory SG or iGrav. Note that at the J9 station in Strasbourg, unlike at AP, the AG is recording in a separate room from that of the SG.

2. Drops or Sets?

Early calibration experiments in Strasbourg (Hinderer et al. 1991) used only 1-day experiments but used both drop and set data, and also considered both the L_1 (minimum absolute error) and L_2 (LSQ) norms when solving for the constants in Eq. (1). It appears the L_1 norm has not been widely used in recent years. Amalvict et al. (2001, 2002), however, showed that with good data there was little difference in the scale factor between drop and set methods, and noted that the errors (which they stated to be standard deviations) in both methods were similar despite the very large difference in the number of SG–AG pairs to be fitted (generally there are about 100 drops for each set) which should result in a smaller formal error using drop data. Although there has been a recent trend in SG–AG calibration processing towards the use of drop data rather than set data, obviously the less numerous AG set values are still less scattered than the drop values.

Several recent papers have covered similar ground to this study, and with which our results are consistent. Tamura et al. (2005) for example used AG drops, and found no evidence for scale factor changes at Esashi (Japan). Wziontek et al. (2006) identified AG offsets at station Bad Homburg (Germany) from calibration experiments using different FG5 instruments, and also treated mainly drops. Meurers (2012) used drops in a comprehensive assessment of many of the factors in SG–AG processing, and Van Camp et al. (2016) also favored using drops, emphasizing the need not just for many drops, by increasing the drop rate, but measuring at high tides to improve accuracy.

We need to be clear about the difference between the two types of measurement. An AG drop results in a trajectory of a falling corner cube, whose flight is sampled by a number of fringe zero-crossing times of which there are a large (many thousand) number of fringes per drop (see e.g., Kren et al. 2016). The variance covariance matrix of the LSQ fit to the fringe crossings yields a statistically determined drop value and a scatter, or standard deviation, σ_d. When drops are processed in sets (often 100 drops, every 10 s) the set mean is the unweighted mean of the accepted drops averaged over a set. One can take the set sigma σ_s as the usual standard deviation of the set mean [Eq. (A2) in the "Appendix"] that reflects the drop to drop scatter (column labeled 'Sigma' in the set text files). This is our choice for most of this paper, except for the final two Figures. Alternatively one may choose the standard error of the set mean (SEM), which is $\sigma_s/\sqrt{(N)}$ where N is the number of drops per set (unweighted), as in Tables 3 and 4. N is frequently close to 100, so the set SEM is about 10 x smaller than σ_s, and given by the column 'Error' in the set file. Drops are accepted or rejected by the g-software based on the usual 3-σ criterion, i.e., a drop outlier is rejected when more than 3-σ from the set mean. When using single drops, more flexibility is available to select drops in the solution, as we will see. Rather than using 'set sigmas', and 'drop sigmas', to avoid any ambiguity we frequently refer to the columns (Sigma, Error) in the drop files and (Sigma, Error) in the set files.

2.1. Tests on Set Data

For reasons that will be clear later, we wish to also find y_0 from the fit, and this requires an initial assessment of the SG mean value x_0; certainly x_0 can be ignored if the only goal is to find β. Various possibilities for the span of SG data were tried: (i) SG values starting at UT 0 the first day, and ending at midnight the last day, (ii) using SG values only at AG times, and (iii) using SG values starting at the first

AG drop time and ending at the last AG drop. The differences in the scale factor were (and expected to be) insignificant, but there was an affect at the μGal level on the AG mean value y_0 in Eq. (4). Obviously option (iii) is the logical choice of the time span for the SG data.

A test was also made to quantify the effect of the SG instrument time delay to the experiment, as mentioned previously. For SG 046 at Apache Point, an observatory-style gravimeter, the nominal time delay (lag) of the system is predominantly that of the GGP1 filter (Hinderer et al. 2015) which is 8.16 s, so this delay has to be incorporated in any calculation that returns the SG value at the AG times. We tested the shift in the scale factor for various time delays in Table 1, showing that for most stations, the effect is negligible even up to 30 s. This confirms similar results of Meurers (2012) and Van Camp et al. (2016).

Again using the AP 2011 data, another test was done to assess the effect of a relative drift between the SG and AG data, i.e., adding the term $\gamma \times t$ in the RHS of Eq. (1). This is not primarily to account for the known SG drift, but allows for other effects occurring preferentially in one of the instruments. The instrument drift of SG046 between 2009 and 2012 was 70 μGal/year, or + 0.192 μGal/day—which is unusually large for an SG. For this reason, the sensor was replaced in 2013 by the manufacturer GWR (Goodkind–Warburton–Reineman, San Diego, California) with a significant decrease in drift. The results are shown in Table 2 where we give the scale factor, the trend, and the AG mean value, all with a time lag of 8.16 s. We have included the errors σ_β to show that although the trend can be larger and of opposite sign than the known SG drift, its effect on β is always smaller than σ_β. The same situation occurs for the AG mean value y_0. Note, however, that σ_β is not determined very accurately by these set solutions at AP. We note that the trend can also be regarded as a diagnostic for possible problems in the data, e.g., the trend for AP2016 is − 0.55 μGal/day, which is sufficient to perturb the offset and SF. In this case, we know the AG data quality for 2016 was rather low.

The argument for using AG set values for SG–AG calibration probably arises because it is the natural choice when determining an AG site value, where the

Table 1

Effect of SG time delay on scale factor (μGal/V)

Time delay (s)	Scale factor β +/0.407	% change
0[a]	− 79.397	0.0
8.16	− 79.399	0.003
15.0	− 79.4015	0.005
30.0	− 79.4027	0.01

AP 2011 experiment, all set data

[a]Reference

Table 2

Effect on the scale factor and base value of adding a trend to the SG–AG fit

AP set experiment	Trend γ (μGal/day)	Scale factor β (μGal/V)	Change in β (μGal/V)	Mean value y_0^a (μGal)	Change in y_0 (μGal)
July 2011	− 0.03 ± 0.22	− 79.39 ± 0.41	+ 0.004	5164.83 ± 0.87	− 0.114
Nov 2013	+ 0.08 ± 0.84	− 94.40 ± 2.43	+ 0.104	5189.66 ± 1.69	+ 0.135
Sept 2014	− 0.36 ± 0.99	− 94.33 ± 1.02	+ 0.098	5161.53 ± 1.40	− 0.417
June 2016	− 0.55 ± 0.78	− 94.96 ± 1.15	+ 0.279	5136.14 ± 2.35	− 1.476

After the July 2011 experiment the sensor at AP was changed, with a new scale factor. Uses JS3 pass 2 method and set data processed with the Sigma column of the set file; values include γ and changes are relative to γ = 0

[a]Add value to 9788050000 μGal

geophysical corrections are applied to get the site gravity. Among past papers, Rosat et al. (2009) used AG set values when doing SG–AG experiments, and there is some benefit in having a set average with a well-defined set sigma (σ_s) used in weighting the fit (as we will see). On the other hand, there are two arguments against using sets, the first being the inability of the set average to precisely track the top and bottom of the semi diurnal tides if the regular set averages are used. To combat this, Meurers (2012) suggested using a moving window average of both AG and SG data to help reduce the AG scatter and yet track the tidal signal more precisely. At each drop time an average of the AG and SG data is taken over the length of a set spanning the drop point, thus keeping the high number of drop values, but reducing the drop to drop scatter. This is related to the second point that in principle set averages, rejecting drop outliers, should work better on AG data where the geophysical corrections (principally tides and atmospheric pressure) have been subtracted, and the corrected signal has only a small scatter. The

situation is different when doing SG–AG calibrations that require the full tidal signal; the AG set averages are biased by the changing level of the large time-varying signal. The importance of this procedure is one of the options we test in our processing. Recent authors have tended to recommend the use of AG drops to generate the scale factor, ignoring the trivial increase in computer time over the set method.

2.2. Initial Attempts at Set and Drop Processing

We start with Fig. 1 showing the fit of SG-to-AG data between July 28 and Aug 3, 2011, which was the span of the first AG measurement at Apache Point Observatory after installation of SG046 in February 2009.

The fit is based on AG set values (100 drops/set, set interval 20 min, drop interval 10 s), where it is seen that the σ_s's are considerably larger during the nighttime hours when the LLR telescope is active for various sky surveys. The set mean can nonetheless be acceptable if poor drops are rejected and de-

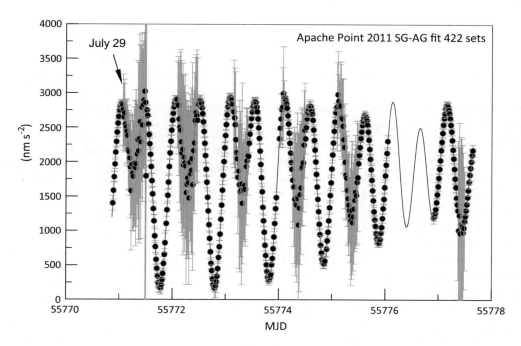

Figure 1
Fit of all Apache Point set data between July 28 (MJD 55771) and Aug 3, 2011. The noisiest data, indicated by large set standard deviations (the Sigma column in the set files), varies according to the telescope viewing schedule

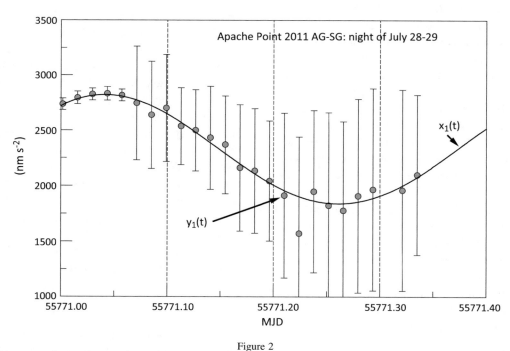

Figure 2
Zoom of Fig. 1 for night of July 28–29 (about the first half day of the experiment) showing the set means are also disturbed by the telescope activity. Labelled are the fitted SG and AG measurements $x_1(t)$ and $y_1(t)$, Eq. (4). Almost all such sets are discarded in the processing

emphasized in the set means by their large inverse variances (i.e., $1/\sigma^2$) in the SF fit. As can be seen in Fig. 2, zoomed to the first night of the calibration, not only are the set errors large, but the set means deviate from the SG values.

There were four SG–AG calibrations done for AP between 2011 and 2016, and for comparison we processed four datasets from ST between 2008 and 2011. We first assessed the histograms of the drop and set σ's (the columns marked 'Sigma' in the FG5 text files) for all datasets used in this study and these are shown in Fig. 3a for AP, and Fig. 3b for ST. For AP the drop sigmas are clearly divided into two groups, for every experiment. There is a tight clustering of values around 16 μGal for AP2011 and AP2013, and similarly around about 25 μGal for AP2014, AP2016. Beyond 30 μGal, the drop sigmas extend to very high values (several 100 μGals) for badly disturbed data, and so we choose a maximum acceptable cutoff value of $\delta g_m = 30$ μGal for all AP data. For the AP set data, however, the sigmas vary from a peak around 5–7 μGal, to a scatter of values up to about 30 μGal, suggesting the latter is also a suitable set cutoff to avoid disturbed data. The situation is different for the ST data where there is very little disturbed data, but the histogram peaks also occur at different values depending on the experiment. The drop sigmas for ST divide between a low of about 10 μGal for ST2010 and ST2011 to between 25 and 30 μGal for the other 2 datasets. The ST set sigmas vary between 5 and about 27 μGal, so to include most data we choose for ST a rather large value of $\delta g_m = 35$ μGal as a cutoff, but this is not a critical value because there are very few sets above 25 μGal.

We processed the four SG–AG experiments at AP using both set and drop data. Initially we used all the set or drop data in the files, without prior selection, and compute the solution relying on the weights (inverse variances from the drop sigmas) to reject large set or drop σ's. This initial processing is called JS0 (selection method 0) and yields the two rows 'JS0' in Table 3. To be explicit this procedure consisted of:

(a) Selecting only those drops or sets that had sigmas below the sigma cutoff discussed above (shown

in Fig. 3) and weighting the AG data according the their inverse variances, and

(b) For the SG data we used 1 min GGP data and interpolated the SG value to the drop time or set time that matched the AG times. This is the procedure discussed in Rosat et al. (2009).

The result is that much data was discarded, depending on the experiment, and in the worst case for the AP2013 experiment less than 30% of the data could be used. Notice in Table 3 that the drop SFs (scale factors) are sometime quite different to the set SFs, whereas the latter (JS0) are generally similar to the other 'better' JS solutions (described below). The reason is that the drop σ's are not necessarily indicative of which drops are far from the SG curve, and so the weighting does not automatically diminish the influence of drop outliers that may have acceptable σ_d's. This is not true of the set σ's, as they are more robust against bad drop data, and sets with high σ's are degraded in the solution against sets with small σ's. Table 3 shows the results for JS1 and JS3 (see later) for passes 1 and 2, but note that pass 2 is not required for some of the set solutions if there are no set residuals outside the 3-σ criterion.

Considering the difficult AP data, namely the lack of agreement between the scale factor for sets and drops, we wondered if it was possible to improve the fit by further selecting the AG drops. Because most drops had about the same σ_d, they all equally contribute to the SG fit, but in reality many drops are far away from the SG curve; perhaps these drops degrade the scale factor fit and could be rejected? It is also clear that drops near the tidal peaks are more important in determining the scale factor that those midway between peaks. Rather than pursue such an approach, we changed strategy and decided to test a number of options presented in the AG–SG processing. For the moment we pass over sections (c) of Tables 3 and 4, and return to them later.

2.3. Improving the Algorithm for Rejecting Bad Data

When collaboration began with the Strasbourg group, and especially after the paper by Calvo et al. (2014), it became clear that better ways to reject bad data had been gaining acceptance, following the

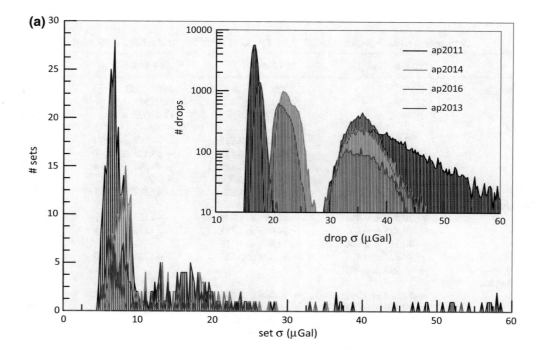

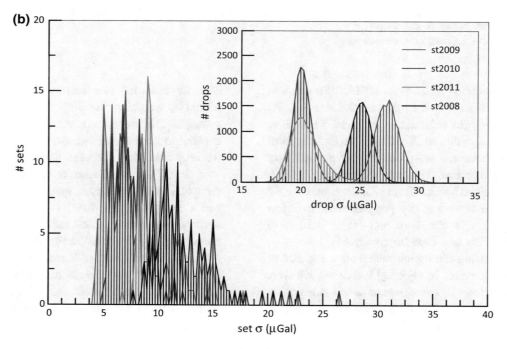

Figure 3

Histograms of standard deviations of drop and set values for 4 datasets from **a** Apache Point and **b** J9 Strasbourg. The AP drop σ's are divided into two groups separated just below 30 μGal for the better data and > 30 μGal for the poor nighttime data. Note that the range of the σ's for data < 30 μGal is similar for both drop and set errors at both stations. One could also plot the set histograms based on the 'Error' column in the set files, in which case the set σ's are about a factor 10 smaller when there are ∼ 100 acceptable drops per set

Table 3

Set versus drop scale factors β (μGal/V) for AP experiments; drop and set δg$_m$ = 30 μGal

Start date	2011 Jul 28	2013 Nov 18	2014 Sept 8	2016 Jun 2
Sets/drops recorded	422/42,200	229/22,900	207/20,700	117/11,700
Sets used/% data	283/67%	64/28%	113/55%	55/47%
Drops used/% data	27,493/65%	6083/27%	11,018/53%	5092/44%
(a) DROP SF's				
JS0	− 79.044 ± 0.092	− 112.13 ± 0.525	− 94.732 ± 0.270	− 96.503 ± 0.245
JS1 pass 1	− 79.342 ± 0.093	− 94.148 ± 0.437	− 94.686 ± 0.261	− 94.930 ± 0.311
JS1 pass 2	− 79.384 ± 0.094	− 94.357 ± 0.445	− 94.541 ± 0.263	− 94.830 ± 0.315
JS3 pass 1	− 79.341 ± 0.093	− 94.451 ± 0.626	− 94.262 ± 0.283	− 94.948 ± 0.313
JS3[a] pass 2	**− 79.389 ± 0.094**	**− 94.420 ± 0.628**	**− 94.317 ± 0.283**	**− 94.847 ± 0.315**
(b) SET SF using σ$_s$				
JS0	− 79.328 ± 0.408	− 94.039 ± 1.240	− 94.123 ± 0.847	− 94.692 ± 1.125
JS1 pass 1	− 79.325 ± 0.405	− 94.001 ± 1.731	− 94.439 ± 0.976	− 94.823 ± 1.149
JS1 pass 2	− 79.348 ± 0.406	− 94.013 ± 1.735	b	b
JS3 pass 1	− 79.373 ± 0.405	− 94.399 ± 2.433	− 94.331 ± 1.016	− 94.911 ± 1.147
JS3 pass 2	− 79.395 ± 0.406	b	b	− 94.958 ± 1.149
(c) SET SF using SEM				
JS0	− 79.325 ± 0.041	− 94.029 ± 0.125	− 94.121 ± 0.085	− 94.674 ± 0.114
JS1 pass 1	− 79.329 ± 0.041	− 93.994 ± 0.175	− 94.431 ± 0.099	− 94.821 ± 0.117
JS1 pass 2	− 79.351 ± 0.041	− 94.006 ± 0.176	b	_b
JS3 pass 1	− 79.378 ± 0.041	− 94.401 ± 0.248	− 94.287 ± 0.162	− 94.957 ± 0.118
JS3 pass 2	− 79.398 ± 0.041	b	b	b

[a]JS3 (bold) for drops is the preferred solution. Sections (a) and (b), drop and set errors from 'Sigma' column of respective text files, (c) set errors from 'Error' column of set files

[b]No pass 2 solution required, same value as pass 1

approach contained in Meurers (2012). We mention again the goal here was to process the AG and SG data using only the information in the FG5 text files, and further to do this on files that come directly from the FG5 without any pre-processing of the AG data using the g-software. This was the case for the AP text files received by mail from NGA, whereas the ST data were already carefully pre-processed to reject bad drops before the text files were written—a significant difference (and improvement).

The following discussion addresses a number of options that we tried in rejecting bad drops, which is the key to getting SFs that are consistent when using both drops and sets. The options are summarized in Table 8 in the "Appendix" with the same abbreviations as here:

1. w_1–w_0: weighting the data when computing means. When the g-software records each drop it is compared to an evolving mean value that eventually becomes the set mean, and there is no chance to weight the drops. But after the drops and sets are recorded one can re-compute the set means by weighting the drops with their σ_d's, in principle this being a better approach. So we decided to base all subsequent processing on just the drop data files, and used the set file data only as a check. We could recover the set file data by computing the unweighted mean of the set mean, and also by adopting the 3-σ rejection of drop outliers. We verified that exactly the same drops were rejected as reported in the set files (columns 'accept', 'reject') and with exactly the same set means. The 3-σ rejection is iterated successively until the number of rejected drops does not change (a maximum of 5 iterations proved adequate). The unweighted solutions are designated 'w_0' and the weighted solutions are 'w_1'. The weighting (or not) is applied to every instance in the program where set means are required; our default preference is to use the weighted option.

2. s_3–s_1: using 3-σ or 1-σ criterion for rejecting outliers. Amalvict et al. (2002) reported on the

Table 4

As Table 3 but for 4 Strasbourg J9 calibrations

Start Date	2008 Dec 17	2009 Jun 16	2010 Jun 10	2011 Oct 11
Sets/drops recorded	143/14,300	167/16,700	168/16,800	164/16,400
Sets used/% data	139/97%	164/98%	161/96%	159/97%
Drops used/% data	13,791/96%	16,478/99%	15,678/93%	15,093/92%
(a) DROP SF's				
JS0	− 79.133 ± 0.390	− 79.154 ± 0.230	− 79.008 ± 0.162	− 78.989 ± 0.248
JS1 pass 1	− 78.951 ± 0.396	− 79.216 ± 0.231	− 78.998 ± 0.162	− 78.930 ± 0.249
JS1 pass 2	− 78.875 ± 0.397	− 79.183 ± 0.232	− 78.999 ± 0.164	− 78.965 ± 0.249
JS3 pass 1	− 78.956 ± 0.396	− 79.212 ± 0.231	− 79.012 ± 0.162	− 78.931 ± 0.248
JS3[a] pass 2	**− 78.877 ± 0.397**	**− 79.183 ± 0.232**	**− 79.003 ± 0.164**	**− 78.969 ± 0.249**
ST processed	− 78.842 ± 0.389	− 79.231 ± 0.230	− 78.549 ± 0.161	− 79.020 ± 0.246
(b) SET SF's using σ_s				
JS0	− 78.841 ± 1.725	− 79.160 ± 0.557	− 78.999 ± 0.574	− 78.892 ± 1.144
JS1 pass 1	− 78.824 ± 1.722	− 79.162 ± 0.556	− 78.986 ± 0.573	− 78.998 ± 1.111
JS1 pass 2	− 78.832 ± 1.722	− 79.156 ± 0.557	− 78.975 ± 0.579	− 78.974 ± 1.112
JS3 pass 1	− 78.856 ± 1.723	− 79.189 ± 0.556	− 79.001 ± 0.573	− 79.037 ± 1.111
JS3 pass 2	− 78.863 ± 1.723	− 79.183 ± 0.556	− 78.991 ± 0.579	− 78.930 ± 1.123
ST processed	− 78.879 ± 1.707	− 79.148 ± 0.554	− 78.975 ± 0.572	− 78.981 ± 1.139
(c) SET SF's using SEM				
JS0	− 78.837 ± 0.173	− 79.160 ± 0.056	− 78.997 ± 0.058	− 78.894 ± 0.115
JS1 pass 1	− 78.825 ± 0.173	− 79.162 ± 0.056	− 78.985 ± 0.058	− 78.928 ± 0.115
JS1 pass 2	− 78.833 ± 0.173	− 79.156 ± 0.056	− 78.978 ± 0.058	− 78.917 ± 0.115
JS3 pass 1	− 78.857 ± 0.173	− 79.188 ± 0.056	− 79.001 ± 0.058	− 78.963 ± 0.115
JS3 pass 2	− 78.864 ± 0.173	− 79.183 ± 0.056	− 78.993 ± 0.058	− 78.954 ± 0.115
ST processed	− 78.879 ± 0.341	− 79.148 ± 0.111	− 78.975 ± 0.114	− 78.981 ± 0.228

[a]JS3 (bold) for drops is the preferred solution. Sections (a) and (b), drop and set errors from 'Sigma' column of respective text files, (c) set errors from 'Error' column of set files

choice of 'n' in using an 'n-σ' selection. They chose 3-σ for drops and 1-σ for sets, so we decided to test both options when rejecting outliers. It is expected that this choice depends on the noise in the experiment; for good data it should make little difference.

3. co–nc data: should we do drop selection on the uncorrected or corrected data? We use 'TLBP' to refer to the geophysical corrections (tide, ocean load, barometric pressure loading, polar motion) that are generally removed to get a regular AG site measurement 'co', as opposed to the 'nc' option that does not remove TLBP when rejecting outliers. This point has been emphasized in the papers by Meurers (2012), Calvo et al. (2014), and Van Camp et al. (2016). As discussed previously, if corrections are not made, the drop rejection is compromised by the inclusion of tides (predominantly) which vary throughout the experiment. Once the drops are rejected, the TLBP corrections can then be reapplied to the accepted AG data for

use in the calibrations. At AP we requested the NGA operator to re-run the four calibration experiments with corrections applied, and based our rejections on those files rather than the uncorrected data. In this case it is necessary to have exactly consistent drop and set files to transfer the accept/reject criteria between corrected and uncorrected versions of the same data. This worked well for the AP data, but unfortunately for the Strasbourg experiments the loss of a computer disk in 2014 with all the processed data meant that we did not have ready access to the corrected text files, although the raw AG files are archived. Up to the present we have not tested the 'co–nc' option for the ST data.

4. p_1–p_2 processing: adding a drop/set rejection of outliers after the first fit was obtained. This turned out to be a significant point. Once pass 1 is made, one has access to the residuals—the deviations between the SG curves and the AG drop or set values. These residuals have a more or less normal

distribution about the SG curve, so it is easy to reject deviations on a 3-σ (or 1-σ) reject criterion; this is equivalent to refining the pass 1 solution based on rejecting outliers, or equivalently choosing drops or sets deviations that are close to the SG curve. For most datasets, even those that had been carefully screened manually using the g-software in Strasbourg, pass 2 found sufficient drops or sets to discard on all the datasets that the solution was noticeably improved. For some of the poorer AP sets up to 900 new drops were rejected and up to 5 more sets. It should be said that all the processing for drop or set SFs up to pass 1 use exactly the same accept/reject drops or sets, but at the last step, the pass 2 solution for drops rejects only drops, and the pass 2 solution for sets rejects only sets, so a small difference appears in the data used for the two types of SF.

An illustration of the effectiveness of this procedure is shown in Fig. 4 for the AP2011 experiment. As seen in Table 3, most calibrations are improved using pass 2 in the sense that the drop and set SFs are brought closer.

In addition to the above options (1)–(4) that could be invoked, we chose 3 additional methods to discard drops/sets. Again these are summarized in Table A1, and described below as a series of processing steps. Within each step we can choose any of the above options.

Step 1: From the drop files with TLBP corrections, compute the set means and implement rejection of drop outliers (with iteration), flag all drops as accept/reject, and save these flags as method 'JS1' (rejection based on set means). For the drop SFs only accepted drops are used, and for the set SF the accepted drops are gathered into sets and the set means are used as the data (as usual). Complete sets are rejected if their $\sigma_s > \delta g_m$ and such sets even with good drops are discarded when doing set SFs. Step 2: From the same drop files, reject drops based on the drop $\sigma > \delta g_m$ (the cut-off σ shown in Fig. 3), flag all drops as accept/reject, and save these flags as method 'JS2' (rejection based on drop σ). The procedure then follows JS1. Step 3: Combine the accept/reject flags from the previous 2 steps, so drops are rejected as method

'JS3' (drop rejection based on both set means and drop σ's).
Step 4: From the drop text files for the calibration with no corrections 'nc', we apply the reject flags on all drops identified from the previous three methods (JS1, JS2, and JS3). This completes the preselection of AG drops.
Step 5: Prepare the SG data in two ways. First we use the old method (JS0) for interpolating the 1 min data to either AG drop times, or AG set times, depending on whether we are using drop or set data; this method is combined with the JS1 and JS2 AG selection above. Second we use the SG 1 s files, select all data spanning the AG drop times, then filter the data to reduce the effect of the microseismic noise; this was suggested explicitly by Van Camp et al. (2016) and turns out to be a valuable suggestion for improving the SFs.

Figure 5 shows a suite of filters, based on the Parzen data window, with lengths from 11 to 501 points and with frequency cutoffs between 3 and 50 s. They are designated sf_1–sf_9, where the last filter sf_9 is the original 1 s to 1 min GGP filter. The effect of applying these filters to the four ST datasets on the calibrations is shown in Fig. 6.

It is clear that there can be a bias in the SF, depending on the data quality, if the SG data are inadequately smoothed. For one dataset, ST2010, the SF using no filter (sf_0) shows a noticeable shift in SF that can be corrected using a filter such as sf_6 (or longer). For this experiment there was a large earthquake that had to be removed from the data, and it was assumed the rest of the data was OK, but the effect of the earthquake carried over unseen in the data. All scale factors using the Calvo et al. (2014) method are obtained from the raw 1 s data without filtering. Here, filtering using sf_6 is used in all the calibration solutions with the 'JS3' designation. It should be pointed out that Meurers (2012) uses the SG 1 min data resampled to the AG drop times in his solution, which is equivalent to the original JS0 processing here. Even though it might be technically better to use filtered raw 1 s data, the SF difference between sf_6-filtered 1 s data and the standard 1 min GGP data is negligible and there appears to be no bias introduced using the 1 min SG data directly. One other advantage of filtering the SG data at AP is a

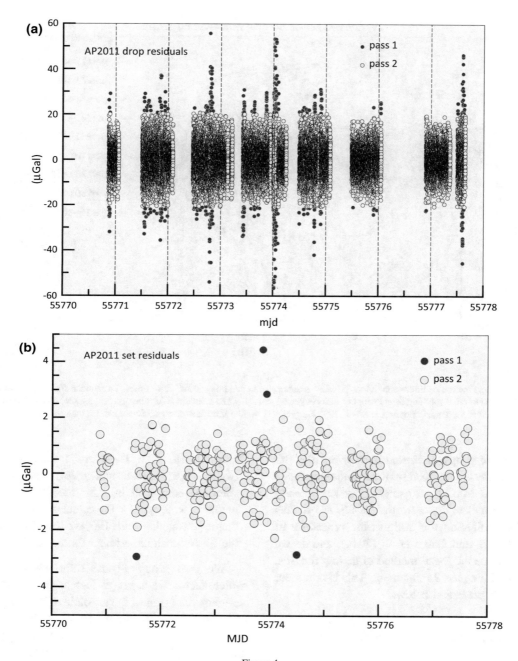

Figure 4
AG residuals **a** drop, and **b** set, after pass 1 and pass 2 of the AP 2011 calibration processed using the JS3 option (see text). The residuals show the departure of the AG values from the SG curve, but these are not related to the drop and set σ's shown in Fig. 3a and b. The second pass successfully removes about 400 drops and 4 sets to improve the solution

noticeable reduction in the amplitude of the telescope glitches that are at the level of 0.5 μGal or less.

The JS1 and JS2 methods combine the different preselection of drops with the SG data interpolated from 1 min data. The JS3 method uses the combined

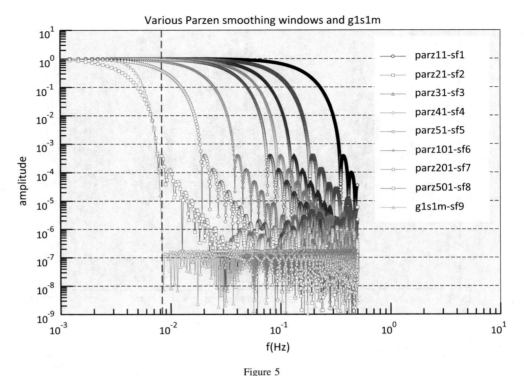

Figure 5

Amplitude transfer function of filters applied to the SG 1 s data prior to sampling at AG drop times. The Parzen filter windows of lengths 11–500 points are simple smoothing filters with cut-offs between 3 and 50 s. The standard GGP filter g_1s_1 m, length 1009 and with a Nyquist cutoff for 1 min sampling at 8.3×10^{-3} Hz (vertical dashed line) is used to decimate the 1 s data to 1 min

preselection of drops, indicated above, but the SG data are preselected at AG drop times and either used directly for the drop SF, or gathered into sets exactly as the AG drops are done, for the set SF. JS3 is thus the only combination that follows the procedure of Meurers (2012) and Calvo et al. (2014), and so we consider it to be the 'best' method of getting the SFs, especially with a pass 2 refinement. This is borne out by the results presented below.

2.4. Results of Tests for the Scale Factors on AP Data

From the previous section we may, therefore, summarize the various tests as follows. For the AP data we have these multiple solutions:

Method JS0: 1 solution for drops, 1 for sets; uses fixed options [w_1, s_3, co (AP) or nc (ST), pass 1] for 2 types of solution

Methods JS1–JS3: There are 2 solutions each for (w_0/w_1, s_3/s_1, co/nc, and p_1/p_2); the number of solutions computed is, therefore (16 options × 3 methods × 2 types = 96 calculations of the scale factors). Together with JS0, we therefore computed the SF for each experiment 98 times.

We then group all the solutions according to which factor we want to isolate (e.g., w_1/w_0) and compute the mean of the absolute difference between the solutions for w_0 and the solutions for w_1 (in units of µGal/V). This provides a metric to judge which of the various options have the most effect on the final scale factors. Some of the more important results are shown in Table 9 ("Appendix") under the 4 methods and 2 types (drops or sets). In this table we have also graded the datasets according to whether they are of high, medium or low quality, based on both the amount of data discarded in Table 3, and on the results in Table 9 themselves. It is important to also evaluate our options on the higher quality datasets

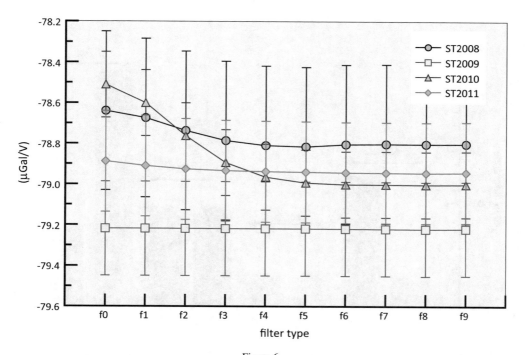

Figure 6

Effect of filtering the 1 s SG data on the scale factor for the Strasbourg datasets. The error bars come from the Sigma column of the set data. The 2010 experiment in particular shows a pronounced bias if the data are not filtered, whereas using filters between f_6 and f_9 suppresses noise at periods < 25 s. Other datasets ST2008 and ST2011 are affected to a lesser extent

that most users will encounter, so we combine all the results with a weighting on a scale of (1 = low, 2 = med, 3 = high) for both the AP and ST datasets combined. The result is shown schematically in Fig. 7 where the two options drops/sets are given with the combined score for the 4 options.

For the w_0/w_1 choice, it depends on whether we are doing drops or sets. For drops the effect is small, but for sets the choice is moderately important, and it seems better to weight the set means as a matter of principle. For the co/nc option, the means differences are surprisingly small, despite the seeming theoretical advantage in rejecting drops on the corrected data (previous discussion), as opposed to the uncorrected data used for the calibration. It is unfortunate that we did not have the corrected ST data at hand to test this result, but it appears that selecting drops directly from the uncorrected files may not in practice give a large SF error. For the s_3/s_1 option the effect is modest for either drops or sets, therefore, retaining the 3-σ outlier is OK for the better data. It is clear from Table 3, however, for the lower quality data there is a

difference (improvement) in choosing a tighter control of outliers using the 1-σ criterion. The final option p_1/p_2, whether or not to have a second pass, makes the biggest difference in determining the SF, especially for the drop solutions. The advantage for sets is much less obvious, as might be expected because the set solutions are weighted more robustly. The number of tests is halved for the ST datasets because there is no corrected data ('co'); thus there are only 49 computations of the various SFs for ST.

2.5. Strasbourg Processing and Calibration Results

We turn from the problematic AG data of AP to station J9 in Strasbourg, which has a very long series of AG–SG calibrations, beginning in the early 1990's (Hinderer et al. 1991; Amalvict et al. 2001, 2002; Rosat et al. 2009; Calvo et al. 2014). We repeat the same calculations as for AP on data for SG CO26 in Strasbourg, using the 4 datasets from 2008, 2009, 2010, and 2011 as shown in Table 4.

119

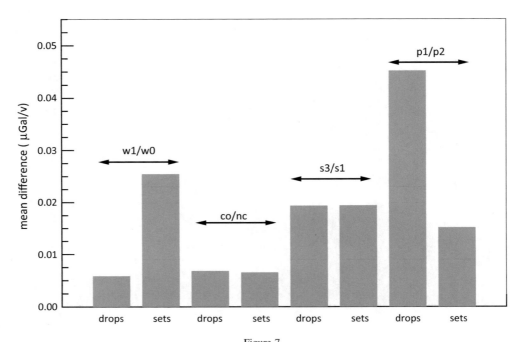

Figure 7
Summary of drop and set tests on options for computing the scale factor, based on Table 9 (see text). The bars represent the mean amplitude of changes in the scale factor over all calculations for both stations AP and ST related to: w_1/w_0—option to weight the set averages, co/nc—option to use corrected or uncorrected AG data, s_3/s_1—option to use 3-σ versus 1-σ rejection of outliers, and p_1/p_2—option to choose a second pass rejecting residual fit outliers. A difference of 0.025 μGal/V is equivalent to a change of 0.03% in the SF

Because of the much better site conditions, the AG data are much cleaner than AP, and with the previously determined cutoff of 35 μGal for both the sets and drops, only a small percentage of the data are excluded. Comparing Tables 4 with 3, we note that the ST scale factors are more consistently determined than at AP, but the error in the SFs is not uniformly better. For example, experiment AP2011 in Table 3 has smaller drop and set errors than any of the data for ST, due probably to the larger number of sets used (283) compared to 139–164 sets for ST. Even so, with careful rejection of the bad data, a satisfactory SF can be obtained, even at AP. As in Table 3, we compare the JS1 and JS3 methods for passes 1 and 2 and note that pass 2 is always required, even though the ST data is much better overall than that at AP. This is a good time to point out that based on set standard deviations, the set SFs in Tables 3(b) and 4(b), have much larger uncertainties than the drop SFs in sections (a) of the Tables, especially for ST2008 and ST2011. The only way the drop and set errors can be comparable, as Amalvict et al. (2002) indicated, is

if we use standard error of the mean (that is to say the standard deviation divided by $\sqrt{(N)}$) as the uncertainty in the set SFs. To show this we recomputed the set SF's using the Error column of the set files, shown in sections (c) of Tables 3 and 4 for the set errors. It is clear that the differences in SF's are very small, and the SF errors are almost exactly a factor 10 smaller than when using the Sigma column. Thus, one can get almost the same result by finding the standard deviation of the SF using the Sigma column (sections (b) of Tables 3 and 4) and dividing by $\sqrt{(N)}$ to get the SEM. The SEM of a weighted mean has the more precise form indicated in the "Appendix" following Eq. (A2).

We see in Fig. 8 a comparison of 7 solutions (JS0, JS1, JS2, and JS3 for drops and sets) for 2 experiments at AP and two from ST, based on sections (a) and (b) of Tables 3 and 4.

The x-axis gives the different solutions comparing pass 1 and 2. Note that the SFs alternate high and low depending on which pass, but with the most important solutions (JS3 pass 2) there is a good

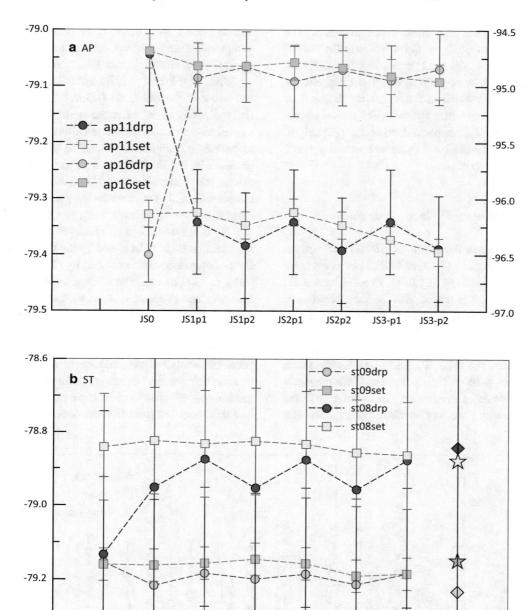

Figure 8

Comparison of scale factors for **a** two AP datasets (2011 left axis, 2016 right axis) and **b** two ST datasets (2008, 2009). All vertical axes are in μGal/V. For each, the drop and set SFs are given for the various methods on the x axis. Abbreviations JS0, JS1, etc., are given in Table 9. Note that the pass 2 solutions are often quite distinct from the pass 1 solutions, and there is a good final convergence between drop and set values for the best datasets (all but AP 2016). The final column in (**b**) shows the drop and set factors obtained by Calvo et al. (2014), diamonds for the drops and stars for the sets (with matching colors). All error bars arise from using the Sigma column in the set files

convergence of the SFs from drops and sets. The original solutions (JS0) are quite different for the AP data, but with the improved processing the results for AP become almost as good as at ST. In Fig. 8b, we also show for comparison the solutions obtained by Calvo et al. (2014) where the drop and set values are more separated. This improvement is due probably to our filtering of the SG 1 s data, as well as using pass 2 to clean up the residuals.

3. Combining Different Scale Factors

Figure 9 shows the result of 51 AG set calibrations from instrument CO26 at J9 in Strasbourg, from 1997 to 2012 (Calvo et al. 2014). The SG sensor did not change over this period, but the data acquisition system electronics changed in 1997/12 (time lag reduced from 36.0 s using the TIDE filter to 17.18 s using the GGP2 filter) and again in 2010/04 when a new GGP filter board (GGP1 filter) was installed with a time lag of 8.16 s. The evolution of the measurements indicates a reduction in scatter of the calibrations with time, but no clear convergence to a

unique value. All these drop scale factors were computed using the drop and set methods in Calvo et al. (2014), similar to our Pass 1 determination.

What is the best way to combine such different estimates of the SG scale factors? The answer partially depends on whether the scale factor should be treated as a constant from one calibration experiment to another. Physicists have long been faced with this problem in the determination of fundamental quantities, for example, the Newtonian gravitational constant G, and in such cases the proper procedure is *conflation,* see "Appendix" Eq. (A1). In the case of an SG, it is assumed (e.g., Hinderer et al. 2015) that the scale factor is determined by the factory magnetic field configuration, or, as stated by Goodkind (1999): "The calibration constant is fixed by the geometry of the coils and suspended mass so that it remains the same if the instrument is turned off and on again no matter how long the time between".

Assuming this is true, the scatter in Fig. 9 must then be attributed to random (and probably also systematic) factors in the experimental setup and environmental noise rather than in the instrument, and this would suggest that conflation is appropriate.

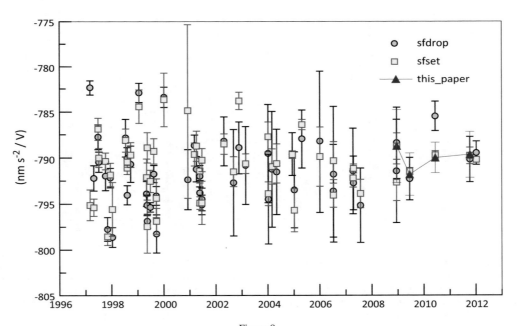

Figure 9
Strasbourg scale factors from sets and drops, 1997–2012. We show set SF's ('sfset') based on the Error column of the set data. Four values denoted by blue triangles ('this_paper') are added from the drop SF's in Table 4(a) for 2008–2011

As discussed in the "Appendix", the calculation of a weighted mean of a series of measurements is unique, but there are two ways to compute its variance, depending on the purpose. One can use the weighted sample variance Eq. (A2), which measures the spread of estimates about the weighted mean. This is appropriate to indicate the scatter of the measurements, but in the case of the SG scale factor it is assumed that the repeated measurements should converge to a unique value, which is the actual calibration. This is indeed the case for SGs, where numerous studies indicate that β can be quite stable. Even the relocation of an SG between two quite different sites does not change the scale factor, as documented by Meurers (2012) for the transition between Vienna and the Conrad Observatory. The SF errors are then appropriately combined by conflation, Eq. (A1).

In principle set and drop scale factors are not independent, but they arise from different procedures and we cannot strictly average them, or conflate them, and they should be treated separately. We apply (A1) to set and drop data independently, and assume the scale factor is not influenced by the changes in electronics, to get the result in Fig. 10.

The initial scale factors prior to 2000 are quite divergent, but each scale factor has more or less stabilized between 2005 and the end of 2012, and the set values are somewhat higher than that used in the Strasbourg data files which are in agreement with the drop value at about -792.00 nm s^{-2}/V. The conflated values are much more revealing of the evolution of the calibration experiments than the scatter plot in Fig. 9, and are a useful way to give a unique value to an SG in a database.

It is to be noted that Eq. (A1) ensures that the eventual error of a long series of scale factor measurements will eventually approach zero, and thus may seem 'unrealistic'. To test this we artificially extended the calibrations at J9 by repeating the same data as shown in Fig. 9 between 2001 and 2012 and adding this series as being qualitatively representative of future (yet to be done) measurements. From the total of 107 calibration experiments (those

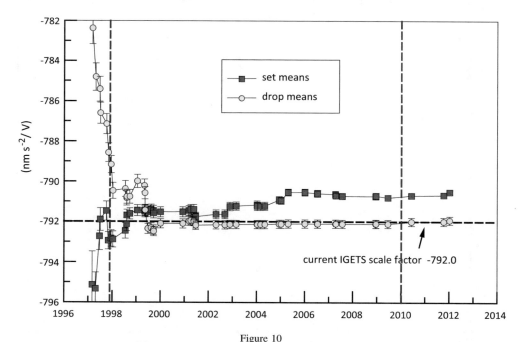

Figure 10

Conflation of the CO26 set and drop scale factors assuming no effect of the time delay (changes at vertical dashed lines). Error bars on the set SF's arise from using the set file Error column, as in Table 4(c), but the drop SF's use the Sigma column of the drop files, as in Table 4(a). The current SF value in the IGETS database is -792.0 ± 1.0 nm s^{-2}/V (0.1%), close to the 2012 conflated drop mean of -791.934 ± 0.195. For sets the SF is slightly reduced to -790.527 ± 0.113

beyond 2012 being repeats) the conflated set error would have dropped slowly from 0.11 nm s^{-2}/V (Fig. 10) to 0.07 nm s^{-2}/V, so the decrease for even a long series of measurements is quite slow.

To be complete, we compare the evolution of the two variances from Eqs. (A1) and (A2) in Fig. 11. The error in the weighted mean (A2) varies somewhat from the scatter of individual scale factor estimates, but does show the same overall downward trend as from conflation (A1). Assuming the actual SF is constant over long time periods, it seems plausible to expect an eventual convergence of the mean and error estimate from Eq. (A1). We also note that conflation can be used for SG scale factors determined using different AG instruments.

4. *The AG Mean Value*

Returning to the basic Eqs. (1)–(3), we note that the AG mean value y_0 includes certain AG static corrections such as the transfer height and gradient effects that are applied to standardize the absolute site level. During the processing of AG data, the operator sets the transfer height and gradient for the experiment. The former is simply a height at which the gravity value is desired for the particular site (frequently ground level), which should be constant from experiment to experiment for consistency. This can be computed from a combination of the actual height for the dropping chamber (in fact the sum of the setup height and a specific height close to 1.2 m given by the manufacturer for each instrument where g is computed from the trajectory over a distance of about 20 cm inside the dropping chamber) and a gradient to be used for the transfer. Ideally, the observed gravity gradient should replace the default $-$ 3.0 µGal/cm (standard free-air gradient).

For various reasons these static corrections (transfer height and gradient) were not always kept constant at Apache Point. In one experiment, the transfer height was set to 0, and the gradient set to $-$ 2.79 µGal/cm, whereas normally we had used 100 or 130 cm for the transfer height and the default gradient. For a long time we did not know the actual gradient below the telescope where the AG

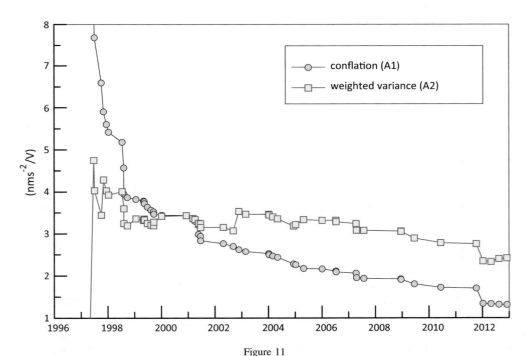

Figure 11

Evolution of the error in the J9 scale factor (Fig. 10) from set data for conflation, Eq. (A1), vs weighted variance, Eq. (A2). Errors are defined in the same way as for Fig. 10

measurements were made. This was eventually measured in 2015 with a Scintrex CG5 both in the cone room (-4.42 μGal/cm) and outside the telescope building at ground level (-3.87 μGal/cm). The gradient resulting from the potential of a homogenous ellipsoidal model at AP is -3.08 (μGal/cm), but cannot be applied to a station at an elevation of 2788 m (referred to WGS84); the difference with the measured value is most likely due to the assumption of a radial Earth model; the discrepancy being due to local topography (including the building) and lateral crustal density anomalies. It then became necessary to adjust not only the transfer height but also the gradient from the values used in the experiment to consistent values. We refer to discussion in the "Appendix" showing how this can be done without access to the g-software.

During a regular AG site measurement, assume $y(t)$ is measured as previously. But this time the geophysical corrections are applied for tides from local gravimetric factors, barometric pressure, and polar motion (TLBP), therefore we write these as:

$$gs(t) = \left[g_{\text{tide}} + g_{\text{press}} + g_{\text{polar}} \right] = gs_0 + gs_1(t) \tag{5}$$

so that during a site measurement, the corrected AG measurements are

$$y_c(t) = y(t) - gs(t). \tag{6}$$

Introducing the mean and time-varying parts from (3) and (5), and recognizing that the corrected gravity $y_c(t)$ should ideally be the site AG value g_0, free of time-varying effects, we find for the constant and time-varying parts

$$g_0 = y_0 - gs_0 = \alpha + x_0 \times \beta - gs_0 \tag{7a}$$

$$0 = y_1(t) - gs_1(t) \tag{7b}$$

Equation (7b) ensures that all time-varying parts of the measured gravity field are accounted for by the time-varying part of $gs(t)$, provided we ignore the errors in the model and other factors such as non-tidal ocean effects and hydrology (but see below).

Equation (7a) shows how to use the mean value y_0 to get g_0, i.e., by subtracting the mean value (or zero level) of the geophysical corrections applied by the g-software. These are the tidal amplitudes with a non-zero mean level, the applied nominal pressure corrections with a specific reference pressure p_0 (calculated for each AG site) and admittance (-0.3 μGal/hPa), and the mean level of polar motion. It was for this reason that we kept track of the constants α, β, and x_0 in the solution of (1).

Normally no other time-varying effects are explicitly involved in the site AG measurements such as local hydrology attraction, non-local hydrology loading, non-tidal ocean effects, and tectonics (see e.g., Pálinkáš et al. 2010); indeed some of these are the target being measured. But many SG users routinely consider such further corrections to their data, so we could assume another model for these: $gh(t) = gh_0 + gh_1(t)$, dominated by hydrology, similar to (6). Adding $gh(t)$ to $gs(t)$ then changes (7a) and (7b) to:

$$g_0 = y_0 - gs_0 - gh_0 \tag{8a}$$

$$y_1(t) = gs_1(t) + gh_1(t) \tag{8b}$$

Again, the time-varying part of the AG measurements can be accounted for in (8b), for which many models exist. It is Eq. (8a) that poses a problem for AG measurements because it is not easy to define the mean level of hydrology gh_0. Unlike the other mean levels, there is no obvious reference level for hydrology; the relative level used for SG studies (e.g., supplied as a loading correction by the EOST/IGETS loading service) is arbitrary and may have no relevance for a particular site. One might use the hydrology levels expected from local environmental parameters (rain, snow, evapotranspiration …) which could define a mean hydrology based, for example, on decades-long averaging, or a reference hydrology level established after a prolonged drought, which might be empirically estimated. This is an interesting unresolved problem that may arise when considering further corrections to the AG site measurements.

Aside from the problem of hydrology, we can still find the site gravity g_0 from Eq. (7a), and compare it with the AG site measurements when re-running the g-software with the geophysical corrections turned on. To estimate gs_0, we turn to the SG-derived version of $gs(t)$ that is readily available at all SG stations due to the need for such modeling, noting this may differ from the FG5 corrections. For example, the SG may provide a superior tidal model (using local gravimetric factors), and we can apply the polar motion as published by IERS instead of predicting the polar motion from a model done in the g-software.

We applied the above method to finding the mean value y_0 and the estimate AG gravity g_0 for the 4 calibrations at AP based on set-derived solutions from the calibration—to be consistent with set means used in the FG5 processing. Table 5 shows the AG mean values from set and drop estimates, estimating g_0 from (5) using the procedure above, and g_0 from a usual FG site measurement.

For AP, the standard transfer height and gradient are (130 cm, -4.42 μGal/cm). We note that the largest component of gs_0 is not necessarily the tide, and TLBP is quite variable over the 4 years. The mean value y_0 comes directly from the fitting (1), with x_0 being found before the fit, after which we read in the corrections $gs(t)$ and find the mean level of the tides, pressure, and polar motion from all the 1 min values coincident with the SG data values. Then $gs(t)$ are splined to the AG set times, and subtracted from the AG values $y(t)$ as in Eq. (6). The corrected AG values $y_c(t)$ allow a weighted mean of the set

values that leads to a 'simulated' FG5 site value denoted by g_1 in Table 5. Alternatively, the mean level of the components of gs_0 are added together and subtracted from y_0 using (7a), giving an AG site value g_0 from y_0, as advertised. Finally the FG5 operator can reprocess the AG data using the g-software with corrections applied to get the actual FG5 site gravity. Some subtleties exist. For example, the SG mean value x_0 and the AG mean value y_0, which can also be obtained directly from the input AG data, must use the highest sampling of the data, either 1 s or 1 min for the SG, and drop data for the AG even though the drop data may be noisy. But the mean TLBP level should be based on the set times, which is based only on good data, and g_0 and g_1 should match the set data as recorded in a site measurement. All the solutions in Tables 5 and 6 are based on the JS3, pass 2 method.

In Table 5, we note that some errors, e.g., the site value g_0 from y_0, seem large, especially for the AP data where there are relatively few TLBP data at set times used in finding gs_0. Thus, quantities derived from gs_0 tend to have larger errors than one might expect. The final two lines of Table 5 show that the difference $g_1 - g_0$ is consistently less than 1 μGal, clearly indicating that Eqs. (7a) and (7b) work in practice for all datasets. The discrepancy between the g_1 and FG5 site values are more variable but still generally within the error bars. The smallest error is on the value g_1 which uses the external corrections directly on the AG data, but other contributions to the

Table 5

Processing of Apache Point AG mean values y_0 from SG to AG calibrations as AG site measurements; all units μGal

	2011 Aug	2013 Nov	2014 Sept	2016 Jun
Means levels				
Tide	-8.2077	18.8580	4.253	-32.4812
Press	-3.9618	-1.7205	-3.030	-3.3393
Polar motion	6.9382	4.5114	4.411	7.8291
gs_0	-5.231 ± 5.003	21.649 ± 4.762	5.635 ± 6.923	-27.991 ± 12.801
AG—9788050000.0				
Mean value y_0	5158.34 ± 0.87	5189.66 ± 1.69	5161.53 ± 1.40	5136.14 ± 2.35
g_0 from y_0 Eq. (7a)	5163.57 ± 5.08	5168.01 ± 5.05	5155.90 ± 7.063	5164.13 ± 13.02
Simulated site value g_1	5164.36 ± 0.88	5167.59 ± 0.90	5155.05 ± 2.71	5164.07 ± 1.676
FG5 site value	5164.11 ± 1.81	5167.45 ± 1.82	5155.29 ± 1.83	5164.39 ± 1.84
$g_1 - g_0$	0.782 ± 5.079	-0.420 ± 5.131	-0.844 ± 7.562	-0.062 ± 13.122
$g_1 -$ FG5	0.246 ± 1.811	0.143 ± 2.029	-0.234 ± 3.271	-0.322 ± 2.489

Table 6

Processing of Strasbourg AG mean values y_0 from SG to AG calibrations as AG site measurements; all units µGal

	2008 Dec 17	2009 Jun 16	2010 Jun 10	2011 Oct 11
Means levels				
Tide	23.8074	23.9927	19.084	22.0829
Press	− 4.1595	− 2.3362	0.194	− 3.4177
Polar motion	0.1435	0.4463	− 1.064	2.5759
gs_0	19.791 ± 3.673	22.103 ± 5.670	18.214 ± 6.070	21.24 ± 4.19
AG—980870000.0				
Mean value y_0	7828.45 ± 1.98	7810.51 ± 0.98	7804.63 ± 1.15	7819.30 ± 1.59
g_0 from y_0 Eq. (7a)	7808.66 ± 4.17	7788.41 ± 5.75	7786.42 ± 6.18	7798.06 ± 4.48
Simulated site value g_1	7809.07 ± 1.27	7789.37 ± 0.36	7797.88 ± 1.28	7797.88 ± 1.28
FG5 site value	7811.62 ± [a]	7793.15 ± 1.46	7789.24 ± 1.51	7800.49 ± 1.30
$g_1 − g_0$	0.412 ± 4.362	0.960 ± 5.766	− 0.547 ± 6.235	− 0.186 ± 4.657
$g_1 −$ FG5	− 2.546 ± 1.265*	− 3.784 ± 1.504	− 3.370 ± 1.729	− 2.612 ± 2.273

[a]FG5 error unavailable for this experiment

total error, i.e., the 'uncertainty', are not added as done within the g-software.

Confirmation of this procedure is provided from the 4 Strasbourg experiments, Table 6. We see that the tide mean value is significantly higher than at AP and does dominate gs_0—it is a significant effect. The overall agreement $g_1 − g_0$ is very close and more consistent than for AP due to the better AG data, but there are discrepancies between g_1 and the standard FG5 set measurement, whose origin may be due to the corrections being either FG5 or 'SG-derived' values.

5. Summary and Conclusions

We show in Table 7, a summary of the amplitudes of the various effects that influence the SG scale factor, according to our estimates. These are taken from the various tables and figures, with additional estimates for the difference between JS0, JS1, and JS3. Notice that the effects are in units of the scale factor (µGal/V), but when translated into percentages the values are close to % errors, e.g., an error of 0.025 in scale factor is equivalent to 0.03%. Certain effects are more important than others, i.e., moving from JS0 to JS3 (using departures from set means which is standard in FG5 processing) when processing drops,

Table 7

Summary of all factors influencing the SG scale factor

Factor	Amplitudes	Effect or decision to be made
Time lag	0.01 for 30 s	Influence of instrumental time lag on SG selection
AG trend	0.01 (µGal/day)	Highly dependent on data
w_1/w_0	0.025 (sets)	Weighting the drops when finding set means
	0.006 (drops)	
co/nc	0.0067	Correcting the AG values for TLBP prior to drop and set selection
s_3/s_1	0.0194	Using a 1-σ versus 3-σ rejection of drops and sets
p_2/p_1	0.045 (drops)	Second pass to eliminate residual outliers far from the SG curve
	0.015 (sets)	
SG 1 s	≤ 0.5	Filtering of 1 s SG data prior to selection, highly data dependent
JS1/JS0	0.529 (drops)	Rejecting outliers from set means vs rejection based on drop δg_m
	0.038 (sets)	
JS3/JS2	0.005 (drops)	For drops, spline-interpolate SG values vs sampling SG value at drop time; for
	0.059 (sets)	sets, gathering SG drops at AG set times vs interpolation of 1 min data

See Table 8 for abbreviations. Values in (µGal/V) except where indicated; a single entry applies to both drops and sets

but this is less important for sets. Two factors have been improved over the processing of Calvo et al. (2014), namely smoothing the SG 1 s data, and doing a second pass—especially in the case of drop SFs. For set SFs the dominant effects are to weight the drops when finding set means, and JS3–JS2—gathering the SG data at AG times into sets rather than interpolating SG 1 min data to AG set times. Any difference below 0.01 μGal/V (such as having to make corrections to AG data before selecting drops), is considered a minimal effect, but we still recommend processing using JS3.

In addition, we have shown that the SG data should be selected at beginning and ending at the AG drop times, and especially for the AG mean value (if used) it can be important to include a trend to account for a drift in one of the instruments but not the other. The SG electronics time delay, which ideally should also be included, has almost negligible effect. Another feature of our study is that we use only the drop text files, because we can compute everything from them, including all the set processing required for a set SF. We do not need special pre-processing of the recorded data using the g-software to reject drops, although this can of course be done by groups that have the facilities and manpower. In terms of choosing whether to report drop or set SFs, both should be computed. Where there is a discrepancy it is likely that the set value may be less affected by bad data. On the other hand if the values are close, the drop SF is preferred as its error is statistically better defined in the sense one does not have to choose between using the 'Sigma' and 'Error' columns of the set data as errors.

We also recommend the use of conflation to combine different estimations of the SF for a particular SG, as this is the best way to characterize the SF for stations in a database. Finally, for users who do not have the g-software, or are reluctant to spend the effort to use it for their calibrations, we have shown it is possible to turn an SG calibration experiment into an AG site measurement at a site by subtracting the geophysical TLBP corrections from the AG mean value. Also it might be useful on occasion to determine the internal distance parameter D, through Eq. (A5), for an FG5 to enable a precise conversion of an AG mean value from one gradient to another.

Acknowledgements

Two anonymous reviewers provided many insightful comments and criticisms that allowed us to rethink some of our initial results and ideas, from which we trust the reader will benefit. We also thank Bruno Meurers as editor for excellent suggestions throughout the reviewing process. We are deeply indebted to R. David Wheeler, from the National Geospatial-Intelligence Agency stationed at Holloman AFB, New Mexico, for making the difficult AG measurements in the cone room at Apache Point Observatory. DC benefitted from very useful discussions with Hartmut Wziontek (BKG, Leipzig) and Derek Van Westrum (National Geodetic Survey, Boulder). The work was done as a subcontract originating from the pioneering work of Tom Murphy (UCSD, California) on the APOLLO LLR system, and his effort to install an SG to improve the LLR; funding came from Grants 10-APRA10-0045 (NASA), PHY-1068879 and 10322410-SUB (NSF). The Strasbourg SG data is available at https://doi.org/10.5880/igets.st.11.001.

Appendix

Weighted Mean and Variance

There are two approaches to finding the variance of weighted samples, assuming N samples of a quantity x_i, each assumed to have a Gaussian probability distribution with a standard deviation σ_i. Interpreting the weight of each sample as its inverse variance, $w_i = 1/\sigma_i^2$, the mean (x_m) and variance σ^2 of the resulting combination are given by

$$x_m = \Sigma_i(w_i x_i) / V_1; \sigma^2 = 1/V_1 \qquad (A1)$$

where $V_1 = \Sigma_i(w_i)$, assuming the data are uncorrelated. Equation (A1) gives the *variance of the weighted mean*, used in combining different quantities, derived for example from the LSQ inversion or fitting of parameters to data. It is also the error of compound quantities, so that when all weights are equal, $\sigma^2 = (1/N^2) \Sigma_i(\sigma_i^2)$. A second approach is to compute the *unbiased weighted sample variance*

$$\sigma^2 = \Sigma_i\left[w_i(x_i - x_m)^2\right] / \left[V_1 - (V_2/V_1)\right] \qquad (A2)$$

where $V_2 = \Sigma_i \, (w_i^2)$, as given for example in the GNU Fortran library (function *gsl_stats_wvariance* at https://www.gnu.org/software/gsl/manual/gsl-ref.html#Weighted-Samples). The mean x_m remains as in (A1) and the weighted SEM is defined as $\sigma\sqrt{(V_2)}/V_1$. Equation (A2) is used to assess the variance of the data about its mean, and is the weighted version of the usual formula for Gaussian mean and variance. In his useful little book, Topping (1979, Section 43) refers to (A1) as measuring *internal consistency of the data*, (using the errors associated only with each experiment) as opposed the *external consistency* of the data where the errors are determined by the spread of each experiment about a common mean.

Hill and Miller (2011) give the name *conflation* to (A1) and argue this is the correct way to combine different experiments to determine the best value of an unchanging physical quantity. They show that conflation is (a) commutative and associative, (b) iterative, and the (c) conflations of normal distributions are normal. Property (b) is useful as new data can be easily combined by adding to the conflations of the previous datasets. A more mathematically oriented justification can be found in Hill (2011).

AG Transfer Height

AG transfer height and gradient effects were discussed by Niebauer (1989) and then extended by Nagornyi (1995) and Timmen (2003) who provided more instrumental details. From a user's point of view it is not possible to deal with changing the gradient using only the transfer height adjustment provided in the FG5 manual. Denote this transfer height correction by

$$\delta g_1 = - (\text{actual height} - \text{transfer height}) \times \text{gradient}$$
$$= - (\text{AH} - h) \times \Delta g \qquad (A3)$$

where AH is the actual height and h the transfer height. The actual height is the sum of the factory height and the setup height, and these quantities are given in the FG5 project files. But (A3) is insufficient to recover the transfer height correction if the gradient Δg changes. A series of experiments was performed by NGA (National Geospatial-Intelligence Agency) by varying the transfer heights and gradients in the FG5 processing and recording the calculated values from the FG5 merged project files. Starting with $g_c(h, \Delta g)$ as the calculated value, we first

Table 8

Terminology and abbreviations used in the paper

	Options	Comment
TLBP corrections		Tides and ocean loading from local gravimetric factors, barometric pressure, and polar motion; for AG data
AG mean value	y_0	The quantity y_0 in Eqs. (2) and (3)
AG site value	g_0	The result of a normal AG determination with TLBP corrections
Sigma rejection	δg_m	Choice of drop and set sigma rejection level for disturbed data, effective for AP data in rejecting nighttime noise
Type of FG5 file	co/nc	co—corrected for TLBP, as in a normal AG measurement nc—not corrected, as in a calibration experiment
Weighting used for set means	w_1/w_0	w_1—set means are computed with drop errors as weights w_0—no weighting used for set means, as in FG5 files
Reject criterion	s_3/s_1	s_3—3-σ rejection of outliers for set means s_1—1-σ rejection
Treatment of residuals	p_1/p_2	p_1—pass 1: all residuals accepted after pass 1 p_2—pass 2: 3-σ rejection of residual outliers and revised fit
SG smoothing	sf0–9	Application of various smoothing filters on SG 1 s data prior to selection of SG value at AG drop times, see Fig. 5
Mechanism for drop rejection	JS0	JS0: drop selection based only on drop sigma, set rejection based on δg_m; SG 1 min data interpolation
	JS1	JS1: drop selection based on 3 σ rejection of drops when computing set means; SG 1 min data interpolated
	JS2	JS2: drop selection based on rejecting drops with errors > δg_m SG data 1 min interpolation
	JS3	JS3 drop selection based on both JS1 and JS2; SG data from smoothed 1 s data at closest 1 s AG drop times, averaged as for AG sets from gathered drop times

Table 9

Test results for AP and ST scale factors. Shown are mean differences between equivalent calculations (keeping other options constant); values in ($\mu Gal/V$)

Station AP	2011	2013	2014	2016
AG data quality	High	Low	Med	Low
Drops w_1-w_0	0.00167	0.0217	0.00584	0.0218
Sets w_1-w_0	0.00507	0.0210	0.00348	0.0358
Drops co–nc	0.00809	0.0294	0.00716	0.0610
Sets co–nc	0.00646	0.0378	0.00287	0.0608
Drops s_3-s_1	0.00403	0.0375	0.0337	0.0408
Sets s_3-s_1	0.00637	0.0658	0.0337	0.0531
Drops p_2-p_1	0.0507	0.0838	0.0507	0.0727
Sets p_2-p_1	0.0218	0.00330	0.0218	0.0290

Station ST	2008	2009	2010	2011
AG data quality	High	High	High	High
Drops w_1-w_0	0.00359	0.00130	0.00024	0.0118
Sets w_1-w_0	0.0169	0.0102	0.00233	0.105
Drops s_3-s_1	0.0334	0.0222	0.00784	0.00621
Sets s_3-s_1	0.0252	0.0208	0.00723	0.00114
Drops p_2-p_1	0.0766	0.0336	0.00068	0.0383
Sets p_2-p_1	0.00802	0.00452	0.00999	0.0261

subtract the standard correction (A1), and also the calculated gravity for zero gradient at this transfer height $g_c(h, 0)$, thus

$$g_1'(h, \Delta g) = g_c(h, \Delta g) - \delta g_1 - g_c(h, 0) \quad (A4)$$

We established that the left hand side (LHS) is a linear function of the gradient, i.e.,

$$g_1'(h, \Delta g) = D \times \Delta g \quad (A5)$$

where for NGA's FG5-107 used at AP, $D = 8.0258$ cm. Thus

$$g_c(h, \Delta g) - g_c(h, 0) = (h - h_e) \times \Delta g \quad (A6)$$

introducing $h_e = (AH - D) = 122.694$ cm as an effective height, such that when $h = h_e$ the gradient has the least effect on the gravity value. Because it is not possible to find h_e or D from the project files, Pálinkáš et al. (2010) observed "...Some users of the gravity data have no access to the FG5g-software; they cannot accurately correct for the new gradient without knowledge of the effective position. There is even a risk that they will compute the new transfer

correction with respect to the top of the drop, because this is often presented as the instrument's reference height."

This was indeed our earlier experience, and we found h_e by adding one additional step, i.e., re-processing the same FG5 data with zero gradient at the same transfer height, $g_1'(h, 0)$, and subtract this to get $g_1'(h, \Delta g)$ using (A5) to find D. We can then correct for both transfer height and gradient from one AG setup to another using

$$g_c(h_2, \Delta g_2) - g_c(h_1, \Delta g_1) = (\delta g_1 - \delta g_2) + D \\ \times (\Delta g_2 - \Delta g_1)$$

$$(A7)$$

Note the distance D in (A5) remains the same when AH changes, and therefore, needs to be determined only once, whereas the effective height h_e depends on AH. In one of our experiments we start with AH = 130.72 cm, a transfer height of 100 cm, and a gradient of -3.0 $\mu Gal/cm$ but want gravity at a height of 130 cm and a gradient of -4.42 $\mu Gal/cm$; using Eq. (A7) gives -100.37 μGal, but Eq. (A3) gives a value of -90.0 μGal, a difference of more than 10 μGal. Equation (A5) in fact is entirely consistent with Eq. (3) in Pálinkáš et al. (2010) to which the reader should refer for complete details (Tables 8, 9).

REFERENCES

Amalvict, M., Hinderer, J., Boy, J.-P., & Gegout, P. (2001). A three year comparison between a superconducting gravimeter (GWR C026) and an absolute gravimeter (FG5#206) in Strasbourg (France). *Journal of the Geodetic Society of Japan, 47,* 410–416.

Amalvict, M., Hinderer, J., Gegout, P., Rosat, S., & Crossley, D. (2002). On the use of AG data to calibrate SG instruments in the GGP network. *Bull d'Inf Marees Terr, 135,* 10621–10626.

Baker, T. F., & Bos, M. S. (2003). Validating Earth and ocean tide models using tidal gravity measurements. *Geophysical Journal International, 152,* 468–485.

Boy, J.-P., Llubes, M., Hinderer, J., & Florsch, N. (2003). A comparison of tidal ocean loading models using superconducting gravimeter data. *Journal of Geophysical Research, 108*(B4), 2193. https://doi.org/10.1029/2002JB002050).

Calvo, M., Hinderer, J., Rosat, S., Legros, H., Boy, J.-P., Ducarme, B., et al. (2014). Time stability of spring and superconducting gravimeters through the analysis of very long gravity records. *Journal of Geodynamics, 80,* 20–33. https://doi.org/10.1016/j.jog.2014.04.009.

Crossley, D., & Hinderer, J. (2010). GGP (global geodynamics project): An international network of superconducting gravimeters to study time-variable gravity. *IAG Symposia, Gravity, Geoid, and Earth Observation, 135,* 627–635. https://doi.org/10.1007/978-3-642-10634-7_83.

Francis, O. (1997). Calibration of the C021 superconducting gravimeter in Membach (Belgium) using 47 days of absolute gravity measurements. In: *Gravity, Geoid and Marine Geodesy, Tokyo, Japan, IAG Symposium* (Vol. 117, pp. 212–219). Berlin: Springer.

Goodkind, J. (1999). The superconducting gravimeter. *Review of Scientific Instruments, 70*(11), 4131–4152.

Hill, T. (2011). Conflations of probability distributions. *Transactions of the American Mathematical Society, 363*(6), 3351–3372.

Hill, T., & Miller, J., (2011). An optimal method to combine results from different experiments. arXiv:1005.4978v3 [physics.data-an].

Hinderer, J., Crossley, D., & Warburton, R. J. (2015). Superconducting gravimetry. In Gerald Schubert (Ed.), *Treatise on geophysics* (2nd ed., Vol. 3, pp. 59–115). Oxford: Elsevier.

Hinderer, J., Florsch, N., Makinen, J., Legros, H., & Faller, J. (1991). On the calibration of a superconducting gravimeter using absolute gravity measurements. *Geophysical Journal International, 106,* 491–497.

Imanishi, Y., Higashi, T., & Fukuda, Y. (2002). Calibration of the superconducting gravimeter T011 by parallel observation with the absolute gravimeter FG5 210—A Bayesian approach. *Geophysical Journal International, 151,* 867–878.

Imanishi, Y., Nawa, K., Tamura, Y., & Ikeda, H. (2018). Effects of horizontal acceleration on the superconducting gravimeter CT

#036 at Ishigakijima, Japan. *Earth, Planets and Space, 70,* 9. https://doi.org/10.1186/s40623-018-0777-9.

Kren, P., Palinkas, V., & Masika, P. (2016). On the effect of distortion and dispersion in fringe signal of the FG5 absolute gravimeters. *Metrologia, 53*(1), 27–40.

Mäkinen, J., Virtanen, H., Bilker-Koivula M, Ruotsalainen, H., Näränen, J., and Raja-Halli, A. (2015). The effect of helium emissions by a superconducting gravimeter on the rubidium frequency standards of absolute gravimeters In: *International Association of Geodesy Symposia.* New York: Springer. https://doi.org/10.1007/1345_2015_205.

Meurers, B. (2002). Aspects of gravimeter calibration by time domain comparison of gravity records. *Bull. d'Inf Marées Terr, 135,* 10643–10650.

Meurers, B., (2012). Superconducting gravimeter calibration by colocated gravity observations: Results from GWRC025. *International Journal of Geophysics.* https://doi.org/10.1155/2012/954271. https://www.hindawi.com/journals/ijge/2012/954271/.

Nagornyi, V. (1995). A new approach to absolute gravimeter analysis. *Metrologia, 32*(3), 201–208.

Niebauer, T. M. (1989). The effective measurement height of freefall absolute gravimeters. *Metrologia, 26,* 115–118.

Pálinkáš, V., Kostelecký, J., & Simek, J. (2010). A feasibility of absolute gravity measurements in geodynamics. *Acta Geodynamic Geomaterial, 7*(1), 61–69.

Rosat, S., Boy, J.-P., Ferhat, G., et al. (2009). Analysis of a ten-year (1997–2007) record of time-varying gravity in Strasbourg using absolute and superconducting gravimeters: New results on the calibration and comparison with GPS height changes and hydrology. *Journal of Geodynamics, 48*(3–5), 360–365.

Tamura, Y., Sato, T., Fukuda, Y., & Higashi, T. (2005). Scale factor calibration of a superconducting gravimeter at Esashi Station, Japan, using absolute gravity measurements. *Journal of Geodesy, 78,* 481–488.

Timmen, L. (2003). Precise definition of the effective measurement height of free-fall absolute gravimeters. *Metrologia, 40,* 62–65.

Topping, J. (1979). *Errors of observation and their treatment.* New York: Halstead Press.

Van Camp, M., Meurers, B., de Viron, O., & Forbriger, T. (2016). Optimized strategy for the calibration of superconducting gravimeters at the one per mille level. *Journal of Geodesy, 90,* 91–99.

Voigt, C., Förste, C., Wziontek, H., Crossley, D., Meurers, B., Pálinkáš, V., Hinderer, J., Boy, J.-P., Barriot, J.-P., Sun, H. (2016). Report on the data base of the international geodynamics and earth tide service (IGETS). *Scientific Technical Report STR—Data.* Potsdam: GFZ German Research Centre for Geosciences. http://doi.org/10.2312/GFZ.b103-16087.

Warburton, R., & Goodkind, J. (1976). Search for evidence of a preferred reference frame. *Astrophysical Journal, 208,* 881–886.

Wziontek, H., Falk, R., Wilmes, H., & Wolf, P. (2006). Rigorous combination of superconducting and absolute gravity measurements with respect to instrumental properties. *Bull d'Inf Marées Terr, 142,* 11417–11422.

(Received January 24, 2017, revised March 2, 2018, accepted March 8, 2018, Published online March 22, 2018)

Pure Appl. Geophys. 175 (2018), 1727–1737
© 2017 Springer International Publishing
https://doi.org/10.1007/s00024-017-1553-7

Two High-Sensitivity Laser Strainmeters Installed in the Canfranc Underground Laboratory (Spain): Instrument Features from 100 to 0.001 mHz

Antonella Amoruso,[1,2] Luca Crescentini,[2,3] Alberto Bayo,[2] Sergio Fernández Royo,[2] and Annamaria Luongo[3]

Abstract—Two laser strainmeters are being operated in the Canfranc Underground Laboratory (LSC, Central Pyrenees, Spain) at about 350 m depth. One of the two laser strainmeters (GAL16, striking 76°) is located about 670 m from the Spanish entrance of a decommissioned train tunnel, along the side wall of one of the bypasses connecting a recent highway tunnel to the train tunnel. The other strainmeter (LAB780, striking −32°) is located about 780 m from the Spanish entrance of the train tunnel, inside two narrow side halls parallel to and built at the same time as the train tunnel. Their mechanical and optical setups derive from a previous installation at Gran Sasso, Italy, with some changes and improvements. Here we show the main instrument features in the frequency range of 100–0.001 mHz. At frequencies lower than 4 mHz, strain noise compares well with the best laser strainmeters made till now, while at higher frequencies strain noise is higher than at Kamioka, Japan, probably because of frequency instabilities of the laser source. Environmental (air temperature and pressure) effects on measured strain are quite small; thus, signal-to-noise ratio in the tidal bands is unusually high. In particular, diurnal Ψ_1 and Φ_1 tides clearly emerge from noise even using a 2-year-long strain record, giving the opportunity to improve previous determinations of the Free Core Nutation parameters from strain data as soon as more data are acquired. The features of the LSC strainmeters allow investigating the Earth in a very broad frequency range, with a signal-to-noise ratio as good as or better than that of the best laser strainmeters in the world.

Key words: Earth strain, strainmeters, earth tides, free core resonance.

1. Introduction

Continuous measurements of the Earth's crustal deformation at spatial scales ranging meters to kilometers can be carried out using different techniques, e.g., rod or wire extensometers (e.g., Richter et al. 1995), borehole dilatometers (e.g., Sacks et al. 1971), tensor strainmeters (e.g., Gladwin 1984), laser strainmeters (e.g., Levine and Hall 1972; Goulty et al. 1974; Crescentini et al. 1997; Takemoto et al. 2004; Milyukov et al. 2005; Kobe et al. 2016; Agnew and Wyatt 2003), and GPS nets (e.g., Borsa et al. 2014). The terms strainmeters and extensometers are often considered synonyms.

Continuous high-sensitivity strain measurements allow to study a wide range of Earth local and global processes, characterized by time scales ranging seismic waves to tectonic deformation and amplitudes ranging several orders of magnitude. As regards local phenomena in tectonic environments, we can mention long-term strain transients (e.g., Gao et al. 2000), seasonal thermoelastic strain (e.g., Ben-Zion and Allam 2013), the correlation between groundwater flow and deformation (e.g., Amoruso et al. 2011, 2014), slow earthquakes and their dynamics (e.g., Linde et al. 1996; Crescentini et al. 1999; Amoruso et al. 2002), slow diffusive fault slip propagation (e.g., Amoruso and Crescentini 2009a), pre-seismic and post-seismic phenomena (e.g., Johnston et al. 2006; Amoruso and Crescentini 2010; Taka-nami et al. 2013; Canitano et al. 2015), co-seismic steps (e.g., Araya et al. 2010), monitoring of rivers (e.g., Díaz et al. 2014), and loading from nonlinear and minor ocean tides (e.g., Agnew 1981; Amoruso and Crescentini 2016). As regards local phenomena in volcanic environments, high-sensitivity strainmeters have been used to detect and infer volcano dynamics (e.g., Linde et al. 1993; Bonaccorso et al. 2016). Global phenomena include the study of the free

---

[1] Department of Chemistry and Biology, University of Salerno, Fisciano, Italy. E-mail: aamoruso@unisa.it

[2] Laboratorio Subterraneo de Canfranc, Huesca, Spain.

[3] Department of Physics, University of Salerno, Fisciano, Italy.

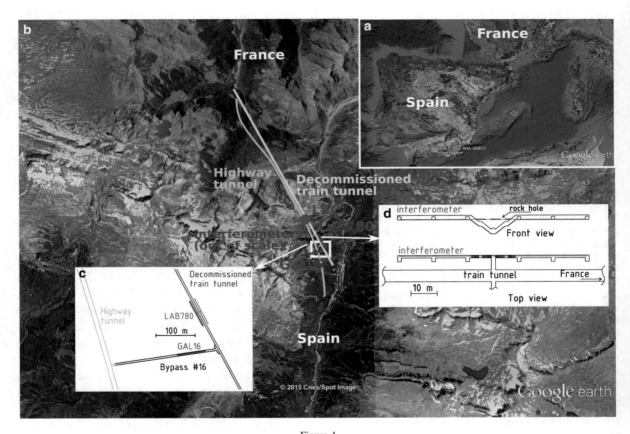

Figure 1
Map of the area surrounding Laboratorio Subterráneo de Canfranc (LSC) and arrangement of the laser strainmeters. **a** Location of LSC in the Central Pyrenees; **b** location of LSC and the strainmeters with respect to the highway and train tunnels; **c** plan of the tunnels where the interferometers (GAL16 and LAB780) have been installed; **d** map and section of the side halls hosting LAB780 interferometer

oscillations of the Earth (e.g., Park et al. 2008; Zürn et al. 2015), the resonant modifications of diurnal tidal waves induced by the free core nutation (e.g., Mukai et al. 2004; Amoruso et al. 2012), and the global deformations of the Eurasian plate (Milyukov et al. 2013).

Even if laser strainmeters are characterized by the best overall performance, exhibiting both very high sensitivity, down to the picostrain level, and long-lasting stability, weeks to years, they are nevertheless not widely used. Some longbase (about 1 km) sur-face-mounted laser strainmeters are operating in California (see, e.g., Agnew and Wyatt 2003), and a few (20–100 m long) underground laser strainmeters are operating inside tunnels (Takemoto et al. 2004; Milyukov et al. 2005; Amoruso and Crescentini 2016; Kobe et al. 2016). The main disadvantages of

longbase surface-mounted laser strainmeters are the cost of the vacuum system and the instability of the end-monuments because of weathering effects (e.g., Agnew 1986). The main disadvantages of under-ground laser strainmeters are the cost of tunnel excavation and the shorter baseline, which reduces sensitivity unless using complex electro-optical setups.

This paper deals with two high-sensitivity laser strainmeters (interferometers) recently installed in Central Pyrenees, Spain, inside the Canfranc under-ground laboratory (LSC, Laboratorio Subterràneo de Canfranc). The instrumental setups and performances are described, by comparing with a previous version of the same instruments and other worldwide laser strainmeters. As usual, here extensions are expressed through the dimensionless ratio $\Delta l/l$, where l is

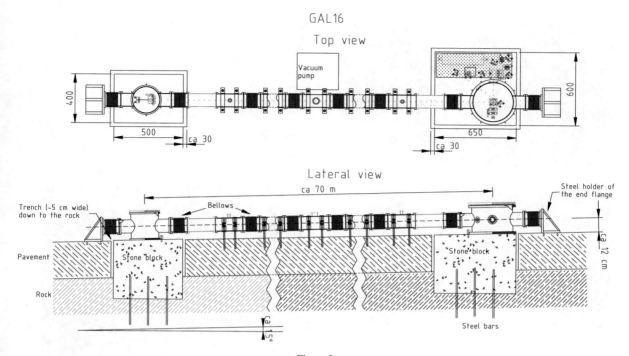

Figure 2
Optical and mechanical setups of GAL16 interferometer. *Top* top view, *bottom* lateral view. Linear dimensions are in millimeters, unless otherwise specified

instrument length, and Δl is positive for an increase in length. We use the symbol nε, nanostrain, for $\Delta l/l = 10^{-9}$. Since each LSC strainmeter is about 70 m long, 1 nanostrain is equivalent to an increase of its length by about 0.07 μm.

2. Instrumental Setup

LSC is located between a highway tunnel connecting Spain to France and a decommissioned train tunnel, 850 m deep under the Mount Tobazo in the Spanish side of the Aragon Pyrenees and about 2400 m from the Spanish entrance of the train tunnel. Although LSC is mainly devoted to particle physics and astrophysics, it also hosts a geodynamical facility (GEODYN project) including a Titan accelerometer, a broad-band Trillium 240 seismometer, and two 70-m-long laser strainmeters (Díaz et al. 2014).

One of the two laser strainmeters (GAL16) is located about 670 m from the Spanish entrance of the train tunnel, along the side wall of one of the bypasses connecting the highway and train tunnels

(Fig. 1). Its azimuth (clockwise from north) is about 76°. The other strainmeter (LAB780, azimuth about −32°) is located about 780 m from the Spanish entrance of the train tunnel, inside two narrow side halls parallel to and built at the same time as the train tunnel. Each of the two collinear halls is about 29 m long and is connected to one end of a single 5-m-long corridor by means of stairs; a closed door separates the other end of the corridor from the train tunnel. The two halls are separated by 15-m-thick rock, which was drilled to allow the strainmeter installation.

The mechanical and optical setups of the laser strainmeters are derived from a previous installation at Gran Sasso, Italy (Crescentini et al. 1997; Amoruso and Crescentini 2009b), with some changes and improvements (Figs. 2, 3). The main differences relate to (1) the reference arm, which now includes a half-wave retarder plate (polarization rotator, see later) and an eighth-wave retarder plate instead of a polarizer and a quarter-wave retarder plate; (2) the endpoint plinths, which now are made of local rock (limestone) instead of reinforced concrete (much

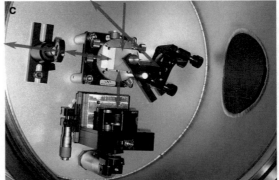

Figure 3
LAB780 interferometer. **a** Optical board including the laser head, two optical isolators, several alignment mirrors, two focusing lenses, one polarizing beam-splitter cube, and two photodiodes; **b** cat's eye of the measurement arm; **c** beam-splitter cube and reference arm; **d** two examples of Lissajous figures

more sensitive to humidity changes); and (3) the size of the vacuum pipes and chambers, which now are much smaller than at Gran Sasso and thus much less affected by any unbalanced force exerted on the end flanges of the vacuum system by atmospheric pressure.

Each interferometer is based on the classical unequal-arm Michelson design and measures distance changes between two about-70-m-distant stone blocks, separated from the floor by means of a narrow trench and fastened to the underlying rock by means of glue and steel bars. Changes in length of the about-70-m-long measurement arm cause changes in phase mismatch between the laser beam traveling along the measurement arm and the laser beam traveling along an about 10-cm-long reference arm. Those phase mismatches, in turn, cause changes in the light intensity measured by output photodiodes PD1 and PD2, which observe the horizontally and vertically polarized components of the laser beam (Fig. 3a). The retroreflector of both the measurement and reference arms consists of a cat's eye (lens and mirror) to minimize the effects of optical misalignments (Figs. 3b, c). As usual, a vacuum system encloses each strainmeter. As laser source, we use the frequency-stabilized He–Ne laser ML1, manufactured by MicroG Lacoste (MicroG Lacoste 2007).

Phase mismatch can be inverted for changes in length of the measurement arm, under the assumption that the length of the reference arm does not change over time. The reference arm includes a $\lambda/8$ retarder plate which makes PD1 and PD2 signals proportional to the sine and cosine of the phase mismatch, respectively (quadrature signals). If PD1 and PD2 were exactly in quadrature and shared the same amplitudes, the graph of PD1 versus PD2 would be a circle. In practice, the graph (Lissajous figure) is an ellipse whose axis lengths and orientations slightly change over time because of mechanical instabilities (Fig. 3d). A baseline-length change of half a wavelength of the laser light (corresponding to about 4.52 nε) produces a complete cycle along the ellipse. Amplitude of Earth

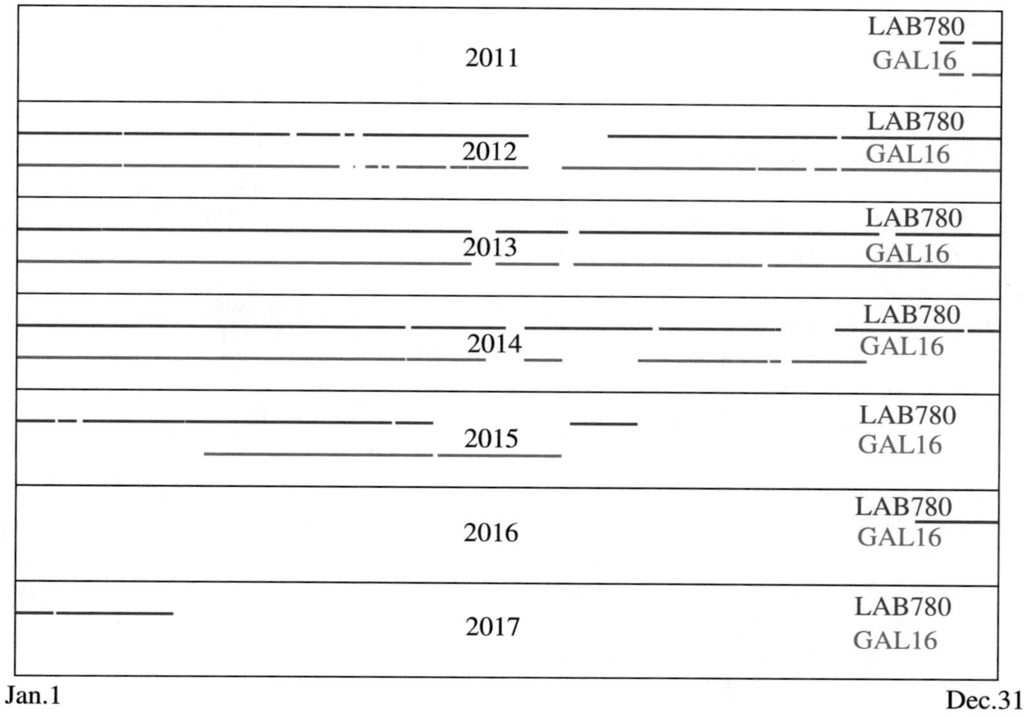

Figure 4
Recording periods of LAB780 (*blue lines*) and GAL16 (*red lines*)

tides is on the order of a couple of tens of nε, and thus, a complete cycle of the Lissajous figure usually takes a few hours, but may be faster during short-lasting phenomena, like oscillations caused by the transit of seismic waves. Since light intensity depends on phase mismatch nonlinearly, PD1 and PD2 signals cannot be low-pass filtered before sampling. To reduce the risk of aliasing, the sample rate is 600 Hz; analog-to-digital conversion is performed with a resolution of 16 bits using a National Instruments USB 6211 digital acquisition module. Every 3 h, the Lissajous figure is nonlinearly inverted for the best-fit ellipse. Then, each experimental point (PD1 and PD2 intensities) is projected on the best-fit ellipse and the phase mismatch retrieved. Because of the current sampling rate, strains faster than $\sim 10^{-6}\,\mathrm{s}^{-1}$ may cause errors in the retrieved sense of revolution (clockwise or counterclockwise) along the Lissajous figure or even missing a complete cycle. To reduce such a risk, the sampling rate can be easily increased up to 2000 Hz; however, this improvement requires using a bigger data storage and faster network connection than the current setup and

has not been necessary till now. Nominal sensitivity, i.e., the minimum strain required to guarantee a change in the digital output signal, is lower than 10^{-4} nε.

Analog-to-digital conversion is triggered by the internal clock of the USB 6211. After every 12,960,000 samples, the data file is closed and a new data file is created. We calibrate the clock of the USB 6211 by comparing the nominal duration of each file, 6 h, with that given by the software clock of the Linux system running on the acquisition PC. This in turn is synchronized with internet time servers through NTP (Network Time Protocol); its precision, as given by the Linux command "ntpq -c rl," results on the order of 1 μs. A simple linear regression between nominal file opening times and real ones during one month gives data timing which is correct within ~ 0.05 s if the linear drift of the USB 6211 clock is corrected for (as we actually do) and within few millisecond if also a quadratic part is removed.

We also monitor air and rock temperature, barometric pressure, pressure inside the vacuum pipe, the laser head temperature, and the null signal of the laser stabilization feedback circuit. These two last signals help checking the laser source for proper functioning.

3. Instrument Performances

Both strainmeters became operational in December 2011. Unfortunately, data acquisition suffered several, sometimes very long, interruptions, because of power outages, failures of the vacuum pumps and the acquisition PCs, and malfunctioning of the laser thermostatting circuits (Fig. 4).

Most problems probably arose from voltage spikes following power supply restoration and the huge amount of dust in bypass #16. Both strainmeters were temporarily shut down in summer 2015 for laser tube replacement, which has to be done by the manufacturer. Lab780 is again operational from November 29, 2016; GAL16 will be made operational as soon as its laser will come back. During that long interruption, we have improved the protection of the strainmeters from power outages and restorations, replaced vacuum pumps and PCs with more suitable models, and improved air and rock temperature monitoring. We have also changed the mechanical support of the retroreflector of the measurement arm, which is now mounted on a single stainless steel rail (Fig. 3c) to improve its mechanical stability.

In this section, we show instrumental performances at time scales ranging seconds to days, mainly focusing on data recorded by LAB780 after it restarted data acquisition on November 29, 2016.

A short (2 days) record, re-sampled at 1 Hz, is shown in Fig. 5a. As expected, the dominant signals are related to earthquakes and tides. The recorded straingram of the 2016/12/08 M7.8 Solomon Islands earthquake is clearly visible in Fig. 5a and, after removing tides, shown in Fig. 5b. A zoom of the first arrivals is also shown in the inset of Fig. 5b. For an example of general description of the event, log on to http://earthquake.usgs.gov/earthquakes/eventpage/us20007z80#executive. Because of the azimuthal position of the source with respect to LAB780, oscillations are mainly due to body and Rayleigh waves. The spectrum (Power Spectral Density, PSD) of the straingram from 17:45:36 to midnight is shown in Fig. 5c, to be compared with the

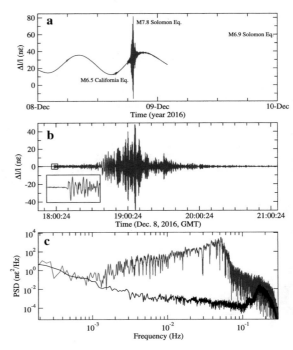

Figure 5
Example of strain signal recorded by LAB780. **a** Raw data re-sampled at 1 sample per second; **b** zoom of plot a during the transit of teleseismic waves produced by a M 7.8 earthquake in the Solomon Islands, after removal of Earth tides from strain; **c** power spectral density of strain recorded from Dec 8, 17:45:36 to 24:00:00 (*blue line*) and from Dec 2, 21:36:00 to Dec 3, 7:12:00 (*black line*)

reference spectrum of the straingram from Dec 2, 21:36:00, to Dec 3, 7:12:00 (black line in Fig. 5c). The reference time window has been selected among the quietest periods preceding the earthquake. The two PSDs are practically the same for frequencies lower than about 0.9 mHz, while the earthquake signal is clearly above noise in the frequency range of 0.9 mHz to around 0.15 Hz.

A longer (15 days) strain record is shown in Fig. 6a; sampling interval is now 1800 s. The solid black line represents tidal residuals, i.e., strain after removing diurnal to sixth-diurnal Earth tides (Amoruso et al. 2000; Amoruso and Crescentini 2016). Tidal residual time series include an almost linear trend and a few-day-long oscillations (solid black line in Fig. 6c, where all time series are shown after removing their linear trends for the sake of clarity). A similar behavior is shared by air temperature inside LAB780 (green line in Fig. 6b).

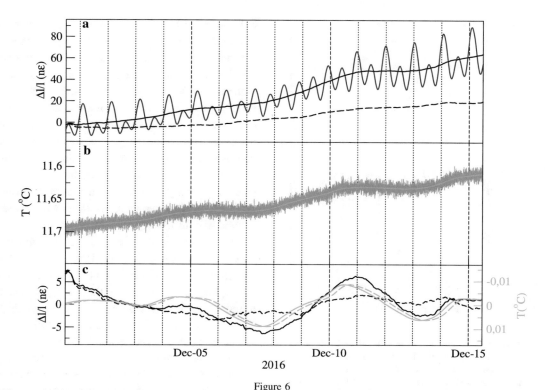

Figure 6

Fifteen-day strain signal recorded by LAB780. **a** Raw data at 1800-s sampling interval (*solid blue line*), tidal residuals (*solid black line*), and tidal residuals corrected for temperature effects (*dashed black line*); **b** LAB780 temperature sampled at 60 s (*solid green line*) and after smoothing by Tikhonov regularization (*solid orange line*); **c** *solid black line*, *solid orange line*, and *dashed black line* as (**a**) after subtracting the best-fitting *straight lines*, *dashed orange line* as *solid orange line* after shifting time by 5 h

Although temperature is very stable, smoothing temperature data evidence the high correlation between long-period tidal residuals and temperature (Fig. 6c). We use Tikhonov regularization (e.g., Wahba 1990) rather than just convolving data with a low-pass filter to avoid border effects in such a short record. Using a standard cross-correlation technique, we find that the maximum correlation is obtained by shifting temperature data by about 5 h (Fig. 6c). As a first attempt to correct strain data for temperature, we multiply the 5-h-shifted temperature by -526 and subtract the resultant time series from strain tidal residuals; the temperature-corrected time series is represented by the dashed black lines in Figs. 6a, c.

During a whole year, temperature cycles over 2 °C in GAL16 and 0.7 °C in LAB780. Diurnal oscillations are smaller than 0.05 °C in GAL16 and below the detection limit in LAB780. Temperature effects on measured strain are significant on time scales longer than several days, but are actually negligible on time scales of one day or shorter. As a consequence, strain amplitude at 1 cpd (cycle per day) is very small and in agreement with computations for S_1 tide (Amoruso and Crescentini 2016). The around-1-cpd amplitude spectrum of about 2 years of LAB780 strain is shown in Fig. 7. It is worth noting that Ψ_1 and Φ_1 tides are clearly visible above noise despite the short length of the time series. Amplitudes of P_1, S_1, K_1, Ψ_1, and Φ_1 from different tidal analyses of the same data are also shown.

Although in principle barometric pressure may also affect strain, mainly at 1 cpd and its multiples or during the transit of atmospheric fronts, till now we could not find any clear, significant effect on LSC strainmeter data.

The capability of investigating the Earth is limited by background noise. Figure 8 shows PSD of background noise down to 10^{-6} Hz; frequencies higher than 0.1 Hz are not shown, because of the large variability over time related to microseisms. For the sake

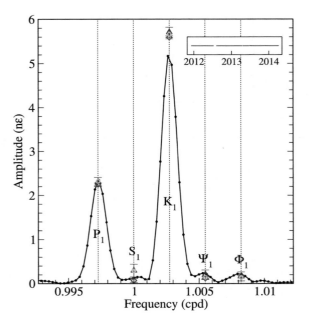

Figure 7

Amplitude spectrum of around 2-yr time series (Dec 2011–March 2014, see *inset*) from LAB780. No pressure or temperature correction was applied to data. Symbols give amplitudes of P_1, S_1, K_1, Ψ_1, and Φ_1 from different tidal analyses of the same data: *green circles*, our own code (Amoruso et al. 2000); *blue triangles*, ETERNA 3.4 (Wenzel 1996); *red diamonds*, VAV05 (Venedikov et al. 2005)

of comparison, Fig. 8 also shows PSDs published by others. Noise at Gran Sasso (GSBA and GSBC), Italy (Amoruso and Crescentini 2009b) and at Queensbury Tunnel (QT), UK (Beavan and Goulty 1977) is somewhat higher than at the other sites. At frequencies lower than about 4 mHz, noise at Poorman Mine (PM) and Piñon Flat Observatory (PFO), USA (Berger and Levine 1974), Kamioka Observatory (K), Japan (Takemoto et al. 2004), and LSC is similar and approximately inversely proportional to the square of frequency. LAB780 PSD seems even lower when temperature effects are subtracted from strain (cyan curve in Fig. 8), but longer records are required to confirm or reject this statement. At frequencies higher than about 4 mHz, LSC noise is approximately inversely proportional to frequency and becomes higher than PFO and PM ones, while K noise is by far the lowest. Since the strainmeter reference arm is much shorter than the measurement arm, strain $\Delta l/l$ in the measurement arm and fractional changes in wavelength (or frequency, $\Delta f/f$) of the laser light are numerically equivalent, i.e., $\Delta l/l = \Delta f/f$. If LSC noise

at frequencies higher than about 4 mHz is due to laser frequency fluctuations, their Allan variance is approximately independent of the time interval over which the laser frequency variation is measured (Rutman and Walls 1991). Computed Allan variance is about 3 kHz, i.e., an order of magnitude lower than the one given by the laser manufacturer when room temperature cycles over 0.4 °C, but this is not surprising, because of the much better temperature stability in GAL16 and LAB780. We cannot check if laser instability really causes the noise floor from 3 to 100 mHz because we would need a much stabler laser source, like an iodine-stabilized one.

An evident feature in Fig. 8 is the much lower noise at LSC than at Gran Sasso, despite the very similar instrumental setups. Gran Sasso strain data and atmospheric pressure inside the hosting tunnels were found to be highly correlated at frequencies higher than about 0.1 mHz (Crescentini et al. 1997). To avoid air entering from the nearby highway tunnel, the underground Gran Sasso laboratories are closed by means of two main doors and a ~ 50 Pa overpressure between inner and outer air is dynamically maintained. Consequently, inner air pressure is highly sensitive to traffic in the highway tunnel and to openings of the main doors thus causing a complex pressure pattern inside the laboratories, which is probably the origin of the noise excess evident in Gran Sasso spectra. Moreover, the endpoints of the measurement arms of both Gran Sasso interferometers were located close to areas where human activity was (and still is) present and strain noise clearly shows daily and weekly modulation. At lower frequencies, Gran Sasso strain is also highly correlated with the regional aquifer dynamics (Amoruso et al. 2014).

4. Conclusions

The features of the LSC strainmeters allow investigating the Earth in a very broad frequency range, with a signal-to-noise ratio not worst than the best laser strainmeters in the word. Unfortunately, we cannot use a single He Ne laser for both interferometers because of their mutual distance. Thus, also the difference between the two recorded extensions is affected by the laser wavelength fluctuations, which

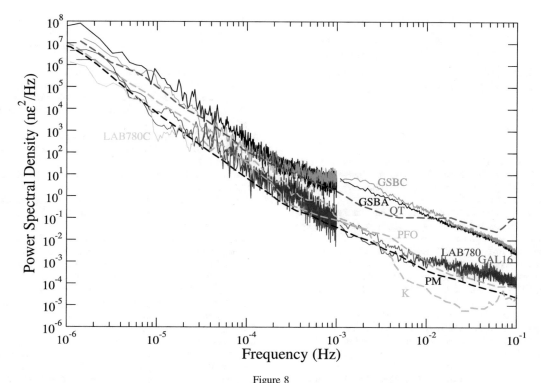

Figure 8

Power spectral density of strain noise from 0.001 to 100 mHz. *Solid green* and *black lines* (GSBC and GSBA), 45-day-long detided (i.e., after removal of Earth tides) strain records collected in Autumn 2007 at Gran Sasso, Italy (Amoruso and Crescentini 2009b); *dashed red line* (QT), Queensbury Tunnel, UK (Beavan and Goulty 1977); *dashed orange line* (PFO), Piñon Flat Observatory, USA (Berger and Levine 1974); *dashed black line* (PM), Poorman Mine, USA (Berger and Levine 1974); *dashed green line* (K), Kamioka Observatory, Japan (Takemoto et al. 2004); *solid red line*, GAL16 (24-day-long detided strain record collected in June 2013); *solid blue line*, LAB780 (20-day-long detided strain record collected in December 2016, *solid black line* in Fig. 6a); *solid cyan line* (LAB780C), LAB780 data corrected for temperature (*dashed black line* in Fig. 6a)

are probably the dominating source of noise at frequencies higher than about 4 mHz. The very low noise at the Kamioka observatory in the same frequency band is probably due to the superior stability of the laser they use. Also wire strainmeters can reach very low noise (e.g., Zürn et al. 2015), but their stability at lower frequencies is not as good.

As regards the diurnal tidal band, Ψ_1 and Φ_1 clearly emerge from noise even using a 2-year-long strain record. This feature is quite uncommon for strain measurements and may lead to improve previous determinations of the Free Core Nutation parameters from strain data (Amoruso et al. 2012).

At frequencies lower than 4 mHz, strain noise is very similar for quite different laser strainmeters, based on different mechanical and optical setups, located from ground surface to about 1 km depth at different distances from the sea, and using different laser

sources. This common feature seems related to some Earth process, but its origin is unclear. In case of gravimeters, the main cause of noise at frequencies lower than 2 mHz is density changes in the atmosphere (e.g., Tanimoto et al. 2015), but we have not yet found a clear correlation between barometric pressure and strain at LSC. Correcting LAB780 strain for temperature effects seems to lower noise below that of the other laser strainmeters at frequencies lower than about 1 cpd; however, the required accuracy of temperature recordings is available only after Nov 29, 2016, and this prevents us from drawing final conclusions.

Acknowledgments

We are grateful to Verdiana Botta for her help in the early stage of the LSC strainmeter installation and data processing. This is a contribution of the Geodyn project (MICINN ICTS2009-33). Figures were produced using the Grace plotting tool (http://plasma-gate.weizmann.ac.il/Grace/) and the Inkscape vector graphics editor (https://inkscape.org/).

REFERENCES

Agnew, D. C. (1981). Nonlinearity in rock: evidence from earth tides. *Journal of Geophysical Research, 86,* 3969–3978.

Agnew, D. C. (1986). Strainmeters and tiltmeters. *Reviews of Geophysics, 24,* 579–624.

Agnew, D.C., & Wyatt, F. K. (2003). Long-base laser strainmeters: A review, Scripps Institution of Oceanography Technical Report, Permalink. http://escholarship.org/uc/item/21z72167.

Amoruso, A., Botta, V., & Crescentini, L. (2012). Free Core Resonance parameters from strain data: sensitivity analysis and results from the Gran Sasso (Italy) extensometers. *Geophysical Journal International, 189,* 923–936. doi:10.1111/j.1365-246X.2012.05440.x.

Amoruso, A., & Crescentini, L. (2009a). Slow diffusive fault slip propagation following the 6 April 2009 L'Aquila earthquake, Italy. *Geophysical Research Letters, 36,* L24306. doi:10.1029/2009GL041503.

Amoruso, A., & Crescentini, L. (2009b). The geodetic laser interferometers at Gran Sasso, Italy: Recent modifications and correction for local effects. *Journal of Geodynamics, 48,* 120–125. doi:10.1016/j.jog.2009.09.025.

Amoruso, A., & Crescentini, L. (2010). Limits on earthquake nucleation and other pre-seismic phenomena from continuous strain in the near field of the 2009 L'Aquila earthquake. *Geophysical Research Letters, 37,* L10307. doi:10.1029/2010GL043308.

Amoruso, A., & Crescentini, L. (2016). Nonlinear and minor ocean tides in the Bay of Biscay from the strain tides observed by two geodetic laser strainmeters at Canfranc (Spain). *Journal of Geophysical Research: Oceans, 121,* 4873–4887. doi:10.1002/2016JC011733.

Amoruso, A., Crescentini, L., Martino, S., Petitta, M., & Tallini, M. (2014). Correlation between groundwater flow and deformation in the fractured carbonate Gran Sasso aquifer (INFN underground laboratories, central Italy). *Water Resources Research, 50,* 4858–4876. doi:10.1002/2013WR014491.

Amoruso, A., Crescentini, L., Morelli, A., & Scarpa, R. (2002). Slow rupture of an aseismic fault in a seismogenic region of Central Italy. *Geophysical Research Letters, 29*(24), 2219. doi:10.1029/2002GL016027.

Amoruso, A., Crescentini, L., Petitta, M., Rusi, S., & Tallini, M. (2011). Impact of the 6 April 2009 L'Aquila earthquake on groundwater flow in the Gran Sasso carbonate aquifer, Central Italy. *Hydrological Processes, 25,* 1754–1764. doi:10.1002/hyp.7933.

Amoruso, A., Crescentini, L., & Scarpa, R. (2000). Removing tidal and atmospheric effects from Earth deformation measurements. *Geophysical Journal International, 140,* 493–499.

Araya, A., Takamori, A., Morii, W., Hayakawa, H., Uchiyama, T., Ohashi, M., et al. (2010). Analyses of far-field coseismic crustal deformation observed by a new laser distance measurement system. *Geophysical Journal International, 181,* 127–140.

Beavan, J., & Goulty, N. R. (1977). Earth strain observations made with the Cambridge laser strainmeter. *Geophysical Journal of the Royal Astronomical Society, 48,* 293–305.

Ben-Zion, Y., & Allam, A. A. (2013). Seasonal thermoelastic strain and postseismic effects in Parkfield borehole dilatometers. *Earth Planet Science Letters, 379,* 120–126. doi:10.1016/j.epsl.2013.08.024.

Berger, J., & Levine, J. (1974). The spectrum of Earth strain from 10^{-8} to 10^2 Hz. *Journal of Geophysical Research, 79,* 1210–1214.

Bonaccorso, A., Linde, A., Currenti, G., Sacks, S., & Sicali, A. (2016). The borehole dilatometer network of Mount Etna: A powerful tool to detect and infer volcano dynamics. *Journal of Geophysical Research, 121,* 4655–4669. doi:10.1002/2016JB012914.

Borsa, A. A., Agnew, D. C., & Cayan, D. R. (2014). Ongoing drought-induced uplift in the western United States. *Science, 345*(6204), 1587–1590. doi:10.1126/science.1260279.

Canitano, A., Hsu, Y. J., Lee, H. M., Hsin-Ming, A. T. Linde, & Sacks, S. (2015). Near-field strain observations of the October 2013 Ruisui, Taiwan, earthquake: source parameters and limits of very short-term strain detection. *Earth Planet and Space, 67,* 1–15. doi:10.1186/s40623-015-0284-1.

Crescentini, L., Amoruso, A., Fiocco, G., & Visconti, G. (1997). Installation of a high-sensitivity laser strainmeter in a tunnel in central Italy. *Review of Scientific Instruments, 68*(8), 3206–3210.

Crescentini, L., Amoruso, A., & Scarpa, R. (1999). Constraints on slow earthquake dynamics from a swarm in Central Italy. *Science, 286,* 2132–2134.

Díaz, J., Ruíz, M., Crescentini, L., Amoruso, A., & Gallart, J. (2014). Seismic monitoring of an Alpine mountain river. *Journal of Geophysical Research: Solid Earth, 119,* 3276–3289. doi:10.1002/2014JB010955.

Gao, S. S., Silver, P. G., & Linde, A. T. (2000). Analysis of deformation data at Parkfield, California: Detection of a long-term strain transient. *Journal of Geophysical Research, 105*(B2), 2955–2967. doi:10.1029/1999JB900383.

Gladwin, M. T. (1984). High precision multi component borehole deformation monitoring. *Review of Scientific Instruments, 55,* 2011–2016.

Goulty, N. R., King, G. C. P., & Wallard, A. J. (1974). Iodine stabilized laser strainmeter. *Geophysical Journal of the Royal Astronomical Society, 39,* 269–282.

Johnston, M. J. S., Borcherdt, R. D., Linde, A. T., & Gladwin, M. T. (2006). Continuous borehole strain and pore pressure in the near field of the 28 September 2004 M 6.0 Parkfield, California, earthquake: Implications for nucleation, fault response, earthquake prediction, and tremor. *Bulletin of the Seismological Society of America, 96,* S56–S72.

Kobe, M., Jahr, T., Pöschel, W., & Kukowski, N. (2016). Comparing a new laser strainmeter array with an adjacent, parallel running quartz tube strainmeter array. *Review of Scientific Instruments, 87,* 034502. doi:10.1063/1.4942433.

Levine, J., & Hall, J. L. (1972). Design and operation of a methane absorption stabilized laser strainmeter. *Journal of Geophysical Research, 77,* 2595–2609.

Linde, A. T., Agustsson, K., Sacks, I. S., & Stefansson, R. (1993). Mechanism of the 1991 eruption of Hekla from continuous borehole strain monitoring. *Nature, 365,* 737–740. doi:10.1038/365737a0.

Linde, A. T., Gladwin, M. T., Johnston, M. J. S., Gwyther, R. L., & Bilham, R. G. (1996). A slow earthquake sequence on the San Andreas Fault. *Nature, 383,* 65–68. doi:10.1038/383065a0.

MicroG Lacoste (2007). ML-1 Polarization-Stabilized HeNe Laser, brochure. http://www.microglacoste.com/pdf/ml1-brochure.pdf. Accessed 8 March 2017.

Milyukov, V. K., Klyachko, B. S., Myasnikov, A. V., Striganov, P. S., Yanin, A. F., & Vlasov, A. N. (2005). A laser interferometer-deformograph for monitoring the crust movement. *Instruments and Experimental Techniques, 48*(6), 780–795.

Milyukov, V., Mironov, A., Kravchuk, V., Amoruso, A., & Crescentini, L. (2013). Global deformations of the Eurasian plate and variations of the Earth rotation rate. *Journal of Geodynamics, 67,* 97–105.

Mukai, A., Takemoto, S., & Yamamoto, T. (2004). Fluid core resonance revealed from a laser extensometer at the Rokko-Takao station, Kobe, Japan. *Geophysical Journal International, 156,* 22–28.

Park, J., Amoruso, A., Crescentini, L., & Boschi, E. (2008). Long-period toroidal earth free oscillations from the great Sumatra–Andaman earthquake observed by paired laser extensometers in Gran Sasso, Italy. *Geophysical Journal International, 173,* 887–905. doi:10.1111/j.1365-246X.2008.03769.x.

Richter, B., Wenzel, H.-G., Zürn, W., & Klopping, F. (1995). From Chandler wobble to free oscillations: comparison of cryogenic gravimeters and other instruments in a wide period range. *Physics of the Earth and Planetary Interiors, 91,* 131–148.

Rutman, J., & Walls, F. L. (1991). Characterization of frequency stability in precision frequency sources. *Proceedings of the IEEE, 79,* 952–960.

Sacks, I. S., Suyehiro, S., Evertson, D. W., & Yamagishi, Y. (1971). Sacks-Evertson strainmeter, its installation in Japan and some preliminary results concerning strain steps. *Papers in Meteorology and Geophysics, 22,* 195–207.

Takanami, T., Linde, A. T., Sacks, S. I., Kitagawa, G., & Peng, H. (2013). Modeling of the post-seismic slip of the 2003 Tokachi-oki earthquake M 8 off Hokkaido: Constraints from volumetric strain. *Earth Planet and Space, 65*(731), 731–738. doi:10.5047/eps.2012.12.003.

Takemoto, S., Araya, A., Akamatsu, J., Morii, W., Momose, H., Ohashi, M., et al. (2004). A 100 m laser strainmeter system installed in a 1 km deep tunnel at Kamioka, Gifu, Japan. *Journal of Geodynamics, 38,* 477–488.

Tanimoto, T., Heki, H., & Artru-Lambin, J. (2015). Interaction of solid earth, atmosphere, and ionosphere. In G. Schubert (Ed.), *Treatise on geophysics* (2nd ed., Vol. 4, pp. 421–443). Oxford: Elsevier.

Venedikov, A. P., Arnoso, J., & Vieira, R. (2005). New version of program VAV for tidal data processing. *Computers & Geosciences, 31,* 667–669. doi:10.1016/j.cageo.2004.12.001.

Wahba, G. (1990). *Spline models for observational data.* Philadelphia: SIAM.

Wenzel, H. G. (1996). The nanogal software: Earth tide data processing package ETERNA 3.30. *Bulletin d'Information des Marées Terrestres, 124,* 9425–9439.

Zürn, W., Ferreira, A. M. G., Widmer-Schnidrig, R., Lentas, K., Rivera, L., & Clévédé, E. (2015). High-quality lowest-frequency normal mode strain observations at the Black Forest Observatory (SW-Germany) and comparison with horizontal broad-band seismometer data and synthetics. *Geophysical Journal International, 203,* 1787–1803.

(Received December 22, 2016, revised April 7, 2017, accepted April 18, 2017, Published online April 24, 2017)

Pure Appl. Geophys. 175 (2018), 1739–1753
© 2017 Springer International Publishing AG
https://doi.org/10.1007/s00024-017-1651-6

Pure and Applied Geophysics

CrossMark

Non-Tidal Ocean Loading Correction for the Argentinean-German Geodetic Observatory Using an Empirical Model of Storm Surge for the Río de la Plata

F. A. Oreiro,[1,2] H. Wziontek,[3] M. M. E. Fiore,[1,2] E. E. D'Onofrio,[1,2] and C. Brunini[4,5]

Abstract—The Argentinean-German Geodetic Observatory is located 13 km from the Río de la Plata, in an area that is frequently affected by storm surges that can vary the level of the river over ±3 m. Water-level information from seven tide gauge stations located in the Río de la Plata are used to calculate every hour an empirical model of water heights (tidal + non-tidal component) and an empirical model of storm surge (non-tidal component) for the period 01/2016–12/2016. Using the SPOTL software, the gravimetric response of the models and the tidal response are calculated, obtaining that for the observatory location, the range of the tidal component (3.6 nm/s^2) is only 12% of the range of the non-tidal component (29.4 nm/s^2). The gravimetric response of the storm surge model is subtracted from the superconducting gravimeter observations, after applying the traditional corrections, and a reduction of 7% of the RMS is obtained. The wavelet transform is applied to the same series, before and after the non-tidal correction, and a clear decrease in the spectral energy in the periods between 2 and 12 days is identify between the series. Using the same software East, North and Up displacements are calculated, and a range of 3, 2, and 11 mm is obtained, respectively. The residuals obtained after applying the non-tidal correction allow to clearly identify the influence of rain events in the superconducting gravimeter observations, indicating the need of the analysis of this, and others, hydrological and geophysical effects.

Key words: Non-tidal ocean loading, superconducting gravimeter, storm surge model, Argentinean-German Geodetic Observatory, Río de la Plata.

1. Introduction

The Argentinean-German Geodetic Observatory (AGGO) is a joint initiative between the Consejo Nacional de Investigaciones Científicas y Técnicas (CONICET) from Argentina and the Bundesamt für Kartographie und Geodäsie (BKG) from Germany. The observatory is strategically situated in the Southern Hemisphere, which makes it a key piece in the Global Geodetic Observing System (GGOS), and has several geodetic techniques: GNSS, VLBI, SLR, absolute gravity, and Superconducting Gravity (SG). Since AGGO is located near the Rio de la Plata, the water height variations that occur in the river affect the observatory. This variation can be considered through the ocean loading, which is a geophysical effect that affects the observations of all the mentioned geodetic techniques, and can be decomposed in a tidal and non-tidal component (Petrov 2015). The high sensitivity of the SG is a very useful tool for testing and validating correction models related to this effect (Virtanen 2004), and the gravity observations of the SG can be used to test corrections for other co-located geodetic observations such as GNSS, VLBI, and SLR (Virtanen and Mäkinen 2003). The superconducting gravimeter SG038 was installed in AGGO on December 16th, 2015 and has made uninterrupted gravity observations since then (Wziontek et al. 2016; Tocho 2016). The traditional processing of SG data usually considers the atmospheric mass changes, the Earth's tides, the pole tide, and the ocean tide loading. Then, the residuals are analyzed to obtain information about other geophysical signals, such as water storage changes or non-tidal ocean loading. In recent years, there has been increased interest on the effects of the non-tidal variation in ocean loading, including ocean circulation and special

[1] Facultad de Ingeniería, Instituto de Geodesia y Geofísica Aplicadas, Universidad de Buenos Aires, Las Heras 2214, 3rd Floor, Buenos Aires, Argentina. E-mail: foreiro@fi.uba.ar; fernandooreiro@yahoo.com.ar

[2] Servicio de Hidrografía Naval, Ministerio de Defensa, Buenos Aires, Argentina.

[3] Bundesamt für Kartographie und Geodäsie (BKG), Leipzig, Germany.

[4] Argentinean-German Geodetic Observatory, CONICET, La Plata, Argentina.

[5] Facultad de Ciencias Astronómicas y Geofísicas, Universidad Nacional de La Plata, La Plata, Argentina.

occasions such as storm surges (Nordman et al. 2015). In this paper, we analyze the incidence of the non-tidal ocean loading at AGGO, which is caused by seafloor pressure variations that relate directly to the water response to atmospheric pressure and wind stress (Geng et al. 2012) and by direct mass effects. This loading can affect the gravity observations in the microgal range, and several mm in the vertical and horizontal displacements. Fratepietro et al. (2006) showed that a storm surge of 2 m in the southern North Sea produces vertical displacements up to -30 mm and increases of gravity up to 80 nm/s^2 in the coastal areas. Geng et al. (2012) studied the loading effects caused by a strong storm surge in the southern North Sea on 2007, and estimated the loading displacements at coastal stations on the order of 40 mm in the vertical direction and over 5 mm in the horizontal direction. Nordman et al. (2015) analyzed the influence of non-tidal loading of the Baltic Sea using GNSS coordinates, and showed that the loading effects should be considered in geodetic measurements, especially near the coast. Virtanen and Mäkinen (2003) calculated that a uniform layer of water covering the whole Baltic Sea increases the gravity in Metsähovi geodetic observatory by 31 nm/s^2 per 1 m of water and the vertical deformation by -11 mm. AGGO is located 13 km from the Río de la Plata, in an area that is frequently affected by storm surges that can vary the level of the river over ± 3 m.

The Río de la Plata is one of the biggest estuaries of the world (Fig. 1) with approximately 35,000 km^2, a water depth between 5 and 15 m (Guerrero et al. 1997), and a width variation from 2 to 220 km. According to Comisión Administradora del Río de la Plata (1989), D'Onofrio et al. (1999) and Dragani and Romero (2004), three regions can be identify in the river: the inner region, the middle region, and the exterior region (Fig. 1).

The hydrological regime of the Río de la Plata is highly influenced by the tidal wave progressing from the Atlantic Ocean, and the circulation is sensitive to the complicated geometry and bathymetry of the estuary (Simionato et al. 2004a). The astronomical tide in the Río de la Plata is mixed mainly semidiurnal type (SHN 2017), and the range is about 1.44 m at the mouth, but in the interior areas, it can reach 0.40 m (D'Onofrio et al. 2009). The main semidiurnal constituent (M2) takes

about 12 h to reach the interior limit, so a full cycle of the M2 component is present at every moment inside the Río de la Plata (D'Onofrio et al. 2009, 2010; Simionato et al. 2004b). The astronomical tide is affected by meteorological forcing and by seasonal variations of the river. Meteorological forcing produces extreme high tides (positive storm surge) or extreme low tides (negative storm surge) according to the direction and intensity of the wind. Positive and negative storm surges are defined by the difference between the observed levels and the corresponding predicted tide, and in the region, they have been studied intensively by several authors (e.g., Balay 1961; D'Onofrio and Fiore 2003; D'Onofrio et al. 2008; Escobar et al. 2004; Campetella et al. 2007; Fiore et al. 2009, Pousa et al. 2013) whom indicate that positive and negative storm surges are more harmful if they coincide with a high-water period or a low one, respectively. The inner and middle regions of the Río de la Plata usually register higher storm surges values than the exterior region due to the funnel like shape of the river. The highest positive surge measured at Buenos Aires city occurred on April 15, 1940 when the water level up to 4.44 m over tidal datum, and the surge was 3.24 m (D'Onofrio et al. 2008). The water height value is the maximum level recorded since the beginning of systematic tidal measurements in 1905. On the contrary, the lowest observed level occurred on 29 May 1984 (-3.66 m), and on that occasion, the surge was -4.61 m. In opposition to the astronomical tide, that allow identifying regions of high tide and low tide simultaneously in the river, the storm surge can generate an increase in water level over the whole river at the same time.

In this study, we compare the gravimetric response of the non-tidal ocean loading model from the Río de la Plata with the residuals of the SG at AGGO, for the period 01/2016–12/2016, and show the improvement in terms of reduction of RMS of gravity residuals when subtracting the non-tidal gravimetric response from the river. We also compare tidal vs non-tidal components to identify which component is more influent in the Río de la Plata. With the purpose of obtaining the gravimetric response of the river, water-level information from seven tide gauge stations located in or nearby the Río de la Plata are used to calculate an empirical model of water heights (tidal + non-tidal component) and an

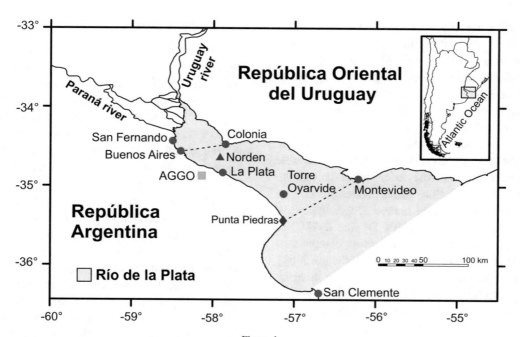

Figure 1
Geographical location of Rio de la Plata estuary, and delimitation of the inner, middle, and outer regions. Tide gauge stations used for the empirical models are represented by *dots*, AGGO location is shown with a *square*, and Norden station, used for the validation of the models, is marked with a *triangle*

empirical model of storm surge (non-tidal component). Then, using the SPOTL software (Agnew 2012), the gravimetric response of each model is calculated for the AGGO location, and compared with the gravimeter residuals after applying corrections for Earth tide, atmospheric effects, pole tide, and ocean tide loading. Using the same software, the vertical and horizontal displacements are also calculated for the indicated period. Although several authors have shown the relation between the storm surges and the SG data (Fratepietro et al. 2006; Boy et al. 2009; Virtanen and Mäkinen 2003), the aim of this work is to quantify the influence of this geophysical signal for the newly installed SG038 and for the co-located instruments at AGGO.

2. Data

2.1. Superconducting Gravimeter Data

The superconducting gravimeter (SG) is the most sensitive and stable spring-type relative gravimeter, where the mechanical spring is replaced by a magnetic levitation of a superconducting hollow sphere in the field of superconducting, persistent current coils (Goodkind 1999; Hinderer et al. 2007). The SG shows only a low and almost linear instrumental drift, providing unequaled long-term instrumental stability, and enables the highest sensitivity. The levitated sphere as the basic element of the sensor is moved from its initial position by gravity changes or inertial accelerations. The displacement is detected by a capacitance bridge and compensated by the feedback force generated by an additional superconducting coil. The required voltage represents the gravity signal and is recorded by a high precision digital voltmeter.

The time series acquired by SG038 at AGGO with 1-s temporal resolution was filtered and reduced to a sample rate of 10 s to eliminate microseismic background noise. The signal was calibrated using a scale factor of -736.5 nm/s^2/V as determined from comparisons with numerous absolute gravimeter measurements at the previous location of the instrument in Concepcion, Chile. A time lag of the whole system including the analogue low-pass filter of −8.3 s was applied as determined from step response experiments (Antokoletz et al. 2017). During preprocessing, disturbances and one step due to a planned power interruption on November 1, 2016 were corrected after removal of a first tidal model (Antokoletz et al. 2017) and atmospheric effects using the local air pressure record and a nominal admittance factor of 3.0 nm/s^2/hPa. Both effects were restored afterwards and the series was further down-sampled to 1 h resolution. Atmospheric effects were now corrected based on the numerical weather model ICON of the Germany Weather Service (DWD) as provided by the atmospheric attraction computation service Atmacs of BKG (Klügel and Wziontek 2009). The local air pressure record was used to improve the temporal resolution of this correction. By this, atmospheric effects are eliminated more efficiently and independently from the measured gravity time series, especially during extreme weather events. This is important as the gravity effects under consideration are related to such situations.

An extensive tidal analysis utilizing the updated and enhanced versions of ETERNA V60 (Schüller 2015), partitioning 49 wave groups and including a hypothesis-free modeling of degree three constituents of the tidal potential, was performed. In this way, the effects of solid Earth tides and ocean tide loading starting from diurnal tides were removed, leaving almost no energy in the tidal spectrum. The fortnightly and monthly tides are with amplitudes of less than 1.5 nm/s^2 extremely low at the latitude of the station and cannot be analyzed from a 1-year record. Instead, the tidal model was completed based on the non-hydrostatic model of Dehant et al. (1999), including annual and semiannual components. The gravity effects caused by the variable Earth rotation and changes of its rotational axis (pole tide and length-of-day variations) were corrected based on the EOP C04 series of IERS. Since the instrumental drift of SG038 is not determined so far by absolute gravity observations, an overall linear trend of 184 nm/s^2/a was removed from the residual time series, which includes possible long-term gravity changes. The remaining fluctuations cover a range of 80 nm/s^2.

2.2. Sea-Level Data

Digital sea-level records from December 2015 to December 2016 at five tidal stations belonging to República Argentina and two to República Oriental del Uruguay were used in this work to obtain the empirical models of water level and storm surge (Fig. 1). Table 1 shows the location of the stations and the sampling interval of the measurements. The Argentinian stations are controlled and operated by the Argentinian Servicio de Hidrografía Naval and the tide gauges Colonia and Montevideo, located in Uruguay, are operated by Administración Nacional de Puertos de Uruguay and by Comisión Administradora del Río de la Plata (CARP), respectively. Besides the mentioned tide gauges, CARP's Norden station is used to validate the empirical models (Table 1; Fig. 1). All sea-level data were processed following D'Onofrio (1984), and since the needed sampling interval of the tide gauges data for the empirical models is 5 min, the series that have a bigger sampling interval are interpolated using a cubic spline interpolation. All observed heights are referred to the mean sea level (MSL) using the available information from the tidal datum of each tide gauge.

Storm surge models were developed using storm surge series for each location. These series were derived from the sea-level data by subtracting the astronomical tide prediction to the observations. The amplitudes and epochs used for each prediction were

Table 1

Location and sampling interval of the water-level gauges used for the empirical models (*) and for the validation of the models

Location	Latitude	Longitude	Sampling interval (min)
San Fernando*	−34°26′00″	−58°30′00″	60
Buenos Aires*	−34°33′45″	−58°24′00″	60
La Plata*	−34°50′00″	−57°53′00″	60
Torre Oyarvide*	−35°06′00″	−57°08′00″	60
San Clemente*	−36°21′30″	−56°42′03″	60
Colonia*	−34°28′30″	−57°51′00″	6
Montevideo*	−34°54′30″	−56°13′03″	5
Norden	−34°37′55″	−57°55′40″	1

The asterisks indicate locations used in the empirical models provided by the Argentinean Servicio de Hidrografía Naval.

2.3. Satellite Altimetry data

Corrected Sea Surface Heights (CorSSH) from Archiving, Validation, and Interpretation of Satellite Oceanographic data (AVISO) (http://www.aviso.oceanobs.com) are used for the period 12/2015–11/2016. For this period, 1319 CorSSH observations registered with Jason-2 (J2) altimeter were obtained for the Río de la Plata (Fig. 3). Due to the launch of Jason-3 altimeter, J2 modified its orbit in October 2016 to an intermediate orbit, obtaining CorSSH of the original orbit until October 2016, and CorSSH of the intermediate orbit from October 2016 to November 2016. This modification favors the validation of the models, since it allows to obtain a better spatial distribution of the data. Following Oreiro et al. (2016), and to make comparable the CorSSH data with the empirical models, the ocean tide (OT) and the dynamic atmospheric correction (DAC), provided in the CorSSH files, were restored to the CorSSH data without modifying the rest of the corrections. Since global ocean tide models do not have good performance in the Río de la Plata, for the comparison of the storm surge models, astronomical tide obtained from the SEAT model (D'Onofrio et al. 2012) was subtracted from the CorSSH with the restored OT and DAC corrections. In both cases, the CorSSH were referred to the mean sea level, which is the datum of the empirical models.

3. Methodology

3.1. Calculation of the Gravimetric Response at AGGO

For the ocean loading calculation of the Río de la Plata, the "nloadf" package of the SPOTL software (Agnew 2012) is used, since it has been developed for this kind of loading calculation. This package allows obtaining the gravity response of a regular grid for a given position, where the height of the water must be stored for each cell of the grid. As the variation of the water height analyzed in this paper is not only

harmonic, this calculation must be performed for each time step for the entire period, resulting in a time series of gravimetric response of the Rio de la Plata at AGGO location. A regular grid of $0.005° \times 0.005°$ (approximately 500 m $\times$ 500 m) that covers the entire Río de la Plata is generated between the coordinates 54.925°W–58.5°W and 33.8°S–36.38°S. The cells are classified in "wet" considering whether they are covered by water or "dry" otherwise. The grid has 370,172 cells (716 columns $\times$ 517 rows), where 125,304 cells correspond to "wet" and 244,868 cells correspond to "dry". Each "dry" cell is assigned to zero in the grid, while for each "wet" cell, the water height (or storm surge) is calculated every hour, for the entire period. The Gutenberg-Bullen Earth model A was used for the convolution through the selection of the Green's functions included in the "nloadf" package that were computed from Farrell (1972). The density of the water used is 1000 kg m^{-3}, and the origin of the coordinate system for the computation is assumed to coincide with the center of mass of the solid Earth.

3.2. Calculation of the Water and Storm Surge Heights

From the observed water level in the Río de la Plata, two empirical models of the water height of the river and two empirical models of storm surge are developed. One of each kind of the models uses information from seven tide gauge stations inside or nearby the river (Fig. 1), and the other two models use exclusively the observations of the station closest to the observatory (La Plata). These last models are developed to compare the gravimetric responses, and evaluate the differences between using all available water height information and the observations from the closest station to the observatory.

The propagation of the observed water level and the storm surge through the river is carried out considering the advance of the M2 tidal constituent provided by the cotidal chart calculated by D'Onofrio et al. (2012). Figure 2 shows the agreement of the high water and low water between the seven stations that provide information, by adding the time lag according to the information of the mentioned cotidal

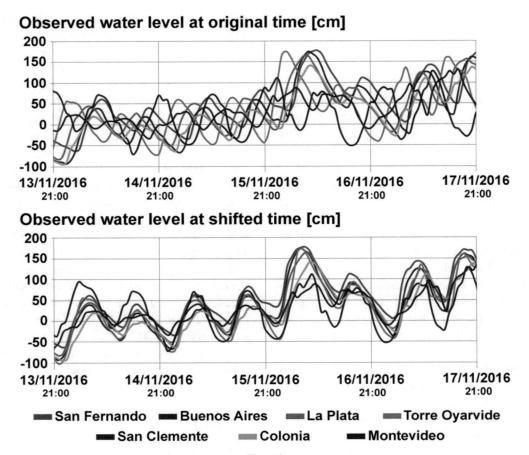

Observed water level at original time [cm]

Observed water level at shifted time [cm]

■■ San Fernando   ■■ Buenos Aires   ■■ La Plata   ■■ Torre Oyarvide
■■ San Clemente   ■■ Colonia   ■■ Montevideo

Figure 2
Agreement of high tides and low tides between the seven tide gauges used for the development of the empirical models. The *upper figure* shows the water levels at their original time, and the *lower figure* shows the water levels with the shifted time, according to the M2 cotidal chart and the speed of the constituent

chart, and the speed of the M2 constituent (28.9841042°/h).

To determine the height of the water (or the storm surge) considering the advance of the tide in the river, the M2 cotidal chart is use to classify the "wet" cells of the grid. The cotidal chart indicates the regions, where the epoch of the tide (high tide, low tide, etc.) occurs at the same time. A 5-min time classification is applied to the cells, determining regions, where the epoch of the tide occurs at the same time. Figure 3 shows this classification, allowing to easily identify the curvature of the M2 cotidal chart. The cells which are located in the same region as La Plata station are classified as 0 min, defining the origin or time reference. Then, the cells of the regions located

towards the outer edge of the river are classified by increasing (+) the time offset in 5-min intervals, while the cells of the regions located towards the inner edge of the river are classified by reducing (−) the time offset by the same amount. With this procedure, all cells are classified according to the time lag to La Plata tide gauge station, and the value assign to a cell indicates the time needed to obtain the tide epoch of that cell in La Plata station.

3.3. Empirical Models Using Observations from the Closest Tide Gauge Station to AGGO

Using the classification of the time lag of the cells to La Plata station, an empirical water height model

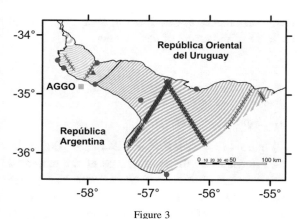

Figure 3
CorSSH data obtained in the Río de la Plata for the period 12/2015–11/2016 are shown with *crosses*. The AGGO location is marked with a *square*, Norden station with a triangle and the seven tide gauges used for the empirical models are marked with *circles*. The *black line* crossing the river is indicating the region of 0-min time frame, corresponding to the classification of the cells. From the *black line*, towards the outer edge of the river, the *lines* indicate the regions that increase in time steps of 5-min intervals, and towards the inner part of the river, the *lines* indicate the regions decreasing in time of 5-min intervals

(EWH_LaPlata) and an empirical storm surge model (ESS_LaPLata) are developed, using only observations from this station. For this, all cells classified as zero time lag at a particular time step are assigned to the corresponding water height (storm surge) at station La Plata. The remaining cells are assigned to the height (storm surge) observed at station La Plata at the time corresponding to the time step plus the time lag obtained from the classification described above. Although this model is an extremely simplified approximation of the water height (storm surge), since it does not consider the variation of the meteorological conditions in the whole river nor the difference of height between the shores, among other things, its calculation is straightforward as it uses information from a single tide gauge station only. In addition, since La Plata station is closest to AGGO, the observations of this station are of great importance to describe the water height of the cells closest to the observatory.

3.4. Empirical Models Using Observations from 7 Tide Gauge Stations

To obtain a more precise description of the water height and storm surge in the Río de la Plata, an empirical water height model (EWH_7TG) and an empirical storm surge model (ESS_7TG) are developed using information from six tide gauge stations around or within the Río de la Plata, and a station located near the outer edge of the river (Fig. 1). To obtain the height of the water (storm surge) of a cell at a particular time, different calculations are developed according to their relative location related to the tide gauges. In regions where the cells coincide with the location of a tide gauge station, the heights (storm surge) of these cells are taken over from the corresponding station. For the rest of the cells, the water height (storm surge) is calculated for respective time, taking the number of neighbor stations into account: if the cell is located between two stations or if the cell is located between a station and a border of the river. In the latter case, the height of the water (storm surge) is assigned in the same way described before, shifting the observed values at the nearest station in time. The water height of the cells between two stations is obtained using the information of the stations involved, following the procedure described below:

1. The values from the stations closest to the analyzed cell are repositioned considering the displacement of the M2 constituent, represented by its cotidal chart, to approximately match the time of occurrence of the high water and the low water.
2. A linear interpolation of the time-shifted series of the stations is performed, to obtain an interpolated series for each region between the stations. For example, if there are nine regions between the stations involved, then nine intermediate series must be obtained.
3. Finally, the intermediate series are time-shifted according to the time lag between each region and the stations, to use an interpolated series in each region.

Although this methodology allows to model the progress of the high and low tides along the river, the calculation of the time offset and its subsequent interpolation is done separately among the Argentinian stations (San Clemente, Torre Oyarvide, La Plata, Buenos Aires, and San Fernando), and the stations at the Uruguayan shore line (Montevideo,

Colonia) as the height of the water (storm surge) is not the same at both margins. In this way, a model is obtained for each of the shorelines of the river. Then, the water height of the cells in the same region is calculated through a linear interpolation, using the calculated values from both shorelines, taking into account the distance to each side.

3.5. Validation of the Empirical Models

The empirical models developed in this work aim to obtain the gravimetric response generated by tidal and non-tidal ocean loadings and by non-tidal ocean loading only, through the modeling of water height and storm surge throughout the Rio de la Plata. Although the models generated are not intended to describe the water levels of the river in detail, the heights that they provide are expected to be accurate enough to obtain an adequate gravimetric response. To validate the models and to obtain a statistical error of the calculated heights of the water, CorSSH from J2 altimeter and observations of the tide gauge Norden, which was not used for the development of empirical models, are used (Fig. 3). The water heights (storm surge) of the empirical models are compared with the water heights (storm surge) observed in Norden and the CorSSH, and the standard deviation of the differences is calculated to obtain a statistical estimate of the error of each model.

By the above-described empirical models, the gravimetric response generated by tidal and non-tidal ocean loadings, or by non-tidal ocean loading only, should be obtained from the water height and storm surge distribution throughout the Rio de la Plata. Although the created models are not intended to describe the water level of the river in detail, the heights that they provide are expected to be accurate enough to infer the gravimetric response precisely enough. To validate the models and to obtain a statistical uncertainty estimate for the calculated grids, CorSSH values from J2 altimeter and observations from the tide gauge Norden, which were not used for the development of empirical models, are used (Fig. 3). The water heights (storm surge) of the empirical models are compared with the observations at Norden and with the CorSSH values, and the RMS

of the differences is calculated to obtain a statistical estimate of the errors of each model.

4. Results

4.1. Empirical Models

Figure 4 shows for February 24th, 2016 00:00, the water heights and the storm surge of the empirical models using observations from seven tide gauge stations and from La Plata station, allowing to identify some differences between the models. For example, in the models obtained using La Plata data only, regions of cells with the same time interval have the same height of the water, or storm surge, throughout the entire region. On the other hand, in the models that use information of the seven tide gauges, the height of the water or the storm surge changes within the same region of cells with the same time interval. In addition, the empirical storm surge model obtained from La Plata observations shows negative values that do not agree with the storm surge from the tide gauges of the exterior region of the river.

4.2. Validation of the Empirical Models

Figure 5 shows the differences between CorSSH data and the empirical water height models EWH_7TG and EWH_LaPlata and the differences between the CorSSH storm surge and the empirical storm surge models ESS_7TG and ESS_LaPlata. The biggest differences visible in Fig. 5 correspond to the models based on the observations at the La Plata station only. These discrepancies can also be characterized in the standard deviation of the differences of each model, where for the water height models, the standard deviation is 13.3 and 21.9 cm for EWH_7TG and EWH_LaPlata, respectively, and for the storm surge models, the standard deviation of the differences is 13.1 and 24.4 cm for ESS_7TG and ESS_LaPlata, respectively.

The standard deviation of the differences of the models using Norden station data is 4.4 and 7.4 cm for the water height models EWH_7TG and EWH_-LaPlata, respectively, and 5.9 and 12.2 cm for the storm surge models ESS_7TG and ESS_LaPlata,

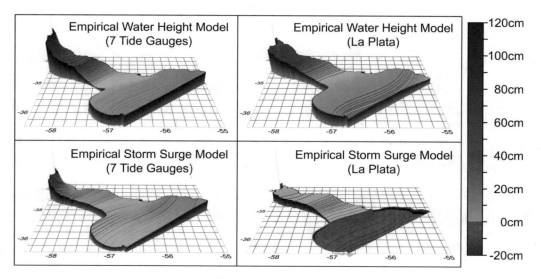

Figure 4
Empirical water height and storm surge models for the Río de la Plata for February 24th, 2016 00:00

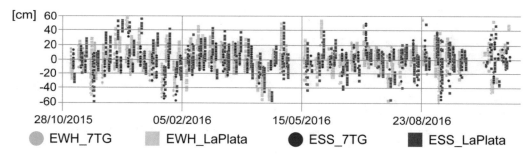

Figure 5
Differences between CorSSH data and CorSSH storm surge, and empirical water height and storm surge models

respectively. The standard deviation found in this comparison is considerably smaller than the ones obtained in the comparison with CorSSH data, probably because Norden station is close to the stations used to generate the empirical models in that region.

The standard deviation of the differences found in the water height and the storm surge models is similar to the water height differences between model vs observations found by other authors, whom made similar comparisons with SG data (Boy et al. 2009; Geng et al. 2012). There is no doubt that the performance of the empirical models can be improved, especially in the middle and exterior regions; however, for the purpose of this work, the

standard deviation of the differences found is considered acceptable for the latter calculation.

4.3. Gravimetric Response of the Empirical Models

Figure 6 shows the gravimetric response of the water height and storm surge empirical models calculated using SPOTL, and Fig. 7 shows the four response series overlapped in a shorter period, where the differences between the models can be better distinguished. All four models have similar gravimetric response in a range from -10 to 20 nm/s^2. The minimum and maximum differences in the gravimetric response between water height models are -3.7 and 6.4 nm/s^2 and between the storm surge

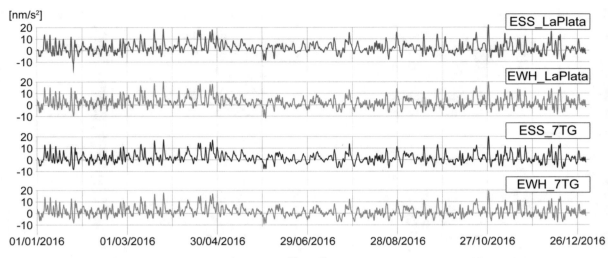

Figure 6
Gravimetric responses of the water height and storm surge empirical models calculated using SPOTL

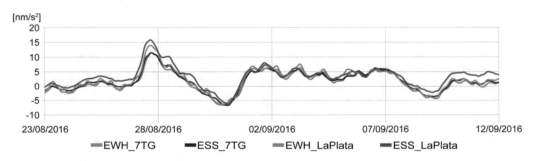

Figure 7
Detail of gravimetric responses of the water height and storm surge empirical models calculated using SPOTL

models are -4.7 and 4.6 nm/s^2. The RMS of the difference of the water height models is 1.0 nm/s^2 and between the storm surge models is 1.4 nm/s^2.

The RMS of the differences between the water height model and the storm surge model is 0.7 and 1.7 nm/s^2 for the seven tide gauges and La Plata models, respectively. The results show that all four gravimetric response series are very similar to each other, with an RMS of the differences smaller than 5% of the range.

Considering the source origin of the data (La Plata station, seven tide gauge stations), the RMS of the differences between the water height model and the storm surge model is 0.7 and 1.7 nm/s^2 for the seven tide gauges and La Plata models, respectively. The results found and show that the four gravimetric

responses are very similar to each other, with RMS of the differences smaller than 5% of the range.

A tidal model can be obtained from the subtraction of a storm surge (non-tidal) model from a water height model. For the Río de la Plata, this calculation can also be applied for the gravimetric response of the calculated empirical models, where the result is the gravimetric response of the tidal component of the loading. Figure 8 shows the gravimetric response of the tidal component of the loading subtracting ESS_7TG model from the EWH_7TG model. This subtraction was verified by subtracting the empirical models cell by cell, and then calculating the gravimetric response using SPOTL. The gravimetric response of the tidal component of the loading has a range of approximately 12% of the range of the

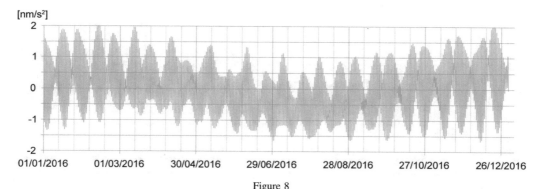

Figure 8
Gravimetric response of the tidal component of the loading, obtained by subtracting ESS_7TG model from EWH_7TG model

non-tidal component for the analyzed period. This difference in range can be explained considering the characteristics of the astronomical tide in the Río de la Plata. The amplitude of the astronomical tide in the river is smaller than 1.5 m, and a full cycle of the M2 constituent is always present in the river, meaning that a low tide and a high tide are present in different regions at the same time. These results show that the non-tidal component defines most of the behavior of the loading in the Río de la Plata.

4.4. Comparison Between the Gravity Residuals and the Gravimetric Response of the Empirical Models

The gravimetric response of the ESS_7TG model was selected to be compared with the gravity residuals of the SG, as during processing of the SG data, the harmonic analysis eliminated the tidal component of the loading, and during the validation process, model ESS_7TG has shown a better performance than the ESS_LaPLata model. Figure 9 shows the comparison of the gravimetric response of the ESS_7TG model with the SG data after applying the Earth tide model (which already includes tidal loading), atmospheric corrections and removing pole tide effect for the whole period, and the final residuals after subtracting the storm surge gravimetric response from the measured gravity residuals.

Although the reduction of the RMS of the series after subtracting the gravimetric response of the storm surge model is minor (15.0 vs 13.9 nm/s^2), the coincident peaks between both series indicate the influence of the storm surge in the SG observations. To make this influence more evident, a wavelet analysis is performed, since it is one of the most appropriate spectral analysis techniques to describe the variability of non-stationary data series. The analysis is applied to the residual series of the SG before and after subtracting the storm surge gravimetric response (Fig. 10) using as mother wavelet the Morlet wavelet. The series without the storm surge influence shows a clear decrease in the spectral energy in the periods between 2 and 12 days, in comparison with the gravity residuals. These periods coincide with the periods of the storm surges in the Río de la Plata.

4.5. Vertical and Horizontal Displacements Using the Empirical Models of the Río de la Plata

Using the SPOTL software, the horizontal and vertical displacements, due to the tidal and non-tidal loading of the Río de la Plata, can be estimated with the empirical models described in this paper. Figure 11 shows North, East, and Up displacements found for the four models in the period 01/2016–12/2016, and Table 2 shows the minimum and maximum values for each direction, for all models.

The biggest displacements correspond to Up direction, where the differences can be easily identify in the minimum displacement values. For the analyzed period, the RMS of the differences between EWH_7TG and EWH_LP models is smaller than 0.4 mm for all the directions, and between ESS_7TG and ESS_LP is smaller than 0.6 mm for all directions.

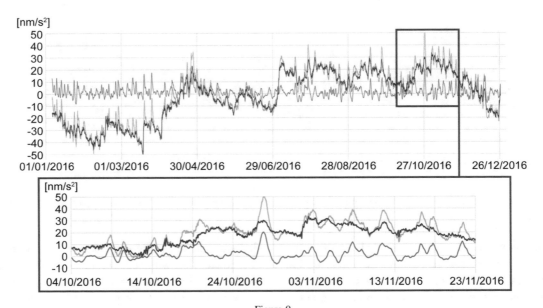

Figure 9
Gravimetric response of the ESS_7TG model (*blue line*), SG data after correcting for Earth tides, pole tide, tidal loading, and atmospheric pressure (*orange line*), and the difference between both series (*purple line*). The *upper figure* shows the comparison for the whole period and the *bottom figure* for a 50 days period

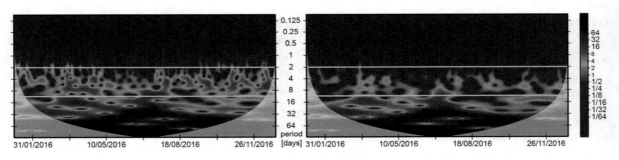

Figure 10
Wavelet power spectrum of the residual series of the SG, before (*left*) and after (*right*) subtracting the storm surge gravimetric response of the Río de la Plata

In addition, for the whole period, the maximum difference between the water height models is 2.4 mm corresponding to up direction, and between the storm surge models is 1.8 mm corresponding to up direction too.

5. Conclusions and Discussion

The empirical models used in this paper allow reducing the variability of the gravity residuals observed by the SG038 located at AGGO by removing the gravity response of the storm surge in the Río de la Plata. The reduction of the residuals is effective at periods between 2 and 12 days, which is coincident with the storm surge periods. According to the validation, the empirical models developed with information from all available tide gauges represented the water height and the storm surge of the river more accurately and, therefore, the derived gravimetric response as well. Although the RMS of the differences from Norden station and CorSSH

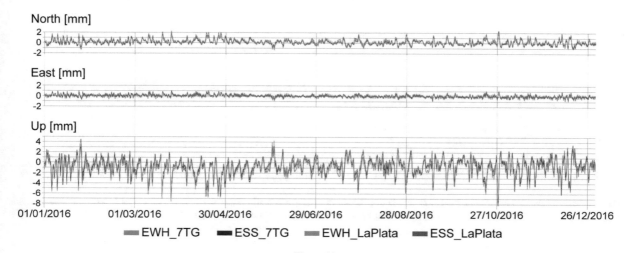

Figure 11

Horizontal and vertical displacements of the SG due to the tidal and non-tidal loadings of the Río de la Plata, obtained from the empirical models EWH_7G, ESS_7G, EWH_LaPlata, and ESS_LaPlata

Table 2

Minimum and maximum values of displacements for directions North, East, and Up obtained from the empirical models EWH_7G, ESS_7G, EWH_LaPlata, and ESS_LaPlata

Empirical model	Maximum negative displacement (mm)			Maximum positive displacement (mm)		
	North	East	Up	North	East	Up
EWH_7TG	−0.9	−0.9	−7.2	2.1	1.2	3.3
ESS_7TG	−0.9	−0.9	−7.6	2.3	1.0	3.1
EWH_LP	−1.4	−0.8	−7.6	2.3	1.0	4.3
ESS_LP	−1.4	−0.7	−8.2	2.5	1.1	4.5

comparisons confirms the better performance of these empirical models, the gravimetric response of the models using La Plata station only also allows reducing the RMS of the gravity residuals significantly. This can be explained by the close distance of the La Plata station to AGGO and to the minor variations of the water height in the vicinity of the inner part of the river.

The tidal and non-tidal loading effects of the river have different ranges; the tidal component is approximately 12% of the non-tidal component. In addition, the tidal component can be eliminated using harmonic analyses, since the frequencies of the oceanic tide are included in the frequencies of the Earth tide. If this technique is used, the storm surge model should be used to reduce the gravity time series after applying the conventional corrections. Otherwise, if the water height model is used, the gravimetric response of this model should be subtracted from the observed series before the harmonic analysis, to avoid a duplicate reduction of the tidal component.

The vertical and horizontal displacements found due to the storm surge are smaller than 1 cm, but will influence all geodetic space techniques installed at AGGO and should, therefore, be corrected to remove the non-tidal signal in the measurements. The observations from the closest tide gauge to AGGO, La Plata station, can provide essential information to model the gravimetric response of the storm surge and should be considered as significant auxiliary information for the observatory.

The results found in this paper will allow a better description of the vertical and horizontal displacements needed for other geodetic techniques, and also the modeling of geophysical signals not considered herein, such as rain and/or local ground water. Figure 12 shows the gravity residuals obtained after subtracting the storm surge gravimetric response, and the rainfall measured by the meteorological station SADL, located at the La Plata Aerodrome, close to AGGO. In Fig. 12, after each rain episode, a

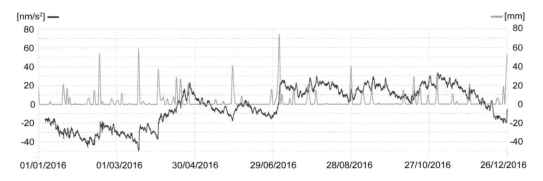

Figure 12

Gravimetric residues (in nm/s²) obtained after subtracting the storm surge gravimetric response of ESS_7TG model (*purple*) and the rain fall (in mm) measure by the meteorological station SADL (*green*)

sudden increase in the gravity time series can be identified, followed by a gradual decrease until the next rain episode. A clear example of this behavior happened on July 4, when an important rain event occurred, and the gravity residuals increased by almost 35 nm/s² within a few days. The analysis of this, and others, hydrological and geophysical effects will be subject developed to in further future work, but is strongly supported by this study, as gravity effects are only separable by precise modeling.

References

Agnew, D. C. (2012). SPOTL: some programs for ocean-tide loading. *SIO Technical Report, Scripps Institution of Oceanography.*

Antokoletz, E., Wziontek, H., & Tocho, C. (2017). First six months of Superconducting Gravimetry in Argentina. In *Proceedings of the International Symposium on Gravity, Geoid and Height Systems 2016, International Association of Geodesy Symposia,* submitted.

Balay, M. (1961). *El Río de la Plata entre la atmósfera y el mar.* Buenos Aires, Argentina: Servicio de Hidrografía Naval.

Boy, J. P., Longuevergne, L., Boudin, F., Jacob, T., Lyard, F., Llubes, M., et al. (2009). Modelling atmospheric and induced non-tidal oceanic loading contributions to surface gravity and tilt measurements. *Journal of Geodynamics, 48*(3–5), 182–188.

Campetella, C. M., D'Onofrio, E. E., Cerne, B. S., Fiore, M. E., & Possia, N. E. (2007). Negative storm surges in the port of Buenos Aires city. *International Journal of Climatology, 27*(8), 1091–1101.

Comisión Administradora del Río de la Plata. (1989). *Estudio para la evaluación de la contaminación en el Río de la Plata* (p. 422). Buenos Aires: Comisión Administradora del Río de la Plata, Montevideo.

D'Onofrio, E. E. (1984). *Desarrollo de un nuevo sistema de procesamiento de información de marea. Informe Técnico N°25/84,* Departamento Oceanografía, Servicio de Hidrografía Naval. 167 pág.

D'Onofrio, E.E., & Fiore, M.M.E. (2003). Estimación de niveles extremos en el Puerto de Buenos Aires contemplando el ascenso del nivel medio. In *Paper presented at V Jornadas Nacionales de Ciencias del Mar.* Mar del Plata, Argentina.

D'Onofrio, E., Fiore, M., Di Biase, F., Grismeyer, W., & Saladino, A. (2010). Influencia de la marea astronómica sobre las variaciones del nivel del Río Negro en la zona de Carmen de Patagones. *Geoacta, 35*(2), 92–104. (**ISSN 1852-7744**).

D'Onofrio, E. E., Fiore, M. M. E., & Pousa, J. L. (2008). Changes in the regime of storm surges at Buenos Aires, Argentina. *Journal of Coastal Research, 24*(1A), 260–265.

D'Onofrio, E. E., Fiore, M. M. E., & Romero, S. I. (1999). Return periods of extreme water levels estimated for some vulnerables areas of Buenos Aires. *Continental Shelf Research, 19,* 1681–1693.

D'Onofrio, E., Oreiro, F., Di Biase, F., Grismeyer, W., & Fiore, M. (2009). Estudio de las principales ondas componentes de la marea astronómica en el Río de la Plata. In *Resúmenes VII Jornadas Nacionales de Ciencias del Mar,* Bahía Blanca, Provincia de Buenos Aires, Argentina (Poster).

D'Onofrio, E., Oreiro, F., & Fiore, M. (2012). Simplified empirical astronomical tide model—An application for the Río de la Plata Estuary. *Computers & Geosciences, 44,* 196–202. doi:10.1016/j.cageo.2011.09.019.

Dehant, V., Defraigne, P., & Wahr, J. M. (1999). Tides for a convective Earth. *Journal of Geophysical Research, 104*(B1), 1035–1058.

Dragani, W., & Romero, S. (2004). Impact of a possible local wind change on the wave climate in the upper Río de la Plata. *International Journal of Climatology, 24,* 1149–1157.

Escobar, G., Vargas, W., & Bischoff, S. A. (2004). Wind Tides in the Río de la Plata Estuary: Meteorological conditions. *International Journal of Climatology, 24,* 1159–1169.

Farrell, W. E. (1972). Deformation of the earth by surface loads. *Reviews of Geophysics, 10*(3), 761–797.

Fiore, M. M. E., D'Onofrio, E. E., Pousa, J. L., Schnack, E. J., & Bértola, G. R. (2009). Storm surges and coastal impacts at Mar del Plata, Argentina. *Continental Shelf Research, 29*(14), 1643–1649.

Fratepietro, F., Baker, T. F., Williams, S. D. P., & Van Camp, M. (2006). Ocean loading deformations caused by storm surges on the northwest European shelf. *Geophysical Research Letters, 33,* L06317. doi:10.1029/2005GL025475.

Geng, J., Williams, S. D. P., Teferle, F. N., & Dodson, A. H. (2012). Detecting storm surge loading deformations around the southern North Sea using subdaily GPS. *Geophysical Journal International, 191*(2), 569–578.

Goodkind, J. M. (1999). The superconducting gravimeter. *Review of Scientific Instruments, 70,* 4131–4152.

Guerrero, R. A., Acha, E. M., Framiñan, M. B., & Lasta, C. A. (1997). Physical oceanography of the Río de la Plata Estuary, Argentina. *Continental Shelf Research, 17*(7), 727–742.

Hinderer, J., Crossley, D., & Warburton, R. (2007). Gravimetric methods—Superconducting gravity meters. In G. Schubert (Ed.), *Treatise on geophysics* (pp. 65–122). Amsterdam: Elsevier.

Klügel, T., & Wziontek, H. (2009). Correcting gravimeters and tiltmeters for atmospheric mass attraction using operational weather models. *Journal of Geodynamics, 48*(3–5), 204–210. doi:10.1016/j.jog.2009.09.010.

Nordman, M., Virtanen, H., Nyberg, S., & Mäkinen, J. (2015). Non-tidal loading by the Baltic Sea: Comparison of modelled deformation with GNSS time series. *GeoResJ, 7,* 14–21.

Oreiro, F., D'Onofrio, E., & Fiore, M. M. E. (2016). Vinculación de las referencias altimétricas utilizadas en las cartas náuticas con el elipsoide WGS84 para el Río de la Plata. *Geoacta, 40*(2), 109–120. **(ISSN 1852-7744)**.

Petrov, L. (2015). The International mass loading service. arXiv preprint 1503.00191.

Pousa, J. L., D'Onofrio, E. E., Fiore, M. M. E., & Kruse, E. (2013). Environmental impacts and simultaneity of positive and negative storm surges on the coast of the Province of Buenos Aires, Argentina. *Environmental Earth Sciences, 68,* 2325–2335.

Schüller, K. (2015). Theoretical basis for Earth Tide analysis with the new ETERNA34-ANA-V4.0 program, Bulletin d'Informations des Marees Terrestres 149, 12024. http://www.eas.slu.edu/GGP/BIM_Recent_Issues/bim149-2015/schuller_theoretical_basis_Eterna-ANA_v4_BIM149.pdf. Accessed 20 Feb 2017.

SHN (Servicio de Hidrografía Naval). (2017). Tablas de Marea, Pub. H- 610. Argentina. Servicio de Hidrografía Naval, Ministerio de Defensa.

Simionato, C. G., Dragani, W., Meccia, V., & Nuñez, M. (2004a). A numerical study of the barotropic circulation of the Río de La Plata estuary: sensitivity to bathymetry, the Earth's rotation and low frequency wind variability. *Estuarine, Coastal and Shelf Science, 61,* 261–273.

Simionato, C. G., Dragani, W., Nuñez, M. N., & Engel, M. (2004b). A set of 3-D nested models for tidal propagation from the Argentinean continental shelf to the Río de la Plata estuary—Part I M2. *Journal of Coastal Research, 20*(3), 893–912.

Tocho, C. (2016). Gravimetría superconductora en Argentina. *Geoacta, 41*(1), 77–78. **(Asociación Argentina de Geofísicos y Geodestas)**.

Virtanen, H. (2004). Loading effects in Metsähovi from the atmosphere and the Baltic Sea. *Journal of Geodynamics, 38,* 407–422.

Virtanen, H., & Mäkinen, J. (2003). The effect of the Baltic sea level on gravity at the Metsähovi station. *Journal of Geodynamics, 35,* 553–565.

Wziontek, H., Nowak, I., Hase, H., Häfner, M., Güntner, A., Reich, M., et al. (2016). A new gravimetric reference station in South America: The installation of the Superconducting Gravimeter SG038 at the Argentinian-German Geodetic Observatory AGGO, EGU General Assembly 2016. *Geophysical Research Abstracts, 18,* EGU2016–12612.

(Received April 26, 2017, revised July 24, 2017, accepted August 11, 2017, Published online August 29, 2017)

Pure Appl. Geophys. 175 (2018), 1755–1763
© 2017 Springer International Publishing
https://doi.org/10.1007/s00024-017-1546-6

Pure and Applied Geophysics

CrossMark

The Improved Hydrological Gravity Model for Moxa Observatory, Germany

A. Weise[1,2] and Th. Jahr[2]

Abstract—The gravity variations observed by the superconducting gravimeter (SG) CD-034 at Moxa Geodynamic
Observatory/Germany were compared with the GRACE results
some years ago. The combination of a local hydrological model of
a catchment area with a 3D-gravimetric model had been applied
successfully for correcting the SG record of Moxa which is especially necessary due to the strong topography nearest to the SG
location. Now, the models have been corrected and improved
considerably by inserting several details in the very near surrounding. Mainly these are: the observatory building is inserted
with the roof covered by a soil layer above the gravity sensor where
humidity is varying, snow is placed on top of the roof and on
topography (steep slope), and ground water is taken into account,
additionally. The result is that the comparison of the corrected
gravity residuals with gravity variations of the satellite mission
GRACE, now using RL5 data, shows higher agreement, not only in
amplitude but also the formerly apparent phase shift is obviously
not realistic. The agreement between terrestrial gravity variations
(SG) and the GRACE data is improved considerably which is
discussed widely.

Key words: Superconducting gravimeter, hydrology, 3D
gravity modeling, local effects.

1. Introduction

Hydrological effects in gravity observations have
been a subject of discussion for several decades, since
the precision and repeatability is in the range of µGal.
Repeated gravity campaigns, time lapse gravity, as
well as time series are affected. One of the first
reporting this effect, especially of soil moisture, was
Bonatz (1967). Lambert and Beaumont (1977) found
effects of ground water variations when seeking for
tectonic gravity signals. Elstner et al. (1978) already
used soil moisture measurements for estimation and
correction of the effect of some 5 µGal (50 nm/s²).
While at times, precipitation data had been used for
statistical estimations, Mäkinen and Tattari (1988)
predicted separately soil moisture and ground water
effects applying the Bouguer plate model and could
verify the calculated effect with repeated gravity
observations. Over years, also statistical methods
have been investigated, e.g. Braitenberg (1999) found
in tiltmeter and extensometer data that the hydrological induced signal showed time variable
correlations, which were not significant over longer
periods or masked by other effects. Especially
regarding repeated campaigns, same seasons are
chosen for measurements to reduce the effect of
hydrological impact. Hydrological effects can be
amplified where strong topography occurs near the
stations, for example at volcanoes (Jentzsch et al.
2004).

At the latest, general knowledge concerning
seasonal changes of continental hydrology in gravity
is widely spread since the GRACE project. Amongst
others, this gives information about weak regions in
continental hydrology models, e.g. with the 'nulltest'
in arid regions (Hinderer et al. 2016). On the other
hand, comparisons with terrestrial gravity observations from superconducting gravimeters (SG) show
local effects at some stations. While Crossley et al.
(2014) favor the Bouguer plate model for reduction
of local hydrological effect where necessary at stations of superconducting gravimeters, in contrast,
Weise et al. (2009, 2012) found that this holds only
for flat and homogeneous surroundings like at
Strasbourg station. But where strong topography
leads to considerable effects physical modeling is
needed, e.g. for stations like Moxa and Vienna.
Naujoks et al. (2010) developed a successful correction model of local hydrological effect first

---
[1] *Present Address*: Leibniz Institute for Applied Geophysics,
Hanover, Germany. E-mail: adelheid.weise@liag-hannover.de
[2] Institute of Geosciences, Friedrich Schiller University Jena,
Jena, Germany.

combining hydrological and 3D-gravimetrical modeling for the SG station Moxa. After reduction the agreement of the SG time series with the result from GRACE for this region had been approved considerably (Weise et al. 2009, 2012).

During the last 3 years. the time series of the existing hydrological model were extended in period, and the combined hydrological–gravimetrical modeling could be clearly enhanced by inserting, especially some important details into the gravimetric model.

2. Hydrology Around Moxa Observatory

The hydrological situation around the Geodynamic Observatory Moxa is characterized by the fact that precipitation and topography cause mass movements above and below the sensor-level of the SG CD-34 (see Fig. 2 in Naujoks et al. 2010). In addition, different flow-paths are active for water transport from the hills to the valley and into the small creek Silberleite.

The local hydrological model of the catchment area of the creek Silberleite, around the observatory Moxa (see Fig. 4 in Naujoks et al. 2010) consists of hydrological responds units (HRUs) according to several soil parameters, discharging downhill along the given fluid paths. In front of the observatory building the runoff of the creek Silberleite is observed at the small weir (Fig. 1b). In addition, several divers installed in shallow drill-holes distributed over the model area are measuring soil moisture changes (Naujoks et al. 2010).

3. Hydrological Gravity Modeling

The hydrological modeling is based on the conceptual model J2000 (Krause 2001; Eisner 2009) and exemplarily shown for three horizons in Fig. 1a. The HRUs are defined according to different soil parameters, layers, permeability, drain information, and slope, among others. The variations of water content in the HRUs are modeled which cause density changes, due to three main storages: surface water

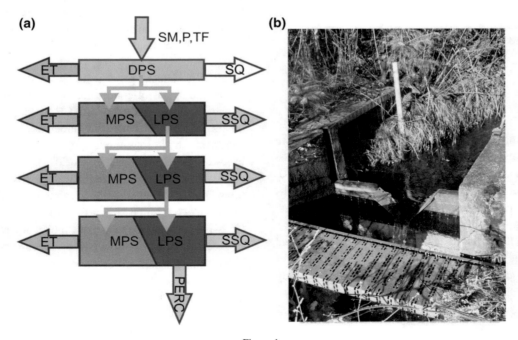

Figure 1
a Strategy of hydrological modeling (after Eisner 2009), b small weir in front of the Geodynamic Observatory Moxa, where the runoff of the catchment area of the creek Silberleite is recorded

(DPS), mid pore storage (MPS), and large pore storage (LPS), (s. Fig. 1a). Input data are temperature, precipitation, humidity, wind velocity, and radiation which together result in the input-storages defined by snowmelt (SM), precipitation (P), and throughfall (TF)/interception. The modeling provides the surface runoff (SQ), subsurface flow (SSQ), evapotranspiration (ET), and flow into groundwater (PERC). The main result of the model is the time dependent amount of stored water content (water column) in different water storages (layers) for all HRUs as time series. The final comparison of modeled vs. observed runoff gives an estimation of the quality of the modeling.

The main storages are large and middle pore storage, snow, interception storage, depression storage, quick and slow ground water storage. The temporal variation of the sum of all storages reflects seasonal variations in the range of 120–160 mm WC (mm water column, Fig. 2), with a maximum of 220 mm WC in 2011 due to extreme snow fall. The modeling period is 2004–2014.

In the next step, a 3D-gravity model was developed on the base of Bouguer gravity values and geological underground information (Naujoks 2008; Naujoks et al. 2010) using the software IGMAS (Schmidt et al. 2011) for describing the geological structures. Subsequently, the time series of the lateral distribution of water content, converted to density changes, go into the 3D-gravimetric modeling. Here, the units of different density are situated parallel to the topography, with a depth of 2–4 m for the soil layer which is underlain by a layer of 8–16 m where mainly the larger and long-term groundwater variation is supposed to take place. The direct vicinity of the observatory is much higher resolved (Fig. 3). The 3D-gravimetric model around the observatory (2×2 km^2) has the SG in the center. The distance of the vertical planes is 5 m in the central part for high resolution of a few meters and increasing distances between the planes outwards. The model comprises 36 vertical planes connected by triangulation and 92 hydrological active units (Naujoks 2008).

The most recent finding based on the work of Naujoks et al. (2010) is—beyond the correction of the conversion to density: the considerably more detailed modeling of the immediate SG-vicinity and

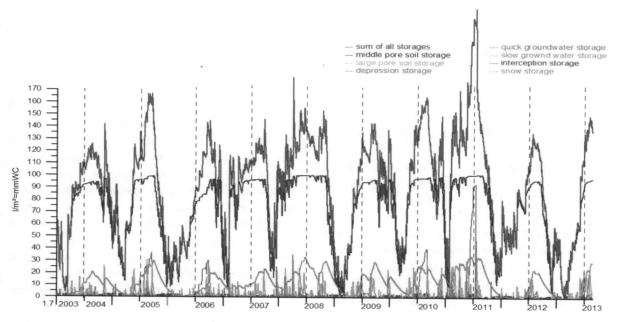

Figure 2
Temporal change of model content in water storages

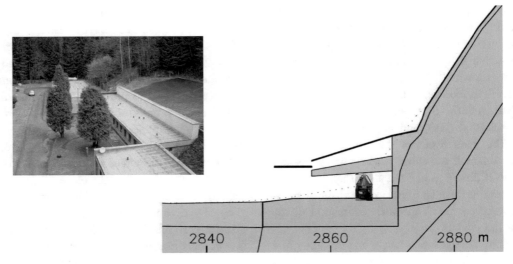

Figure 3
Improved realistic detail from gravimetric 3D-model of vertical plain through the observatory building

observatory building is crucial and leads to improved results (Weise and Jahr 2015):

- Roof:
  1. snow is stored on the roof of the observatory and on the ground above the SG (Fig. 3),
  2. with respect to the plastic cover soil humidity in the roof coverage is reduced to 30% (not zero),

- building acts as a shield which leads to reduced variation of humidity in clefts beneath the observatory and SG, estimated with 10%,
- topography next to the SG has been improved (especially steep slope),
- geometry next to SG has been adapted more specific, including pillar height, thickness of basement, height of rooms, cover layer in the roof,
- constant thickness of units next to the SG is set for exact conversion of water storage to density change,
- slow ground water has been included (up to 30 mm WC seasonal). The equivalent density change has been included in the gravimetric model partly in the soil layer and partly in the deeper layer. The impact in the seasonal amplitude of gravity change at the SG site is about 20–30%. The apparent phase shift against the satellite data in former results vanished after including the slow ground water.

The results of the hydrological modeling are inserted into the gravimetric model by continuously changing densities dependent on the water content. In time steps of 1 h the gravity-effect due to water mass changes in the upper model units is calculated for the location of the SG resulting in the time series of the hydrological induced gravity effects for correction of the local hydrological effect in the SG recording.

4. Results: SG-Residuals and Comparison with GRACE

The result of the modeled gravity effect due to local hydrological variations (Fig. 4, total in blue) has the clear seasonal amplitude (peak-to-peak) of 30–45 nm/s^2, up to 75 nm/s^2 in the begining of 2011 caused by a high snow layer. Maxima in summer and minima in winter prove the dominating water masses above the SG sensor. The separation of particular contributions of the nearest areas (Fig. 4) demonstrates that the nearest area of <75 m radius is responsible for ∼80% (red) of the local hydrological effect, whereof the most considerable and effective parts are the slope area above the observatory (∼60%, green) and the roof (mostly snow, ∼24%, brown) while the roof had not been considered before in the modelling of Naujoks et al. (2010).

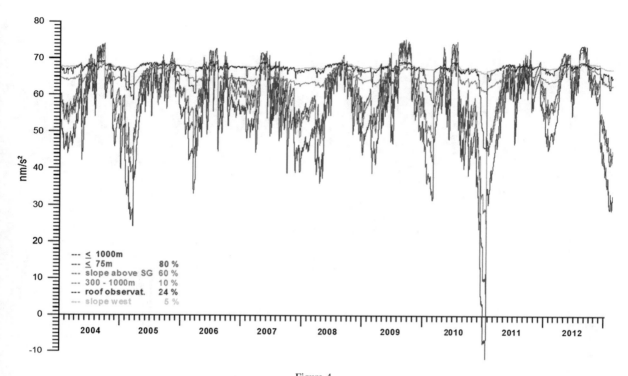

Figure 4

Contributions to local combined hydrological and 3D-gravimetrical modeling from several parts of the observatory surrounding, *blue* sum of the total modeled local hydrological gravity effect

Figure 5 illustrates the observed time series of gravity variations (monthly samples, black), reduced for tides, 3D atmospheric masses, and drift. The atmospheric correction from Atmacs (Atmospheric attraction computation service, ATMACS 2017) is used with the temporal resolution of 3- and 6-hourly samples which has been adapted to hourly samples by interpolation applying the local observed air pressure and an adapted regression coefficient. The reduced gravity variations are in the range of 20 nm/s^2, up to maximum 57 nm/s^2, without clear seasonal content. After subtracting the modeled local hydrological gravity effect (blue) the resulting gravity variations (red, monthly samples) show a seasonal signal in the range of 30–40 nm/s^2 with maxima in winter and minima in summer as expected from water storage changes in global continental hydrology.

We compare the SG residuals after reduction of local hydrological effect (red in Fig. 5) with gravity variations from the satellite mission GRACE for Moxa area which are computed according to Abe et al. (2012) using monthly sets (RL5) from JPL (Jet Propulsion Laboratory, Pasadena/California) and the spherical harmonic (SH) approach to the degree/order 120 (download from GFZ Information System and Data Center, ISDC, http://isdc.gfz-potsdam.de/, see Abe et al. 2012). The deformation effect is added. The GRACE data are Gauss filtered with 1000 km radius, and additionally DDK1 anisotropic filter from Kusche (2007) was applied, both exemplarily as their suitability was demonstrated in Weise et al. (2012).

The general agreement (Fig. 6) of the corrected terrestrial gravity with gravity variations from GRACE for the Moxa area can be stated in magnitude and in phase and has been improved successfully against Naujoks et al. (2010). Partly agreeing structural details, e.g. with GRACE in winter 2007/8 and 2008/9 (double peaks) and with GLDAS in autumn 2010 and 2011, and in summer 2005, 2009 can support the similar origin of the signal. But also anomalous seasonal variations have to be ascertained, mainly from summer 2006 to 2007, and the minimum

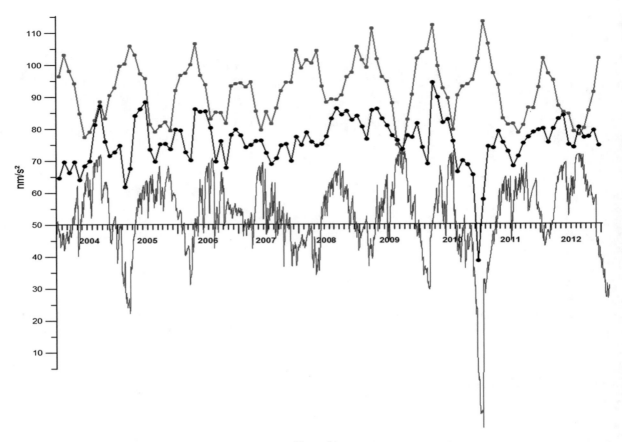

Figure 5
Local hydrological effect 2004–2012 from combined hydrological and 3D-gravity modeling at the SG site (*blue*) with seasonal variations, SG-gravity residuals (monthly) observed (*black*) and after hydrological reduction (*red*)

in summer 2008, with more slight extent in summer to autumn 2007. The reason is not clear, yet. All meteorological and gravity data have been checked. A most possible candidate could be the ground water in the hydrological modeling which still is a great challenge. Here, major work is still waiting to improve the underground information for modeling.

As the modeling and correction of the hydrological effect of the gravity record is mainly successful in the long-term range, we intended to check the possibility of realistic modeling in the short-term period range of hours to days. For several events of strong rain fall the modeling and correction has been checked, exemplarily shown for one event in Fig. 7. The single event is characterized by a nearly immediate and quick gravity decease and followed by a slower gravity increase nearly linear over 1–2 days.

The almost immediate gravity decrease is due to the precipitation which increases soil humidity, which in Moxa occurs also and mainly above the gravity sensor.

Modeling of the first start of simulated gravity decrease is basically agreeing in time, amplitude and in course of process. The subsequent gravity increase due to relaxation is too slow and delayed in the original modeled version (pink in Fig. 7). The main effective process in the model seems to be evaporation. The flow downhill should be driven by quick ground water. The orange model version shows that the flow of the 'quick ground water' downhill, which is not sufficiently fast in the model, could be increased, also by dividing the HRU at the slope. In the red model version the storage capacity of the quick ground water has been reduced, in the orange

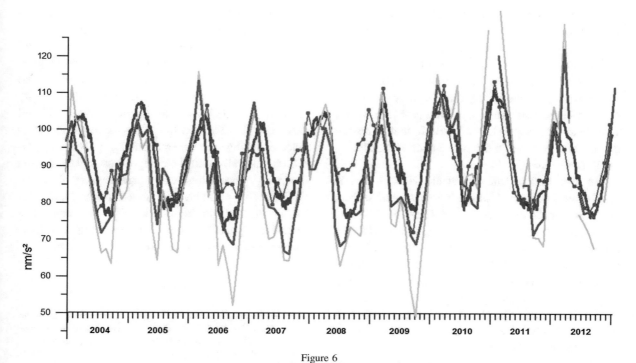

Figure 6
Monthly gravity variations reduced for local hydrological effect (*red*; Fig. 5) compared for Moxa to the global hydrological model GLDAS (*blue*, hourly samples) and to satellite data from GRACE (RL5 JPL) which are Gauss filtered ($R = 1000$ km, *dark green*) and DDK1 anisotropic filter (*light green*, Kusche 2007) is applied

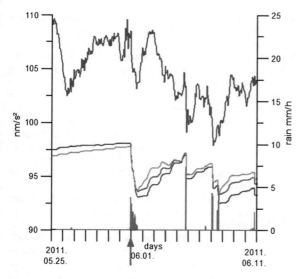

Figure 7
Example of short-term gravity changes after heavy rainfall with three modeled versions of the hydrological induced gravity effect

version, the amount of quick ground water is increased. Both experiments show that the dynamic is not sufficient in the model. Simulating the full

dynamic of flow in the complex situation in the slope through weathered and fissured rock remains a big challenge. Still, there remains a lack of knowledge about the real flow path in the underground and its characteristics. One point is also the steepness of the slope where the realistic ground water modeling seems to be problematic.

5. Discussion and Conclusion

The main improvements against Naujoks et al. (2010) of modeling the local hydrological effect at Moxa observatory which is mainly caused by the near and strong topography are part of the 3D-gravimetric model: Snow on and humidity in the cover layer of the observatory roof above the SG, reduced storage changes beneath the observatory and the SG due to the building acting as a shield, improved geometry and topography nearest to the gravimeter sensor. Especially including the ground water which had been neglected before, lead to higher effects in the

167

modeling and finally to more realistic amplitude in gravity compared to GRACE. With these improvements the agreement with the GRACE time series is considerably higher, not only in amplitude but also in phase which seemed to be a problem before. This enhancement is mainly regarding long-term gravity changes in the range of months and probably weeks. This result also supports the position, that terrestrial gravity mainly 'sees' the same temporal variation as satellites as GRACE do.

As discussed in Crossley et al. (2014) a certain part of the local effect which is representative for a larger region is included in the GRACE data and should, in best case, remain in or should be added to the terrestrial gravity record after subtraction of a local effect. This will be small for the area around Moxa observatory due to a thin soil layer on the rocks of "Thüringer Schiefergebirge". We estimate, that the part subtracted from below the sensor is <5% of the whole modelled effect, and thus perhaps within the uncertainty. Up to now, it is not solved how to estimate a realistic value for the smooth effect representing a wider area and to be added after reduction of the local effect due to topography and an underground location of the SG. On the other hand, if we would leave out the part from below the SG in the reduction model we would neglect the fact that also the building is working as a shield. This point remains an open question in this work.

In the higher frequency range of days to hours the comparison of the hydrological effect from the model combination with the observed gravity time series from the SG shows on the one hand a certain improvement but on the other hand also the obviously limited possibilities of the modeling. Flow processes of high dynamics within the deeper underground and the slope next to and under the observatory seem to remain a big challenge to be modeled in a realistic way.

Acknowledgements

The investigations were carried out in the frame of the project JA-542-24-1, funded by the German Research Foundation (DFG), which is gratefully acknowledged. We also thank our project partners Daniel Hagedorn (PTB, Braunschweig) and Ronny Stolz (IPHT, Jena) and the colleagues from the Arbeitskreis Geodäsie/Geophysik for the fruitful discussions about the reduction of hydrological effects in SG recordings. Further, we thank Maiko Abe, GFZ Potsdam, for calculating the gravity variations of GRACE for Moxa location with all her expertise. To keep the Geodynamic Observatory Moxa running, the superconducting gravimeter SG-CD034 as well as all the components of weather and hydrology recordings, was guaranteed by Wernfrid Kühnel and Matthias Meininger; many thanks for doing this job in highest quality. We appreciate valuable suggestions improving the manuscript from two reviewers, one anonymous and N. Florsch.

REFERENCES

Abe, M., Kroner, C., Förste, C., Petrovic, S., Barthelmes, F., Weise, A., et al. (2012). A comparison of GRACE-derived temporal gravity variations with observations of six European superconducting gravimeters. *Geophysical Journal International, 191*(2), 545–556.

ATMACS, Atmospheric attraction computation service. (2017). http://atmacs.bkg.bund.de/ in 3 or 6 hr. Accessed Jan 2015.

Bonatz, M. (1967). Der Gravitationseinfluß der Bodenfeuchtigkeit. *ZfV, 92,* 135–139.

Braitenberg, C. (1999). Estimating the hydrologic induced signal in geodetic measurements with predictive filtering methods. *Geophysical Research Letters, 26*(6), 775–778.

Crossley, D. J., Boy, J.-P., Hinderer, J., Jahr, T., Weise, A., Wziontek, H., et al. (2014). Comment on: 'The quest for a consistent signal in ground and GRACE gravity time-series', by Michel Van Camp, Olivier de Viron, Laurent Metivier, Bruno Meurers and Olivier Francis. *Geophysical Journal International, 199,* 1811–1817.

Eisner, S. (2009). Analyse der Bodenwasserdynamik im hydrologischen Modell J2000 und Multi-Response-Validierung am Beispiel des Einzugsgebiets der Silberleite (Moxa), Dipl. Thesis, Chem.-Geowiss. Fak., Friedrich Schiller University Jena, p 109 **(unpublished)**.

Elstner, C., Harnisch, G., & Altmann, W. (1978). Ergebnisse präzisionsgravimetrischer Messungen auf der W-E-Linie der DDR 1970-1976. *Vermessungstechnik, 26,* 12–14.

Hinderer, J., Boy, J.-P., Humidi, R., Abtout, A., Issawy, E., Radwan, A., & Hamoudi, M. (2016). Ground-satellite comparison on time variable gravity: Issues and on-going projects for the null test in arid regions, IAG Symposium, Prague June 2015 **(in press)**.

Jentzsch, G., Weise, A., Rey, C., & Gerstenecker, C. (2004). Gravity changes and internal processes: Some results obtained from observations at three volcanoes. *PAGEOPH, 161*(7), 1415–1431.

Krause, P. (2001). Das hydrologische Modellsystem J2000: Beschreibung und Anwendung in großen Flußeinzugsgebieten, in

Schriften des Forschungszentrums Jülich: Reihe Umwelt/Environment, vol. 29, Forschungszentrum Jülich.

Kusche, J. (2007). Approximate decorrelation and non-isotropic smoothing of time-variable GRACE-type gravity field models. *Journal of Geodesy, 81*(11), 733–749.

Lambert, A., & Beaumont, C. (1977). Nanovariations in gravity due to seasonal groundwater movements: Implications for the gravitational detection of tectonic movements. *Journal of Geophysical Research Atmospheres, 82*(2), 297–306.

Mäkinen, J., & Tattari, S. (1988). Soil moisture and ground water: Two sources of gravity variations. *BGI, 62,* 103–110.

Naujoks, M. (2008). Hydrological information in gravity: Observation and modelling. Dissertation at Chem.-Geowiss.-Faculty of Friedrich Schiller University, Jena.

Naujoks, M., Kroner, C., Weise, A., Jahr, T., Krause, P., & Eisner, S. (2010). Evaluating local hydrological modelling by temporal gravity observations and a gravimetric three-dimensional model. *Geophysical Journal International, 182,* 233–249.

Schmidt, S., Plonka, C., Götze, H.-J., & Lahmeyer, B. (2011). Hybrid modelling of gravity, gravity gradient and magnetic fields. *Geophysical Prospecting, 59*(6), 1046–1051.

Weise, A., & Jahr, T. (2015). News from the local hydrological correction of the SG gravity record at Moxa, Present. IUGG Prague, IAG-Session, June 23–28, 2015.

Weise, A., Kroner, C., Abe, M., Creutzfeldt, B., Förste, C., Güntner, A., et al. (2012). Tackling mass redistribution phenomena by time-dependent GRACE- and terrestrial gravity observations. *Journal of Geodynamics, 59–60,* 82–91.

Weise, A., Kroner, C., Abe, M., Ihde, J., Jentzsch, G., Naujoks, M., et al. (2009). Gravity field variations from superconducting gravimeters for GRACE validation. *Journal of Geodynamics, 48,* 325–330.

(Received December 15, 2016, revised March 15, 2017, accepted March 30, 2017, Published online April 9, 2017)

Pure Appl. Geophys. 175 (2018), 1765–1781
© 2018 Springer International Publishing AG, part of Springer Nature
https://doi.org/10.1007/s00024-018-1860-7

Pure and Applied Geophysics

CrossMark

Cansiglio Karst Plateau: 10 Years of Geodetic–Hydrological Observations in Seismically Active Northeast Italy

BARBARA GRILLO,[1] CARLA BRAITENBERG,[1] ILDIKÓ NAGY,[1] ROBERTO DEVOTI,[2] DAVID ZULIANI,[3] and PAOLO FABRIS[3]

Abstract—Ten years' geodetic observations (2006–2016) in a natural cave of the Cansiglio Plateau (Bus de la Genziana), a limestone karstic area in northeastern Italy, are discussed. The area is of medium–high seismic risk: a strong earthquake in 1936 below the plateau (M_m = 6.2) and the 1976 disastrous Friuli earthquake (M_m = 6.5) are recent events. At the foothills of the karstic massif, three springs emerge, with average flow from 5 to 10 m^3/s, and which are the sources of a river. The tiltmeter station is set in a natural cavity that is part of a karstic system. From March 2013, a multiparametric logger (temperature, stage, electrical conductivity) was installed in the siphon at the bottom of the cave to discover the underground hydrodynamics. The tilt records include signals induced by hydrologic and tectonic effects. The tiltmeter signals have a clear correlation to the rainfall, the discharge series of the river and the data recorded by multiparametric loggers. Additionally, the data of a permanent GPS station located on the southern slopes of the Cansiglio Massif (CANV) show also a clear correspondence with the river level. The fast water infiltration into the epikarst, closely related to daily rainfall, is distinguished in the tilt records from the characteristic time evolution of the karstic springs, which have an impulsive level increase with successive exponential decay. It demonstrates the usefulness of geodetic measurements to reveal the hydrological response of the karst. One outcome of the work is that the tiltmeters can be used as proxies for the presence of flow channels and the pressure that builds up due to the water flow. With 10 years of data, a new multidisciplinary frontier was opened between the geodetic studies and the karstic hydrogeology to obtain a more complete geologic description of the karst plateau.

Key words: Geodesy, karst hydrogeology, GPS, tilt measurements.

1. Introduction

The study region is located in northeastern Italy, in the seismically active area of the karstic Cansiglio Plateau. The present seismicity of NE Italy is well manifested toward the Friuli region, whereas toward the western sector a relative calmness is found. This picture emerges when considering the local seismicity recorded since the 1976 disastrous Friuli earthquake, certainly biased by the post-seismic sequence of this event. The western sector was hit in 1936 by the destructive Cansiglio earthquake, showing that the seismic potential is high in the entire region, reaching also farther west to the eastern Venetian sector. For this reason, 10 years ago it was decided to monitor the deformation of the area, installing two Zöllner-type Marussi tiltmeters in a natural cavity at 25 m depth (Bus de la Genziana). They are operating continuously since 2005 (Braitenberg et al. 2007).

During this period, we proposed an interdisciplinary study of karstic aquifers using hydrogeological data, tiltmeters and GPS observations (Grillo et al. 2011). During the year 2010, two data acquisition campaigns have been carried out to integrate the research started with the tiltmeter recordings: hydrogeological flow measurements (level, conductivity and temperature) in two principal springs and the installation of a small geodetic GPS network. The geodetic campaign extended from May to October 2010 measuring two GPS benchmarks in the neighborhood of the FReDNet permanent station CANV (950 m a.s.l; Zuliani 2003; Zuliani et al. 2009), the results of which demonstrated the presence of an observable hydrologic signal in the GPS time series (Devoti et al. 2015).

[1] Department of Mathematics and Geosciences, University of Trieste, Via Weiss, 1, 34100 Trieste, Italy. E-mail: berg@units.it
[2] Centro Nazionale Terremoti, Istituto Nazionale di Geofisica e Vulcanologia, Via di Vigna Murata, 605, 00143 Rome, Italy.
[3] Centro di Ricerche Sismologiche - CRS, OGS - Istituto Nazionale di Oceanografia e di Geofisica Sperimentale, Via Treviso, 55, 33100 Cussignacco, Udine, Italy.

To monitor the underground hydrodynamics of Bus de la Genziana and to correlate it with the geodetic recordings, for the first time a multiparametric logger (temperature, stage, electrical conductivity) was installed in the siphon located at the bottom of the Genziana cave (587 m deep) from March 2013 to December 2016 (Grillo and Braitenberg 2015).

The decade-long tilt and GPS observations in the karstic area allows us to univocally characterize the expected deformation signal in relation to the rainfall and spring discharge from the karstic plateau. The results can be of relevance for other karstic areas worldwide (see, e.g., Longuevergne et al. 2009; Tenze et al. 2012; Braitenberg 1999b, c), because hydrologic karst systems have often common characteristics, such as prominent epikarst, a well-developed network of deep channels in which water is efficiently drained toward the base level, and a less important matrix-flow component. The knowledge of the geometry of the cave network is important, because boreholes to the channels produce drinking water provision. An improved knowledge of the relation between the karstic water flows and the geodetic signal will allow the use of the geodetic measurement as a hydrologic investigation tool in the future. The main features of the geodetic measurements are furthermore generally comparable to the signals of analogous instrumentation operating in non-karstic environs affected by fissures and faults. The understanding of the physical relations between extensometer, tilt and gravity observations and the hydrology is fulfilled also elsewhere, e.g., in the station Moxa, Germany (Jahr 2017). Here, induced pore pressure variations through pumping and injection experiments were made to test the poroelastic deformation models. The aim is to use the physical relations to plan hydro-deformation experiments to recover in situ rock physical properties (Wang and Kümpel 2003; Jahr et al. 2008).

2. Geographical and Geological Setting

The Cansiglio–Cavallo Plateau is a karstic massif situated in the Forealps (Prealpi Carniche) (Fig. 1), which stretches forward as a mountainous block toward the plain. It is divided between two regions, Veneto on the west and Friuli-Venezia-Giulia on the east and three provinces Pordenone, Treviso and Belluno (towns outside the orientation map of Fig. 1). Its maximum height above mean sea level is 2200 m, and it has two plateaus of medium height of 1000 m, the Cansiglio and the Piancavallo Plateau.

A geological description of Cansiglio Plateau is discussed in Cancian and Ghetti (1989). The outcropping rocks range in age from Upper Jurassic to Paleocene and are mainly composed of carbonates (Fig. 1). The eastern area is characterized by a thick succession of Cretaceous peritidal carbonates (Cellina limestones), while the central western part is characterized by slope breccia deposits (Fadalto Formation and Mount Cavallo Formation), all capped by basinal marly carbonates (Scaglia Formation).

Notoriously, the zone is of medium–high seismic risk: in recent history, we recall the strong earthquake of 1936, which according to the magnitude scale used was quantified as having $M_s = 5.8$ (magnitude of surface wave) or $M_m = 6.2$ (macroseismic magnitude) (Pettenati and Sirovich 2003). 65 km to the northeast the Friuli earthquake hit on 6 May 1976 with $M = 6.4$, with relevant aftershocks (Pondrelli et al. 2001). The present local seismicity (OGS-RSC Working Group 2012) is shown in Fig. 2. The seismicity is monitored with a local network due to the presence of a natural gas storage concession. Details on the relation between gas pumping and seismicity can be found in Priolo et al. (2015). The magnitudes range from -1.8 to 4.5, with average magnitude of 0.6 in the period 2012–2017. Clockwise from upper left, the figure displays the seismicity for the period 1 January 2012–31 March 2017. Then a depth profile along the Caneva line and two parallel profiles orthogonal to the thrust are traced. On the seismicity map, the gray star shows the position of the Genziana station, the smaller black dots show locations of local towns, and the three red dots show the three springs marked in Fig. 1. A distinct fault plane (brown line overlain on seismicity depth plot in Fig. 2c) is delineated by the seismicity in the southern profile. In the northern profile, this fault is no longer seen (Fig. 2). The Cansiglio Plateau has very few seismic events, displaying a seismic gap, which could be interpreted as due to the fact that it is a rigid block

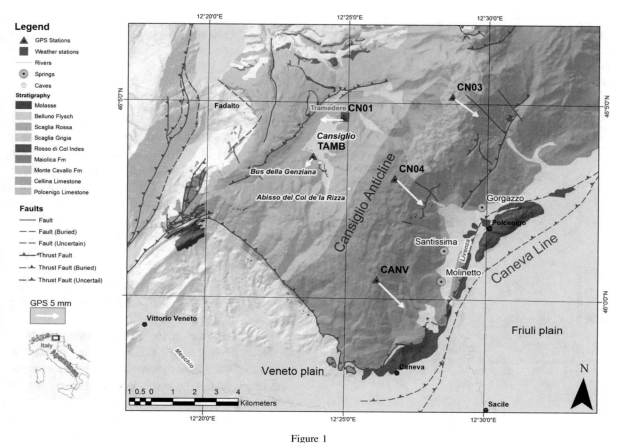

Figure 1

Localization of the study area. Red triangles: location of GPS stations; white arrows show horizontal movement of GPS after strong rainfall (details in Devoti et al. 2015). Genziana cave: tiltmeter station. Green dots: the locations of the three karstic springs forming the Livenza River. Red and blue: dominant regional thrusts. Also shown is the geological setting of the study area with the principal structural lineaments (Design by A. Riva). Surface projection of the Genziana cave sketched in black

with little seismicity. The plateau is bounded at the southwest and southeast by regional thrust faults (Cavallin 1980). The Caneva Line thrust faulting resulted in a relatively wide asymmetric anticline (Cansiglio Anticline) in the hanging wall, with steep dipping beds near the main fault. The main thrust plane is also associated with minor faults developing a quite wide cataclastic zone, about 500 m in width. The main karst springs emerge in the lower limb of the anticline, where the Mesozoic limestones (Upper Jurassic Polcenigo limestone and Cretaceous Cellina limestones) are in tectonic contact with the Cenozoic and Quaternary impermeable units of the footwall.

The River Livenza flows down from the southeastern slope of the carbonatic Massif of Cansiglio–Cavallo. It is supplied by three main springs: the Gorgazzo, which has a recharge basin of 170 km^2, the Santissima of 500 km^2 and the Molinetto of 230 km^2. All three have an average flow from 5 to 10 m^3/s (Cucchi et al. 1999).

The massif is characterized by a markedly deep karst, with about 200 caves and clear karst surface morphology. Although the annual mean precipitation is about 1800 mm, the Cansiglio Plateau for the time being has no surface runoff, but acts like an endorheic basin with a pronounced system of underground drainage through caves. Essentially, two noteworthy caves are considered: the Bus de la Genziana (Fig. 1) with a maximum depth of 587 m and a development of 8 km, and the Abisso Col de la Rizza, the deepest cavity of the area, reaching 800 m below the surface. Morphologically, all caves have a complex tunnel

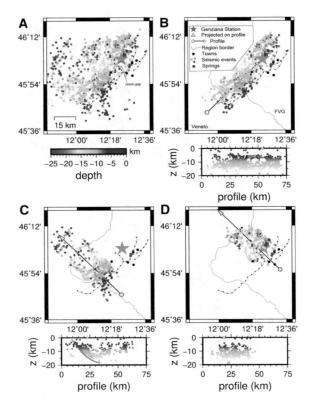

Figure 2
Present local seismicity (OGS-RSC Working Group 2012) for the period 1 January 2012 to 31 March 2017. **a** Epicenters with colors representing depth; following three graphs. **b–d** Seismic events along the different profiles, with depth distribution for the profile shown below the map. The distance along the section extends from arrowtail to arrowhead of the profile

system, including shafts, halls, canyons, meanders and sometimes are also well decorated (Grillo 2007).

The hydrologic connection between the Cansiglio Plateau and two of the three main Friulian sources, the Santissima and Molinetto origins of the Livenza River, have been demonstrated by recent tracing examinations (Filippini et al. 2016).

3. Geodetic and Hydrologic/Environmental Data

3.1. Climate Data

Rainfall, snow, atmospheric pressure and air–temperature data were obtained from the ARPA Veneto (regional Meteorological Service and Snow and Avalanches Service), considering the weather station Cansiglio–Tramedere (1028 m a.s.l.) for

hourly sampling. It is placed at about 8 km north from the CANV station and 2.5 km from the tiltmeter location.

3.2. Hydrological Monitoring

Multiparametric loggers (CTD-Diver, Schlumberger Water Services) have been installed in 2010 in the springs Santissima and Gorgazzo located in Polcenigo at the foothills of the massif, while a barometer (Baro-Diver, Schlumberger Water Services) was placed close to one of the springs for barometric correction of the stage data. The pressure is used to reduce the stage observations for atmospheric pressure variations using the inverse barometric hydrologic response. These instruments record variations in water level (h), temperature (T) and specific electrical conductivity (EC) every hour.

Since the year 2013, in the cave Bus de la Genziana another Diver CTD has been installed to record the water flow dynamics below the tiltmeters, with hourly sampling, at the depth of 587 m from the surface, or 433 m above sea level (Fig. 3a, b).

The hydrometric station of the Livenza River is located at the foothills of the massif in the city of Sacile, about 10 km south from the springs area. The data consist in the river level measurement with hourly sampling and are available through the monitoring network of the Civil Protection of Friuli Venezia Giulia.

3.3. Marussi Tiltmeters

Since 2005, the University of Trieste runs a tiltmeter station in the natural cave "Bus de la Genziana" 25 m below the entrance (Grillo et al. 2011). The Genziana tiltmeters are horizontal pendulums with Zöllner-type suspension and described in detail (Braitenberg 1999a; Zadro and Braitenberg 1999). They are sturdy instruments due to their relatively big size (0.5 m tall) and stable mount, inside a cast iron bell resting on compact rock. The iron bell is sustained by three supports placed on the solid rock. The horizontal arm rotates in the horizontal plane around a subvertical axis, the rotation angle being picked up by a magnetic transducer. The

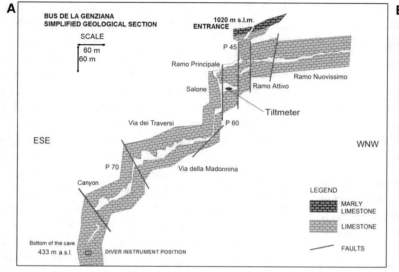

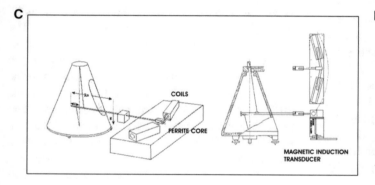

Figure 3

The Genziana cave and instrumentation: **a** Simplified geological section of the Genziana cave and **b** picture of Diver instrument located at the bottom of the cave. **c** Schematic design of the Marussi tiltmeter. The digital acquisition occurs by means of a magnetic induction transducer. **d** The pair of tiltmeters in the Bus de la Genziana–Pian Cansiglio, located 25 m below the surface (Photo: Barbara Grillo)

pendulum is sensitive to a tilting of the ground at 90° with respect to the off-vertical angle of the pendulum's rotation axis (see Fig. 3c) plus a mass attraction effect. The digital data acquisition has a sampling rate of 1 h and uses an inductive transducer. The resolution of the tiltmeter is near 5 nrad, the value corresponding to the unit value of the digitizing process. Sign convention is positive for east- and northward tilting of the pendulum (tilting down).

3.4. GPS Observations

Two permanent GPS stations are available on the Cansiglio Massif, and five temporary stations had been installed to outline the area affected by the hydrologic deformation. The permanent stations are the following: the Caneva station (CANV), belonging to the FReDNet (http://frednet.crs.inogs.it; Battaglia et al. 2003), and the Tambre station (TAMB), owned by the Regione Veneto authority (http://147.162.229.63/Web). The CANV station is located at 800 m a.s.l. at the southern margin of the Cansiglio–Cavallo Plateau at a distance of about 8 km from the Genziana cave and is established on a reinforced concrete pillar anchored on bedrock. The TAMB station is located on the plateau at about 400 m from the entrance of the Genziana cave, settled on the roof of a stone-made house. Five monitoring stations were set up along the southern margin of the plateau and on top of Mt. Pizzoc in the southwestern edge of the plateau and were measured occasionally in the last years (CN01, CN03 and CN04 in Fig. 1).

4. Description of Geodetic Observations in Relation to Hydrologic Data

4.1. Characteristic Signals of Tiltmeter Observations

Interpretation of the tilt signal must take into account both environmental factors and deformation, since the tilting can be provoked by thermal expansion, deformation due to loading or pressure variations, as well as being sensitive to earth tides, tectonic stresses and tectonic deformation (Jahr 2017). The environmental generative sources are temperature, superficial and underground water flow and snow (Jahr et al. 2008), and atmospheric pressure gradient. Our objective is to be able to identify the different signals and quantify the source. The Cansiglio station started recording in December 2005, including an instrumental adjustment and testing period. The useful recordings for representing the tectonic movement start from 13 February 2006. The long-term tilting direction is principally southward, with a slowdown in 2008 (Grillo et al. 2011), relative stability in 2009, an episode of northward movement until 2012 and a return to the southward movement which is still ongoing (Fig. 4a, b). In all graphs, increase of tilt numbers corresponds to downward tilting of the platform on which the tiltmeter stands toward north and east, respectively. The amount of tilting is such that the mass effect on the deviation of the vertical, which would also generate a tilt signal, is negligible.

The contribution of the barometric effect is negligible in comparison to the hydrologic effect, and correlation to atmospheric pressure gradient changes could not be found (Kroner et al. 2005; Boy et al. 2009), probably masked by the hydrologic effect.

The tiltmeter time series are highly correlated with the rainfall data of the Cansiglio weather station (station Tremedere ARPA Veneto) and to the stage data of River Livenza, as can be seen in Figs. 4 and 6.

Both EW and NS components have a seasonal periodical variation, as can be seen after detrending and low pass filtering the time series. By least squares adjustment, a third-order polynomial is fitted to the NS and EW components separately and subtracted; then, the outcome of this operation is low pass filtered by a running average of 10 days. The final residual is the signal that most clearly identifies the hydrologically induced tilting (Fig. 5). In another tilt station (Grotta Gigante station) installed in a karst plateau (Classical Karst, straddling the Italian–Slovenian border; outside the map of Fig. 1), the spectral analysis had demonstrated that the yearly and seasonal signal has both thermal (365, 183 day period) and rainfall (365, 183, 171 days) spectral components, so it is generated by the superposition of both effects (Tenze et al. 2012). Also in Genziana, an annual and semi-annual signal can be identified, although not regularly repeating itself (Fig. 5). The different amplitudes of the long-term movement of the NS and EW components cannot be due to the mountings or instability of the location, because the two instruments are identical and there are no visible differences in the site on which the instruments stand (Fig. 3d). Mutual swap of the position of the two instruments confirmed the southward polarized movement. The tilt effect of barometric changes has been shown to be induced by the horizontal gradient of air pressure (Kroner et al. 2005; Boy et al. 2009).

4.2. Discussion of Hydrogeological Campaign Compared to the Tiltmeter Observations

The Cansiglio karst aquifer is a mature pre-alpine deep karst, characterized by high permeability and rapid runoff through enlarged fractures and caves, although mitigated by the existence of a base flow component through a network of smaller channels (Grillo 2007). The underground karst phenomenon is mainly developed in Monte Cavallo limestone with a complex of caves 600–800 m deep, controlled by the geological–structural setting. The aquifer has a high conductivity and high vulnerability (Cucchi et al. 1999).

Starting in the year 2010, a few hydrological campaigns limited in time were made studying the Gorgazzo (50 m a.s.l.) and Santissima (30 m a.s.l.)

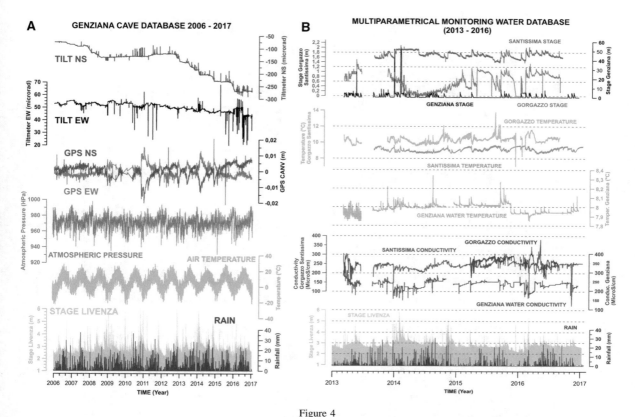

Figure 4

The 10 years of tilt in Genziana cave, hourly rainfall and Livenza River stage. **a** The tilting signal of the two components of the tiltmeter compared with the stage data of Livenza River and the rainfall measured at the Cansiglio weather station. Please notice the different scale factors used for representing the EW and NS components. We note a small eastward movement and a nearly seasonal signal in the component EW, and a southward movement in the component NS, which could be of tectonic origin. The hydrologic signal manifests itself as spikes corresponding to rain events, and also as slow movement in correlation with the runoff curves of the aquifer (Grillo et al. 2011). **b** The hydrologic, temperature and electrical conductivity data of the two springs (Gorgazzo and Santissima springs, and the measurements below Genziana cave. Also shown is the stage of the River Livenza and hourly rainfall

springs (see data in Fig. 4b). These studies confirmed what previous studies had shown: the Gorgazzo spring is characterized by a drainage through a well-developed fracture network, because it has highly variable conductivity and sudden and abundant rate of flow, which normalizes after a few hours and thus falls within the classic case of upward siphons (electrical conductivity between 150 and 370 µS/cm and temperature values of 10–11.5 °C). The Santissima spring has electrical conductivity values between 220 and 300 µS/cm and temperature values of 8–9.5 °C, typical of a less well-developed karst network draining with mixing of waters (Grillo 2007). The system of this spring is different because it has smaller water level fluctuations with longer

times of decay. This, in terms of chemical and physical properties, is in support of a water reservoir of significant extensions feeding the Santissima spring (A.R.P.A. F.V.G. 2006; Filippini et al. 2016). For a limited number of months, the hydrological studies of the two springs overlap with the tiltmeter recordings. As seen in Fig. 6, the tiltmeter signal is associated with the stage signals of the Gorgazzo and Santissima springs, which confirms that the deformation monitored with the tiltmeters is induced mainly by the incoming rainwater and the drainage to the karst conduct system. The impulsive deformation at rainfall onset is highly amplified, and the subsequent slow decay has a similar time evolution to the Gorgazzo although shorter, but the

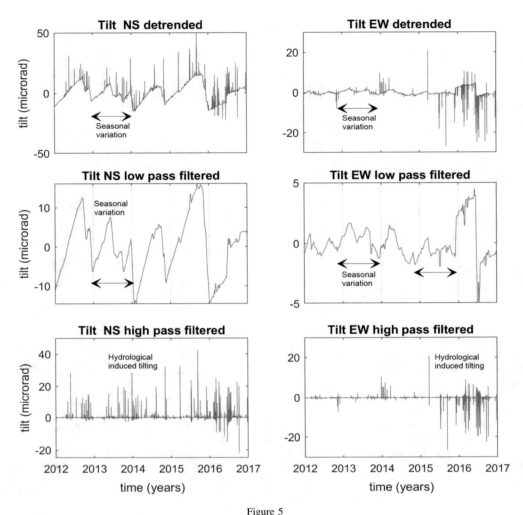

Figure 5
The seasonal and hydrological signal in tilt. The figure shows the detrended, seasonal (low pass filtered) and hydrological (high pass filtered) tilt signal for station Bus de la Genziana. Time window, years 2012–2017

amplitude of the spring is small compared to the onset of deformation. Similar relations hold for the level variations of the Santissima spring.

The level in the siphon, recorded far below the tiltmeters with the CTD data logger, gives definitive clues about the dynamic source of the tilting. Data are available for 4 years, starting 2013, and only partly overlap with the data available for the two springs (Fig. 7a). In contrast to the records of the two springs and River Livenza, the water level is zero, except after rainfalls. This is because the water channels always filled with water are supposed to be at a lower level than the siphon. The water level in the siphon

rises abruptly by 4–5 m after a few millimeters of rain: a statistical analysis has shown that when the rainfall is above 12 mm/h, the water level will rise by more than 5 m. In three extreme cases, the monitored water level had reached 27 m (16 May 2013, 31 mm of rainfall in 5 h, and 25 h for level decay) and 50 m (31 January 2014 and 26 December 2013) with heavy rainfall, lasting for 43 and 79 h, respectively. The velocity of maximum rise was 7 m/h. The characteristic response is an impulsive signal and near to linearly decaying discharge (Fig. 7b). The complete recession of the event lasts up to 10 days.

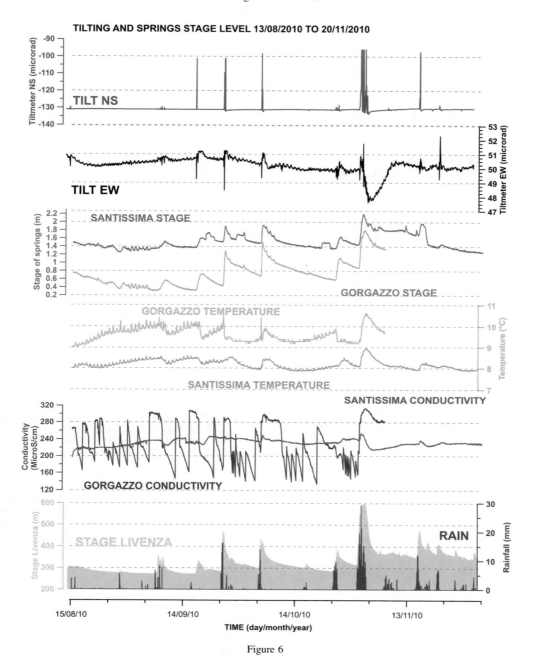

Figure 6
Comparing the river level, the spring level (Santissima 30 m a.s.l., Gorgazzo 50 m a.s.l.) and the tiltmeter signals at 1000 m a.s.l. (EW, NS) from 15 August to 30 November 2010: the spring levels correlate very well with the tiltmeter data. The tilt transients last some days like the groundwater runoff signals

The electrical conductivity values are on average of 230 µS/cm, while the parameter reduces down to 150 µS/cm when the karstic system fills due to rainfall. The water temperature variations are well correlated to the conductivity and also have the characteristic impulsive increase with slow recovery. The medium temperature is 8 °C, with variations limited to − 0.15° to +0.30°. Increase of temperature correlates with reduction of electrical conductivity

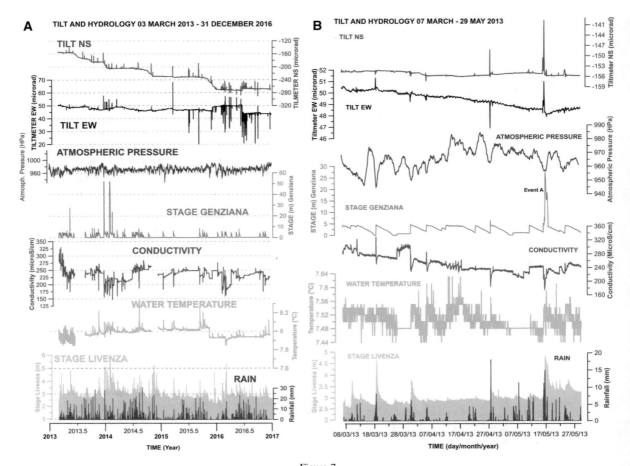

Figure 7
Comparison between tiltmeter data and the CTD multiparametric station below the cave, next to other ambient and hydrologic parameters.
a Entire time series. Tiltmeter data, atmospheric pressure, hourly rainfall and river series, the recordings of the underground Genziana flow
monitoring, conductivity and water temperature. The water level below the cave Genziana rises quickly to 4–5 m with a few mm of rain: it
had reached 27 m in one event and had risen over 50 m two times (compare text). **b** Significant time window cut-out from (**a**), period 7 March
to 29 May 2013. On tilt, we note the earth tides and the impulsive signals marking the onset of the underground water level rise. Event A
shows the time of 25 m water level rise below the tiltmeter station (compare text)

and with onset of rainfall (Grillo and Braitenberg
2015).

These parameters show that the hydrologic system
is affected by mixing of new infiltration water, which
is very fast in the inflow, but slower in the outflow, so
the water level abruptly increases and then slowly
decays.

It is to be expected that the tilting correlates to the
filling of the siphon below the instruments. This is
most convincingly seen in Fig. 7b, where in a time
window of 2.5 months data for the tilt, siphon water
height, level of River Livenza and rainfall are shown.
The river level must be at least 3.3 m high to flood

the siphon, and at all times that the river rises above
this threshold level the siphon is filled. Every time the
siphon is filled, there is a tilting event, with 2 μrad
tilting for a rise of 5 m in the siphon. The tilting has a
short duration and marks the filling stage of the
siphon. The marking of the filling stage is demonstrated best by analyzing the time derivative of the
siphon, as shown in the next figure. Only in rare cases
in which the siphon is filled above the 5 m mark do
the tiltmeters also have a slow recovery of deformation, as at the event A in Fig. 7b.

To further illustrate the tiltmeter response during
the rising stage in the siphon, we calculate the first

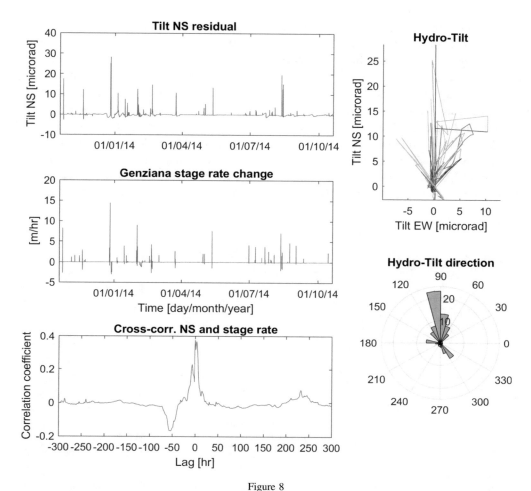

Figure 8
Tilting signal induced by the rapid stage rise in the siphon below the tiltmeter. Upper left: detrended and high pass filtered NS tilt component. Middle left: time derivative of water level in Genziana cave. Lower left: cross-correlation function between tiltmeter and stage rate in siphon. Negative lag: siphon rate delayed with respect to tilt. Upper right: hodograph of tilting in the 12 h following the rapid stage rise in the Genziana cave. Lower right: polar histogram of orientations of hydrologic-induced tilt signal

time derivative of the stage and compare it to the tilt records, calculating the cross-correlation function between the two. A negative lag corresponds to a delay in the siphon The cross-correlation function is maximum (cross-correlation coefficient 0.38) for zero delay between the diver stage rate and the tiltmeter records, and has a smaller but well-developed negative extreme at − 54 h delay of the diver change rate with respect to the tilting.

We further select the time of each steep rise in stage and a time window covering the successive 12 h. We draw the tilt hodograph for each of these events and calculate the azimuth (positive from EW anticlockwise) of the tilting at the time of maximum tilting in each of these windows (Fig. 8).

4.3. Model for Explaining the Diver and Tiltmeter Observations

The physical model that explains the tilt observations must comply with the observations. We have no direct observations of the shafts, so the physical model we propose here is a conjecture. To justify the model, we summarize the responses of the diver and the tilt to variations in rainfall and the Livenza River, and then formulate a structural model that explains the observations. The conceptual model can be used for a numerical simulation of the hydraulic flows and the induced deformation, which requires a separate study based on the present one.

The characteristics of the diver observations below the tilt station are twofold: a general maximum height of 5 m water height and linear level decrease, accompanied by a tilt signal limited mainly to the fast water level increase. A possible model could be made of a shaft of which the lower part evolves to tight fissures filled with silt and mud (see Fig. 9). The shaft

collects the rain water from above and fills up with water up to 5 m maximum height, above which the water drains entering a free lateral channel. The shaft quickly fills up due to the rain and slowly releases the water through the tight fissures. The permeability of the filled fissures is inversely proportional to the water height, so as to cancel the height-proportional term in the Darcy flow rule, and the outgoing flux is at constant rate, as is also the decrease of water level above the diver. Due to the presence of the fissure filling, with viscous rheology, the stress acting on the

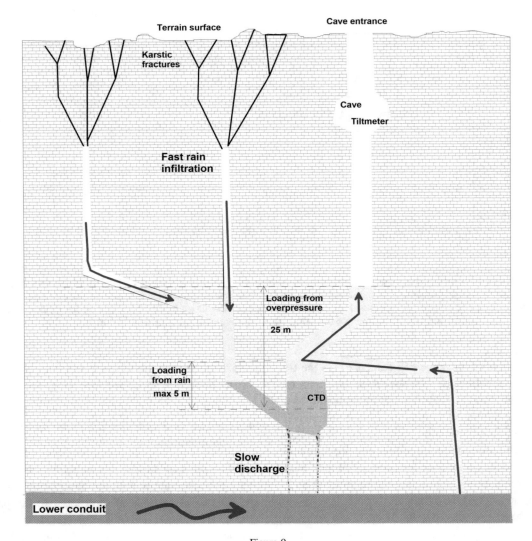

Figure 9
Cartoon illustrating the mechanism that explains the upper limit of 5 m water level and linear level decrease in case of standard rainfalls, and the uprising to over 25 m level in case of exceptionally strong rainfalls that produce an overpressure in the lower conduit (see text)

fissure walls is proportional to the first time derivative of the pressure increase due to the water level height change. Therefore, the movement of the tiltmeters is perceptible only in the impinging phase, whereas during outflow the time derivative is too low to generate an observable tilt signal. This is documented in Fig. 8, where the NS tilting is compared to the first time derivative of the diver. It can be verified that the strong peaks in the time derivative of the diver are well correlated to the peaks of the NS component. The tilting occurs in both NS and EW components for all cases in which the velocity of the increasing diver level overcomes a threshold value (here chosen as of 1 m/h, but also other values can be chosen and the results are not affected). In rare cases of strong rainfall, the flow in the channel system below the cave is such that the water pressure is increased, and the vertical shaft of the diver is filled by a lateral arm we conjecture must be connected with the channel system below the diver. The shaft of the diver acts as a pressure gauge, with a fast increase of the water level by up to 25 m (see Event A in Fig. 7) governed by the pressure pulse in the channel. The level decays after the pressure events are different from the common linear decay described above, because it no longer represents the mud-filled shafts and their permeability, but the flow in the open channel system, with a different flow dynamic. The tilt response is proportional to a superposition of viscous and elastic time evolution.

4.4. Discussion of GPS Observations in Relation to Subsurface Hydrologic Flows

The hydrologic-induced deformation has also been detected with GPS stations installed on the karstic plateau. The results of a dedicated campaign were discussed in Devoti et al. (2015) and it was demonstrated that the horizontal movement of the stations after rainfall is outward toward the margins of the plateau and orthogonal to the prevailing fracture direction (Devoti et al. 2016). The signals were interpreted as due to filling and subsequent pressure exerted by vertical fractures.

In Fig. 4a, we show the updated time series (updated with respect to the paper Devoti et al. 2015) of the permanent station CANV, which has a consistent pattern of deformation correlated to the rainfall (Fig. 10), with a fast displacement (1–2 days) of up to 15 mm and a subsequent slow rebound following the rainfall event (lasting 2–3 weeks). The signals were interpreted as due to the percolation of rainwater filling subvertical fractures resulting in a lateral pressure stress and subsequent elastic opening of the fractures, followed later on by closure after draining.

On comparing the tiltmeter's signal recorded at the Genziana station with the local rainfall series, the Livenza River gauge height, and the time series of the GPS stations located on the southern slopes of the Cansiglio Massif, a clear correspondence with the water runoff is seen (Fig. 10).

The hydrologic-induced deformation was particularly evident during the flood of the Livenza River between October 31 and November 3, 2010 (Fig. 10) (Grillo 2010). In those days, a total amount of 520 mm of rain fell with a peak 29 mm/h. The GPS station and the tiltmeters recorded the displacement and deformation induced by the water influx into the system instantly, and also due to the underground water runoff during the flood. The flood resulted in a GPS maximum displacement of 10 mm to the east and 15 mm to the south. The tiltmeters show a 1 μrad variation toward east and then a continuous westward drift of about 3 μrad during the rainy days, recovering the initial easterly position in the following week. The NS component provided initial complex variations during the rainy period, drifting first to the north and then slowly to the south and recovering the initial position in the following week (see inset in Fig. 10).

5. Discussion

We observed a variety of signals in and around the caves of the Cansiglio Plateau, ranging from standard hydrological signals such as rainfall, stage data, temperature and electrical conductivity, to GPS observations and measurements of tilt in the cave Bus de la Genziana. From this variety of observations, we show how the tiltmeters respond to the fast infiltration, related to the amount and duration of the rainfall, and to a lesser extent to the phreatic discharge, because of lower pressure change. In contrast,

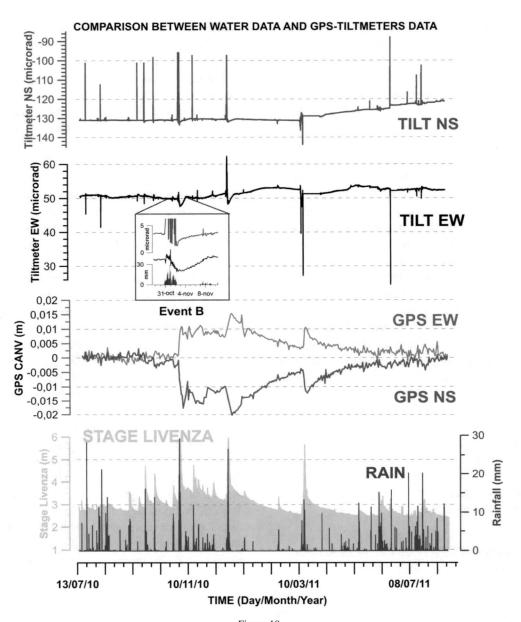

COMPARISON BETWEEN WATER DATA AND GPS-TILTMETERS DATA

Figure 10

GPS, tiltmeter observations, hourly rainfall and stage of the River Livenza during flood events. During the event of the Livenza flood from 31 October to 3 November 2010 (Event B in figure), the total amount of rain was of 520 mm. The interdisciplinary comparison shows a clear correspondence in the recordings. For the GPS NS component, the shift is 16 mm to south and the tilt is 35 μrad southward; for the EW component, it is 12 mm to the east recorded by the GPS and 3 μrad to the east for the tiltmeter. The inset shows a zoom of the EW, NS tilt and rainfall for the period 29 October–10 November 2010. The colors of the curves in the inset are those of the full graph

the horizontal displacement from GPS does not show any reaction from the fast infiltration, but records the slower phreatic discharge. The amplitude relation between the GPS movement and the amount of rainfall is approximately linear, accounting for 3 mm displacement every 100 mm cumulative rainfall for strong events. The prevailing movement is horizontal along a stable direction 130–160° north azimuth for stations at the eastern part of the plateau. The mutual correlation is significant, with a high correlation coefficient of 0.92 (Devoti et al. 2015). The direction of movement points to the area of the two springs mentioned above (Fig. 7) and is directed orthogonal to the frontal thrust systems of the Cansiglio Massif. The high correlation with the rainfall suggests that the deformation of the GPS, like the tiltmeters, is caused by subsurface hydrologic processes in the karstic vadose zone rather than with deep-rooted water table variations in the phreatic zone, as pointed out by Devoti et al. (2015).

A geological and structural survey revealed that the whole southern Cansiglio slope is affected pervasively by structural features related to the Cansiglio thrust and by other NNW–SSE and NNE–SSW conjugate faults and fractures perpendicular to the Cansiglio southern slope (Fig. 11). These structural features are responsible for rock mass weakening that leads to large-scale gravitational instability and favors the nucleation of karstic landforms (dolines and swallow holes) (Devoti et al. 2016).

The region to which the Cansiglio carbonatic massif belongs is subjected to compressional stress-oriented SSE–NNW with subhorizontal angle, as has been deduced from focal mechanisms (Bressan et al. 2003). This direction is nearly compatible with southward tilting that has been observed over the past 10 years. Different studies (Davis et al. 1983; Yeats 1986; Huang and Johnson 2016) have proposed the model in which an anticline on an active thrust fault

shows growth and increase of the anticline topographic slopes. A southward tilting, such as the one documented by the tiltmeters, could be interpreted by this model as being the expression of the southward movement of the massif along the two thrust faults that mark the margin of the southern border of the massif (Fig. 1) and are responsible of the southward tilting of the slope facing the thrust fault.

6. Conclusions

We describe an active slope deformation monitored with GPS and tiltmeter stations in a karstic limestone plateau in southeastern Alps (Cansiglio Plateau), one of the most interesting karstic areas of northeastern Italy. Considering the geophysical and hydrogeologic setting, the geodetic tilt station located in Bus de la Genziana is situated in a strategic position. The tiltmeters record crustal movements: long-term tectonic movement and hydrogeological information (epikarst fast infiltration and phreatic slow unloading).

The long-term tilt has a definite southward direction, with episodes of stability and northward movement. The tilting reflects the actual tectonic situation in northeast Italy, which shows the convergence of the Adriatic and Euroasiatic plates.

The long-term movement is overprinted by fast tilting induced by rainfall and a slower movement due to the phreatic water discharge. GPS observations

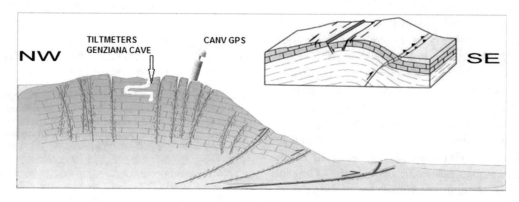

Figure 11

Cansiglio Plateau model: the slope is affected pervasively by structural features (faults, thrusts, fractures). The tiltmeters and GPS observations record two different movements: the epikarst fast infiltrations and the phreatic discharge (modified after Devoti et al. 2016). The long-term tilting could be due to a tilting of the slope facing the thrust, due to the growth of the anticline

record the slower movement in the horizontal components as well, and movement is principally horizontal in the SSW direction and back, and can reach the order of 1 cm during the hydrologic-induced movement. The recovery of the deformation following the rain can last up to 2 weeks. The interpretation of the different time constants is a fast water pressure change in the epikarst and a slower water pressure buildup at deeper karst levels, where water fills the karstic channels, with local level increments of up to 50 m. We can approximate the behavior of the Cansiglio Massif as a network of drainage channels with dominant preferential directions, flowing in the karst vadose reticulum with different trigger times depending on the amount and duration of rainfall. The long-term deformation pattern revealed by geodetic instruments probably reflects the discharge of the karst aquifer, a first impulsive reaction due to rapid and turbulent flow in the conduit network, followed by a slow discharge in the porous matrix (pores and fissures).

The direct link between the aquifer system cycles and the induced surface deformation provides interesting insights into karst-style hydrological processes that could also be relevant in the assessment of hydrologic hazards. The GPS and the tilt observations are complementary and sensitive enough to study and monitor the effects of water infiltration in karst systems.

Considering that the southern Alpine front accommodates a compression of a few millimeters per year and that the area is known to be in Zone 2 (defined as: Zone 2: Municipalities in this area may be affected by quite strong earthquakes, Protezione Civile 2015) at medium to high seismic risk, we are asking how and if these sudden shifts due to hydraulic load may affect the geophysical and geodynamic context of the Cansiglio area. The analysis of time series of the permanent GPS at Caneva suggests an elastic response of a hydro-structure with a drainage system directed along NW–SE, parallel to the direction of the complex headwater of Polcenigo–Caneva. Furthermore, an improved knowledge of the relation between the karstic water flows and the geodetic signal will allow the use of the geodetic measurement as a hydrologic investigation tool in the future.

Acknowledgements

We thank Alberto Casagrande for the constant engagement and for the precious collaboration; A.R.P.A. Veneto Centro Meteorologico of Teolo for providing the meteorological data; Ivano di Fant (Ufficio Idrografico del Servizio Gestione delle Risorse Idriche—Friuli Venezia Giulia) for the hydrometrical data of Livenza River; Dr. Alberto Riva for providing the geological map; OGS-RSC Working Group for publishing the local seismicity on their homepage; Dr. Paola Favero and the Comando Forestale of Pian Cansiglio for the hospitality and the collaboration; the local cavers (in particular, the speleologists of Unione Speleologica Pordenonese CAI Pordenone) for the support and collaboration in the installation and sampling of the diver in the bottom of Bus de la Genziana. We acknowledge the use of the GMT software (Wessel et al. 2013). We thank Georg Kaufmann and an anonymous reviewer for meticulous reviews.

References

A.R.P.A. F.V.G. (2006). Rilevamento dello stato dei corpi idrici sotterranei della Regione Friuli Venezia Giulia (Survey on the state of underground hydrologic units of the Friuli Venezia Giulia Region). Final Report, pp. 68–71. Regione Autonoma Friuli Venezia Giulia.

Battaglia, M., Zuliani, D., Pascutti, D., Michelini, A., Marson, I., Murray, M. H., et al. (2003). Network assesses earthquake potential in Italy's Southern Alps. EOS, 84(28), 262–264.

Boy, J.-P., Longuevergne, L., Boudin, F., Jacob, T., Lyard, F., Llubes, M., et al. (2009). Modelling atmospheric and induced non-tidal oceanic loading contributions to surface gravity and tilt measurements. Journal of Geodynamics. https://doi.org/10.1016/j.jog.2009.09.022.

Braitenberg, C. (1999a). The Friuli (NE Italy) tilt/strain gauges and short term observations. Annali di Geofisica, 42, 1–28. https://doi.org/10.4401/ag-3745.

Braitenberg, C. (1999b). Estimating the hydrologic induced signal in geodetic measurements with predicitive filtering methods. Geophysical Research Letters, 26, 775–778.

Braitenberg, C. (1999c). The hydrologic induced strain—A review. Marees Terrestres Bulletin D'Informations, 131, 1071–1081.

Braitenberg, C., Grillo, B., Nagy, I., Zidarich, S., & Piccin, A. (2007). La stazione geodetico—geofisica ipogea del Bus de la Genziana (1000VTV)—Pian Cansiglio. Atti e Memorie della C.G.E.B., S.A.G. CAI, Trieste, Italia, 41, 105–120.

Bressan, G., Bragato, P. L., & Venturini, C. (2003). Stress and strain tensors based on focal mechanisms in the seismotectonic framework of the Friuli-Venezia Giulia Region (Northeastern

Italy). *Bulletin of the Seismological Society of America, 93,* 1280–1297. https://doi.org/10.1785/0120020058.

Cancian, G., & Ghetti, S. (1989). Stratigrafia del Bus de la Genziana (Cansiglio, Prealpi Venete). *Studi Trentini di Scienze Naturali—Acta Biologica, Trento, 65,* 125–140.

Cavallin, A. (1980). Assetto strutturale del Massiccio Cansiglio—Cavallo, Prealpi Carniche Occ. Atti del 2° Convegno di Studi sul Territorio della provincia di Pordenone (Piancavallo, 19–2 ottobre 1979).

Cucchi, F., Forti, P., Giaconi, M., & Giorgetti, F. (1999). Note idrogeologiche sulle sorgenti del Fiume Livenza. Atti della Giornata Mondiale dell'Acqua "Acque Sotterranee: Risorsa Invisibile", Roma, 23 marzo 1998 (Pubbl. n°1955 del GNDCI, LR49), pp. 51–60.

Davis, D., Suppe, J., & Dahlen, F. A. (1983). Mechanics of fold and thrust belts and accretionary wedges. *Journal of Geophysical Research, 88,* 1153–1172.

Devoti, R., Falcucci, E., Gori, S., Poli, M.E., Zanferrari, A., et al.. (2016). Karstic slope "breathing": Morpho-structural influence and hazard implication. Poster General Assembly EGU, Vienna.

Devoti, R., Zuliani, D., Braitenberg, C., Fabris, P., & Grillo, B. (2015). Hydrologically induced slope deformations detected by GPS and clinometric surveys in the Cansiglio Plateau, southern Alps. *Earth and Planetary Science Letters, 419,* 134–142. https://doi.org/10.1016/j.epsl.2015.03.023.

Filippini, M., Casagrande, G., Fiorucci, A., Gargini, A., Grillo, B., Riva, A., et al. (2016). Geological and hydrogeological investigations for the design of a multitracer test in a major karst aquifer (Cansiglio-Cavallo, Italian Alps). *Rendiconti online della Società Geologica Italiana, 39*(suppl. 1). ISSN 2035-8008.

Grillo, B. (2007). Contributo alle conoscenze idrogeologiche dell'Altopiano del Cansiglio. *Atti e Memorie della C.G.E.B., S.A.G. CAI, Trieste, Italia, 41,* 5–15.

Grillo, B. (2010). *Applicazioni geodetiche allo studio dell'idrogeologia del Cansiglio. AA. 2009–2010.* Master Thesis, Environmental Sciences University of Trieste. Edizioni Accademiche Italiane. ISBN 978-3-639-66321-1.

Grillo, B., & Braitenberg, C. (2015). Monitoraggio delle acque di fondo del Bus de la Genziana (Pian Cansiglio, Nord-Est Italia). *Atti e Memorie della C.G.E.B., S.A.G. CAI, Trieste, Italia, 46,* 3–14.

Grillo, B., Braitenberg, C., Devoti, R., & Nagy, I. (2011). The study of karstic aquifers by geodetic measurements in Bus de la Genziana station—Cansiglio plateau (North-Eastern Italy). *Acta Carsologica.* https://doi.org/10.3986/ac.v40i1.35.

Huang, W. J., & Johnson, K. M. (2016). A boundary element model of fault-cored anticlines incorporating the combined mechanisms of fault slip and buckling. *Terrestrial, Atmospheric and Oceanic Sciences (TAO).* https://doi.org/10.3319/TAO.2015.06.18.01(TT).

Jahr, T. (2017). Non-tidal tilt and strain signals recorded at the Geodynamic Observatory Moxa, Thuringia/Germany. *Geodesy and Geodynamics,.* https://doi.org/10.1016/j.geog.2017.03.015.

Jahr, T., Jentzsch, G., Gebauer, A., & Lau, T. (2008). Deformation, seismicity, and fluids: Results of the 2004/2005 water injection experiment at the KTB/Germany. *Journal of Geophysical Research, 113*(B11), 410.

Kroner, C., Jahr, T., Kuhlmann, S., & Fischer, K. D. (2005). Pressure-induced noise on horizontal seismometer and strainmeter records evaluated by finite element modeling. *Geophysical Journal International, 161*(1), 167–178. https://doi.org/10.1111/j.1365-246X.2005.02576.x.

Longuevergne, L., Florsch, N., Boudin, F., Oudin, L., & Camerlynck, C. (2009). Tilt and strain deformation induced by hydrologically active natural fractures: Application to the tiltmeters installed in Sainte-Croix-aux-Mines observatory (France). *Geophysical Journal International, 178,* 667–677. https://doi.org/10.1111/j.1365-246X.2009.04197.x.

OGS-RSC Working Group. (2012). Rete Sismica di Collalto. http://rete-collalto.crs.inogs.it, https://doi.org/10.7914/sn/ev.

Pettenati, F., & Sirovich, L. (2003). Source inversion of intensity patterns of earthquakes: A destructive shock in 1936 in the northeast Italy. *Journal of Geophysical Research, 109,* B10309. https://doi.org/10.1029/2003JB002919.

Pondrelli, S., Ekström, G., & Morelli, A. (2001). Seismotectonic re-evaluation of the 1976 Friuli, Italy, seismic sequence. *Journal of Seismology, 5,* 73–83. https://doi.org/10.1023/A:1009822018837.

Priolo E., Romanelli, M., Plasencia Linares, M. P., Garbin, M., Peruzza, L., Romano, M. A., et al. (2015). Seismic monitoring of an underground natural gas storage facility: The Collalto Seismic Network. *Seismological Research Letters, 86*(1), 109–123 + esupp. https://doi.org/10.1785/0220140087

Protezione Civile. (2015). Retrieved October 10, 2017, from http://www.protezionecivile.gov.it/jcms/en/classificazione.wp.

Tenze, D., Braitenberg, C., & Nagy, I. (2012). Karst deformations due to environmental factors: Evidences from the horizontal pendulums of Grotta Gigante, Italy. *Bollettino di Geofisica Teorica ed Applicata, 53,* 331–345. https://doi.org/10.4430/bgta0049.

Wang, R., & Kümpel, H.-J. (2003). Poroelasticity: Efficient modeling of strongly coupled, slow deformation processes in multilayered half-space. *Geophysics, 68*(2), 1–13. https://doi.org/10.1190/1.1567241.

Wessel, P., Smith, W. H. F., Scharroo, R., Luis, J. F., & Wobbe, F. (2013). Generic Mapping Tools: Improved version released. *Eos Transactions American Geophysical Union, 94,* 409–410.

Yeats, R. S. (1986). Active faults related to folding. In R. E. Wallace (Ed.), *Active tectonics* (pp. 63–79). Washington, DC: National Academy Press.

Zadro, M., & Braitenberg, C. (1999). Measurements and interpretations of tilt-strain gauges in seismically active areas. *Earth Science Reviews, 47,* 151–187. https://doi.org/10.1016/S0012-8252(99)00028-8.

Zuliani, D. (2003). FReDNet: una rete di ricevitori GPS per la valutazione del potenziale sismico nelle Alpi sudorientali italiane. GNGTS, Atti del 22° Convegno Nazionale, Roma, 18–20 November 2003. https://doi.org/10.13140/rg.2.1.1350.1845.

Zuliani, D., Priolo, E., Palmieri, F., & Fabris, P. (2009). Progetto GPS-RTK: una rete GPS per il posizionamento in tempo reale nel Friuli Venezia Giulia. *Dimensione Geometra, 12,* 30–36.

(Received November 14, 2017, revised March 28, 2018, accepted March 30, 2018, Published online April 7, 2018)

Pure Appl. Geophys. 175 (2018), 1783–1792
© 2017 Springer International Publishing AG
https://doi.org/10.1007/s00024-017-1585-z

❙ Pure and Applied Geophysics

Earth Tide Analysis Specifics in Case of Unstable Aquifer Regime

Evgeny Vinogradov,[1] Ella Gorbunova,[1] Alina Besedina,[1] and Nikolay Kabychenko[1]

Abstract—We consider the main factors that affect underground water flow including aquifer supply, collector state, and distant earthquakes seismic waves' passage. In geodynamically stable conditions underground inflow change can significantly distort hydrogeological response to Earth tides, which leads to the incorrect estimation of phase shift between tidal harmonics of ground displacement and water level variations in a wellbore. Besides an original approach to phase shift estimation that allows us to get one value per day for the semidiurnal M_2 wave, we offer the empirical method of excluding periods of time that are strongly affected by high inflow. In spite of rather strong ground motion during earthquake waves' passage, we did not observe corresponding phase shift change against the background on significant recurrent variations due to fluctuating inflow influence. Though inflow variations do not look like the only important parameter that must be taken into consideration while performing phase shift analysis, permeability estimation is not adequate without correction based on background alternations of aquifer parameters due to natural and anthropogenic reasons.

Key words: Unstable confined aquifer, tidal analysis, hydrogeological response to seismic waves, phase shift, underground inflow.

1. Introduction

Underground fluid motion is often considered as an indicator of Earth crust deformation (Kissin 2015), so permanent hydrogeological monitoring is a significant tool for saturated rock state diagnosis.

Precise hydrogeological monitoring is usually performed in seismoactive regions such as Japan, Kamchatka, Eastern Taiwan, etc. to study co- and postseismic earthquakes' influence on aquifers (Liu

et al. 2006; Kocharyan et al. 2011; Xue et al. 2013; Shi et al. 2015) and to analyze informativeness of wellbores that are being used for hydrogeodeformation observation (Kopylova et al. 2007). Researchers identify several main areas of aquifers' permeability to study. The first group involves studies of fractured aquifer type by harmonic analysis of Earth tides and barometric pressure (Rahi and Halihan 2013) with special emphasis on estimation of aquifer protectability (Hussein et al. 2013). Other studies include the assessment of filtration characteristics of saturated rocks and fault affected zones in terms of quazistatic poroelastic theory (Rojstaczer 1988; Furbish 1991; Ritzi et al. 1991). The same approach can be useful to analyze massif deformation mode (Cutillo and Bredehoeft 2011; Burbey et al. 2012; Lai et al. 2013; Xue et al. 2013), and hydraulic fracturing effectiveness (Burbey and Zhang 2010). Allegre et al. (2016) proved that tidal analysis prevails over pumping tests for permeability estimation within fault affected zones. Phase shift in tidal waves as an indicator of permeability change due to seismic influence was considered in Elkhoury et al. (2006), Doan et al. (2006) and Kitagawa et al. (2011).

Though precise hydrogeological monitoring is rather effective for the estimation of reservoir properties in the case of fractured collectors, it is essential to take into consideration background parameters that can lead to non-stationary flow and incorrect results. In this paper, we present the results of the precise monitoring of underground water level in platform conditions within the Moscow region. We studied the phase shift of Earth tide waves M_2 between water level variations and ground displacement, and considered the main factors that can affect them, including high inflow periods, distant earthquakes, and geological conditions.

[1] Institute of Geosphere Dynamics, Russian Academy of Sciences, 38 Leninskiy Prosp., Bldg. 1, Moscow 119334, Russian Federation. E-mail: gian.vin@gmail.com

2. Observational Area and In Situ Data

We review the water level change as a response to Earth tides in an open well that was drilled within the territory of the Mikhnevo Geophysical Observatory (54.96 N. lat. 37.77 E. long., central part of the Russian Platform, 80 km to the south of Moscow; "Mikhnevo" GPO, international code—MHV). An experimental area is located in the southern part of the Moscow artesian basin that represents a complex multilayer system of monoclinal bedding aquifers and aquicludes that gently slope in a north-easterly direction (Fig. 1).

Ground waters are partly developed in quaternary sand layers. Upper aquifers are supplied at watershed areas and discharged within a local erosive river system. According to drilling tests, unconfined Lopasny (C_2lp) and Narsky (C_2nr) aquifers are spread in the Middle carboniferous Kashira deposit and separated by Middle carboniferous Khatun horizon. Underlying confined Aleksin aquifer (C_1al) belongs to the Lower carboniferous Tarussa-Oksky aquiferous complex (C_1ok-tr) and is bounded by two aquicludes. Due to different depths of uneven-aged aquifers' levels we consider them as disconnected. Long-standing hydrogeological monitoring data confirm different conditions of underground water regime formation.

Oka river valley lies 14 km to the south of "Mikhnevo" GPO and partly erodes the top of the Tarussa–Oksky aquiferous complex and is considered as a contour of permanent pressure head. The Aleksin aquifer is tapped in the interval of 92–115 m; major permeable fractured and cavernous zones were found at the depths of 100.5–101.7 and 104.5–105.5 m. The pressure head reaches 25 m within the observation site. Observational area hydrogeological characteristics are presented in detail in Besedina et al. (2016).

Hydrogeologically, "Mikhnevo" GPO research area is undisturbed and distinctively affected by seasonal supply. The water level normally rises during flash flood periods in spring and fall and drops in summer and winter. Amplitude of annual alterations varies from 1.2 to 2.6 m. The cross-correlation coefficient between water level in the wellbore and Oka river surface (according to Serpukhov hydrological station) does not exceed 0.4 and obtains its maximum value in 2012–2013 with a time shift of 30–40 days (Fig. 2). Significant distance from the wellbore to the river valley in comparison to Aleksin aquifer thickness and the self-restrained strike of aquicludes makes it reasonable to consider this aquifer as uniform and confined.

Precise observation of groundwater level change in the well was conducted using a LMP308i submersible precision digital level sensor with a 2-mm accuracy and sampling rate of 1 Hz. The Vontage Pro-2 digital meteorological station is installed at the surface for precision registration of barometric pressure with an accuracy of 0.1 hPa.

Vertical displacement of the ground surface was theoretically calculated with the ETERNA 3.0 program (Wenzel 1994). In Besedina et al. (2015) we used Fourier analysis to determine the main types of tidal waves. The most stable is M_2 wave that primarily contributes a semidiurnal component. Main diurnal waves are lunar O_1 and lunar-solar K_1 waves. The most significant wave in barometric pressure is S_2 that exists in water level variations, but is much less than other waves. K_1 wave is hardly seen.

We have been monitoring water level in the well since March 2008 but due to recurrent technical problems there were a lot of long-term disruptions in data. In this paper we consider three long spaces of continuous observation: January 2011–May 2013, July 2013–May 2014, and August 2014–January 2016. All the handling was the same for every part of data.

3. Research Methods

Figure 3 shows spectra of water level variation data for the time interval from January 2011 to May 2013. The relative power of different waves corresponds to a confined aquifer type (Rahi and Halihan 2013) that correlates with early described hydrogeological investigations.

To estimate the barometric effect on water level variations in the range of Earth tides we evaluated coherence functions between barometric pressure, ground displacement, and water level (Fig. 4). Initial data from January 2011 to May 2013 were divided into eight segments weighted by Hamming window

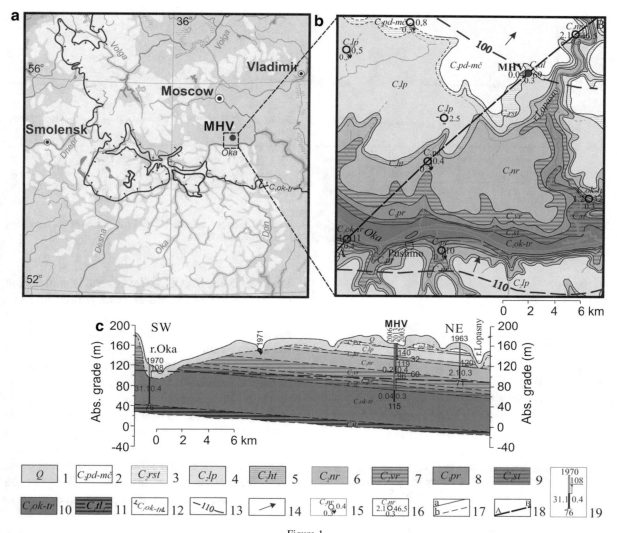

Figure 1

Experimental area location (**a**), hydrogeological scheme (**b**) and cross section (**c**): *1* quaternary deposits (sand, loamy sand, loam, shale), *2, 4, 6, 8, 10* carboniferous aquifers, *3, 5, 7, 9, 11* aquicludes; *12* Taryssa-Oksko (Aleksin) aquifer spread contour *13* Taryssa-Oksko (Aleksin) aquifer hydroisohype in absolute grade (m) *14* Taryssa-Oksko (Aleksin) main direction (*arrow*); *15* the gravity spring, *above* geological index and hydrogeological survey date at the cross section, *below* mineralization (g/l), on the *right* discharge (l/s); *16* the well, *above* geological index, *below* mineralization (g/l), on the *left* specific capacity (l/s), on the *right* level (m); *17* spread boundaries of aquifers and aquicludes: established **a** *solid line*, anticipated *dashed line*; *18* the line of cross section; *19* the well at the cross section, *above* drilling date, *below* deep (m), on the *left* specific capacity (l/s), on the *right* level (abs.grade, m) and mineralization (g/l), the *arrow* head, *blue band* length of open interval. The *red circle marks* Geophysical Observatory "Miknevo". Hydrogeological scheme and cross section are based on State Geological Map. Scale 1: 200,000. Moscow series. Sheet # N-37-VIII. Moscow. 1982 (in Russian)

with the overlap of 50%. The coherence function values between water level variations and ground displacement for the tidal waves frequencies is about 0.95–0.98 while between water level and barometric pressure it is close to 1 for all the range except tidal where it dramatically drops. Coherence function values at the M_2 and O_1 lunar-type waves are beneath 0.2, at the K_1 lunar-solar wave—beneath 0.3. Whereupon the coherence function between the water level and the atmospheric pressure is close to 1 for the S_2 solar-type wave since barometric pressure has a strong harmonic at this frequency as well. We avoid

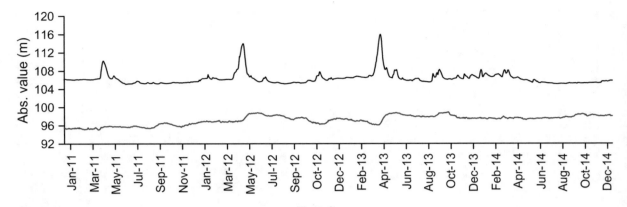

Figure 2

Water level in the wellbore (*blue line*) and the Oka river level (*black line*). The Oka river data are provided by the funds of the Hydrology Department of the Central Hydrometeorological Office

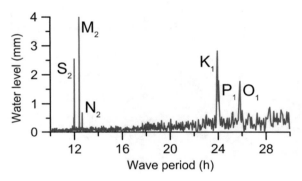

Figure 3

Spectra of water level variations in a wellbore tapped Aleksin aquifer. Data for the time interval of August, 2014–January, 2016 inclusive, sample rate $F_S = 1/300$ Hz

K_1 wave because of thermal effects and core nutation (Doan et al. 2006). M_2 wave is the most stable and makes the biggest contribution to the semidiurnal tidal components (Besedina et al. 2015) so our further analysis is based on this harmonic.

We used two different methods to calculate phase shift between M_2 components in water level variations and ground displacement. In the first standard technique, we performed analysis in the frequency domain. We decimated both ETERNA and water level data by the order of 300 and then filtered them by first-order Butterworth digital filter in the range of 12.32–12.52 h, i.e. app. 10 min apart from semidiurnal M_2 wave with period $T = 12.4206$ h (Melchior 1966). According to the majority of papers that deal with tidal waves (Doan et al. 2006; Xue et al. 2013;

Lai et al. 2013, etc.) a 28-day interval is sufficient to distinguish M_2 and S_2 waves (with periods of 12.42 and 12.00 h accordingly) and minimize spectral leakage. So, we divided long data spaces into four-week parts with a two-week overlap and calculated spectra of every part. M_2 wave frequency corresponds to 54th harmonic—we take the largest value of the two closest harmonics to $n_F = \frac{N}{F_S \cdot \tau(M_2)} = \frac{8064}{12 \cdot 12.42} \approx 54.1$, where N is the number of counts, F_S is a sample rate, and $\tau(M_2) \approx 12.42$ hours is the period of M_2 wave. The resulting M_2 phase of water level variation was subtracted from the ETERNA data to find phase shift. The shortcoming of this approach is that we can find an average phase shift of a two-week term to have rather precise values.

The second approach enables us to find out a phase shift value for every wave period (in practice we took one value per day for simplicity). Here we use initial data with sample rate $F_S = 1$ Hz that are filtered in the same range of 12.32–12.52 h. As far as spectra harmonic is a complex number $A + jB$ the corresponding wave can be calculated as $(A \cdot \cos(\omega t) - B \cdot \sin(\omega t) \cdot {}^2/_N$, where $\omega = {}^{2\pi}/_f$ is an angular frequency. To estimate phase shift we can draw an ellipse "ground displacement–water level variation" equation $\left(\frac{y}{B}\right)^2 + \left(\frac{x}{A}\right)^2 - \frac{2xy\cos(\varphi)}{A \cdot B} = \sin^2(\varphi)$, where $x = A\sin(\omega t)$ and $y = B\sin(\omega t - \varphi)$ are signals along axes, and φ is a required phase shift (Fig. 4). Putting $x = 0$ we get $\varphi = \arcsin\left(\frac{y_1}{B}\right)$. Ground displacement and water level variations are close to antiphase and if a point on the ellipse moves

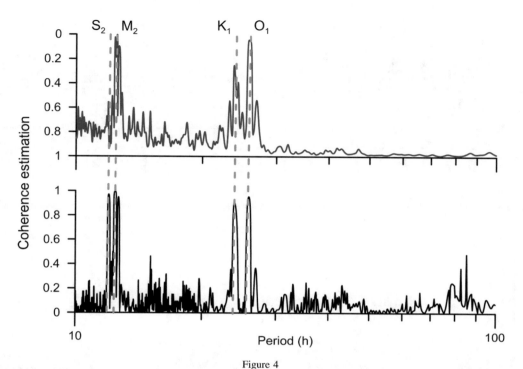

Figure 4
Coherence function between data rows: *top* water level and barometric pressure; *bottom* water level and theoretical ground displacement.
Dashed lines correspond to tidal waves periods

clockwise then phase shift $d\varphi = -180\circ -\varphi$, in case of counterclockwise direction $d\varphi = -180 \circ +\varphi$ (Fig. 5).

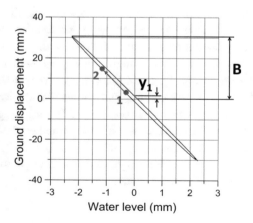

Figure 5
Ground displacement–water level variations (data are filtered in the range of 12.32–12.52 h, round the M_2 wave): y_1 corresponds to the moment, when water level shift equals zero, B is the maximum daily ground displacement. Clockwise consecutive movement (from moment 1 to moment 2) corresponds to leading ground displacement relatively to water level

Both methods give similar results (Fig. 6) so we chose the second method due to the better resolution (one item per day). Henceforward we change water level variations to minus water level variations to exclude permanent ($-180°$) item and consider phase shift alterations around null value.

4. Measurement Results

We performed a comprehensive analysis of phase shifts between ground displacement and water level variations for M_2 tidal wave, high flow rates, and distant earthquakes' impact.

In the case of a stable aquifer, as was described in Hsieh et al. (1987), phase shift between water level variations and pressure head disturbance (that is always in antiphase with vertical ground displacement) must be negative; thus positive values mark terms when aquifer cannot be considered in a frame of this theory. It means that it was affected by some endogenic or exogenous impact.

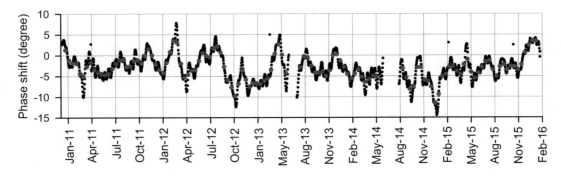

Figure 6

Phase shifts found in two ways: *red crosses* spectra calculating of 28-day spaces; *black dots* daily values estimated from ellipses

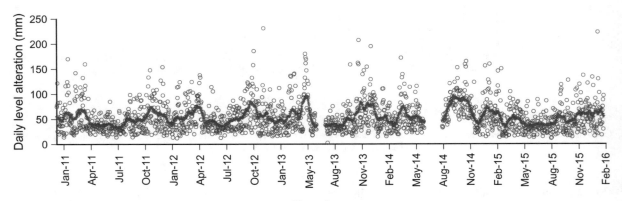

Figure 7

Daily level alterations of the water level. *Blue circles* initial daily level variation. *Red circles* values smoothed by running window of 28 days

We estimated daily water level alteration as the difference between daily minimum and maximum values. As the tidal component variation is a rather slow process, we smoothed alteration data by a running window of 28 days (Fig. 7). For convenience, we duplicate smoothed data in Fig. 8 to compare high flow periods with phase shift of M_2 wave. The purple line corresponds to the median value of smoothed data that is equal to 51.35 mm (initial data median value equals to 45.1 mm). We discard days when level variation exceeds this value (red areas in Fig. 8), which allows us to retain periods of time when the aquifer is not affected by heavy inflow that can distort data. Blue arrows correspond to distant earthquakes that were evidently seen in the water level data.

Figure 9 shows distribution of phase shifts between ground displacement and water level variations. Red rectangles correspond to all the 1766 days from spaces under consideration and striped blue rectangles are plotted on the base of 883 days that lie outside high inflow periods of time (red areas in Fig. 8). Solid lines mark corresponding normal Gauss distribution. Plot shows that the average value stays the same ($\varphi_{full} = -3.0°$, $\varphi_{corrected} = -2.95°$) while standard deviation of corrected data is about 20% smaller $\sigma_{full} = 3.3°$, $\sigma_{corrected} = 2.7°$, so corrected range $\varphi \pm \sigma$ stays negative though there still are some positive values.

5. Discussion

We recorded the hydrological response of the aquifer in a geodynamically stable platform to tides.

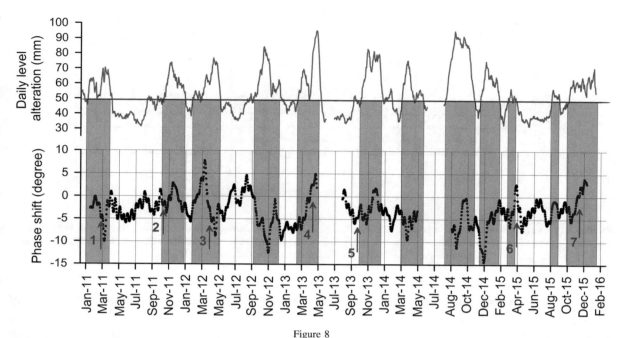

Figure 8

Phase shifts (*black dots*) between ground displacement and water level variations for M_2 wave. Daily water level alteration (*red line*, similar to one in Fig. 5). *Red areas* correspond to days with high inflow. *Blue arrows* show distant earthquakes: *1* March 11, 2011, $M_w = 9.1$; *2* October 23, 2011, $M_w = 7.1$; *3* April 11, 2012, $M_w = 8.6$; *4* April 16, 2013, $M_w = 7.7$; *5* September 24, 2013, mb $= 7.7$; *6* April 25, 2015, $M_w = 7.9$; *7* December 7, 2015, $M_w = 7.2$. *Purple line* corresponds to median value of smoothed data

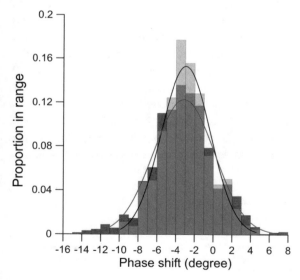

Figure 9

Distribution of phase shift between ground displacement and water level variations. *Red rectangles* correspond to phase shift estimated by initial data, *red solid lines* normal distribution with $\mu = -3.0$, $\sigma = 3.3$. *Striped blue rectangles* correspond to corrected phase shift by elimination of high inflow periods, *black solid lines* normal distribution with $\mu = -2.95$, $\sigma = 2.70$

Consider phase shift dependence on aquifer's transmissivity using the method offered by Hsieh et al. (1987). Our aquifer and wellbore parameters are as follows: radius of well $r_w = 0.059$ m, well casing $r_c = 0.0635$ m, aquifer specific storage $S = 2.3 \times 10^{-4}$, and M_2 wave period $\tau \approx 44715$ s. Figure 10 shows estimated phase shift as a function of transmissivity. Excluding phase shift values on days when inflow was too high allows us to reduce statistically possible transmissivity range to approximately 1.6–30 m²/day while initial range lasts from 1.3 to infinity. Though range is still rather wide, calculated average transmissivity value $T_\varphi \approx 3$ m²/day is close to the one that was estimated after pumping tests $T_{pumping} \approx 4 \pm 2$ m²/day (green area in Fig. 10). Hence according to tidal variations from January 2011 to January 2016 permeability of the unevenly fractured collector belongs to the range of 8.2×10^{-14} to 15.4×10^{-13} m².

Tidal analysis shows that permeability of different zones of enhanced fracturing and faults varies in a wide range (Xue et al. 2013, 2016; Allegre et al.

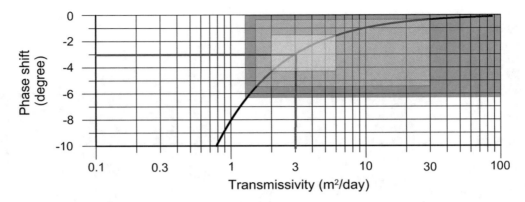

Figure 10

Relation between aquifer transmissivity and phase shift in the range of tidal waves. *Black line* theoretical curve (Hsieh et al. 1987), *red line marks* mean value of phase shift distribution. Big area corresponds to the estimation based on all the values of phase shift, *medium* after elimination of high inflow periods, *small* as estimated after pumping tests

2016). It is rather low for sandstones— $5.6–9.0 \times 10^{-14}$ m^2 in the zone of a regional fault— and rises up to $26–33 \times 10^{-14}$ m^2 in the rock massif and in the affected zone of poorly exposed faults (Allegre et al. 2016). In the close zone of the active San Andreas fault permeability of fractured gabbro can change in the wide range from $1.5–2.2 \times 10^{-15}$ to 1.6×10^{-14} m^2 within the small area about 0.1 km^2 (Xue et al. 2016). Thus permeability of "Mikhnevo" GPO fractured carbonate collector $k = 8.2 \times 10^{-14}$ to 15.4×10^{-13} m^2 lays within data range of other wellbores in different regions and in our opinion corresponds to dynamic variations of aquifer specific storage due to Earth tides.

Several papers (Brodsky et al. 2003; Elkhoury et al. 2006; Shi et al. 2015; etc.) consider near- and far-field earthquakes as one of the main causes of aquifer permeability change. Thus responses to large earthquakes with seismic energy density exceeding 0.76×10^{-4} J/m^3 were found during the monitoring of the wellbore near the Wenchuan fault (Xue et al. 2013). Overall permeability of brecciated rocks of the fault zone changed from 0.95 to 1.75×10^{-15} m^2 while the underground water level was irregularly recovering by 16 m during a year and a half.

According to "Mikhnevo" GPO we considered earthquakes with seismic energy density above 0.7×10^{-4} J/m^3 that took place during the observation time (Table 2). We took moment magnitudes M_w and corresponding epicentral distances r from the Global CMT Catalog (http://www.globalcmt.org/ CMTsearch.html) and used empirical scaling relation $M_w = 2.7 + 0.69\lg e + 2.1\lg r$ (Wang 2007) to determine the seismic energy density of earthquakes. All events showed in Table 2 except the 11th March 2011 Tohoku earthquake had the same order of energy density as those from Xue et al. (2013) (Table 1). We visually determined hydrogeological responses only for seven earthquakes with $e \geq 2.17 \times 10^{-4}$ J/m^3. However, for them we did not detect phase shift changes during high (arrows 1–4 in Fig. 8) and low inflow rate (arrow 5–6 in Fig. 8). Subject to significant background phase shift variations it makes it obvious that seismic impact on aquifers in terms of tidal influence must be considered only in accordance with the geological and hydrogeological structure of observational area.

Table 1

Parameters of earthquakes caused aquifer response within the Wenchuan earthquake fault zone (Xue et al. 2013)

Date	R, km	M_w	e, 10^{-4} J/m^3
2010-02-26	2492	7.0	0.78
2010-04-06	3258	7.8	5.01
2010-04-13	703	6.9	26.44
2010-12-21	3906	7.4	0.76
2011-03-11	3632	9.1	275.51
2011-03-24	1211	6.8	3.62

Table 2

Parameters of earthquakes caused aquifer response t the GFO "Mikchnevo"

Date	R, km	M_w	e, 10^{-4} J/m^3
2011-03-11	7446	9.1	30.99
2011-10-23	1865	7.1	2.65
2012-04-11	7645	8.6	5.39
2013-04-16	3564	7.7	2.73
2013-05-24	6491	8.3	3.26
2013-09-24	3830	$m_b = 7.7$	2.19
2014-05-24	1864	6.9	1.36
2015-04-25	4783	7.9	2.17
2015-12-07	3206	7.2	0.71

6. Conclusions

Based on the results above we suggest that due to possible strong instability of aquifers and, therefore, phase shift intensity, it is crucially important to estimate background oscillations of this parameter before using it for determining aquifer response to dynamic impact. One of the possible causes for phase shift range growth could be inflow rise due to either natural or anthropogenic reasons like flash floods, aquifer supply change or pumping. We still do not know how significant the level change must be to cause tidal form distortion and what other reasons should be taken into consideration, but insist on the importance of this factor for tidal analysis of aquifer state. Either high inflow periods of time should be excluded from consideration or corresponding corrections must be done. Special attention must be paid to the estimation of possible range of phase shift variations for different geological and hydrogeological conditions.

Acknowledgements

This work was supported by Project no. 16-17-00095 of the Russian Science Foundation (precise monitoring of water level variations, data pre-handling) and Project no. 0146-2015-0008 of the Russian Academy of Science (final data handling and analysis).

REFERENCES

Allegre, V., Brodsky, E., Xue, L., Nale, S.M., Parker, B.L., Cherry, J.A. (2016). Using earth-tide induced water pressure changes to measure in situ permeability: A comparison with long-term pumping tests. *Water Resources Research.* doi:10.1002/2015WR017346.

Besedina, A. N., Vinogradov, E. A., Gorbunova, E. M., Kabychenko, N. V., Svintsov, I. S., Pigulevskiy, P. I., et al. (2015). The response of fluid-saturated reservoirs to lunisolar tides: Part 1. Background parameters of tidal components in ground displacements and groundwater level. *Izvestiya, Physics of the Solid Earth, 51*(1), 70–79.

Besedina, A., Vinogradov, E., Gorbunova, E., & Svintsov, I. (2016). Chilean earthquakes: Aquifer responses at the Russian platform. *Pure and Applied Geophysics, 173*(4), 1039–1050. doi:10.1007/s00024-016-1256-5.

Brodsky, E. E., Roeloffs, E., Woodcock, D., Gall, I., & Manga, M. (2003). A mechanism for sustained groundwater pressure changes induced by distant earthquakes. *Journal of Geophysical Research, 108*, 2390.

Burbey, T. J., Hisz, D., Murdoch, L. C., & Zhang, M. (2012). Quantifying fractured crystalline-rock properties using well tests, earth tides and barometric effects. *Journal of Hydrology, 414–415,* 317–328.

Burbey, T. J., & Zhang, M. (2010). Assessing hydrofracing success from Earth tide and barometric response. *Ground Water, 48*(6), 825–835.

Cutillo, P. A., & Bredehoeft, J. D. (2011). Estimating aquifer properties from the water level response to Earth tides. *Ground Water, 49*(4), 600–610.

Doan, M.L., Brodsky, E.E., Priour, R., Signer, C. (2006). *Tidal analysis of borehole pressure—a tutorial.* Schlumberger Oil Field Services Research. December 20, 2006.

Elkhoury, J.E., Brodsky, E.E., Agnew, D.C. (2006). Seismic waves increase permeability. *Nature, 441* (Supplementary Material for Nature manuscript 2005-11-13339 Seismic Waves Increase Permeability). doi:10.1038/nature04798.

Furbish, D. J. (1991). The response of water level in a well to a time series of atmospheric loading under confined conditions. *Water Resources Research, 27*(4), 557–568.

Hsieh, P. A., Bredehoeft, J. D., & Farr, J. M. (1987). Determination of aquifer transmissivity from Earth tide analysis. *Water Resources Research, 23*(10), 1824–1832. doi:10.1029/WR023i010p01824.

Hussein, M. E. A., Odling, N. E., & Clark, R. A. (2013). Borehole water level response to barometric pressure as an indicator of aquifer vulnerability. *Water Resources Research, 49,* 102–7119. doi:10.1002/2013WR014134.

Kissin, I. G. (2015). *Fluids in the Earth crust: geophysical and tectonic aspects.* Moscow: Nauka. **(in Russian)**.

Kitagawa, Y., Itaba, S., Matsumoto, N., & Koizumi, N. (2011). Frequency characteristics of the response of water pressure in a closed well to volumetric strain in the high-frequency domain. *Journal of Geophysical Research Letters, 116,* B08301. doi:10.1029/2010JB007794.

Kocharyan, G. G., Vinogradov, E. A., Gorbunova, E. M., Markov, V. K., Markov, D. V., & Pernik, L. M. (2011). Hydrologic response of underground reservoirs to seismic vibrations. *Izvestiya, Physics of the Solid Earth, 47*(12), 1071–1082.

Kopylova, G. N., Kulikov, G. V., & Timofeev, V. M. (2007). Estimation of state and outlook of hydrogeodeformation monitoring of seismo active regions of Russia. *Subsurface exploring and protection, 11,* 75–83. **(in Russian)**.

Lai, G., Ge, H., & Wang, W. (2013). Transfer functions of the well-aquifer systems response to atmospheric loading and Earth tide from low to high-frequency band. *Journal of Geophysical Research: Solid Earth, 118,* 1904–1924. doi:10.1002/jgrb.50165.

Liu, C., Huang, M.-W., & Tsai, Y.-B. (2006). Water level fluctuations induced by ground motions of local and Teleseismic earthquakes at two wells in Hualien, Eastern Taiwan. *TAO, 17*(2), 371–389.

Melchior, P. (1966). *The Earth tides.* Oxford: Pergamon Press.

Rahi, K. A., & Halihan, T. (2013). Identifying aquifer type in fractured rock. Aquifers using Harmonic Analysis. *Ground Water, 51*(1), 76–82.

Ritzi, R. W., Soroosshian, S., & Hsieh, P. A. (1991). The estimation of fluid flow properties from the response of water levels in wells to the combined atmospheric and Earth tide forces. *Water Resources Research, 27*(5), 883–893.

Rojstaczer, S. (1988). Intermediate period response of water levels in wells to crustal strain: sensitivity and noise level. *Journal of Geophysical Research, 93*(B11), 13619–13634.

Shi, Z., Wang, G., Manga, M., & Wang, C.-Y. (2015). Mechanism of co-seismic water level change following four great earthquakes—insights from co-seismic responses throughout the Chinese Mainland, Earth Planet. *Science Letters, 430,* 66–74.

Wang, C.-Y. (2007). Liquefaction beyond the near field. *Seismological Research Letters, 78*(5), 512–517.

Wenzel, H. G. (1994). Earth tide analysis package ETERNA 3.0. *BIM, 118,* 8719–8721.

Xue, L., Brodsky, E. E., Erskine, J., Fulton, P. M., & Carter, R. (2016). A permeability and compliance contrast measured hydrogeologically on the San Andreas fault. *Geochemistry, Geophysics, Geosystems.* doi:10.1002/2015GC006167.

Xue, L., Li, H.-B., Brodsky, E. E., Xu, Z.-Q., Kano, Y., Wang, H., et al. (2013). Continuous permeability measurements record healing inside the Wenchuan earthquake fault zone. *Science, 340,* 1555–1559.

(Received December 16, 2016, revised May 25, 2017, accepted May 31, 2017, Published online June 9, 2017)

Pure Appl. Geophys. 175 (2018), 1793–1804
© 2017 Springer International Publishing
https://doi.org/10.1007/s00024-017-1554-6

Pure and Applied Geophysics

Analyses of a 426-Day Record of Seafloor Gravity and Pressure Time Series in the North Sea

S. Rosat,[1] B. Escot,[1] J. Hinderer,[1] and J.-P. Boy[1]

Abstract—Continuous gravity observations of ocean and solid tides are usually done with land-based gravimeters. In this study, we analyze a 426-day record of time-varying gravity acquired by an ocean-bottom Scintrex spring gravimeter between August 2005 and November 2006 at the Troll A site located in the North Sea at a depth of 303 m. Sea-bottom pressure changes were also recorded in parallel with a Paroscientific quartz pressure sensor. From these data, we show a comparison of the noise level of the seafloor gravimeter with respect to two standard land-based relative gravimeters: a Scintrex CG5 and a GWR Superconducting Gravimeter that were recording at the J9 gravimetric observatory of Strasbourg (France). We also compare the analyzed gravity records with the predicted solid and oceanic tides. The oceanic tides recorded by the seafloor barometer are also analyzed and compared to the predicted ones using FES2014b ocean model. Observed diurnal and semi-diurnal components are in good agreement with FES2014b predictions. Smallest constituents reflect some differences that may be attributed to non-linearity occurring at the Troll A site. Using the barotropic TUGO-m dynamic model of sea-level response to ECMWF atmospheric pressure and winds forcing, we show a good agreement with the detided ocean-bottom pressure residuals. About 4 hPa of standard deviation of remaining sea-bottom pressure are, however, not explained by the TUGO-m dynamic model.

Key words: Seafloor gravimeter, seafloor pressure measurements, inverted-barometer response, oceanic tides, dynamic response of the oceans.

1. Introduction

There are very few available long records of time-varying gravity on the seafloor compared to land-based continuous gravity measurements. Seafloor observations are usually limited to spatial gravity surveys of short duration (e.g., Ballu et al. 1998; Zumberge et al. 2008). From the first marine gravity

measurements (Beyer et al. 1966) to the latest ones (Sasagawa et al. 2008), the seafloor gravity precision has gained a factor of hundred. The necessity of precise models of the time-varying tidal signals for reservoir monitoring has motivated the installation of long-term seafloor gravity and pressure observations in the North Sea, in the frame of the Troll A gas-reservoir monitoring program (Sasagawa et al. 2003). From August 14th, 2005 to November 3rd, 2006 an ocean-bottom Scintrex gravimeter (SN970439) and a Paroscientific quartz pressure sensor (model 31K, SN 74329) were measuring at a depth of 303 m on the North Sea floor (60.64227°N, 3.72417°E; see Fig. 1) nearly continuously (Sasagawa et al. 2008). The sensors are mounted in a single frame carried by a remotely operated vehicle (ROV). The instrument is called ROVDOG for remotely operated vehicle-deployed deep-ocean gravimeter.

The records contain two time-series of gravimetric and pressure data initially sampled at 1 s then decimated to 1 min after applying an anti-aliasing low-pass filter: the first from August 14th, 2005 to March 5th, 2006 and the second from March 25th, 2006 to November 3rd, 2006. The gap between the two series is due to mechanical maintenance (Sasagawa et al. 2008). In Fig. 2, we have plotted the seafloor gravity and pressure measurements before and after tidal analyses. As any relative Scintrex instrument, the seafloor gravimeter is affected by a strong non-linear drift. We decided to high-pass filter the gravity records with a cut-off period of 10 days in order to remove this trend before performing the analyses.

In a first part, we perform a noise level comparison with standard land-based gravimeters and barometer. Then, we compare the tidal analysis results with the predicted solid and oceanic tides. Thanks to the seafloor barometric records, we have a

---

Tides and non-tidal loading (Bruno Meurers, David Crossley).

[1] Institut de Physique du Globe de Strasbourg, UMR 7516, Université de Strasbourg/EOST, CNRS, 5 rue Descartes, 67084 Strasbourg, France. E-mail: srosat@unistra.fr

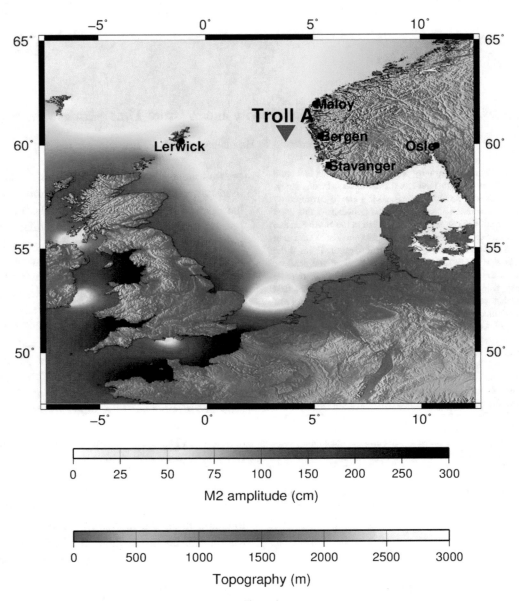

Figure 1

Map of the Troll A site where the seafloor gravimeter and barometer were deployed from August 2005 to November 2006. The amplitude of M_2 tide from FES2014b ocean model is also plotted in cm

direct measurement of oceanic tides that we compare with FES2014b (Carrère et al. 2015) tidal heights. Finally, we investigate the in situ dynamics of oceans by comparing the sea-bottom pressure data with the TUGO-m 2D (Toulouse Unstructured Grid Ocean model 2D, ex-MOG2D) modeling of the dynamic ocean response to ECMWF pressure and winds forcing (Carrère and Lyard 2003).

2. Noise Level Analysis

Knowledge of the noise level at a site is fundamental in the search for small signals or to infer the quality of a site or the precision of an instrument. With one instrument at a site it is not possible to separate the contribution of the instrument itself from the environmental noise. Here we have the possibility

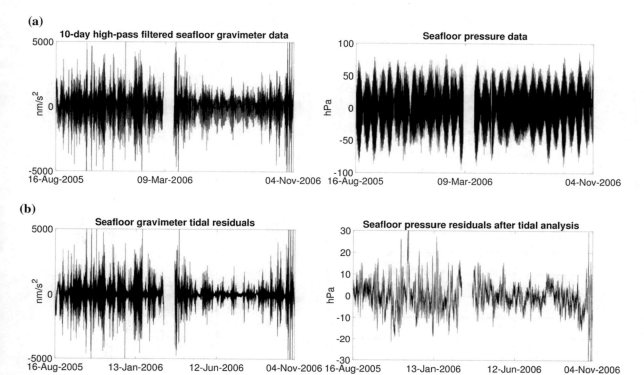

(a)

Figure 2

Seafloor records: **a** gravimetric records after a high-pass filtering with a cut-off period of 10 days (*left plot*) and pressure records (*right plot*); **b** gravimetric (*left plot*) and pressure (*right plot*) residuals after tidal analyses with ETERNA software

to compare the noise level of a relative Scintrex gravimeter installed on the sea bed with one installed on land. We may assume that the instrumental noise from one Scintrex to the other one is not so much different, nevertheless the noise coming from the ocean is expected to be much larger. The question is by how much a seafloor gravimeter is noisier than a land-based gravimeter. In that purpose of quantifying the noise, we compute power spectral densities (PSDs) using a Welch's overlapped segment averaging estimator on data sampled at 1 min. These PSDs are obtained from fortnightly segments in order to represent the noise level at seismic and sub-seismic frequencies. For each fortnightly segment, the Welch periodogram is computed using a Hanning window on daily sections with a 50% overlap, leading to an average of 15 modified periodograms to produce a PSD estimate. Then, we compute the median of the distribution of PSDs obtained from 115 fortnightly

segments (only 13 segments for the Scintrex CG5 at J9 since the available time-record is shorter). The median PSD means that the noise does not exceed this level 50% of the time.

The noise level of the seabed gravimetric record is compared in Fig. 3 with the noise levels of the superconducting gravimeter (SG-C026) and the Scintrex CG5 that were recording at the J9 gravimetric observatory of Strasbourg (France) (Rosat et al. 2015). The seismological NLNM reference noise model (Peterson 1993) and the more recent Global Seismic Network (GSN) noise model (Berger et al. 2004) are also plotted for reference. The NLNM corresponds to the lowest envelope of the PSDs computed from the GSN stations. The GSN noise model that Berger et al. (2004) have developed more recently is based on 118 GSN stations and is more complete than NLNM since they computed not only the minimum, but also the percentiles of PSD

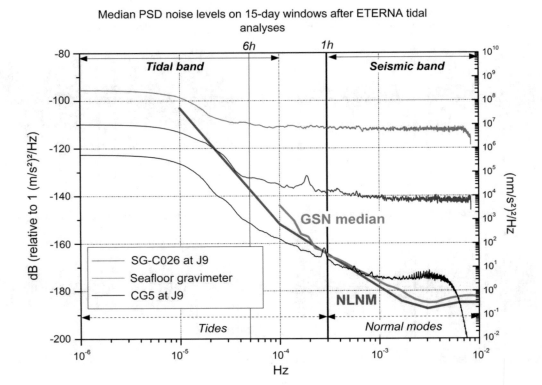

Figure 3

Median noise levels computed from fortnightly time windows for the SG-C026 and CG5 land gravimeters recording at the J9 gravimetric observatory of Strasbourg (France) and for the seafloor gravimeter recording at Troll A site. The PSDs were computed on the gravity residuals after a tidal analysis with ETERNA software. The low noise model (NLNM, in *red*) of Peterson (1993) and the median GSN noise model (in *gray*) of Berger et al. (2004) are also plotted for reference

distributions. Note that we used the median noise level from Berger et al. (2004), but those were computed from daily time-series while we used fortnightly ones to represent the noise level at sub-seismic frequencies. We can see on Fig. 3 that the median PSD for the SG-C026 follows well the NLNM while records from the seabed gravimeter are 50 dB noisier and the ones from the land CG5 are 20 dB noisier at 1-h period. Note that the CG5 gravimeter was not operating at the time the seafloor gravimeter was recording while the SG was, so the environmental noises (mostly oceanic noise) seen by the SG and the CG5 at J9 are not the same. However, we also see, like in a previous study (Rosat et al. 2015), that the CG5 is 20 dB noisier than the SG-C026 at 1-h period. The instrumental drifts of the land gravimeters were removed before the tidal analysis by fitting a linear trend.

The observed median noise level for the seafloor gravimeter is around -110 dB (or 10^7 $(nm/s^2)^2/$ $Hz = 0.1$ $mGal^2/Hz$) in the sub-seismic band between 10^{-4} Hz and the Nyquist frequency of 8.33 10^{-3} Hz. Sasagawa et al. (2003) had shown that the noise level at the Troll site was between $2\ 10^{-2}$ and 0.2 $mGal^2/Hz$ at 0.02 Hz, respectively, during low and high swell states. Assuming the PSD remains flat towards higher frequencies, the value we have obtained in Fig. 3 lies in between but closer to high swell state, i.e., when significant wave heights can be observed (3–3.5 m). As shown in Table 1, the Troll A site indeed experienced some high swells during this period.

We can also compare the seabed pressure measurements from the Paroscientific model 31 K with the land-based barometric records at J9 from a Druck DPI145 barometer. To be able to compare with the

Table 1

Meteorological data at the Troll A platform between October, 27 and November, 1 2006

Date	Wind speed (m/s)	Seafloor pressure variation (hPa)	Wave ocean height (m)	ECMWF + TUGO-m pressure variation (hPa)	ECMWF pressure variation (hPa)
2006/10/27 03 h	20.6	−28.5	6	−24.3	−31.7
2006/10/27 18 h	6.2	41.3	1.7	23.4	6.24
2006/10/28 15 h	5.7	−9.6	2.1	−6.22	−4.7
2006/10/29 18 h	9.8	18.6	2.6	9.6	3.18
2006/10/31 15 h	25	−16.1	7.5	−11.8	−31.7
2006/11/01 15 h	7.7	42.4	2.1	22.45	14.1

NLNM and with the seafloor gravimeter too, we compute the Newtonian attraction of the corresponding air column for the land barometer or water column for the seabed barometer, using the cylinder formula

$$A = -2\pi G \rho_w \left[2z - h - \sqrt{r^2 + z^2} + \sqrt{r^2 + (z - h)^2} \right],$$

(1)

where ρ_w is the seawater density (or air density for the land barometer), z is the height of the point inside the cylinder located on the axial symmetric axis at which the attraction is computed (in our case $z = 0$), r is the radius of the cylinder and h is the height of the water column. We choose a value of r much larger than the water height so that we tend towards the Bouguer approximation of $2\pi G \rho_w h$.

The median PSDs for the ocean-bottom barometer are plotted in Fig. 4 with the seismological noise models and with the median PSD of the seafloor gravimeter.

It appears that the seabed pressure measurement is less noisy than the seabed gravimeter. It is known that bottom pressure is less sensitive to internal tides hence mostly records the barotropic two-dimensional water column (Ray 2013), while the gravimeter will integrate the three-dimensional water mass changes. Besides, the gravity residuals

plotted in Fig. 1 illustrate the additional noise present in these records. We can also see that the seafloor barometer is 10 dB noisier than the land barometer at 1-h period. A 10 dB difference corresponds to a factor 3 in amplitude, meaning that with the land barometer we can detect signals that are 3 times smaller than with the seabed barometer. Since the barometers are different, we cannot infer the part due to the instrumental noise from the contribution to the environmental noise, but we can assume that most of the noise difference is coming from the ocean noise, since ocean noise is larger in shallow waters than in the deep ocean. It is also larger at sites near the sea and J9 is located far, about 500 km, from the North Sea.

3. Tidal Analysis Results

In this part we focus on the tidal analysis of the seafloor records. The seabed barometric records clearly show the oceanic tidal signal (see Fig. 2). These records can be analyzed to retrieve the tidal amplitudes and phases of the oceanic waves. For that we need to convert the seafloor barometric data from pressure variations ΔP to equivalent tidal height changes ξ using the hydrostatic equation:

$$\Delta P = \rho_w g \xi,$$

(2)

where g is the mean gravity value and ρ_w the mean density of the water column. This simple and widely employed conversion may, however, be inadequate (Ray 2013). We have supposed a constant value of 1025 kg/m^3 for seawater density, but this simplification is questionable since it depends on temperature and salinity. Indeed compression of the water column should be considered by adding a correcting factor so that $\xi = \Delta P (1 - gH/2c^2)/(g\rho_w)$ with c the speed of sound and H the ocean depth (Ray 2013). This factor represents a 1% adjustment in deep water but less in shallow seas, where the conversion factor approaches unity. Since our site is at 303 m depth in the North Sea and according to Fig. 6 of Ray (2013), we can consider the usual Eq. (2) to convert bottom pressure into equivalent sea surface height with a negligible error. Moreover, seawater temperature profiles were collected to improve the seawater density model, but

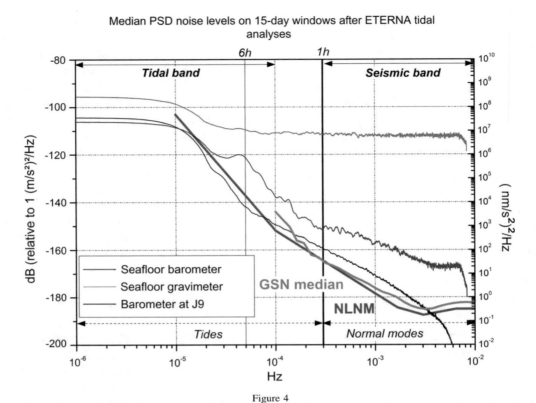

Figure 4
Median noise levels computed from fortnightly time windows for the land Druck DPI145 barometer recording at the J9 gravimetric observatory of Strasbourg (France) and for the seafloor barometer recording at Troll A site. The barometric pressures were converted into gravitational acceleration using the Bouguer approximation for comparison with the NLNM and with the median PSD of the seafloor gravimeter. The PSDs were computed on the residuals after a tidal analysis with ETERNA software for the seafloor gravimeter and barometer. The low noise model (NLNM, in *red*) of Peterson (1993) and the median GSN noise model (in *gray*) of Berger et al. (2004) are also plotted for reference

it turned out to be too small to affect the conversion from pressure to depth (Sasagawa et al. 2003).

The tidal analysis is performed with the ETERNA3.4 software (Wenzel 1996) which performs least-squares fit to tides and instrumental drift to give residual gravity and a polynomial drift function. The two time-series of, respectively, 203 and 223 days of seafloor records are analyzed as two separate blocks within ETERNA. For each block we remove a linear instrumental drift and use the HW95 tidal potential development (Hartmann and Wenzel 1995). The ETERNA analysis enables to retrieve the gravimetric δ factors and the κ phase shifts of the selected waves with respect to the tidal potential for gravimetric records and the amplitudes and phase shifts of the selected waves for tidal height (pressure)

records. We show the results of the δ factors for the diurnal waves obtained from the seafloor gravimeter on Fig. 5a. As for land gravimeters, we can observe an enhancement of the Ψ_1 wave due to the resonance with the free core nutation (FCN). However, the error bars are too large to attempt any retrieval of the FCN frequency from these data. These gravimetric factors represent the complete Earth's response to tidal forcing, that is to say the solid Earth tides and the ocean tides that we can compare to predicted ones. On Fig. 5, we have also plotted the predicted solid tides for an inelastic Earth model using the Dehant et al. (1999) model (noted hereafter DDW) in addition to the oceanic tides computed from FES2014b ocean model. For the gravimetric effects of the oceanic tides, we have taken into account the elastic

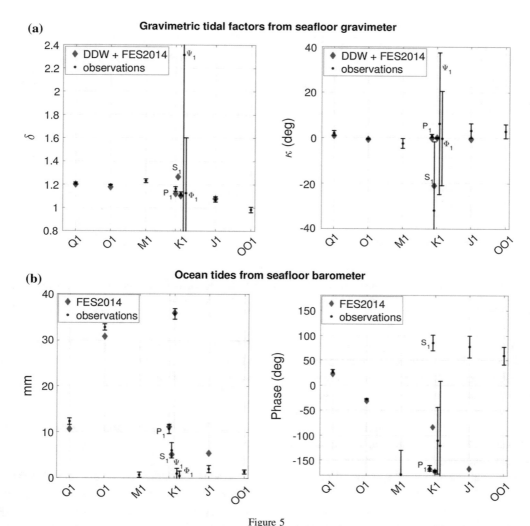

Figure 5

a Gravimetric amplitude factors for the 10 main diurnal waves obtained with ETERNA software by tidal analysis of the seafloor gravimeter data: amplitude factor δ (*left plot*) and phase shifts κ (*right plot*) with respect to the predicted amplitude and phase for a solid Earth. Predicted tides are also plotted in *red diamonds*. They include the solid tides for a DDW (Dehant et al. 1999) inelastic Earth model and the total gravimetric effect (Newtonian attraction, elastic loading and local effect using a Bouguer approximation) from the oceanic tides using FES2014b model; **b** ocean tides amplitudes (*left plot*) and phases (*right plot*) for the 10 main diurnal waves obtained with ETERNA software by tidal analysis of the seafloor barometer data converted into ocean tide heights. The predicted amplitudes and phases for FES2014b ocean tide model are plotted in *red diamonds*. The HW95 tidal potential development (Hartmann and Wenzel 1995) was used as driving potential for all tidal analyses

loading, the direct Newtonian attraction and since the location is below the sea, we have added the local contribution from a Bouguer plate using the Eq. (1) and with FES2014b tidal amplitudes. Note that the δ amplitude factor for S_1 is out of range in Fig. 5a with a value close to 6 since it contains the atmospheric thermal effect which is much larger than the gravitational S_1 tide.

The seafloor pressure data are converted into water heights using Eq. (2) and then analyzed with ETERNA as ocean tide heights. The resulting amplitudes and phases with respect to the HW95

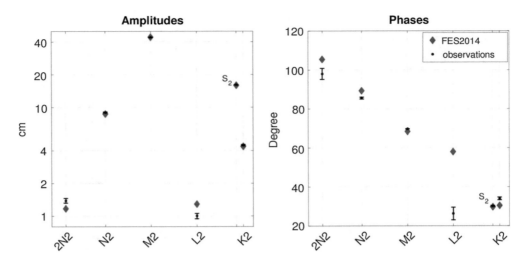

Figure 6
Amplitudes and phases for the semi-diurnal tidal waves analyzed with ETERNA from the seafloor barometer. The predicted amplitudes for FES2014b ocean tide model are plotted in *red diamonds*

(Hartmann and Wenzel 1995) driving tidal potential are plotted in Fig. 5b. We can compare them with the FES2014b (Carrère et al. 2015) ocean tidal model, in terms of amplitude and phase as shown in Fig. 5b for the diurnal waves and in Fig. 6 for the semi-diurnal waves. The observed amplitudes and phases for the diurnal waves are in good agreement, within the error bars, with the predicted FES2014b model except for J_1 component. The frequency of J_1 is exactly equal to the difference of M_2 and Q_1 frequencies, so some potential non-linear interaction may affect this small component. Atmospheric tides perturb the sea-bottom pressure by adding a 1–2 hPa pressure to the measured one and by inducing a dynamic ocean response, the so-called radiational tide (Ray 2013). The S_1 amplitude in FES2014b is very close to the observed one, but the phase is very different. Large phase difference for small-amplitude waves like J_1 and S_1 is not surprising (e.g., Ray and Egbert 2004). The S_2 tide is in good agreement between FES2014b and the sea-bottom observation with a difference of 1.7 mm meaning that the barometric tide over the oceans and the radiational tide are correctly modeled in FES2014b. As for the semi-diurnal waves, the largest difference occurs for the small L_2 constituent. For M_2, the FES2014b amplitude is 44.147 cm and the observed M_2 amplitude from analysis of sea-bottom pressure records is 43.854 cm, resulting in a

difference of 3 mm, comparable to the difference obtained by Ray (2013) from worldwide sea-bottom pressure measurements.

4. Ocean Response to Atmospheric Pressure Changes

Under the influence of the atmospheric forcing, the ocean will respond in different ways depending on the time and space scales of the forcing (e.g., Ponte 1993). When the atmospheric pressure variations occur on time-scales larger than 7 days, the ocean has enough time to balance this forcing by a water level change. This is the classical inverted barometer (IB) model, where a 1 hPa of atmospheric pressure change will be compensated by a 1 cm sea-level variation, and thus no pressure change occurs at the sea bottom and ocean currents are negligible (Ponte et al. 1991; Wunsch and Stammer 1997; Egbert and Ray 2003). This IB static hypothesis was validated at long periods for instance by surface gravity observations (e.g., Bos et al. 2002; Boy et al. 2002). It is, however, well known that the sea-level response to air pressure changes does not always act like an ideal inverted barometer, particularly at periods shorter than 3 days and at high latitudes (Carrère and Lyard 2003; Boy and Lyard 2008). One

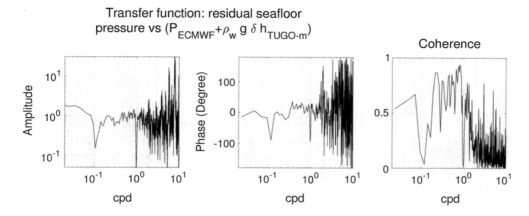

Figure 7

Transfer function (amplitude and phase) and magnitude squared coherence between residual seafloor pressure and TUGO-m (ex-MOG2D) sea-level response to ECMWF pressure and winds after subtraction of the IB response. Frequency unit is in cycle per day (cpd)

major reason for this dynamic response of the oceans at periods longer than 10 days is the wind effects which prevail, particularly at high latitudes and in the tropics (e.g., Fukumori et al. 1998). At seasonal timescales, the oceanic circulation is mostly forced by thermal effects, and the oceans cannot be considered as barotropic (Boy and Lyard 2008). Development of satellite altimetry like TOPEX-POSEIDON has enabled to quantify such deviations from the IB model (e.g., Ponte and Gaspar 1999).

In order to take into account the dynamic response of the oceans, a barotropic non-linear model based on 2-dimension gravity waves model, called TUGO-m 2D (Toulouse Unstructured Grid Ocean model 2D, ex-MOG2D), has been developed by Lynch and Gray (1979) and Carrère and Lyard (2003). The governing equations of the model are the classical shallow water continuity and momentum equations. TUGO-m 2D is a barotropic model that represents the dynamic response of the oceans to atmospheric winds and pressure forcing from ECMWF pressure (noted P_{ECMWF}). The total ocean-bottom pressure should be given by:

$$P_{ECMWF} + \rho_w g \delta h_{TUGO-m}, \qquad (3)$$

where δh_{TUGO-m} is the predicted dynamic sea-level change.

After removing the tidal effects, the sea-bottom pressure changes still include effects of the atmospheric mass changes (atmospheric pressure and winds), oceanographical components (e.g., seiches, waves), density variations of the water column due to changes of temperature and salinity and some possible seafloor displacement due to crustal deformation. To infer the part coming from the atmospheric pressure and winds, we first check the coherence between the TUGO-m sea-level changes from which we removed the hydrostatic load due to average height of the water column responding as an IB to the ECMWF pressure changes using Eq. (3) and the seafloor pressure residuals after ETERNA tidal analysis. As argued in Sect. 3, we neglect the changes of temperature and salinity. The magnitude squared coherence is represented in Fig. 7 and is given by:

$$C_{xy} = |P_{xy}|^2 / (P_{xx} P_{yy}),$$

where x is the dynamic part of the ocean response ($P_{ECMWF} + \rho_w g \delta h_{TUGO-m}$) and y the seafloor pressure residuals after ETERNA tidal analysis. P_{xx} is the PSD estimate of x, P_{yy} is the PSD estimate of y and P_{xy} is the cross-PSD between x and y. The PSD is obtained using the Welch's averaged, modified periodogram method, i.e., the signals x and y are divided into eight sections with 50% overlap and tapered with a Hamming window. For each section a modified periodogram is computed and the eight periodograms are averaged.

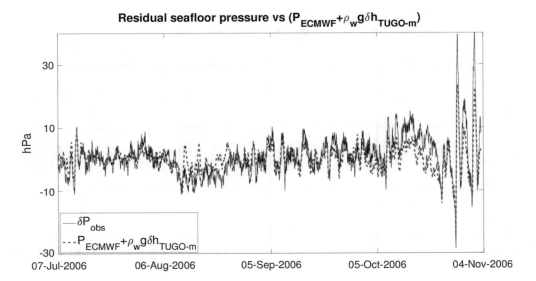

Figure 8

A 4-month comparison of the TUGO-m (ex-MOG2D) dynamic response of the oceans (after subtraction of the IB response from ECMWF pressure and winds forcing) and seafloor pressure residuals

The transfer function between the modeled ocean response and seafloor pressure residuals is computed in the same way by:

$$T_{xy}(f) = P_{yx}(f)/P_{xx}(f).$$

The transfer function (amplitude and phase) is also plotted on Fig. 7.

Between 1 and 10 days the coherence is large (about 0.9), the phase of the transfer function is zero and the admittance is close to one. This means that the dynamic TUGO-m barotropic response of the ocean agrees with the sea-bottom pressure residuals as expected. At the period of 1 day, there is also a drop of coherence, associated with a large phase and an admittance close to zero. This is due to the removal of S_1 tide during the tidal analysis of pressure data while it was not removed from the TUGO-m model. Between about 0.5 and 0.9 day, the coherence is large again, with a phase close to zero and an admittance close to one again. There is a good agreement between sea-bottom observations and TUGO-m. At periods shorter than 0.5 day, the coherence becomes weak as TUGO-m is not given for higher frequencies. Around a period of 10 days, we have a drop of coherence between the ocean dynamic response and the seafloor pressure residuals, with a very small admittance (amplitude of the transfer function) of 0.15 and a large phase close to $-100°$. Indeed at periods longer than 10 days, the barotropic hypothesis within TUGO-m may not be appropriate anymore and thermal forcing becomes important.

An illustration of the good agreement between seafloor pressure residuals and the dynamic prediction is plotted in Fig. 8 with a stormy event that occurred at the end of October and early November 2006. The time scale concerned by this phenomenon is less than a day, so out of the IB approximation. We can see that there is no time-shift between the predicted TUGO-m dynamic response and the observed pressure changes and their amplitudes are very similar. The standard deviation of the difference is close to 4 hPa (4 cm) while the pressure data after tidal removal had a standard deviation of 5.5 hPa, so one part of the signal is coming from the dynamic ocean response to ECMWF pressure and winds, and the remaining comes from other meteorological effects that we cannot identify yet.

Wind speed and pressure variations associated to Fig. 8 are given in Table 1. The meteorological data were provided by the Norwegian

Meteorological service and the wave ocean heights come from radar data recorded at the Troll A platform. We can see from Table 1 that wind strength is correlated with sea level, with the pressure variations recorded by the seafloor barometer, and with ECMWF and ECMWF + TUGO-m predictions. Indeed, strong winds fit with low pressure and weak winds fit well with high pressure. Hence the event of Fig. 8 illustrates a dynamic ocean response driven by winds that is well modeled in TUGO-m. Boy and Lyard (2008) have also shown a good agreement between tide gauge measurements and TUGO-m sea surface height variations during two large storm surges in the North Sea. We observe, however, a 4-hPa difference in standard deviation between the ocean-bottom pressure data and TUGO-m model. The time-scale of this storm is too short to consider a thermal forcing as a possible explanation for the remaining seafloor pressure. The spatial resolution of TUGO-m is 0.25° which may be not precise enough to represent local effects at the Troll A site that may induce additional sea-bottom pressure.

5. Conclusion

Observed ocean tides at Troll A site with the seafloor barometer are in good agreement with FES2014b ocean tide model for the diurnal and semi-diurnal waves. Some non-linearity may explain the noticed differences for some small constituents like J_1. After removing tides, the standard deviation of the barometric residuals at Troll A decreased from 34 to 5.5 hPa. After removing the TUGO-m dynamic response of the oceans to ECMWF pressure and winds forcing, the standard deviation further decreased to nearly 4 hPa (equivalent to 4 cm of seawater height). The in situ seafloor pressure observations clearly detect the part due to the dynamic response of the oceans to atmospheric pressure and winds forcing as modeled by TUGO-m. Other sources of sea-bottom pressure changes have, however, not yet been identified to explain the remaining sea-bottom pressure signal.

Acknowledgements

The authors acknowledge the suggestions of two anonymous reviewers that contributed to significantly improve this manuscript. We also thank David Crossley for his re-reading and suggested corrections to this paper. We are grateful to Glenn Sasagawa as a data contributor and to Ola Eiken for their useful information about the data and about the site measurement. This seafloor dataset was acquired by Scripps, under research contract with Statoil. Statoil is recognized as a sponsor of the data collection, and also for help with connecting cables and recording on the Troll A platform. Historical wind speed data were provided by the Norwegian Meteorological service at http://www.yr.no/place/Norway/Hav/Troll_A/ and the wave ocean heights come from radar data recorded on the Troll A platform. The General Mapping Tools (Wessel and Smith 1998) was used for plotting the map.

REFERENCES

Ballu, V., Dubois, J., Deplus, G. C., Diament, M., & Bonvalot, S. (1998). Crustal structure of the Mid-Atlantic Ridge south of the Kane fracture zone from seafloor and sea surface gravity data. Journal of Geophysical Research, 103, 2615–2631.

Berger, J., Davis, P., & Ekström, G. (2004). Ambient Earth noise: a survey of the Global Seismographic Network. Journal of Geophysical Research, 109, B11307. doi:10.1029/2004JB003408.

Beyer, L. A., von Huene, R. E., McCulloh, T. H., & Lovett, J. R. (1966). Measuring gravity on the sea floor in deep water. Journal of Geophysical Research, 71(8), 2091–2100.

Bos, M. S., Baker, T. F., Rothing, K., & Plag, H.-P. (2002). Testing ocean tide models in the Nordic seas with tidal gravity observations. Geophysical Journal International, 150, 687–694.

Boy, J.-P., Gegout, P., & Hinderer, J. (2002). Reduction of surface gravity data from global atmospheric pressure loading. Geophysical Journal International, 149, 534–545.

Boy, J.-P., & Lyard, F. (2008). High-frequency non-tidal ocean loading effects on surface gravity measurements. Geophysical Journal International, 175, 35–45.

Carrère, C., & Lyard, F. (2003). Modeling the barotropic response of the global ocean to atmospheric wind and pressure forcing—comparisons with observations. Geophysical Research Letters, 30(6), 1275. doi:10.1029/2002GL016473.

Carrère, L., Lyard, F., Cancet, M., A. Guillot, 2015. FES 2014, a new tidal model on the global ocean with enhanced accuracy in shallow seas and in the Arctic region, *Geophysical Research Abstracts, 17*, EGU2015-5481-1.

Dehant, V., Defraigne, P., & Wahr, J. M. (1999). Tides for a convective Earth. *Journal of Geophysical Research, 104,* 1035–1058.

Egbert, G. D., & Ray, R. D. (2003). Deviation of long period tides from equilibrium: kinematics and geostrophy. *Journal of Physical Oceanography, 33,* 822–839.

Fukumori, I., Raghunath, R., & Fu, L.-L. (1998). Nature of global large-scale sea level variability in relation to atmospheric forcing: a modeling study. *Journal of Geophysical Research, 103*(C3), 5493–5512.

Hartmann, T., & Wenzel, H.-G. (1995). The HW95 tidal potential catalogue. *Geophysical Research Letters, 22*(24), 3553–3556.

Lynch, D. R., & Gray, W. G. (1979). A wave equation model for finite element tidal computations. *Computers & Fluids, 7,* 207–228.

Peterson J., 1993. Observations and Modelling of Seismic Background Noise. Open-File Report 93-332. U.S. Department of Interior, Geological Survey, Albuquerque, NM.

Ponte, R. M. (1993). Variability in a homogeneous global ocean forced by barometric pressure. *Dynamics of Atmospheres and Ocean, 18,* 209–234.

Ponte, R. M., & Gaspar, P. (1999). Regional analysis of the inverted barometer effect over the global ocean using TOPEX/POSEIDON data and model results. *Journal of Geophysical Research, 104*(C7), 15587–15601.

Ponte, R. M., Salstein, D. A., & Rosen, R. D. (1991). Sea level response to pressure forcing in a barotropic numerical model. *Journal of Physical Oceanography, 21,* 1043–1057.

Ray, R. D. (2013). Precise comparisons of bottom-pressure and altimetric ocean tides. *Journal of Geophysical Research Oceans, 118,* 4570–4584. doi:10.1002/jgrc.20336.

Ray, R. D., & Egbert, G. D. (2004). The global S_1 tide. *Journal of Physical Oceanography, 34,* 1922–1935.

Rosat, S., Calvo, M., Hinderer, J., Riccardi, U., Arnoso, J., & Zürn, W. (2015). Comparison of the performances of different Spring and Superconducting Gravimeters and a STS-2 Seismometer at the Gravimetric Observatory of Strasbourg, France. *Studia Geophysica et Geodaetica, 59,* 58–82.

Sasagawa, G. S., Crawford, W., Eiken, O., Nooner, S., Stenvold, T., & Zumberge, M. A. (2003). A new seafloor gravimeter. *Geophysics, 68,* 544–553.

Sasagawa, G., Zumberge, M., & Eiken, O. (2008). Long-term seafloor tidal gravity and pressure observations in the North Sea: testing and validation of a theoretical tidal model. *Geophysics, 73*(6), 143–148.

Wenzel, H. G. (1996). The Nanogal Software: earth tide data processing package ETERNA 3.30. *Bull. Inf. Marées Terrestres, 124,* 9425–9439.

Wessel, P., & Smith, W. H. F. (1998). New, improved version of the Generic Mapping Tools released. *EOS Transactions American Geophysical Union, 79*(47), 579.

Wunsch, C., & Stammer, D. (1997). Atmospheric loading and the oceanic "inverted barometer" effect. *Reviews of Geophysics, 35,* 79–107.

Zumberge, M., Alnes, H., Eiken, O., Sasagawa, G., & Stenvold, T. (2008). Precision of seafloor gravity and pressure measurements for reservoir monitoring. *Geophysics, 73*(6), 133–141.

(Received December 15, 2016, revised April 18, 2017, accepted April 19, 2017, Published online April 25, 2017)

Pure Appl. Geophys. 175 (2018), 1805–1822
© 2018 The Author(s)
This article is an open access publication
https://doi.org/10.1007/s00024-018-1814-0

Pure and Applied Geophysics

Multichannel Singular Spectrum Analysis in the Estimates of Common Environmental Effects Affecting GPS Observations

MARTA GRUSZCZYNSKA,[1] SEVERINE ROSAT,[2] ANNA KLOS,[1] MACIEJ GRUSZCZYNSKI,[1] and JANUSZ BOGUSZ[1]

Abstract—We described a spatio-temporal analysis of environmental loading models: atmospheric, continental hydrology, and non-tidal ocean changes, based on multichannel singular spectrum analysis (MSSA). We extracted the common annual signal for 16 different sections related to climate zones: equatorial, arid, warm, snow, polar and continents. We used the loading models estimated for a set of 229 ITRF2014 (International Terrestrial Reference Frame) International GNSS Service (IGS) stations and discussed the amount of variance explained by individual modes, proving that the common annual signal accounts for 16, 24 and 68% of the total variance of non-tidal ocean, atmospheric and hydrological loading models, respectively. Having removed the common environmental MSSA seasonal curve from the corresponding GPS position time series, we found that the residual station-specific annual curve modelled with the least-squares estimation has the amplitude of maximum 2 mm. This means that the environmental loading models underestimate the seasonalities observed by the GPS system. The remaining signal present in the seasonal frequency band arises from the systematic errors which are not of common environmental or geophysical origin. Using common mode error (CME) estimates, we showed that the direct removal of environmental loading models from the GPS series causes an artificial loss in the CME power spectra between 10 and 80 cycles per year. When environmental effect is removed from GPS series with MSSA curves, no influence on the character of spectra of CME estimates was noticed.

Key words: Multichannel singular spectrum analysis, seasonal signals, GPS, environmental loading models.

1. Introduction

Seasonal changes are a component part of the Global Positioning System (GPS) position time series, especially the vertical direction (Blewitt and Lavallée 2002; Collilieux et al. 2007). In most cases, those variations result from real geophysical phenomena which deform the Earth's surface. They are broadly explained and modelled by environmental loading effects (van Dam and Wahr 1998; Jiang et al. 2013). Atmospheric (van Dam and Wahr 1987), hydrological (van Dam et al. 2001) and non-tidal ocean (van Dam et al. 2012) loadings are the most important contributors of seasonal variations to many GPS stations in different parts of the world. The appropriate models can be removed directly from the GPS position time series to reduce the influence they might have on the observed displacements. This approach has proved to decrease the root mean square (RMS) of the GPS position time series (Jiang et al. 2013). However, according to Santamaría-Gomez and Mémin (2015), such an approach reduces only the amplitude of white noise of the GPS position time series. Klos et al. (2017) showed that the direct removal of environmental loading models from the GPS observations causes the evident change in the power spectrum density of noise for frequencies between 4 and 80 cycles per year (cpy).

Beyond real geophysical origins, seasonal changes in the GPS position time series can be also generated by systematic errors (Ray et al. 2008) or by spurious effects (Penna et al. 2007). Both can influence permanent stations individually or be similar for stations situated not far from each other.

As shown by Dong et al. (2002), both the GPS position time series and the environmental loading

Electronic supplementary material The online version of this article (https://doi.org/10.1007/s00024-018-1814-0) contains supplementary material, which is available to authorized users.

[1] Faculty of Civil Engineering and Geodesy, Military University of Technology, Warsaw, Poland. E-mail: marta.gruszczynska@wat.edu.pl

[2] Institut de Physique du Globe de Strasbourg, UMR 7516, Université de Strasbourg/EOST, CNRS, Strasbourg, France.

The original version of this chapter was revised. The correction to this chapter is available at https://doi.org/10.1007/978-3-319-96277-1_23

models contain the common seasonal signal which characterizes the data from a certain region of the world. Freymueller (2009), Tesmer et al. (2009) and Bogusz et al. (2015a) proposed to employ stacking and clustering methods to estimate regional mean annual oscillations from the time series and to group them from a regional signal. They all proved that the neighbouring stations are characterized by a similar seasonal signal.

Few methods such as Singular Spectrum Analysis (SSA), Wavelet Decomposition (WD) and Kalman Filter (KF) have been already used to retrieve station-dependent time-varying curves from the GPS position time series (Chen et al. 2013; Gruszczynska et al. 2016; Didova et al. 2016; Klos et al. 2018a). However, neither of them is able to separate real and spurious effects from the GPS data. Lately, multi-channel singular spectrum analysis (MSSA) was proposed by Walwer et al. (2016) for stations situated close to each other. They demonstrated that MSSA is able to model the common seasonal signal in the GPS position time series. Zhu et al. (2016) used the MSSA approach to investigate the inter-annual oscillations in glacier mass change estimated from gravity recovery and climate experiment (GRACE) data in Central Asia. Gruszczynska et al. (2017) compared the MSSA-, SSA- and least-squares estimation-derived seasonal signals. They showed that the seasonal signals detected by MSSA are not affected by noise as much as the SSA-derived oscillations. It was explained by the fact that SSA estimates individually fitted curve for the analysed station, while MSSA takes into account the common effects which are observed by few neighbouring GPS stations. As the noise is mainly a station-specific signal, MSSA curves will not be affected by it as much as SSA curves.

To indicate the signals which arise from real geophysical effects, Klos et al. (2017) proposed a two-stage solution based on Improved Singular Spectrum Analysis (ISSA, Shen et al. 2015). When applied for loading models from an individual station, the time-varying seasonal signals from environmental atmospheric, hydrological and non-tidal ocean loadings were extracted, causing the character of the stochastic part characterized a power-law noise (e.g. Williams 2003; Bogusz and Kontny 2011;

Santamaría-Gomez et al. 2011; Klos et al. 2016; Klos and Bogusz 2017) remained intact. In this research, we assumed that the GPS position time series should not be affected by the high frequency part of environmental loadings series as we did not know if GPS observations to a large extent are influenced by the environmental effects. Therefore, we focused on the common environmental effect in a form of time-varying annual curve which is observed at the GPS permanent stations and proposed its modelling without any alteration in the character of the stochastic part of the time series.

Beyond seasonal signals, the GPS-derived series are also characterized by common mode error (CME), being a sum of the systematic errors (Wdowinski et al. 1997; Nikolaidis 2002; King et al. 2010). Mismodeling of the earth orientation parameters (EOPs), mis- or un-modelled large-scale atmospheric and hydrologic effects or small scale crust deformations, all increase spatial correlations between individual series. CME can be easily estimated using stacking (Wdowinski et al. 1997; Nikolaidis 2002), spatial filtering (Márquez-Azúa and DeMets 2003) or orthogonal transformation functions. The latter is considered to be the most effective in reflecting the real nature of CME (Dong et al. 2006). Yuan et al. (2008) proved that a direct removal of surface mass loadings can significantly reduce the power-law noise. However, they did not investigate to what extent the properties of CME are affected. As was shown by Klos et al. (2017) a part of the power is removed when real geophysical effects are considered by direct subtraction of environmental loading models. We presumed that the CME values may also be affected by such removal.

In this paper, we proposed the spatial analysis of environmental loading models based on MSSA to extract the common annual signal for 16 different sections related to the climate zones (i.e. equatorial, arid, warm, snow, polar) and continents. We then modelled with MSSA the time-varying seasonal signal from environmental loading models and subtracted them from the GPS height time series. This was aimed to remove real geophysical changes from the GPS data leaving the noise character of time series unchanged. The benefits of this approach were presented for CME estimates which should not include the environmental effects, as they were removed by the MSSA approach and, most importantly, no influence on CME character should be noticed.

2. Methodology

In this section, we provided a detailed description of the data and the methodology we used. Data included the environmental loading models and the GPS height time series collected at 229 stations from around the globe. The division into 16 different sections according to the climate classification is also presented.

2.1. Data

We employed the GPS position time series from 229 stations distributed globally (Fig. 1). Daily time series were derived from network solution

(Rebischung et al. 2016) produced by the International GNSS Service (IGS). They contributed to the latest realization of the International Terrestrial Reference System (namely ITRF2014; Altamimi et al. 2016). We selected the vertical components which were not shorter than 10 years. To remove offsets, we used the epochs defined by IGS in station log-files with the manual inspection of the series. Outliers were removed with the Interquartile Range (IQR) approach.

Under the Synthesis Report about Climate Change published by Intergovernmental Panel on Climate Change (IPCC), a region's climate is generated by the system, which has five main components: atmosphere, hydrosphere, cryosphere, lithosphere, and biosphere (AR4 SYR Synthesis Report Annexes, http://www.ipcc.ch, retrieved on 2017-06-28). According to this report, we investigated environmental loading effects which may be correlated within climate zones. Therefore, 229 stations were divided into sixteen sections (Table 1, and Table S1

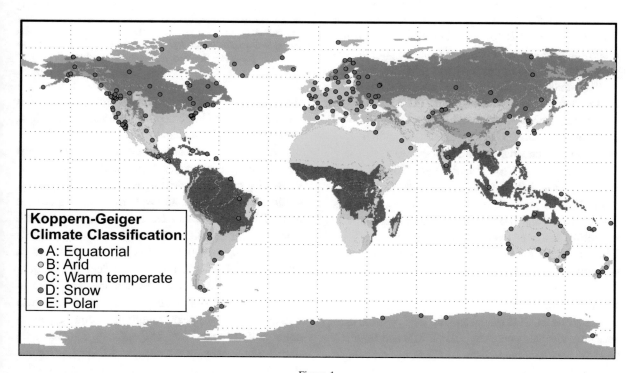

Figure 1

World Map of Köppen–Geiger Climate Classification with the GPS permanent stations considered in this analysis plotted in grey. The A: Equatorial, B: Arid, C: Warm temperate, D: Snow and E: Polar climate zones are presented in red, yellow, green, turquoise and blue, respectively

213

Table 1

Number of stations included in a particular section, i.e. a part of continent that is assigned by climate zone

Number of section	Climate zone	Continent	Number of stations
1	A: Equatorial	Asia	6
2		Australia	6
3		South America	7
4	B: Arid	Asia	6
5		Australia	3
6		North America	13
7	C: Warm	Asia	11
8		Australia	24
9		Europe	30
10		North America	33
11		South America	6
12	D: Snow	Asia	17
13		Europe	18
14		North America	25
15	E: Polar	Antarctica	13
16		North America	11

The full list of stations is available in Table S1 in Supplementary materials. The symbols and names of the sections are used throughout the entire paper, as, e.g. A: Equatorial Asia or D: Snow Europe

in Supplementary Materials) which are associated with these zones and also with different continents. We used the Köppen–Geiger Climate Classification (Rubel and Kottek 2010) to divide stations according to similar conditions. Then, the time-varying curves were modelled separately for each section.

The atmospheric, hydrological and non-tidal ocean loading models in Centre-of-Figure (CF) frame were employed. Atmospheric loadings were determined from ERA interim (ECMWF Reanalysis) model (Dee et al. 2011). Non-tidal ocean loadings were estimated from Estimation of the Circulation and Climate of the Ocean version 2 (ECCO2) (Menemenlis et al. 2008) ocean bottom pressure model. Hydrological loading (soil moisture and snow) was estimated from modern era-retrospective analysis (MERRA) land model (Reichle et al. 2011). These environmental loadings were developed at the Ecole et Observatoire des Sciences de la Terre (EOST) loading service available at http://loading.u-strasbg.fr/.

Since the ECCO2 model and the GPS position time series are sampled every day, the ERAIN and MERRA models were decimated into a daily sampling using a low-pass filter. One of the requirements of the MSSA approach is a common time span of data sets; therefore, we selected a period from 1st January 1994 to 14th February 2015 to be common for all stations. In this

Figure 2 ▶

The selected environmental loading models presented for various sections considered in this research. ERAIN, MERRA and ECCO2 models are shown in blue, brown and violet, respectively. A clear time variability in each series may be observed. Obviously, the spread and amplitudes of ERAIN, MERRA and ECCO2 curves depend on the sections which are considered

study, we focused on the vertical component for which the environmental loading is the most significant (Dach et al. 2011; van Dam et al. 2012).

Figure 2 presents series of environmental loadings for selected stations from each section to show the time variability we may be dealing with. Depending on the section considered, the atmospheric, hydrological and non-tidal ocean effects can be larger than others. According to BAKO (Cibinong, Indonesia), ALIC (Alice Springs, Australia), BREW (Brewster, United States) and TUCU (San Miguel de Tucuman, Argentina) stations, we may notice that the amplitude of MERRA model significantly varies over time with maximum peak-to-peak amplitude being equal to about 5, 7, 10 and 4 mm, respectively. Similarly, we may see in atmospheric loading at ACOR (A Coruna, Spain) and SEAT (Seattle, United States) stations that the peaks differ by 5 and 4 mm, respectively. DARW (Darwin, Australia) station is characterized by ECCO2 model with time-varying amplitude of about 2 mm.

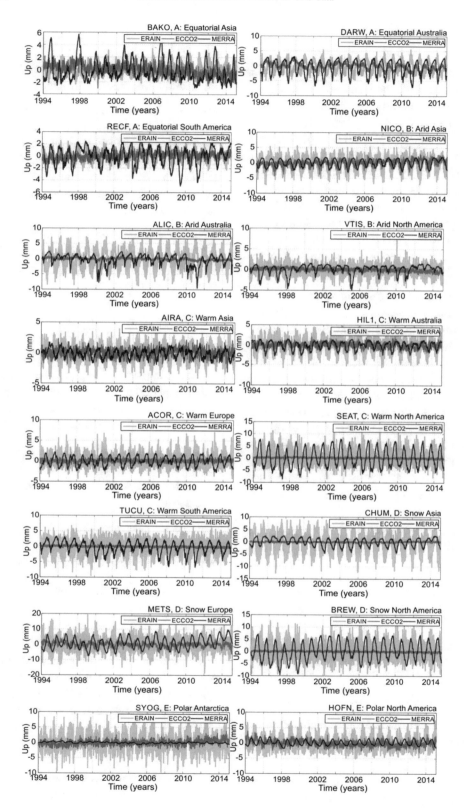

2.2. Multichannel Singular Spectrum Analysis (MSSA)

Multichannel (or multivariate) singular spectrum analysis (MSSA) is a method that allows analysing spatial and temporal correlations between different time series (Broomhead and King 1986a, b; Allen and Robertson 1996; Ghil et al. 2002). The common modes of spatio-temporal variability of a set of time series are described by empirical basic functions. MSSA, similarly to its univariate equivalent method named SSA, consists of two stages: decomposition of the grand-covariance matrix and reconstruction of interesting component. Trend, seasonalities, and noise common for all series are computed in the undermentioned steps:

Step 1. Application of the Embedding Procedure to Estimate the Full Augmented Trajectory Matrix. Firstly, the sliding window of M-length which will be moved across the entire dataset has to be chosen. To determine annual signals we focused on 3-, 4- and 6-year sliding windows, depending on the series. In the embedding procedure, a trajectory matrix was computed (Eq. 1). Each row relates to observations which were included in the sliding window of size M. This window is shifted until the last observation N is reached. This trajectory matrix has a dimension of $N' \times M$, where $N' = N - M + 1$ (Broomhead and King 1986a, b; Allen and Robertson 1996; Ghil et al. 2002):

$$\tilde{X}_l = \begin{pmatrix} X_l(1) & X_l(2) & \cdots & X_l(M) \\ X_l(2) & X_l(3) & \cdots & X_l(M+1) \\ \vdots & \vdots & \cdots & \vdots \\ X_l(N'-1) & \vdots & \cdots & X_l(N-1) \\ X_l(N') & X_l(N'+1) & \cdots & X_l(N) \end{pmatrix}.$$

$$(1)$$

Then, the multichannel trajectory matrix $\tilde{\mathbf{D}}$ is estimated as:

$$\tilde{\mathbf{D}} = \left(\tilde{\mathbf{X}}_1, \tilde{\mathbf{X}}_2, \ldots, \tilde{\mathbf{X}}_L \right) \qquad (2)$$

where L is the number of time series included in the analysed dataset. In our research, L was equal to 229 (total number of stations).

Step 2. Estimation of the Grand Lag-Covariance. The grand lag-covariance matrix is defined as:

$$\tilde{\mathbf{C}}_D = \frac{1}{N'} \tilde{\mathbf{D}}^t \tilde{\mathbf{D}} \qquad (3)$$

and was computed in this research using the BK algorithm (Broomhead and King 1986b).

Step 3. Decomposition of Grand Lag-Covariance Matrix to Determine Eigenvalues and Eigenvectors. The grand lag-covariance matrix is diagonalized using singular value decomposition (SVD) in order to compute eigenvalues λ_k and eigenvectors $\mathbf{E}^k$ also known as empirical orthogonal functions (EOFs).

Step 4. Determination of k-th Principal Component (PC) as Single-Channel Time Series. The consecutive PCs A^k are computed with the eigenvectors estimated in step 3:

$$A^k(t) = \sum_{j=1}^{M} \sum_{l=1}^{L} X_l(t+j-1) E_l^k(j). \qquad (4)$$

The frequency of a particular PC is determined using a periodogram. Any two PCs relating to the same frequency as well as the eigenvectors which correspond to those PCs are in quadrature.

Step 5. Computation of the Reconstructed Components (RC) of Frequency of Interest. The k-th RC at time t for time series l is defined as (Plaut and Vautard 1994):

$$R_l^k(t) = \begin{cases} \frac{1}{M} \sum_{j=1}^{M} A^k(t-j+1) E_l^k(j) & \text{for} \quad M \leq t \leq N-M+1 \\ \frac{1}{t} \sum_{j=1}^{M} A^k(t-j+1) E_l^k(j) & \text{for} \quad 1 \leq t < M-1 \\ \frac{1}{N-i+1} \sum_{j=1-N+M}^{M} A^k(t-j+1) E_l^k(j) & \text{for} \quad N-M+2 \leq t \leq N \end{cases}.$$

$$(5)$$

Depending on the frequency which interests us, various components can be reconstructed. It is worth emphasizing that summing all RCs, the original time series is reconstructed with no loss in information.

3. Results

In the following section, we presented the results of research on time-varying seasonal signals estimated with MSSA, separately for environmental loadings in each considered section. Then, we analysed the residuals of the GPS height time series obtained by subtraction of the common time-varying annual curves from data. Finally, we estimated the CME values to decide on the efficiency of the proposed approach. We aimed to propose MSSA as an alternative method to remove the common environmental effect from the GPS position time series without affecting their stochastic characters.

3.1. Common Seasonal Signals Estimated from Environmental Loadings

Common seasonal signals were estimated with MSSA from environmental loadings separately for sixteen considered sections (Fig. 3) and then summed to each other for the time span of 1.01.1994-14.02.2015. Residuals of the GPS height time series were produced by de-trending each of them and by removing the common annual signal for epochs corresponding to GPS observations. These residuals were then subjected to CME estimates. In this research, we intentionally focused on annual period, as the percentage of total variance of time series explained by modes of annual signal is much higher than the variance explained by any other pair of modes (Gruszczynska et al. 2016).

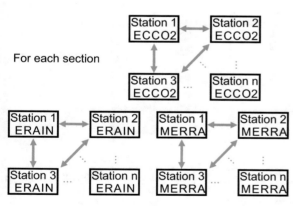

Figure 3
An idea of MSSA employed for a particular section and environmental model. The common annual signal was estimated based on time series included in the considered section, separately either for each section or each loading model. Then, the common annual signals estimated for ERAIN, MERRA and ECCO2 were summed to each other to reveal the common environmental effect which affects the GPS position time series

The principal components which correspond to the same frequency describe the common character of the employed time series. In other words, each individual PC constitutes a pattern of the common signal. PCs and eigenvectors of the annual period were combined to estimate the k-th RC to determine common oscillations for stations in a particular section, not of the individual station. The percentage of variance explained by the common signal is strictly related to the variance of entire time series estimated including trend, noise and other seasonal signals. It can be only interpreted in the context of all the time series common components variability. As an example, we can provide a series with the significant amplitude of the common annual signal, and the common trend, noise, and other seasonalities being also significant. Then, the contribution of the common annual signal is comparatively low in relation to the remaining components identified through the MSSA procedure. On the contrary, if the variance of trend, noise and other seasonalities is small, the total variance of data explained by the annual curve will be significant and large.,

From Fig. 4, we can notice that for ERAIN loading model, two first PCs are always related to common annual signals explaining between 4 and 80% of the total variance of data. If the percentage of variance explained by the annual signal is relatively low, the higher is the variance that represents the sum of the other time series components (trend, noise, and other seasonalities). For 7 sections, i.e. B: Arid North America, C: Warm Europe, C: Warm North and South America, D: Snow Europe, D: Snow North America and E: Polar Antarctica, the variance explained by the annual signal is relatively low in comparison to the variance explained by other modes and varies between 4 and 8%. For other sections, this variance is much higher and ranges 31-80%.

For MERRA loading model, the two first PCs always combine to annual signal. The variance explained by it varies between 31 and 83% of total variance of data. Zones as B: Arid Australia, B: Arid North America and C: Warm Australia are characterized by a high contribution of lower modes into the total percentage of variance equal to nearly 70% at maximum. For other sections, this percentage

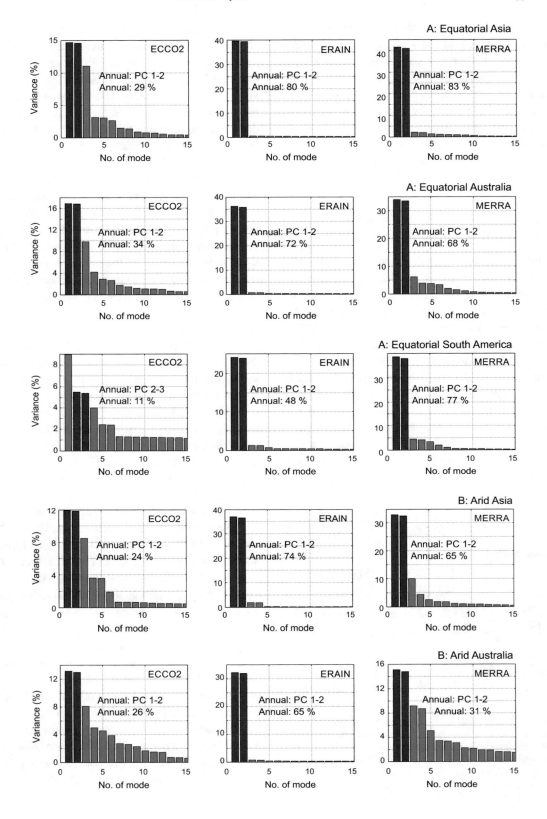

◀Figure 4
Percentage of variance explained by individual modes for environmental loadings for all sections we analysed. The modes which were used to reconstruct the annual changes are marked in red. Also, the percentage of variance explained by annual signal is given in each plot

explains between 60 and 83% and obviously is location dependent.

For ECCO2 loading model, the 2nd and 3rd PCs from sections: A: Equatorial South America, C: Warm North America, D: Snow Asia, and D: Snow North America are related to the common annual signal, while for 3 sections, namely: B: Arid North America, E: Polar Antarctica and E: Polar North America—the 3rd and 4th. For all these cases, first PCs explain the long-term non-linear trend. As an example, for E: Polar Antarctica section, the ECCO2 series are characterized by large non-linearity mainly between 2006 and 2015, while for stations included in E: Polar North America section, the non-linearity from 2000 onwards is hardly seen, as the variance explained by this non-linear trend is comparable to the annual signal. The long-term trend seen in the ECCO2 model may result from the Boussinesq approximation used to create this model (Ponte et al. 2007). For remaining 9 sections, annual signals were detected in the 1st and 2nd PCs with the amount of variance explained by annual curve being significantly higher than the variance explained by other modes.

Figure 5 presents the spatial distribution of the percentage of variance which is explained by a common annual signal in the vertical direction for ECCO2, ERAIN and MERRA model, respectively, for individual stations situated in different sections.

For the non-tidal ocean loading model, the common annual signal identified using MSSA accounts for an average of 16% of the total variance of data. The highest contribution of the annual signal was noticed for stations situated in A: Equatorial Australia section, reaching 34%. In Europe, the annual signal explains approximately 26%, while in North and South Americas 6 and 9%, respectively. Common annual oscillations identified using MSSA for atmospheric loading account for an average of 24% of the total variance of data with the highest percentage equal to 80, 72 and 74%, respectively, for

stations situated in South Asia, in North Australia and in Asia. The percentage of variance related to annual oscillation explains approximately 4% of the total variance for stations in Europe. The annual signal estimated using MSSA approach for the hydrological loading models explains around 68% of the total variance, proving that local hydrological plays a significant role in the observed signals. We noticed that stations situated in Europe are strongly affected by annual oscillations from hydrological loading, which accounts for 80%. Those results may be further used for other research related for example to the climate studies, but it falls out of the scope of this paper.

Then, we estimated the RCs for ERAIN, MERRA and ECCO2 models for all examined stations. Figure 6 reveals the original MERRA time series for stations from A: Equatorial Asia section and common seasonal signal derived by the MSSA approach. Due to the non-parametric character of MSSA, we were able to estimate common seasonal pattern which is not constant over time. From Fig. 6, we can notice that the maximum amplitude estimated for A: Equatorial Asia was equal to 5.9 mm at the beginning of 2007, while the smallest peak was equal to 4.8 mm in 2002. We compared the common annual signals estimated with MSSA with those determined with LSE separately for each loading model, obtaining maximum difference in peak-to-peak oscillations of 2.2 mm (or 77% in other words) at maximum.

Finally, the common seasonal signal estimated from ERAIN, MERRA, and ECCO2 with MSSA algorithm was subtracted from the GPS height time series. Due to numerical artefacts, the environmental loadings do not account for entire seasonal oscillation observed by the GPS time series, as pointed the Introduction section. Therefore, we modelled the remaining annual oscillations from the residual time series using the LSE approach. The median amplitude of the residual annual oscillations estimated for the 229 GPS stations was equal to 1.7 mm. They arose from the fact that the common annual oscillation estimated with MSSA did not reflect the entire seasonal changeability of series due to its spatial pattern. These residual annual oscillations were

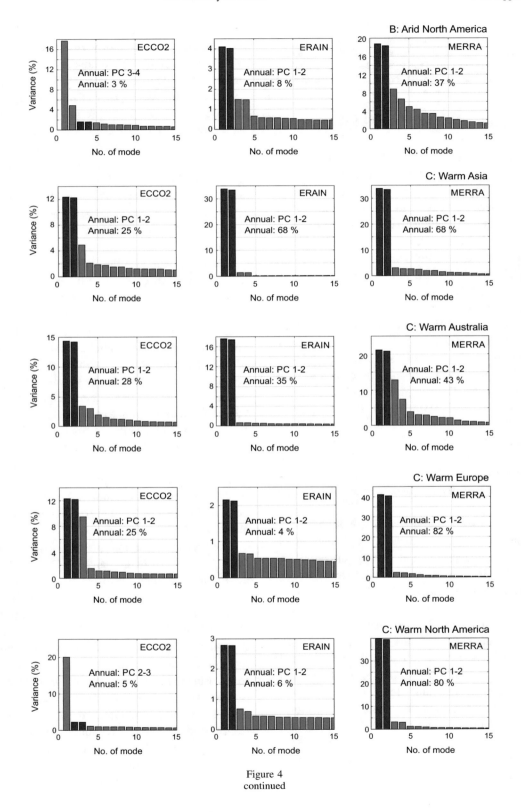

Figure 4
continued

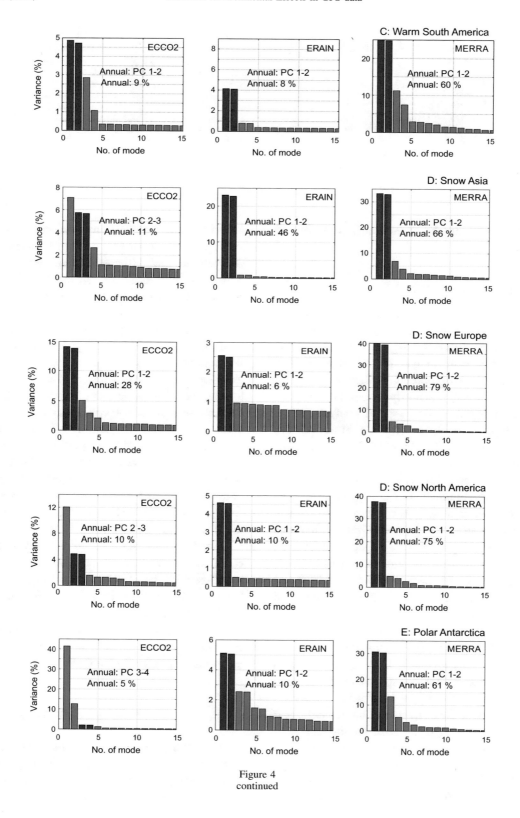

Figure 4
continued

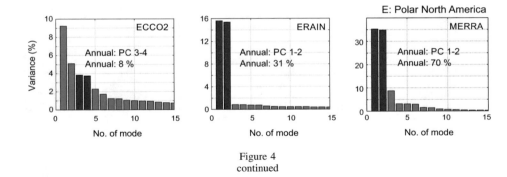

Figure 4
continued

subtracted to estimate the series submitted for further analysis.

3.2. Common Mode Errors

In order to show the main advantage of estimating the seasonal signal with MSSA, we determined the Common Mode Error (CME) with the use of the Principal Component Analysis (PCA) for incomplete GNSS time series as proposed by Shen et al. (2013). This procedure was previously successfully applied by the authors for spatio-temporal analysis of the GPS position time series (Bogusz et al. 2015b; Gruszczynski et al. 2016). First, CME was estimated for residuals after MSSA curves were removed. Then, CME was estimated for residuals after direct subtraction of environmental loading models. The second approach is nowadays widely used to remove the environmental effects from the GPS position time series, while first is a novelty introduced in this paper.

Figure 7 presents the selected stacked power spectral densities of height time series CME determined for individual sections after environmental effect was removed using MSSA-derived annual curve as well as removed directly from time series. These results confirm that although the direct subtraction of environmental loading models from the GPS position time series may help in removing the loading effect and in reducing the RMS values, it causes a change in the character of the stochastic part of time series leading to an artificial subtraction of some power in CME. On the other hand, we may suppose that some part of the Common Mode Error observed by the GPS sensors arises from environmental influence. However, it has not yet been considered if the GPS records are sensitive to environmental influence in the frequency bands between 10 and 80 cpy. If not, we should not artificially influence the character of CME by removal of entire environmental loadings.

From Fig. 7, we may notice that the character of CME is very similar for both cases for the B: Arid Australia section. For A: Equatorial Australia, C: Warm South America, D: Snow Europe, and E: Polar North America, we can observe that environmental loading models remove lots of power in a frequency band between 10 and 80 cpy. Especially, zones as C: Warm South America and D: Snow Europe are affected by a large cut in the power, and, therefore, a change in a character of CME. When MSSA curves which reveal the common geophysical signal are removed from the GPS position time series, the change in power is not observed, which means that the character of CME remains intact.

Based on the results, we concluded that the common geophysical signal can be successfully modelled with the MSSA method and then removed from the GPS position time series. We showed that the CME values are not affected by MSSA estimates, which makes it to be an effective approach to investigate and/or subtract a common large-scale environmental effect in the GPS position time series.

4. Discussion and Conclusions

The MSSA approach is a non-parametric method that is able to investigate simultaneously the spatial and temporal correlations for analysing the dependence between any geodetic time series. This method

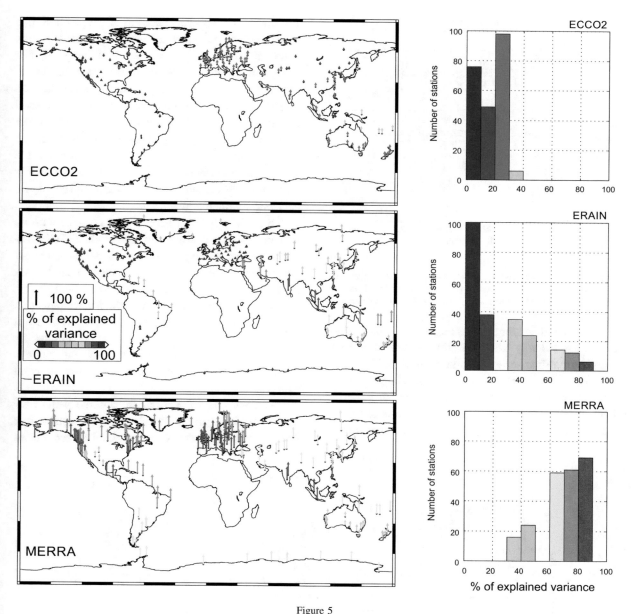

Figure 5
Percentage of variance explained by common annual signal in the vertical component of environmental loadings

provides the opportunity to determine a signal which is common for stations included in the analysis. In this research, we focused on the common annual signal as environmental loadings from atmosphere, hydrosphere and non-tidal ocean can similarly affect stations situated in a specific area. As it has been already proven, the GPS position time series are influenced by real geophysical changes, systematic errors and spurious effects (Dong et al. 2002; Ray

et al. 2008; Collilieux et al. 2010), which may be observed as seasonal curves with amplitudes changing over time (Gegout et al. 2010; Bennet 2008; Davis et al. 2012; Chen et al. 2013). If we do not assume the time variability of seasonal variations, this can be transferred to the reliability of station's velocity estimates (Klos et al. 2017).

The recent researches (Santamaría-Gomez and Mémin 2015; He et al. 2017) confirmed that

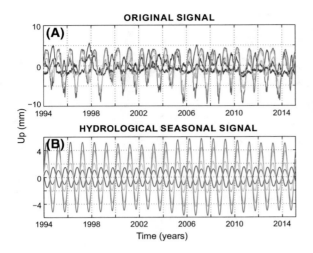

ORIGINAL SIGNAL

HYDROLOGICAL SEASONAL SIGNAL

Figure 6
Original height time series **a** and common annual signal **b**, estimated from MERRA model for stations included in A: Equatorial Asia. Different colours denote different time series

environmental loading models should be considered before the velocity of GPS station is estimated from the position time series. Collilieux et al. (2012) used the loading models to mitigate the aliasing effect in the GPS technique during the frame transformation. Klos et al. (2017) proved that a direct removal of loadings causes a reduction in the RMS values, but it

also changes the power spectrum of the position time series for frequencies between 4 and 80 cpy, removing a part of the power. Following up, this change in the power spectrum can cause an underestimation of the uncertainty of station's velocity. As we still do not know if GPS senses the environmental effects in the entire frequency range, i.e. we are not sure if the high frequency changes of GPS are due to environmental loadings, we should not remove the environmental effects directly from the GPS position time series. What should be aimed at is the stochastic part which remains intact.

To retrieve the real geophysical changes with no influence on the character of the residual GPS position time series, we proposed to model and subtract the common annual part from environmental loadings. In this way, we remove the geophysical signals of common origin which may affect the changes in the positions of GPS permanent stations. We used a set of the IGS ITRF2014 stations and discussed the amount of variance explained by individual models. Afterwards, the common annual curve was removed from the heights.

We determined the common seasonal pattern for environmental loadings which are not constant over time. For example, for A: Equatorial Asia section we

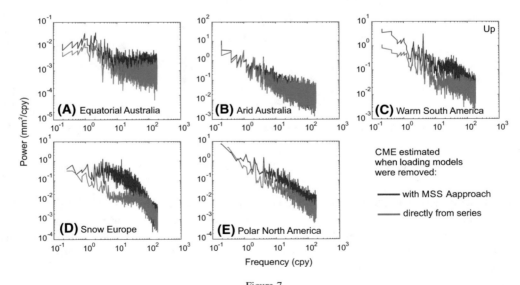

Figure 7
Stacked power spectral densities (PSDs) estimated for CME of residuals of the GPS height time series after the influence of environmental loadings was removed using MSSA curve (in red) and directly from series (in blue). The CME was stacked within each of the considered sections. All PSDs were estimated using Welch periodogram (Welch 1967)

may make a misfit of 2.2 mm if we assume the time constancy. Annual signal in the MERRA model accounts for 68% of the total variance. It proves that hydrological series are strongly affected by common annual signal. For the ERAIN model, the annual curve explains 24% of the total variance on average, while for the ECCO2 model—only 16%.

The highest contribution of the annual changes into ECCO2 loading model reached 40% for stations situated in Indonesia. This was previously noticed by van Dam et al. (2012), who emphasized that both the RMS values and maximum predicted radial surface displacement are also much larger for Indonesia than they are for any other regions of the world. Also, the region of North Sea is affected by large contribution of non-tidal ocean loading, as stated before by Williams and Penna (2011). They noticed that the removal of ECCO2 model from the GPS position time series can reduce the variance of the series of the same amount as the atmospheric loading does. In our analysis, we observed that the contribution of annual curve for European stations for ECCO2 is much higher (40%) than it is for ERAIN model (10%).

As was previously noted by van Dam et al. (2012), the most of the power in the non-tidal ocean loading records is at the annual period with other frequencies being not as powerful as annual signal. This was observed in our research as well, as for almost all sections considered, the annual signal explains most of the variance of time series. We also found that most of the power in the ERAIN atmospheric model is cumulated at 1 cpy excluding sections B: Arid North America, C: Warm Europe, D: Snow Europe and E: Polar Antarctica. The annual signal dominates also in the MERRA hydrological model, explaining up to 90% of the total variance of the height time series.

Van Dam et al. (2012) found the largest reduction in the RMS values for Asian stations after atmospheric non-tidal loading was incorporated. In our analysis, we noticed that Australian, Indonesian and Asian stations are characterized by the largest contribution of annual signal in the total variance of ERAIN model. If the correlation between the GPS height time series and ERAIN model is high, as showed by van Dam et al. (2012), we expect that the annual curves will constitute a good approximation of

annual signal found in the GPS position time series. For MERRA model, we compared our results with Tregoning et al. (2009), who found a large RMS reduction of the GPS height time series, when these were corrected for elastic deformation using continental water load estimates derived from GRACE. We found that the percentage of variance explained by annual signal correlates well with the RMS reduction they presented, especially for areas of North America, South America and Northern Australia. For those areas, the reduction in RMS they found varied from − 9 to − 3 mm, while the percentage of variance explained by common annual signal we estimated varies between 30 and 50%. The RMS reduction for stations situated in Central and Southern Europe reached the values between 3 and 9 mm as reported by Tregoning et al. (2009). For these stations, the percentage of variance we explained by common annual signal is in the range of 80–90%. The above result means that the reduction in RMS values which was found before can arise from the annual signal present in hydrological signal, which dominates the loading effect.

We found the evident long-term trends for ECCO2 model, confirming the findings of van Dam et al. (2012). This trend explains the majority of variance of ECCO2 model for most of sections considered. According to Ponte et al. (2007), the long-term trends in the ECCO2 model may arise from the Boussinesq approximation which is employed to compute the model. Van Dam et al. (2012) emphasized that a real long-term trend in non-tidal ocean loading can arise from a trend in freshwater fluxes, trends in the atmospheric forcing as well as long-term climate variability.

After the common annual seasonal curve was modelled with MSSA and removed from the GPS position time series, the residual station-specific annual curve was removed with LSE, as it may affect the character of stochastic part of the GPS data (Bos et al. 2010; Bogusz and Klos 2016; Klos et al. 2018b). We showed that the amplitudes of residual oscillations do not exceed 2 mm at maximum. These findings were consistent with results obtained by Klos et al. (2017). In their paper, the authors analysed the residual annual oscillation after the SSA curve was modelled for environmental loading models and then removed from the GPS position time series, but on a station-by-station basis. This means that those models underestimate the seasonalities observed by the GPS system. So, probably, the remaining part of signal being present in the seasonal frequency bands arises from the systematic errors which are not of geophysical origin.

The approach based on removing the common environmental effect with the MSSA method we introduced was compared to the widely applied removing the loading models directly from the GPS position time series. The advantages were shown using the CME estimates. When a common annual signal was removed from the GPS position time series, neither removal of power nor artificial reduction in the CME was noticed. In this way, we remove the geophysical annual effects that disturb stations included in a particular section with no influence on the high frequency part of the spectra. We observed that direct subtraction of environmental loading models causes an evident change in the PSD for frequencies between 10 and 80 cpy. Removing the MSSA-derived curves from environmental loading models causes no influence on the stochastic part of the GPS time series. So, a change in the noise character and CME estimates noted before by Yuan et al. (2008), when the environmental loading models were removed directly from series, arose from the artificial cut in the power of residuals of the GPS position time series.

The two-step solution we propose allows us to consider the real geophysical effects arising from environmental loading models and to model the time-varying common seasonal signal using the MSSA approach with no artificial change of the stochastic part of the GPS height time series we analysed. However, this approach can be successfully applied for any other type of geodetic observations, in which we may expect a common influence of different type.

Acknowledgements

We would like to thank Giuliana Rossi and other anonymous reviewer for their suggestions and comments improving the manuscript. This research was financed by the National Science Centre, Poland, Grant no. UMO-2017/25/B/ST10/02818 under the leadership of Prof. Janusz Bogusz. World Map of Köppen-Geiger Climate Classification was downloaded from http://koeppen-geiger.vu-wien.ac.at/shifts.htm. GPS time series were accessed from http://acc.igs.org/reprocess2.html on 2016-05-05. Environmental loadings were downloaded from EOST loading service: http://loading.u-strasbg.fr/ on 2017-01-10. Maps were drawn in the generic mapping tool (GMT) (Wessel et al. 2013). We modified Matlab-based algorithms written by Eric Breitenberger which were downloaded from https://pantherfile.uwm.edu/kravtsov/www/downloads/KWCT2014/SSAMATLAB/mssa/. This research was partially supported by the PL-Grid Infrastructure, grant name "mssagnss".

REFERENCES

Allen, M. R., & Robertson, A. W. (1996). Distinguishing modulated oscillations from coloured noise in multivariate datasets. *Climate Dynamics, 12,* 772–786. https://doi.org/10.1007/s003820050142.

Altamimi, Z., Rebischung, P., Métivier, L., & Collilieux, X. (2016). ITRF2014: A new release of the international terrestrial reference frame modelling nonlinear station motions. *Journal of Geophysical Research: Solid Earth, 121*(8), 6109–6131. https://doi.org/10.1002/2016JB013098.

Bennet, R. A. (2008). Instantaneous deformation from continuous GPS: Contributions from quasi-periodic loads. *Geophysical Journal International, 174*(3), 1052–1064. https://doi.org/10.1111/j.1365-246X.2008.03846.x.

Blewitt, G., & Lavallée, D. (2002). Effect of annual signals on geodetic velocity. *Journal of Geophysical Research, 107*(B7), ETG 9-1–ETG 9-11. https://doi.org/10.1029/2001jb000570.

Bogusz, J., Gruszczynska, M., Klos, A., & Gruszczynski, M. (2015a). Non-parametric estimation of seasonal variations in GPS-derived time series. In: van Dam T. (Eds.) *REFAG 2014. International Association of Geodesy Symposia,* 146. Cham: Springer. https://doi.org/10.1007/1345_2015_191

Bogusz, J., Gruszczynski, M., Figurski, M., & Klos, A. (2015b). Spatio-temporal filtering for determination of common mode error in regional GNSS networks. *Open Geosciences, 7*(1), 140–148. https://doi.org/10.1515/geo-2015-0021.

Bogusz, J., & Klos, A. (2016). On the significance of periodic signals in noise analysis of GPS station coordinates time series. *GPS Solutions, 20*(4), 655–664. https://doi.org/10.1007/s10291-015-0478-9.

Bogusz, J., & Kontny, B. (2011). Estimation of sub-diurnal noise level in GPS time series. *Acta Geodynamica et Geomaterialia, 8*(3), 273–281.

Bos, M. S., Bastos, L., & Fernandes, R. M. S. (2010). The influence of seasonal signals on the estimation of the tectonic motion in short continuous GPS time-series. *Journal of Geodynamics, 49*(3–4), 205–209. https://doi.org/10.1016/j.jog.2009.10.005.

Broomhead, D.S., & King G.P. (1986a). On the qualitative analysis of experimental dynamical systems. In Adam Hilger (Ed.) Nonlinear Phenomena and Chaos, 113–144. Bristol.

Broomhead, D. S., & King, G. P. (1986b). Extracting qualitative dynamics from experimental data. *Physica, 20*(2–3), 217–236. https://doi.org/10.1016/0167-2789(86)90031-X.

Chen, Q., van Dam, T., Sneeuw, N., Collilieux, X., Weigelt, M., & Rebischung, P. (2013). Singular spectrum analysis for modelling seasonal signals from GPS time series. *Journal of Geodynamics, 72*, 25–35. https://doi.org/10.1016/j.jog.2013.05.005.

Collilieux, X., Altamimi, Z., Coulot, D., Ray, J., & Sillard, P. (2007). Comparison of very long baseline interferometry, GPS, and satellite laser ranging height residuals from ITRF2005 using spectral and correlation methods. *Journal of Geophysical Research*. https://doi.org/10.1029/2007JB004933.

Collilieux, X., Altamimi, Z., Coulot, D., van Dam, T., & Ray, J. (2010). Impact of loading effects on determination of the International Terrestrial Reference Frame. *Advances in Space Research, 45*(1), 144–154. https://doi.org/10.1016/j.asr.2009.08.024.

Collilieux, X., van Dam, T., Ray, J., Coulot, D., Métivier, L., & Altamimi, Z. (2012). Strategies to mitigate aliasing of loading signals while estimating GPS frame parameters. *Journal of Geodesy, 86*(1), 1–14. https://doi.org/10.1007/s00190-011-0487-6.

Dach, R., Boehm, J., Lutz, S., Steigenberger, P., & Beutler, G. (2011). Evaluation of the impact of atmospheric pressure loading modelling on GNSS data analysis. *Journal of Geodesy, 85*(2), 75–91. https://doi.org/10.1007/s00190-010-0417-z.

Davis, J. L., Wernicke, B. P., & Tamisiea, M. E. (2012). On seasonal signals in geodetic time series. *Journal of Geophysical Research*. https://doi.org/10.1029/2011JB008690.

Dee, D. P., Uppala, S. M., Simmons, A. J., Berrisford, P., Poli, P., Kobayashi, S., et al. (2011). The ERA-Interim reanalysis: configuration and performance of the data assimilation system. *Quarterly Journal of the Royal Meteorological Society, 137*(656), 553–597. https://doi.org/10.1002/qj.828.

Didova, O., Gunter, B., Riva, R., Klees, R., & Roese-Koerner, L. (2016). An approach for estimating time-variable rates from geodetic time series. *Journal of Geodesy, 90*(11), 1207–1221. https://doi.org/10.1007/s00190-016-0918-5.

Dong, D., Fang, P., Bock, Y., Cheng, M. K., & Miyazaki, S. (2002). Anatomy of apparent seasonal variations from GPS-derived site position time series. *Journal of Geophysical Research, 107*(4), ETG 9-1–ETG 9-16. https://doi.org/10.1029/2001jb000573.

Dong, D., Fang, P., Bock, Y., Webb, F., Prawirodirdjo, L., Kedar, S., et al. (2006). Spatio-temporal filtering using principal component analysis and Karhunen-Loeve expansion approaches for regional GPS network analysis. *Journal of Geophysical Research*. https://doi.org/10.1029/2005JB003806.

Freymueller, J.T. (2009). Seasonal position variations and regional Reference frame realization. In H. Drewes (Ed.), *Geodetic Reference Frames, International Association of Geodesy Symposia*, 134, 191–196. Springer, Berlin. https://doi.org/10.1007/978-3-642-00860-3_30.

Gegout, P., Boy, J.-P., Hinderer, J., & Ferhat, G. (2010). Modeling and observation of loading contribution to time-variable GPS site positions. In S. Mertikas (Ed.), *Gravity, Geoid and Earth Observation, International Association of Geodesy Symposia*, 135, 651–659. Springer, Berlin. https://doi.org/10.1007/978-3-642-10634-7_86.

Ghil, M., Allen, M. R., Dettinger, M. D., Ide, K., Kondrashov, D., Mann, M. E., et al. (2002). Advanced spectral methods for climatic time series. *Reviews of Geophysics*. https://doi.org/10.1029/2000RG000092.

Gruszczynska, M., Klos, A., Gruszczynski, M., & Bogusz, J. (2016). Investigation of time-changeable seasonal components in the GPS height time series: A case study for Central Europe. *Acta Geodynamica et Geomaterialia, 13*(3), 281–289. https://doi.org/10.13168/AGG.2016.0010.

Gruszczynska, M., Klos, A., Rosat, S., & Bogusz, J. (2017). Deriving common seasonal signals in GPS position time series: By using multichannel singular spectrum analysis. *Acta Geodynamica et Geomaterialia, 14*(3), 267–278. https://doi.org/10.13168/AGG.2017.0010.

Gruszczynski, M., Klos, A., & Bogusz, J. (2016). Orthogonal transformation in extracting of common mode errors from continuous GPS networks. *Acta Geodynamica et Geomaterialia, 13*(3), 291–298. https://doi.org/10.13168/AGG.2016.0011.

He, X., Montillet, J.-P., Hua, X., Yu, K., Jiang, W., & Zhou, F. (2017). Noise analysis for environmental loading effect on GPS time series. *Acta Geodynamica et Geomaterialia, 14*(1), 131–142. https://doi.org/10.13168/AGG.2016.0034.

Jiang, W., Li, Z., van Dam, T., & Ding, W. (2013). Comparative analysis of different environmental loading methods and their impacts on the GPS height time series. *Journal of Geodesy, 87*(7), 687–703. https://doi.org/10.1007/s00190-013-0642-3.

King, M., Altamimi, Z., Boehm, J., Bos, M., Dach, R., Elosegui, P., Fund, F., Hernández-Pajares, M., Lavallee, D., Mendes Cerveira, P. J., Penna, N., Riva, R. E. M., Steigenberger, P., van Dam, T., Vittuari, L., Williams, S., Willis, P. (2010). Improved constraints on models of glacial isostatic adjustment: a review of the contribution of ground-based geodetic observations. *Surveys in Geophysics, 31*(5), 465–507. https://doi.org/10.1007/s10712-010-9100-4.

Klos, A., & Bogusz, J. (2017). An evaluation of velocity estimates with a correlated noise: case study of IGS ITRF2014 European stations. *Acta Geodynamica et Geomaterialia, 14*(3), 255–265. https://doi.org/10.13168/AGG.2017.0009.

Klos, A., Bogusz, J., Figurski, M., & Gruszczynski, M. (2016). Error analysis for European IGS stations. *Studia Geophysica et Geodaetica, 60*(1), 17–34. https://doi.org/10.1007/s11200-015-0828-7.

Klos, A., Bos, M. S., & Bogusz, J. (2018a). Detecting time-varying seasonal signal in GPS position time series with different noise levels. *GPS Solutions*. https://doi.org/10.1007/s10291-017-0686-6.

Klos, A., Gruszczynska, M., Bos, M. S., Boy, J.-P., & Bogusz, J. (2017). Estimates of vertical velocity errors for IGS ITRF2014 stations by applying the improved singular spectrum analysis method and environmental loading models. *Pure and Applied Geophysics*. https://doi.org/10.1007/s00024-017-1494-1.

Klos, A., Olivares, G., Teferle, F. N., Hunegnaw, A., & Bogusz, J. (2018b). On the combined effect of periodic signals and colored noise on velocity uncertainties. *GPS Solutions*. https://doi.org/10.1007/s10291-017-0674-x.

Márquez-Azúa, B., & DeMets, C. (2003). Crustal velocity field of Mexico from continuous GPS measurements, 1993 to June 2001: Implications for the neotectonics of Mexico. *Journal of Geophysical Research*. https://doi.org/10.1029/2002JB002241.

Menemenlis, D., Campin, J., Heimbach, P., Hill, C., Lee, T., Nguyen, A., et al. (2008). ECCO2: High resolution global ocean and sea ice data synthesis. *Mercator Ocean Quarterly Newsletter, 31*, 13–21.

Nikolaidis, R. (2002). Observation of geodetic and seismic deformation with the Global Positioning System. Ph.D. thesis. Univ. of Calif., San Diego

Penna, N. T., King, M. A., & Stewart, M. P. (2007). GPS height time series: Short-period origins of spurious long-period signals. *Journal of Geophysical Research*. https://doi.org/10.1029/2005JB004047.

Plaut, G., & Vautard, R. (1994). Spells of low-frequency oscillations and weather regimes in the northern hemisphere. *Journal of the Atmospheric Sciences, 51*(2), 210–236. https://doi.org/10.1175/1520-0469(1994)051<0210:SOLFOA>2.0.CO;2.

Ponte, R., Quinn, K., Wunsch, C., & Heimbach, P. (2007). A comparison of model and GRACE estimates of the large-scale seasonal cycle in ocean bottom pressure. *Geophysical Research Letters*. https://doi.org/10.1029/2007GL029599.

Ray, J., Altamimi, Z., Collilieux, X., & van Dam, T. (2008). Anomalous harmonics in the spectra of GPS position estimates. *GPS Solutions, 12*(1), 55–64. https://doi.org/10.1007/s10291-007-0067-7.

Rebischung, P., Altamimi, Z., Ray, J., & Garayt, B. (2016). The IGS contribution to ITRF2014. *Journal of Geodesy, 90*(7), 611–630. https://doi.org/10.1007/s00190-016-0897-6.

Reichle, R. H., Koster, R. D., De Lannoy, G. J. M., Forman, B. A., Liu, Q., Mahanama, S. P. P., et al. (2011). Assessment and enhancement of MERRA land surface hydrology estimates. *Journal of Climate, 24*, 6322–6338. https://doi.org/10.1175/JCLI-D-10-05033.1.

Rubel, F., & Kottek, M. (2010). Observed and projected climate shifts 1901–2100 depicted by world maps of the Köppen–Geiger climate classification. *Meteorologische Zeitschrift, 19*, 135–141. https://doi.org/10.1127/0941-2948/2010/0430.

Santamaría-Gomez, A., Bouin, M.-N., Collilieux, X., & Woppelmann, G. (2011). Correlated errors in GPS position time series: Implications for velocity estimates. *Journal of Geophysical Research*. https://doi.org/10.1029/2010JB007701.

Santamaría-Gomez, A., & Mémin, A. (2015). Geodetic secular velocity errors due to interannual surface loading deformation. *Geophysical Journal International, 202*, 763–767. https://doi.org/10.1093/gji/ggv190.

Shen, Y., Li, W., Xu, G., & Li, B. (2013). Spatio-temporal filtering of regional GNSS network's position time series with missing data using principle component analysis. *Journal of Geodesy*. https://doi.org/10.1007/s00190-013-0663-y.

Shen, Y., Peng, F., & Li, B. (2015). Improved singular spectrum analysis for time series with missing data. *Nonlinear Processes in Geophysics, 22*, 371–376. https://doi.org/10.5194/npg-22-371-2015.

Tesmer, V., Steigenberger, P., Rothacher, M., Boehm, J., & Meisel, B. (2009). Annual deformation signals from homogeneously reprocessed VLBI and GPS height time series. *Journal of Geodesy, 83*(10), 973–988. https://doi.org/10.1007/s00190-009-0316-3.

Tregoning, P., Watson, C., Ramillien, G., McQueen, H., & Zhang, J. (2009). Detecting hydrologic deformation using GRACE and GPS. *Geophysical Research Letters*. https://doi.org/10.1029/2009GL038718.

van Dam, T., Collilieux, X., Wuite, J., Altamimi, Z., & Ray, J. (2012). Nontidal ocean loading: Amplitudes and potential effects in GPS height time series. *Journal of Geodesy, 86*(11), 1043–1057. https://doi.org/10.1007/s00190-012-0564-5.

van Dam, T., & Wahr, J. M. (1987). Displacements of the Earth's surface due to atmospheric loading: Effects on gravity and baseline measurements. *Journal of Geophysical Research: Solid Earth, 92*(B2), 1281–1286. https://doi.org/10.1029/JB092iB02p01281.

van Dam, T., & Wahr, J. M. (1998). Modeling environmental loading effects: A review. *Physics and Chemistry of the Earth, 23*, 1077–1086.

van Dam, T., Wahr, J., Milly, P. C. D., Shmakin, A. B., Blewitt, G., Lavallée, D., et al. (2001). Crustal displacements due to continental water loading. *Geophysical Research Letters, 28*(4), 651–654. https://doi.org/10.1029/2000GL012120.

Walwer, D., Calais, E., & Ghil, M. (2016). Data-adaptive detection of transient deformation in geodetic networks. *Journal of Geophysical Research: Solid Earth, 121*(3), 2129–2152. https://doi.org/10.1002/2015JB012424.

Wdowinski, S., Bock, Y., Zhang, J., Fang, P., & Genrich, J. (1997). Southern California permanent GPS geodetic array: Spatial filtering of daily positions for estimating coseismic and postseismic displacements induced by the 1992 Landers earthquake. *Journal of Geophysical Research, 102*(B8), 18057–18070. https://doi.org/10.1029/97JB01378.

Welch, P. D. (1967). The use of fast fourier transform for the estimation of power spectra: A method based on time averaging over short. *Modified Periodograms, IEEE Transactions on Audio Electroacoustics, 15*, 70–73. https://doi.org/10.1109/TAU.1967.1161901.

Wessel, P., Smith, W. H. F., Scharroo, R., Luis, J., & Wobbe, F. (2013). Generic mapping tools: Improved version released. *Eos, Transactions, American Geophysical Union, 94*(45), 409–410. https://doi.org/10.1002/2013EO450001.

Williams, S. D. P. (2003). The effect of coloured noise on the uncertainties of rates estimated from geodetic time series. *Journal of Geodesy, 76*(9–10), 483–494. https://doi.org/10.1007/s00190-002-0283-4.

Williams, S. D. P., & Penna, N. T. (2011). Non-tidal ocean loading effects on geodetic GPS heights. *Geophysical Research Letters, 38*, L09314. https://doi.org/10.1029/2011GL046940.

Yuan, L., Ding, X., Chen, W., Guo, Z., Chen, S., Hong, B., et al. (2008). Characteristics of daily position time series from the Hong Kong GPS fiducial network. *Chinese Journal of Geophysics, 51*(5), 1372–1384. https://doi.org/10.1002/cjg2.1292.

Zhu, C., Lu, Y., Shi, H., & Zhang, Z. (2016). Spatial and temporal patterns of the inter-annual oscillations of glacier mass over Central Asia inferred from Gravity Recovery and Climate Experiment (GRACE) data. *Journal of Arid Land, 9*(1), 87–97. https://doi.org/10.1007/s40333-016-0021-z.

(Received November 2, 2017, revised January 30, 2018, accepted February 21, 2018, Published online March 3, 2018)

Pure Appl. Geophys. 175 (2018), 1823–1840
© 2017 Springer International Publishing
https://doi.org/10.1007/s00024-017-1494-1

Pure and Applied Geophysics

Estimates of Vertical Velocity Errors for IGS ITRF2014 Stations by Applying the Improved Singular Spectrum Analysis Method and Environmental Loading Models

Anna Klos,[1] Marta Gruszczynska,[1] Machiel Simon Bos,[2] Jean-Paul Boy,[3] and Janusz Bogusz[1]

Abstract—A reliable subtraction of seasonal signals from the Global Positioning System (GPS) position time series is beneficial for the accuracy of derived velocities. In this research, we propose a two-stage solution of the problem of a proper determination of seasonal changes. We employ environmental loading models (atmospheric, hydrological and ocean non-tidal) with a dominant annual signal of amplitudes in their superposition of up to 12 mm and study the seasonal signal (annual and semi-annual) estimates that change over time using improved singular spectrum analysis (ISSA). Then, this deterministic model is subtracted from GPS position time series. We studied data from 376 permanent International GNSS Service (IGS) stations, derived as the official contribution to International Terrestrial Reference Frame (ITRF2014) to measure the influence of applying environmental loading models on the estimated vertical velocity. Having removed the environmental loadings directly from the position time series, we noticed the evident change in the power spectrum for frequencies between 4 and 80 cpy. Therefore, we modelled the seasonal signal in environmental models using the ISSA approach and subtracted it from GPS vertical time series to leave the noise character of the time series intact. We estimated the velocity dilution of precision (DP) as a ratio between classical Weighted Least Squares and ISSA approach. For a total number of 298 out of the 376 stations analysed, the DP was lower than 1. This indicates that when the ISSA-derived curve was removed from the GPS data, the error of velocity becomes lower than it was before.

Key words: GPS, seasonal signals, singular spectrum analysis, environmental loadings, ITRF2014, dilution of precision.

1. Introduction

Most global navigation satellite system (GNSS) position time series show displacements with annual and semi-annual periodicities (e.g. Blewitt et al. 2001; Collilieux et al. 2007; van Dam et al. 2007). Those periodicities, or so-called seasonal changes, are produced by geophysical effects related to tides (solid Earth, ocean, atmosphere or pole), transportation of mass over the surface of the Earth (atmosphere- and hydrosphere-related non-tidal effects) or bedrock and antenna thermal expansion (Dong et al. 2002). However, the power at certain frequencies results from systematic errors rather than real geophysical effects. In part, these errors are due to the residual diurnal and semi-diurnal tidal signatures that are aliased into a discrete 24-h GPS solution (Dong et al. 2002; Penna and Stewart 2003). Moreover, unmodelled- or mismodelled oscillations can propagate into spurious features with periods that vary from a few weeks to 1 year (Penna et al. 2007). A draconitic year (Agnew and Larson 2007) with a period equal to 351.6 days (Amiri-Simkooei 2013) has also been found in a position time series from International GNSS Service (IGS) stations. It is an artefact of GPS solution as it has not been detected using very long baseline interferometry (VLBI) and satellite laser ranging (SLR) techniques, or in geophysical fluids (Ray et al. 2008). In addition, the quality of the GNSS position time series suffers from errors when considering the double differences and network adjustment including transfer from fiducial stations or inclusion of scale (Tregoning and van Dam 2005). Finally, each type of modelling (satellite orbits, the Earth Orientation Parameters, satellite Antenna Phase Center Variations, mapping functions,

[1] Faculty of Civil Engineering and Geodesy, Military University of Technology, Warsaw, Poland. E-mail: marta.gruszczynska@wat.edu.pl
[2] Instituto D. Luis, University of Beira Interior, Covilhã, Portugal.
[3] Institut de Physique du Globe de Strasbourg, CNRS/ Université de Strasbourg (EOST), Strasbourg, France.

etc.) may introduce artificial oscillations into the time series. The comprehensive contributions of both geophysical sources and errors of models to the observed periodic (annual) changes in position were demonstrated by Dong et al. (2002).

Jiang et al. (2013) corrected the GPS height time series from three various environmental loading models (atmospheric pressure, continental water storage and non-tidal ocean loading) and tested their effectiveness for a set of 233 globally distributed GPS reference stations. They investigated the influence of the loading contributions on the IGS weekly repro-cessed time series and found significant improvement in the values of weighted root mean square (WRMS) error. They also found that the RMSs of the times series after applying loadings show larger variations in the Northern Hemisphere than in the Southern Hemisphere, mainly due to differences in land–ocean distribution.

Previous research into environmental loading models and their contribution to ground displace-ments showed that some power of observed annual variation can be explained by hydrological (e.g. van Dam et al. 2012; Dill and Dobslaw 2013), atmo-spheric (e.g. van Dam and Wahr 1987) and non-tidal oceanic (e.g. Fratepietro et al. 2006) loadings. Non-tidal ocean loading (NTOL) can cause a 5-mm peak-to-peak variation in a vertical direction (van Dam et al. 1997) and can reduce the scatter in the GPS height changes of about 10% when it is subtracted (Zerbini et al. 2004). Van Dam et al. (2001) found that continental water storage (CWS) may explain more than half of the annual signal in vertical dis-placements derived from GPS. Loading deformation of Earth's crust due to the redistribution of air masses can reach 20 mm (Petrov and Boy 2004). As much as 40% of the observed annual variation in vertical direction results from many cumulative factors, such as pole tide effects, ocean tide loading, atmospheric loading, non-tidal oceanic mass and groundwater loading (Dong et al. 2002). Apart from the mass loading due to the atmosphere, continental water storage and non-tidal ocean loading jointly contribute at a rate of about 53% to the annual amplitude of GPS vertical changes (Yan et al. 2009). The effects of non-tidal ocean and atmospheric loading on GPS displacement in vertical direction

are comparable when considered separately and may reduce the GPS RMSs by 30% when they are combined (Williams and Penna 2011). When atmo-spheric tidal model (ATML) is improperly modeled, it may mitigate in the combined GPS solution introducing anomalous signal around draconitic, annual and semi-annual period (Tregoning and Watson 2009). Dach et al. (2011) proved that the repeatability of station coordinates improves at a rate of 20% when the atmospheric pressure loading (APL) is applied directly during the analysis of data and at a rate of 10% when the APL is employed as a post-processing correction comparing to a solution that does not take APL into account.

An accurate removal of the seasonal signals from the position time series is beneficial for the accuracy of the derived velocities. These velocities are used, e.g. to create the horizontal (e.g. Bogusz et al. 2014) and vertical velocity (e.g. Kontny and Bogusz 2012) field, to correct the mean-sea level (MSL) records for vertical land movements (VLM) (Teferle et al. 2002) or to create the kinematic reference frames, as the newest realization of the International Terrestrial Reference System (ITRS)—ITRF2014 (Altamimi et al. 2016). As far as their reliability is concerned, the velocities of GPS sta-tions should be estimated simultaneously with the annual and semi-annual signals, as the seasonal changes will bias the value of velocity when they are not accounted for (Blewitt and Lavallée 2002; Bos et al. 2010). Moreover, all seasonal periods have to be modelled when noise analysis is per-formed afterwards, as they may significantly bias the noise parameters when neglected (Davis et al. 2012; Bogusz and Klos 2016). A seasonal model should also be considered during the reference frame definition, either terrestrial (Freymueller 2009; Collilieux et al. 2012) or celestial (Krásná et al. 2015) and residual annual oscillations are still significant when continental (Kenyeres and Bruyn-inx 2009) and regional networks (Bogusz and Figurski 2014) are processed. All of the aforemen-tioned phenomena will transfer directly to the velocity and its error when they are un- or mis-modelled. Recently, Santamaria-Gomez and Memin (2015) subtracted atmospheric, oceanic and conti-nental water mass loading to access the effect on secular velocities estimated from the short time series. They found that at least 4 years of continuous data (in some regions this should be extended to 10 years) are needed to mitigate the effect of the inter-annual deformations on secular velocities.

Since we are not able to model these seasonal effects perfectly, we have to estimate seasonal curves for each station from the position time series to improve the accuracy of the derived velocities. The most common approach to modelling the seasonal signal on a station-by-station basis is to employ the weighted least-squares (WLS) to estimate the sine curves and then subtract them from the data. Due to the fact that some power of observed seasonal variation can be explained by environmental loadings, one can also subtract them from data to account for a signal that may vary with time and, therefore, cannot be entirely covered by a constant WLS-derived sine curve.

In this paper, we start by estimating the annual and semi-annual amplitudes of environmental (atmospheric, hydrological and ocean non-tidal) loading models (ELM) and of IGS ITRF2014 GPS position time series with the most common WLS approach. Furthermore, we demonstrate that assuming their constancy over time may lead to mismodelled seasonal phenomena. Therefore, we subtract the superposition of ELM from GPS position time series and prove that some power from the frequencies between 2 and 100 cpy is being artificially cut from the GPS data. To avoid this artificial loss in power, we mitigate the combined seasonal effect in the GPS position time series by an Improved Singular Spectrum Analysis (ISSA) approach, which is suitable to time varying effects. We employ the ISSA analysis to estimate the time-varying seasonal curve in environmental loading models and then subtract this curve from the GPS position time series, which results in no loss in power and allows us to model a seasonal curve that varies with time. We show that although the seasonal signal was removed, some residual oscillations still remain, but their time-variability is not as large as the one for loading models. Finally, we assess the estimates of vertical velocity errors for IGS ITRF2014 stations when the entire time-varying real seasonal signal is removed.

2. Data

2.1. GPS

We selected 376 globally distributed GPS stations with data sets no shorter than 10 years to be consistent with Santamaria-Gomez and Memin (2015) statements. The time series from chosen stations contributed to the newest International Terrestrial Reference Frame (ITRF2014, Fig. 1). For each station, the GPS measurements were adjusted in the network solution named "repro2" by the International GNSS Service (IGS) (Rebischung et al. 2016). A detailed description of the processing may be found at: http://acc.igs.org/reprocess2.html.

First, a standard pre-processing for offsets, outliers and gaps was performed. The offsets were removed based on epochs defined by the IGS and also a few offsets that were unreported before were detected with the STARS algorithm (Rodionov and Overland 2005) and removed. A number of no more than 6 offsets were employed for one series. The interquartile range (IQR) approach was employed to remove outliers. Missing values accounted for a maximum of 8% of the entire data.

In this research we only focus on the vertical position time series from the IGS contribution to the

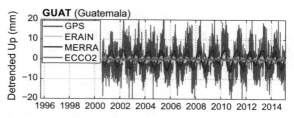

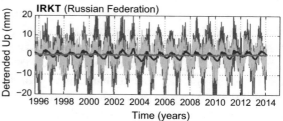

Figure 1
Detrended vertical displacements (in *red*) for time series for GUAT (Guatemala): *upper diagram* and IRKT (Russian Federation): *lower diagram*, permanent stations. Beyond IGS GPS data, the environmental loading models used for research are also plotted: ERAIN in *blue*, MERRA in *brown* and ECCO2 in *violet*

ITRF2014, as the time-variability of height changes is much larger than the ones in the horizontal direction. However, the presented approach can also be successfully applied to the North and East time series.

Figure 1 presents detrended vertical displacements for IRKT (Russian Federation) and GUAT (Guatemala) stations. The IRKT station follows a long-period trend non-linear trend, which can be clearly noticed by changes in maxima around 2002 and between 2010 and 2012. IRKT station is situated within Lake Baykal region. Two large earthquakes have been already registered to affect the area of Irkutsk with magnitudes of M5.9 and M6.3. Those two happened in 1999 and 2008, which in its turn may mean that there is some stress in the rock mass with a very long period. We checked the long-term ISSA curve delivered during analysis and the trend is not linear. However, this non-linear trend did not filter into any other frequencies and did not affect the estimates of ISSA-curve we analysed in this research. In the following analysis, we only focus on seasonal estimates.

2.2. Environmental Loadings

In this research, we used the atmospheric (ATM), hydrological (HYDR) and non-tidal oceanic (NTOL) loading effects on surface displacements using recent global reanalysis models, respectively, ERA (ECMWF Re-Analysis) interim (Dee et al. 2011), MERRA (Modern Era-Retrospective Analysis) land (Reichle et al. 2011) and ECCO2 (Estimation of the Circulation and Climate of the Ocean version 2) (Menemenlis et al. 2008). The induced surface displacements were computed using Green's function formalism (Farrell 1972) based on the Preliminary Reference Earth Model (Dziewonski and Anderson 1981). All loading computations were performed in the center-of-figure (CF) reference frame. For the atmospheric loading computation, we assumed that the ocean responds to pressure variations as an inverted barometer (IB) process. This approximation is assumed to be valid for periods exceeding few weeks. The mass conservation of the ocean is enforced by adding/removing a uniform ocean layer to compensate for the non-zero mean atmospheric

pressure over the oceans (for details see Petrov and Boy 2004). We used a land sea mask at higher resolution (0.10°) than the resolution of the atmospheric model (about 0.70°). We also enforced the total mass conservation of the hydrological model by adding/removing a uniform ocean layer to compensate for any lack/excess of water on the land surface. The ECCO2 model conserves its volume, but not its mass, so we used a similar approach to conserve the ocean mass. For these, we used the ocean model at 0.25° resolution, i.e. higher than the hydrological model (0.66° in longitude and 0.5° in latitude). For more details, see http://loading.u-strasbg.fr/.

In further part of this research, for consistency, we refer to environmental loadings as ERAIN to atmospheric loading, MERRA to hydrological loading and ECCO2 to non-tidal oceanic loading. We also estimate the joint contribution of the above-mentioned environmental models, referred to as a superposition (SUP).

3. Methodology

Initially, we estimated the annual curve along with its three subsequent harmonics (periods of half a year, three and four months) with the WLS approach. These peaks are clearly significant for IGS GPS position time series and environmental loading models (more detailed description in the Sect. 4). We described the observation model as

$$x(t) = x_0 + \sum_{i=1}^{4} A_i \times \sin(\omega_i \times t + \phi_i) + \varepsilon_x(t), \quad (1)$$

where x_0 is the initial value, A is the amplitude of seasonal sine with phase shift ϕ and residuals formed when a model was subtracted are denoted as ε_x (this is also a so-called noise or stochastic part). GPS position time series contain different seasonal signals that may be significant for up to 9 harmonics of a tropical year (Bogusz and Klos 2016). The amplitude of vertical displacements may be as high as between tenths of a millimetre to 8 mm. Beyond the deterministic model, we consider here residuals ε_x which are well approximated by white plus power-law process (e.g. Williams et al. 2004). These reveal the amplitudes of few millimetre and, therefore, may have a strong impact on the linear parameters that are estimated from the GPS position time series (e.g. Klos et al. 2016).

Then, we used the singular spectrum analysis to subtract the time-changeable signals. SSA algorithm, which has been introduced by Broomhead and King (1986) and Allen and Robertson (1996) is a powerful tool of multivariate statistics that is more and more often employed in various areas, such as economics, hydrology, climatology, geophysics or medicine (e.g. Zhang et al. 2014; Wu and Chau 2011; Ghil et al. 2002; Kumar et al. 2015 or Lee et al. 2015). The SSA algorithm involves the decomposition of the time series into a sum of components, each of them having a meaningful interpretation. The SSA approach is a suitable tool to determine time-varying seasonal signals on a station-by-station basis. It is based on the formation of a completely new covariance matrix and then the retrieval of the eigenvalues and eigenvectors (also so-called empirical orthogonal functions—EOFs) by eigenvalue decomposition performed on this new matrix. The eigenvalues and eigenvectors are then sorted in descending order and reconstructed into the reconstructed component (RC) as (Vautard et al. 1992):

$$
RC_i^k = \begin{cases} \frac{1}{i}\sum_{j=1}^{i} a_{i-j}^k E_j^k & 1 \le i \le M-1 \\ \frac{1}{M}\sum_{j=1}^{M} a_{i-j}^k E_j^k & M \le i \le N-M+1 \\ \frac{1}{N-i+1}\sum_{j=1}^{i} a_{i-j}^k E_j^k & N-M+2 \le i \le N, \end{cases}
$$

$$(2)$$

where a is the set of principal components (PC) of the time series, E is the set of eigenvectors retrieved by covariance matrices, M is the length of the window that is sliding through the time series and dividing in this way the time series into a number of N sets. Two approaches to create the covariance matrix can be distinguished: the BK algorithm (Broomhead and King 1986) which creates the covariance matrix by sliding the M-point window through the entire data, and the VG algorithm (Vautard and Ghil 1989) for which the lagged-covariance matrix with a maximum lag of M has the Toeplitz structure.

The SSA approach has been already successfully applied to extract modulated seasonal signals from weekly GPS time series (Chen et al. 2013) or to investigate the inter-annual temporal and spatial variability of atmospheric pressure, terrestrial water storage and surface mass anomalies observed with the GRACE gravity mission in Europe (Zerbini et al. 2013). Most recently, Xu and Yue (2015) investigated the effects of known geophysical sources (atmospheric and hydrological) and stated that they are insufficient to explain the annual variations in GPS observations. They also applied the SSA in combination with the Monte Carlo test to extract time-variable seasonal signals, however, it was concluded that this method was problematic, since by using this approach the SSA-filtered annual signal may contain a signal driven by colored noise. Therefore, the MCSSA signal may artificially remove some power and in consequence, reduce the amplitudes of flicker noise.

The key problem in the SSA approach is the choice of the size of the M-point lag-window. This size depends on the length of time series and frequencies of seasonal signal we intend to model. In this research, based on the tests we performed with the Akaike Information Criterion (AIC, Akaike 1974), we adopted a 2- and 3-year window adaptively, depending on the time series. This will resolve for periods between $M/5$ and M and will result in the determination of the annual and semi-annual oscillations which amplitude changes over time (Gruszczynska et al. 2016).

The SSA approach requires a continuous time series without gaps (Schoellhamer 2001). Shen et al. (2015) proposed an Improved Singular Spectrum Analysis (ISSA) to be appropriate to employ to incomplete time series with gaps. In this research, we analyse the GPS position time series and environmental loading models with the ISSA approach and employ the BK algorithm to create the covariance matrix. A set of frequencies that corresponds to each k-th EOF is estimated in this research with Welch's periodogram (Welch 1967). Considering the length of the tropical year (365.25 days) and the latest determination of the length of draconitic year (351.6 ± 0.2 days according to Amiri-Simkooei, 2013), we would need 25.6 years of data to resolve them with spectral analysis. In this way, when the annual cusp is estimated, we mitigate the combined seasonal effect of a tropical plus draconitic year which in fact contribute the most to total data variance (Blewitt and Lavallée 2002; Bogusz and Figurski 2014). Whenever the annual cusp is being referred to in this research, we mean a combined effect of tropical and draconitic year which we are not able to differentiate due to a limited data span, since the longest time series available were 21.1 years long (YELL—Canada).

By using the ISSA approach, we estimated the annual and semi-annual seasonal signals that have major variability over time. The changeability of higher harmonics of a tropical year is not significant. Therefore, they can be accurately modelled with the WLS approach. Apart from time-changeable curves, we estimated a percentage of the total variance of the time series which is explained by particular ISSA-curve. In this way, we used the ISSA approach to model the seasonal signal (annual + semi-annual) in environmental loadings to estimate the combined effect of atmosphere and hydrosphere (ERAIN + - MERRA + ECCO2). Afterwards, we subtracted this ISSA-curve from vertical GPS position time series. In consequence, we should be able to remove all time-changeable curves that arise from an impact of the environment. However, we realize that the environment is not the only source of seasonal signals in GPS position time series. Therefore, residual oscillations of a year and half-a-year were modelled with WLS.

Finally, the maximum likelihood estimation (MLE) approach was employed to derive the parameters of residuals when the ISSA-curve was subtracted. We used Hector software (Bos et al. 2013) to estimate both the spectral index and amplitude of power-law noise that GPS time series are likely to follow.

4. Results

In this section, we present the results of annual and semi-annual amplitudes estimates. We applied the WLS approach for GPS position time series and a superposition of environmental loadings (SUP). Then, we subtract the SUP directly from GPS vertical

data and show that this approach can change the stochastic part leading in this way to the artificial bias of vertical velocities. Therefore, we apply the ISSA approach to retrieve the seasonal signal from the SUP and subtract it from the GPS position time series. Afterwards, we estimate the vertical velocity errors for selected IGS ITRF2014 stations with the seasonal signal being efficiently removed.

4.1. Seasonal Signal Constant Over Time

In this research, we focused on 376 stations situated around the globe with GPS position time series of the length between 10 and 21 years. In order to assess the impact of environmental loadings on the vertical displacements of ITRF2014 stations, we employed the ERAIN, MERRA and ECCO2 models and summed them into the superposition of environmental effects (SUP). Figure 2 presents the amplitudes of the annual signal for the GPS and SUP data. The mean annual amplitude estimated with the WLS approach for the IGS ITRF2014 solution is equal to 3.5 mm. The highest amplitudes of the annual period are equal to 9.7, 11.3 and 9.7 mm, respectively, and were found for three South American stations BELE, IMPZ and BRAZ (all situated in Brazil). The amplitudes for North American stations are homogenous, aside from a few neighbouring stations of the longitude between 230°W and 250°W and a latitude of 40°–50°N on the coast of the Pacific Ocean. These amplitudes even reach 9.0 mm (WSLR station). In Europe, the amplitudes of annual signal are uniform with values between 0 and 5 mm; however, one exception is noteworthy: station METZ (France) with the annual amplitude for vertical direction equal to 7.3 mm. Inland stations situated in Asia are characterized by a larger amplitude of annual signal than the remaining parts of the world. Five Asian stations are clearly different from the others with higher annual amplitudes: NVSK (Russia), CHAN (China) and three neighbouring stations: IRKT, IRKM and IRKJ (Russia). The amplitudes of the annual curve for NVSK and CHAN are equal to 8.9 and 8.2 mm, while neighbouring stations are characterized by values of 8.5 (IRKT), 8.9 (IRKM) and 4.5 (IRKJ) mm. The annual changes in vertical direction for Australian and Antarctica stations are

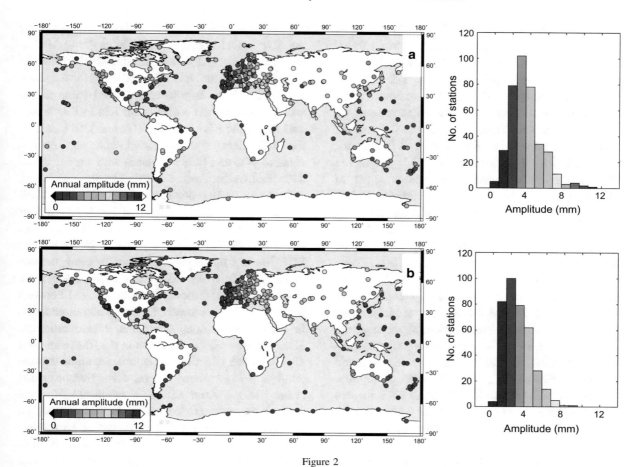

Figure 2

The annual amplitudes estimated with the WLS approach for IGS ITRF2014 series (**a**) and superposition of environmental loading—SUP (**b**). All amplitudes are presented for vertical time series. The histograms of estimated amplitudes are presented on *plots* on the *right side*. The *colours* correspond to the amplitude of the annual signal. The error of estimated amplitudes is close to 0.2 mm for most of the analysed stations

homogenous for all stations with amplitudes between 0 and 4 mm, but again, two exceptions are clearly different from the other amplitudes: DARW and ADE1 (Australia) with amplitudes of 6.9 and 6.2 mm. The median of semi-annual amplitudes (not shown here) estimated for the ITRF2014 GPS position time series was equal to 0.8 mm with a maximum of 2.9 mm for FAIR (USA).

Environmental loadings are characterized by an annual sine curve with a median amplitude of 1.3, 2.0 and 0.3 mm for ERAIN, MERRA and ECCO2, respectively. Asian stations are characterized by the largest annual amplitudes for the ERAIN model. These amplitudes fall between 3 and 5 mm with a maximum of 5.7 mm for NVSK (Russia), while

amplitudes for other stations range between 0 and 2 mm. The annual amplitudes of the MERRA model range from 0 to 8 mm. Considering the Southern Hemisphere, almost all annual sinusoids have an amplitude close to zero and only a few stations differ from this rule. These are BELE, IMPZ, BRAZ, CHPI (all situated in Brazil) and DARW (Australia) with amplitudes of 4.8, 7.8, 5.7, 3.3 and 3.2 mm, respectively. The hydrological model for the Northern Hemisphere is much more varied in amplitude than for the Southern Hemisphere with a maximum for ZWE2 and MDVJ (Russia) with amplitudes around 6 mm. European stations are characterized by amplitudes of the MERRA model between 0 and 6 mm. The amplitudes of the annual curve for the oceanic model range between 0.0 and 1.5 mm with a median of 0.3 mm. The maxima of the annual sinusoid of ECCO2 were found for HELG, VIS0 and TERS (all situated in Europe) with values of 1.5, 1.4 and 1.5 mm, respectively. Beyond the annual signal, environmental models are also characterized by semi-annual curves. These are equal to 0.3 mm for ERAIN, 0.2 mm for MERRA and 0.1 mm for ECCO2. When a combined effect SUP is considered (Fig. 2), the mean annual amplitude is equal to 2.8 mm with a maximum of 8.6 mm for NVSK (Russia). The overall picture of annual amplitudes is homogenous for all stations. Apart from NVSK which has a maximal amplitude, only a few of the 376 stations analysed stand out from others. These are BELE, IMPZ, BRAZ (Brazil), ZWE2, MDVJ, ARTU, NRIL and YAKT (Russia) with amplitudes of 5.3, 8.3, 5.6, 7.1, 7.0 and 7.4 mm, respectively. Apart from the annual curve, environmental loadings are also characterized by a semi-annual signal (not-shown here). Its amplitude ranges between 0.0 and 1.4 mm for ERAIN with a median value of 0.3 mm, between 0.0 and 1.4 mm for MERRA with a median value of 0.2 mm and between 0.0 and 0.4 mm with a median of 0.1 mm for ECCO2. The median of semi-annual signal for SUP is equal to 0.4 mm with a maximum of 1.8 mm for the SCH2 station (Canada).

Then, we subtracted the combined loading model (SUP) from the vertical GPS position time series.

Figure 3 shows a reduction in the RMS value after the SUP model was removed. The majority of GPS stations show a decrease in the RMS value of 10–30% when the loading model was subtracted. RMS fell below 40–50% of its initial value for 40 stations. A maximal reduction in RMS of 45% was noticed for the KHAR station (Ukraine), followed by the IRKM (Russia) station for which a reduction was equal to 44% and BOR1 (Poland) with a reduction of 43%. Stations in Australia and Oceania are characterized by a mean reduction in RMS of 12%, stations in Antarctica of 17%, stations in South America of 13%, stations in North America of 15%, European stations of 20% and Asian stations of 25% when the SUP loading model was applied. A clear latitude dependence in RMS reduction may be noticed for North American stations. Stations situated between 0°N and 32°N are characterized by a lower reduction in RMS than stations situated at higher latitudes. Having compared the reduction in the RMS value for all 376 globally distributed stations, we found that the RMS is being reduced the most for European and Asian stations. After SUP was subtracted from the vertical changes, 16 from 376 stations showed an increase in the RMS value and 8 out of 16 stations showed an increase higher than 1%: PARC (Chile), BSHM (Israel), GUAM (Guam), PDEL (Azores), KIT3 (Uzbekistan), ZHN1 (Hawaii), POHN (Micronesia) and BRMU (Bermuda). According to the

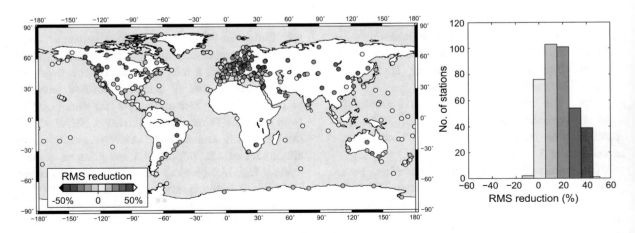

Figure 3
A reduction of RMS after environmental loading models (SUP) were subtracted directly from vertical GPS position time series. A number of 100% means a reduction of the RMS to zero

visual inspection of the time series we found that the increase in the RMS value resulted from the annual signal, the amplitude of which increases in case of a disagreement between the phase of GPS position time series and the loading model employed. A disagreement arises from the fact that for some GPS stations, seasonal frequencies may be affected by artificial phenomena, predominant over the environmental loadings. On the other hand, environmental loadings change their phases for stations located at the seaside, which is why the inland stations show larger reduction of RMS value than the seaside ones. Moreover, other natural phenomena may exist, like sun shine exposure or light/gravity (Neumann 2007; Kalenda and Neumann 2014), which are not entirely covered.

Next step of our research was to produce the Power Spectral Densities (PDSs) of all analysed stations using Welch periodogram. Figure 4 presents two examples. The stochastic behaviour of the hydrological model MERRA is close to random-walk noise. The oscillations of 1-year period and its three harmonics are clearly seen at the PSD. The annual peak of MERRA estimated for the GUAT station is higher than the one for GPS data, while the annual oscillation of MERRA derived for the IRKT station is

as high as one-tenth of the annual peak estimated for GPS. The atmospheric loading ERAIN is characterized by a stochastic process close to the autoregressive model (as was previously noted by Petrov and Boy 2004 or Klos et al. 2017). All stations analysed in this research show a sudden drop in the power of ERAIN for frequencies higher than 30 cpy. The ERAIN for Asian stations is much more prevailing than for other stations. In these cases, the annual peak is as powerful as the annual oscillation estimated for GPS. For stations in Asia, the hydrological model MERRA is not so powerful as it used to be for any other region around the world. Therefore, when it is subtracted, it should not have a significant impact on the GPS data. The oceanic model, ECCO2, is the less powerful for all 376 stations analysed in this research. Its annual peak is almost 300 times smaller than the annual amplitudes estimated for GPS. In this research, when a joint environmental loading impact is being considered as an SUP, ECCO2 makes the smallest contribution to a combined effect.

A power-law behaviour with a spectral index close to flicker noise (Fig. 4) is a good approximation for GPS position time series (e.g. Zhang et al. 2014;

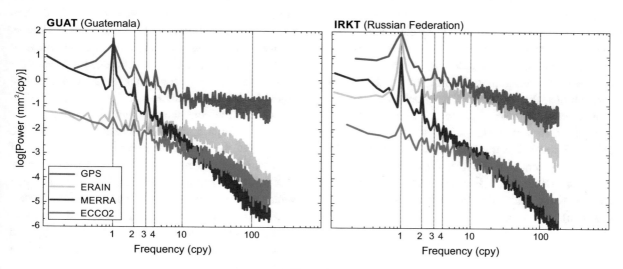

Figure 4

Power Spectral Densities of environmental loadings and GPS data estimated for GUAT (*left*) and IRKT (*right*) stations in the vertical direction. The GPS data (in *red*) follow a power-law behaviour close to flicker noise (slope of −1) with its power decreasing constantly over time. The ERAIN model (in *blue*) is close to autoregressive behaviour with a sudden drop in its power for frequencies higher than 30 cpy. The MERRA model (in *brown*) is characterized by the largest annual cusp and even more powerful than GPS data for GUAT while slightly less powerful than GPS for IRKT. The ECCO2 model (in *violet*) is the least powerful of all environmental models

Williams et al. 2004 or Klos et al. 2016). In cases where a pure white noise is wrongly assumed instead of a power-law noise, the velocity uncertainty is underestimated from 5 to 11 times (Mao et al. 1999). Conversely, if any seasonal signal remains unmodelled, it leads to an additional correlation to the GPS position time series leading in this way to an overestimation of velocity uncertainty which can be then inappropriately interpreted. In this instance too, if any obvious and clear time-variability remains unmodelled, it will additionally bring artificial correlation into GPS residuals leading to overestimation of velocity error. As stated by Santamaria-Gomez and Memin (2015), having subtracted the loading models directly from the GPS position time series, the data variance is reduced which is only a variance of white noise. However, when it is subtracted directly, loading models might influence the stochastic part of GPS data and in this way, change the noise parameters from flicker noise into less correlated white noise or into a non-stationary random-walk noise. If this is not taken into account in the noise modelling, the errors of velocities are going to be under- or over-estimated, respectively, and might be misinterpreted. Having subtracted the SUP model, we examined the noise parameters of residuals with MLE. All stations from a number of the 376 selected in this study, showed a change in spectral index when SUP was removed. A number of 104 from 376 data sets changed the spectral index of a value lower than 0.1 when SUP was removed. For a total of 210 out of 376 stations, the character of the stochastic part was moved towards a random walk with a maximal change of spectral index equal to 0.5 for station NICA (France). For the rest of the 376 stations, the noise parameters changed from flicker noise into less correlated white noise. For this group, the maximal change of spectral index was 0.5 for station SKE0 (Sweden).

4.2. Seasonal Signals Changing Over Time

The seasonal signal observed in the GPS position time series may vary slightly from year to year, since the geophysical causes that absorb the GPS data are not constant over time (Chen et al. 2013; Davis et al. 2012). Whenever the WLS approach is applied, it only estimates sinusoidal curves with amplitudes constant over time. In this section, using the Improved SSA (ISSA), we introduce the seasonal signals that change over time for the SUP loading model. Then, we subtract this curve from the vertical changes for GPS stations. We focus on annual and semi-annual curves, as their time-variability is the highest. Other harmonics of a tropical year may be modelled with the WLS approach afterwards.

To produce an overall view of how much the annual signal contributes to the variance of time series, we estimated the percentage of the total variance explained by the annual curves for ERAIN, MERRA, ECCO2, SUP and GPS, respectively. The annual signal explains on average 23% of the total variance of GPS vertical displacements. Considering IMPZ, BRAZ (Brazil), IRKT (Russia) and LHAZ (China), the annual oscillations, respectively, explain 64, 62, 54 and 53% of total variance. Stations situated in Asia are characterized by the highest contribution of the annual signal to GPS data. The annual signal estimated with ISSA explains on average 23% of the total variance of ERAIN. The greatest percentage of explained variance was found for DARW (82%), KARR (80%) (Australia) and for WUHN (78%) and SHAO (74%) (China). The percentage of total variance explained by the annual signal for ERAIN represents approximately 10% for stations in Europe. Annual oscillations estimated for the hydrological loading (MERRA) account on average for 68% of the total variance. Such a high contribution of the annual signal to the total variance of the loadings demonstrates that the continental hydrosphere may have a significant impact on the observed signals. For the majority of the stations in Europe, the percentage of total variance explained by the annual signal ranges between 83% and a maximum of 91%. The annual signal derived from non-tidal ocean loading models (ECCO2) explains on average 19% of the total variance for 376 selected stations. In Europe, the annual signal accounts for approximately 29% of the total variance of this loading.

Figure 5 shows the percentage of the total variance which can be explained by the annual signal for the SUP environmental model. When environmental loading models are combined into one superposition, the annual signal explains up to 66% of the total

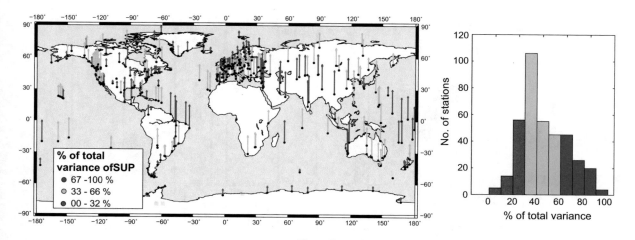

Figure 5
A percentage of the total variance of SUP model explained by the annual signal determined by the ISSA approach

variance of the SUP for Europe, between 30 and 100% for Asia, and up to 30% for North America and up to 100% for South America. The annual curve estimated for Antarctica explains 10% of the SUP variance. The geographical distribution of a total variance explained by SUP means that there is at least one natural process, which is seemingly correlated with atmospheric, hydrologic and/or ocean non-tidal loadings, but varies for specific geographical areas.

Then, the ISSA-curve estimated for SUP was compared to the WLS-curve estimated for the vertical displacements of GPS data. Comparing to the ISSA-curve derived for loading models, the seasonal signal estimated with WLS for GPS position time series is underestimated by a maximum of 4.8 mm for KIT3 station (Uzbekistan).

To date, the GPS position time series were corrected for loading models by direct subtraction of the model from GPS data. Earlier in this chapter, we showed that this might absorb some power in the GPS data, leading in this way to incorrect estimates of velocity errors. In this research, we show a completely new approach to modelling the seasonal curve that comes from the SUP loading model. Time changeable curves are provided by the ISSA approach and then, subtracted from the GPS position time series. In this way, the stochastic properties of the GPS signal remain intact, and seasonal changes that arise from environmental loading are removed. Having removed this curve, one should still model

the annual and semi-annual residual oscillations with the WLS approach, as these result not only from loading models, but also arise from GPS artefacts.

Figure 6 introduces the problem of absorption in the stochastic part when loading models were directly removed from the GPS position time series. The evident change in the power of GPS data for frequencies between 4 and 80 cpy was noticed. However, stations as GUAT, for which the hydrological model MERRA contributes to SUP much more than the atmospheric model ERAIN, an absorption in power is not so obvious. In 272 out of 376 stations examined in this research, we found a change in spectral index of more than 0.1 when SUP was directly removed from the GPS data (see the previous section). Having subtracted the ISSA-curve modelled for SUP from GPS data, we found a maximum change in the spectral index of 0.1 in a power-law character. Also, a seasonal curve of a period of a year and half-a-year was efficiently modelled and its power reduced in the vertical displacements of GPS stations. Although the seasonal signal was removed with the SUP-ISSA curve, we modelled the remaining annual and semi-annual oscillations with the WLS approach (Fig. 7), as these curves may not only arise from the environmental impact. For 122 of the 376 stations examined, residual annual amplitude did not exceed 1 mm. However, it was higher than 4 mm for 9 stations: VAAS (Finland), BRAZ (Brazil), KIR0 (Sweden),

239

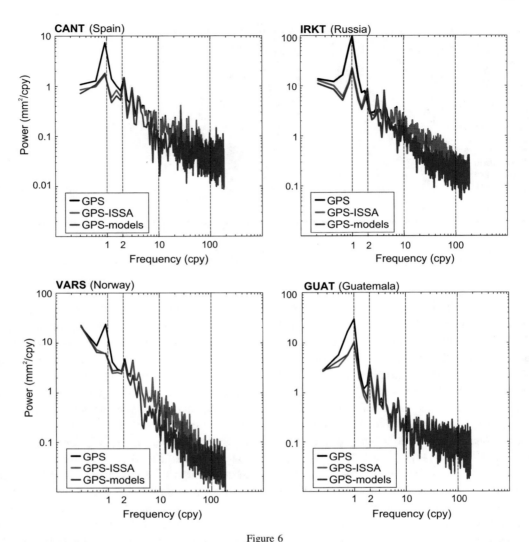

Figure 6
The Power Spectral Densities of vertical displacements of GPS stations (in *black*) plotted against residuals after the ISSA seasonal signal was removed (in *red*) and when we subtracted the superposition of loading models directly from the GPS position time series (in *blue*). Four stations are presented here: CANT (Spain), IRKT (Russia), VARS (Norway) and GUAT (Guatemala). An evident absorption in a power when environmental loadings were directly removed from the GPS data is shown for the first three aforementioned stations. No artificial absorption in power can be seen for residuals after the ISSA-curves were removed. In this instance, a power of annual peak was reduced, but no change in the stochastic behaviour was observed

BAMF (Canada), YSSK (Russia), KIT3 (Uzbekistan), PETS (Russia), BELE (Brazil) and WSLR (Canada). With the exception of the VAAS and KIT3 stations, for the aforementioned stations, the amplitude of the annual signal was reduced by 2-4 mm compared to the WLS amplitude of the GPS data (Fig. 2a). For VAAS and KIT3, the annual signal increased when the ISSA-curve was removed. In the following situations, the ISSA signal estimated for

the SUP model did not correspond in phase to the GPS position time series. Moreover, for the KIT3 station, the combined model SUP is characterized by a semi-annual signal with an amplitude of 1 mm for the period 1995–2000, which then disappears. This semi-annual signal was not observed for the GPS data. This contradiction between the phase and decay of the semi-annual signal caused the annual oscillation to increase when the SUP-ISSA-curve was

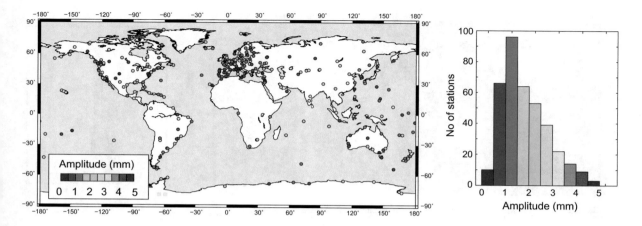

Figure 7
Annual amplitudes (mm) of residual oscillations estimated with the WLS approach for vertical displacements of GPS stations after the seasonal signal was removed with the SUP-ISSA-curve

subtracted from the GPS data. In addition, the growth in the amplitude of the annual signal was also noticed for 9 out of 376 stations analysed: BSHM (Israel), FAIR (USA), GUAM (Guam), LPAL (Spain), MALI (Kenya), MORP (UK), POHN (Micronesia), PTBB (Germany) and SOUF (Guadeloupe). Three of the stations mentioned (KIT3, POHN, BSHM) are ones for which we also noticed an increase in the RMS value.

4.3. Dilution of Precision

We followed the approach of Blewitt and Lavallée (2002) and Bos et al. (2010) and estimated the value of the dilution of precision (DP), i.e. the ratio between two uncertainties of velocity estimated for the various deterministic models considered. In this research, we estimate the DP as the ratio between the uncertainty of the vertical velocity determined twofold. Firstly, the uncertainty was estimated along with the annual period and its three harmonics with the WLS approach($\sigma_{v_{GPS}}$). Then, the uncertainty of vertical velocity was estimated when the seasonal of: SUP-ISSA-curve was removed together with the annual period and its three harmonics (residual oscillations of a year and half-a-year) modelled with the WLS ($\sigma_{v_{GPS-ISSA}}$). This ratio is expressed as:

$$DP = \frac{\sigma_{v_{GPS-ISSA}}}{\sigma_{v_{GPS}}}. \tag{3}$$

The velocity uncertainty is estimated following Bos et al. (2008):

$$\sigma_v \approx \pm \sqrt{\frac{A_{PL}^2}{\Delta T^{2-\frac{\kappa}{2}}} \cdot \frac{\Gamma(3-\kappa) \times \Gamma(4-\kappa) \times (N-1)^{\kappa-3}}{\left[\Gamma\left(2-\frac{\kappa}{2}\right)\right]^2}}, \tag{4}$$

where Γ is the gamma function, N is the number of data in the time series and ΔT is the sampling interval.

Figure 8 presents the results of DP values. 48 out of the 376 stations examined are characterized by a DP value lower than 0.95, which means that the uncertainty of the vertical velocity was lowered by 5%. An improvement greater than 10% was noted for MIKL (Ukraine), RWSN (Argentina), MERI (Mexico), VARS (Norway), BRAZ (Brazil), GAIA (Portugal), TLSE (France) and MTY2 (Mexico). For the aforementioned stations, no loss in power in the GPS position time series was noticed when the SUP-ISSA-curve was removed. Therefore, the improvement in velocity errors arises from more efficient modelling of the seasonal signal. A total of 298 out of the 376 stations analysed was characterized by a DP value lower than 1. This means that the ISSA seasonal curve modelled for the superposition of environmental loadings may explain more variability in the GPS position time series than the WLS model itself. We presume that this is due to site-

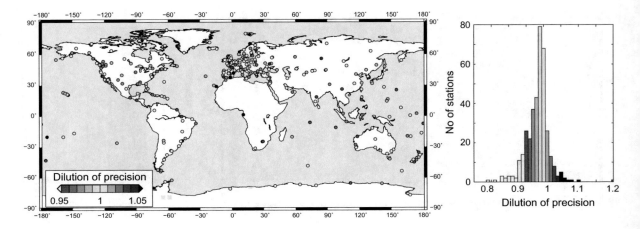

Figure 8
A dilution of precision (DP) of vertical velocities derived for IGS ITRF2014 stations when the ISSA-curve was employed to remove the environment-related seasonal signal. Values lower than 1 indicate that errors of vertical velocity were lower when the ISSA-curve was removed. DP values lower and higher than 5% were marked with *yellow* and *black*, respectively

specific effects recorded by GPS which superimpose the geophysical changes explained by environmental models. In other words, environmental models show time-changeability which cannot be really seen in GPS position time series. This is why the subtraction of ISSA curve is not as efficient as it was supposed to be.

5. *Discussion and Conclusions*

We studied 376 IGS ITRF2014 stations to measure the influence of applying atmospheric, hydrological and oceanic non-tidal loading models on the estimated vertical velocity. These loadings show a dominant annual signal and their superposition shows amplitudes of up to 12 mm. The highest amplitudes were noted for Asian stations but also three South American stations are characterized by amplitudes close to 11 mm, which is in agreement with the findings of Bogusz et al. (2015). We summed the hydrological, atmospheric and non-tidal oceanic model into one superposition (SUP) to examine the impact all models may have on the GPS position time series. For the SUP model, we found the annual amplitude larger than 7 mm for 5 stations in Asia and one station in Brazil.

Looking at each contribution separately, we found that the annual amplitude of the atmospheric loading,

using the ERAIN model, can be as large as 6 mm, especially in Asia. For hydrological loading this value is 8 mm using the MERRA model. The highest amplitudes circa 4–8 mm were delivered for central European, north Asian and the northern group of North American stations. Also, BRAZ and IMPZ (Brazil) are characterized by an annual signal greater than 5 mm. Finally, using the oceanic ECCO2 model, we estimated that 13 out of 376 stations have an annual loading amplitude higher than 1 mm with three maximal amplitudes of 1.5 mm for three stations in Europe: VIS0 (Sweden), TERS (Netherlands) and HELG (Germany). The two latter stations were also noted by Williams and Penna (2011).

After subtracting the SUP directly from the GPS vertical position time series we obtained a reduction of more than 25% in RMS value for the majority of the 376 stations. We found an average reduction in RMS of 17%. The RMS reduction we estimated in this research is very similar to what was shown by Tregoning and van Dam (2005).

Next, we examined the noise properties of these residuals with MLE. Assuming a power-law behaviour, we found a maximum change in the spectral index of 0.5, which would cause an underestimation in velocity error, if this error was to be estimated from residuals after the SUP was removed. This change was mostly caused by the ERAIN model, which is characterized by autoregressive-like behaviour. If ERAIN contributes to SUP more than MERRA, it will cause a change in the noise parameters. On the other hand, if MERRA contributes more than ERAIN, the change in the noise parameters will not be so dramatic, due to its power-law-like character. Beyond a change in the stochastic part, we also obtained a decrease in the annual peak of the GPS position time series.

The loading models do not remove the seasonal signal complete from the GNSS time series and, therefore, we fitted another seasonal signal to the residuals using the ISSA approach and subtracted it from the GPS position time series. The ISSA approach was chosen, as seasonal signals from real geophysical sources are not constant over time. Naturally enough, we still modelled the residual oscillations of a year and half-a-year with the WLS approach, as the seasonal signal in GPS does not only result from environmental loading. We discovered that for environmental loading models, the annual signal explains up to 66% of a total variance of SUP for Europe, between 30 and 100% for Asia, up to 30% for North America and up to 100% for South America. The annual curve estimated for Antarctica explains 10% of the SUP variance. Having removed the ISSA curve from the GPS data, we examined the properties of residuals with MLE. We found a maximal difference in the spectral index between the GPS position time series and the GPS data without ISSA-seasonal of only 0.1. This means that we are able to reduce the seasonal peaks from environmental impact with no significant influence on the stochastic part of the GPS data.

The reliability of the GPS-based velocities determined from the time series are constantly improving due to the ongoing efforts in the refinement of equipment, processing strategies and the models of geophysical phenomena as well. With regard to this, we need to estimate the errors of velocity as reliably as possible. In this research, we estimated the dilution of Precision (DP) of the vertical velocities of the ITRF2014 series. For a total of 298 out of the 376 stations analysed, the DP was lower than 1. This indicates that when the SUP-ISSA-curve was removed from the GPS data, the error of velocity becomes lower than it was before. We were able to reduce the velocity uncertainty with no change in the

stochastic part of data. It means that the change in velocity uncertainty arises only from a proper modelling of time-varying seasonal signal.

Concluding, we propose a two-stage solution of the problem of reliable subtraction of seasonal curves. It is important to implement the environmental loading and model a seasonal curve with an approach that allows for considering their changeability over time. Afterwards, the analytical methods still need to be implemented to remove the remaining seasonal oscillations. These two steps may provide an estimation of the vertical velocities that are used for kinematic reference frames and geodynamical interpretations as well as their reliable uncertainty assessment.

Acknowledgements

Anna Klos, Marta Gruszczynska and Janusz Bogusz are financed by the Polish National Science Centre, Grant No. UMO-2014/15/B/ST10/03850. Machiel Simon Bos is financially supported by Portuguese funds through FCT in the scope of the Project IDL-FCT-UID/GEO/50019/2013 and Grant Number SFRH/BPD/89923/2012. Jean-Paul Boy is partly funded by CNES (Centre National d'Etudes Spatiales), through its TOSCA program. Loading time series used here are available at EOST/IPGS loading service (http://loading.u-strasbg.fr). Maps and charts were plotted in the Generic Mapping Tool (Wessel et al. 2013). IGS time series were accessed from ftp://igs-rf.ensg.eu/pub/repro2.

REFERENCES

Agnew, D. C., & Larson, K. M. (2007). Finding the repeat times of the GPS constellation. *GPS Solutions, 11*(1), 71–76. doi:10.1007/s10291-006-0038-4.

Akaike, H. (1974). A new look at the statistical model identification. *IEEE Transactions on Automatic Control, 19*(6), 716–723. doi:10.1109/TAC.1974.1100705.

Allen, M. R., & Robertson, A. W. (1996). Distinguishing modulated oscillations from coloured noise in multivariate datasets. *Climate Dynamics, 12,* 772–786. doi:10.1007/s003820050142.

Altamimi, Z., Rebischung, P., Métivier, L., & Collilieux, X. (2016). ITRF2014: A new release of the International Terrestrial Reference Frame modelling nonlinear station motions. *Journal of Geophysical Research: Solid Earth.* doi:10.1002/2016JB013098.

Amiri-Simkooei, A. R. (2013). On the nature of GPS draconitic year periodic pattern in multivariate position time series. *Journal of Geophysical Research: Solid Earth, 118*(5), 2500–2511. doi:10.1002/jgrb.50199.

Blewitt, G., & Lavallée, D. (2002). Effect of annual signals on geodetic velocity. *Journal of Geophysical Research, 107,* ETG 9-1–ETG 9-11. doi:10.1029/2001JB000570.

Blewitt, G., Lavallée, D., Clarke, P., & Nurutdinov, K. (2001). A new global mode of Earth deformation: seasonal cycle detected. *Science, 294,* 2342–2345. doi:10.1126/science.1065328.

Bogusz, J., & Figurski, M. (2014). Annual signals observed in regional GPS networks. *Acta Geodynamica et Geomaterialia, 11*(2), 125–131. doi:10.13168/AGG.2014.0003.

Bogusz, J., Gruszczynska, M., Klos, A., & Gruszczynski, M. (2015). Non-parametric estimation of seasonal variations in GPS-derived time series. *Springer IAG Symposium Series, 146.* Springer Berlin Heidelberg, doi:10.1007/1345_2015_191.

Bogusz, J., & Klos, A. (2016). On the significance of periodic signals in noise analysis of GPS station coordinates time series. *GPS Solutions.* doi:10.1007/s10291-015-0478-9.

Bogusz, J., Klos, A., Grzempowski, P., & Kontny, B. (2014). Modelling velocity field in regular grid on the area of Poland on the basis of the velocities of European permanent stations. *Pure and Applied Geophysics, 171*(6), 809–833. doi:10.1007/s00024-013-0645-2.

Bos, M. S., Bastos, L., & Fernandes, R. M. S. (2010). The influence of seasonal signals on the estimation of the tectonic motion in short continuous GPS time-series. *Journal of Geodynamics, 49,* 205–209. doi:10.1016/j.jog.2009.10.005.

Bos, M. S., Fernandes, R. M. S., Williams, S. D. P., & Bastos, L. (2008). Fast error analysis of continuous GPS observations. *Journal of Geodesy, 82,* 157–166. doi:10.1007/s00190-007-0165-x.

Bos, M. S., Fernandes, R. M. S., Williams, S. D. P., & Bastos, L. (2013). Fast error analysis of continuous GNSS observations with missing data. *Journal of Geodesy, 87*(4), 351–360. doi:10.1007/s00190-012-0605-0.

Broomhead, D. S., & King, G. P. (1986). Extracting qualitative dynamics from experimental data. *Physica, 20*(2–3), 217–236. doi:10.1016/0167-2789(86)90031-X.

Chen, Q., van Dam, T., Sneeuw, N., Collilieux, X., Weigelt, M., & Rebischungc, P. (2013). Singular spectrum analysis for modelling seasonal signals from GPS time series. *Journal of Geodynamics, 72,* 25–35. doi:10.1016/j.jog.2013.05.005.

Collilieux, X., Altamimi, Z., Coulot, D., Ray, J., & Sillard, P. (2007). Comparison of very long baseline interferometry, GPS, and satellite laser ranging height residuals from ITRF2005 using spectral and correlation methods. *Journal of Geophysical Research, 112,* B12403. doi:10.1029/2007JB004933.

Collilieux, X., van Dam, T., Ray, J., Coulot, D., Métivier, L., & Altamimi, Z. (2012). Strategies to mitigate aliasing of loading signals while estimating GPS frame parameters. *Journal of Geodesy, 86,* 1–14. doi:10.1007/s00190-011-0487-6.

Dach, R., Boehm, J., Lutz, S., Steigenberger, P., & Beutler, G. (2011). Evaluation of the impact of atmospheric pressure loading modelling on GNSS data analysis. *Journal of Geodesy, 85*(2), 75–91. doi:10.1007/s00190-010-0417-z.

Davis, J. L., Wernicke, B. P., & Tamisiea, M. E. (2012). On seasonal signals in geodetic time series. *Journal of Geophysical Research, 117*(B1), B01403. doi:10.1029/2011JB008690.

Dee, D. P., Uppala, S. M., Simmons, A. J., Berrisford, P., Poli, P., Kobayashi, S., et al. (2011). The ERA-Interim reanalysis: configuration and performance of the data assimilation system. *Quarterly Journal of the Royal Meteorological Society, 137,* 553–597. doi:10.1002/qj.828.

Dill, R., & Dobslaw, H. (2013). Numerical simulations of global-scale high-resolution hydrological crustal deformations. *Journal of Geophysical Research: Solid Earth, 118,* 5008–5017. doi:10.1002/jgrb.50353.

Dong, D., Fang, P., Bock, Y., Cheng, M. K., & Miyazaki, S. (2002). Anatomy of apparent seasonal variations from GPS-derived site position time series. *Journal of Geophysical Research, 107*(B4), 2075. doi:10.1029/2001JB000573.

Dziewonski, A. M., & Anderson, D. L. (1981). Preliminary reference Earth model. *Physics of the Earth and Planetary Interiors, 25,* 297–356. doi:10.1016/0031-9201(81)90046-7.

Farrell, W. E. (1972). Deformation of the earth by surface loads. *Reviews of Geophysics, 10*(761–797), 1972.

Fratepietro, F., Baker, T. F., Williams, S. D. P., & Van Camp, M. (2006). Ocean loading deformations caused by storm surges on the northwest European shelf. *Geophysical Research Letters, 33,* L06317. doi:10.1029/2005GL025475.

Freymueller, J.T. (2009). Seasonal position variations and regional Reference frame realization. In H. Drewes (Ed.), *Geodetic reference frames,* International Association of Geodesy Symposia 134 (pp. 191–196). Springer Berlin Heidelberg, doi:10.1007/978-3-642-00860-3_30.

Ghil, M., Allen, M. R., Dettinger, M. D., Ide, K., Kondrashov, D., Mann, M. E., et al. (2002). Advanced spectral methods for climatic time series. *Reviews of Geophysics, 40,* 1-1–1-41. doi:10.1029/2000RG000092.

Gruszczynska, M., Klos, A., Gruszczynski, M., & Bogusz, J. (2016). Investigation of time-changeable seasonal components in the GPS height time series: a case study for Central Europe. *Acta Geodynamica et Geomaterialia, 13*(3), 281–289. doi:10.13168/AGG.2016.0010.

Jiang, W., Li, Z., van Dam, T., & Ding, W. (2013). Comparative analysis of different environmental loading methods and their impacts on the GPS height time series. *Journal of Geodesy, 87,* 687–703. doi:10.1007/s00190-013-0642-3.

Kalenda, P., & Neumann, L. (2014). The Tilt of the Elevator Shaft of Bunker Skutina. *Transactions of the VSB: Technical University of Ostrava., 60*(1), 55–61. doi:10.22223/tr.2014-1/1978.

Kenyeres, A., & Bruyninx, C. (2009). Noise and Periodic Terms in the EPN Time Series. In H. Drewes (Ed.), *Geodetic reference frame*: International Association of Geodesy Symposia 134 (pp. 143–148). Springer Berlin Heidelberg, doi:10.1007/978-3-642-00860-3_22.

Klos, A., Bogusz, J., Figurski, M., & Gruszczynski, M. (2016). Error analysis for European IGS stations. *Studia Geophysica et Geodaetica, 60,* 17–34. doi:10.1007/s11200-015-0828-7.

Klos, A., Hunegnaw, A., Teferle, F. N., Abraha, K. E., Ahmed, F., & Bogusz, J. (2017). Noise characteristics in zenith total delay from homogeneously reprocessed GPS time series. *Atmospheric Measurement Techniques Discussion.* doi:10.5194/amt-2016-385. **(in review)**.

Kontny, B., & Bogusz, J. (2012). Models of vertical movements of the Earth crust surface in the area of Poland derived from leveling and GNSS data. *Acta Geodynamica et Geomaterialia, 9*(3), 331–337.

Krásná, H., Malkin, Z., & Böhm, J. (2015). Non-linear VLBI station motions and their impact on the celestial reference frame and Earth orientation parameters. *Journal of Geodesy, 89,* 1019–1033. doi:10.1007/s00190-015-0830-4.

Kumar, A., Walia, V., Arore, B. R., Yanh, T. F., Lin, S.-J., Fu, Ch-Ch., et al. (2015). Identifications and removal of diurnal and semidiurnal variations in radon time series data of Hsinhua monitoring station in SW Taiwan using singular spectrum analysis. *Natural Hazards, 79*(1), 317–330. doi:10.1007/s11069-015-1844-1.

Lee, T. K. M., Lim, J. G., Sanei, S., & Gan, S. S. W. (2015). Advances on singular spectrum analysis of rehabilitative assessment data. *Journal of Medical Imaging and Health Informatics, 5*(2), 350–358. doi:10.1166/jmihi.2015.1399.

Mao, A., Harrison, C. H. G. A., & Dixon, T. H. (1999). Noise in GPS coordinate time series. *Journal of Geophysical Research, 104*(B2), 2797–2816. doi:10.1029/1998JB900033.

Menemenlis, D., Campin, J., Heimbach, P., Hill, C., Lee, T., Nguyen, A., et al. (2008). ECCO2: High resolution global ocean and sea ice data synthesis. *Mercator Ocean Quarterly Newsletter, 31,* 13–21.

Neumann, L. (2007). Static pendulum with contactless 2D sensor measurements opens the question of gravity dynamic and gravity noise on earth's surface. *Physics Essays, 20*(4), 535. doi:10.4006/1.3254506.

Penna, N. T., King, M. A., & Stewart, M. P. (2007). GPS height time series: Short-period origins of spurious long-period signals. *Journal of Geophysical Research, 112*(B2), B02402. doi:10.1029/2005JB004047.

Penna, N. T., & Stewart, M. P. (2003). Aliased tidal signatures in continuous GPS height time series. *Geophysical Research Letters, 30*(23), 2184. doi:10.1029/2003GL018828.

Petrov, L., & Boy, J.-P. (2004). Study of the atmospheric pressure loading signal in VLBI observations. *Journal of Geophysical Research, 109,* B03405. doi:10.1029/2003JB002500.

Ray, J., Altamimi, Z., Collilieux, X., & van Dam, T. (2008). Anomalous harmonics in the spectra of GPS position estimates. *GPS Solutions, 12*(1), 55–64. doi:10.1007/s10291-007-0067-7.

Rebischung, P., Altamimi, Z., Ray, J., & Garayt, B. (2016). The IGS contribution to ITRF2014. *Journal of Geodesy, 90*(7), 611–630. doi:10.1007/s00190-016-0897-6.

Reichle, R. H., Koster, R. D., De Lannoy, G. J. M., Forman, B. A., Liu, Q., Mahanama, S. P. P., et al. (2011). Assessment and enhancement of MERRA land surface hydrology estimates. *Journal of Climate, 24,* 6322–6338. doi:10.1175/JCLI-D-10-05033.1.

Rodionov, S., & Overland, J. E. (2005). Application of a sequential regime shift detection method to the Bering Sea ecosystem. *ICES Journal of Marine Science, 62,* 328–332. doi:10.1016/j.icesjms.2005.01.013.

Santamaria-Gomez, A., & Memin, A. (2015). Geodetic secular velocity errors due to interannual surface loading deformation. *Geophysical Journal International, 202,* 763–767. doi:10.1093/gji/ggv190.

Schoellhamer, D. H. (2001). Singular spectrum analysis for time series with missing data. *Geophysical Research Letters, 28,* 3187–3190. doi:10.1029/2000GL012698.

Shen, Y., Peng, F., & Li, B. (2015). Improved singular spectrum analysis for time series with missing data. *Nonlinear Processes in Geophysics, 22,* 371–376. doi:10.5194/npg-22-371-2015.

Teferle, F. N., Bingley, R. M., Dodson, A. H., Penna, N. T., & Baker, T. F. (2002). Using GPS to separate crustal movements and sea level changes at tide gauges in the UK. In H. Drewes, A. H. Dodson, L. P. S. Fortes, L. Sanchez, & P. Sandoval (Eds.), *Vertical reference systems* (pp. 264–269). Heidelberg: Springer.

Tregoning, P., & van Dam, T. (2005). Effects of atmospheric pressure loading and seven-parameter transformations on estimates of geocenter motion and station heights from space geodetic observations. *Journal of Geophysical Research, 110,* B03408. doi:10.1029/2004JB003334.

Tregoning, P., & Watson, C. (2009). Atmospheric effects and spurious signals in GPS analyses. *Journal of Geophysical Research, 114,* B09403. doi:10.1029/2009JB006344.

van Dam, T., Collilieux, X., Wuite, J., Altamimi, Z., & Ray, J. (2012). Nontidal ocean loading: amplitudes and potential effects in GPS height time series. *Journal of Geodesy, 86,* 1043–1057. doi:10.1007/s00190-012-0564-5.

van Dam, T., & Wahr, J. M. (1987). Displacements of the Earth's surface due to atmospheric loading: effects on gravity and baseline measurements. *Journal of Geophysical Research: Solid Earth, 92*(B2), 1281–1286. doi:10.1029/JB092iB02p01281.

van Dam, T., Wahr, J., Chao, Y., & Leuliette, E. (1997). Predictions of crustal deformations and of geoid and sea-level variability caused by oceanic and atmospheric loading. *Geophysical Journal International, 129,* 507–517. doi:10.1111/j.1365-246X.1997.tb04490.x.

van Dam, T., Wahr, J., & Lavallee, D. (2007). A comparison of annual vertical crustal displacements from GPS and Gravity Recovery and Climate Experiment (GRACE) over Europe. *Journal of Geophysical Research, 112,* B03404. doi:10.1029/2006JB004335.

van Dam, T., Wahr, J., Milly, P. C. D., Shmakin, A. B., Blewitt, G., Lavallée, D., et al. (2001). Crustal displacements due to continental water loading. *Geophysical Research Letters, 28*(4), 651–654. doi:10.1029/2000GL012120.

Vautard, R., & Ghil, M. (1989). Singular Spectrum Analysis in nonlinear dynamics, with applications to paleoclimatic time series. *Physica D: Nonlinear Phenomena, 35,* 395–424. doi:10.1016/0167-2789(89)90077-8.

Vautard, R., Yiou, P., & Ghil, M. (1992). Singular spectrum analysis: A toolkit for short, noisy chaotic signals. *Physica D: Nonlinear Phenomena, 58,* 95–126. doi:10.1016/0167-2789(92)90103-T.

Welch, P. D. (1967). The use of fast fourier transform for the estimation of power spectra: a method based on time averaging over short, modified periodograms. *IEEE Transactions on Audio Electroacoustics, 15,* 70–73. doi:10.1109/TAU.1967.1161901.

Wessel, P., Smith, W. H. F., Scharroo, R., Luis, J., & Wobbe, F. (2013). generic mapping tools: improved version released. *Eos, Transactions, American Geophysical Union, 94*(45), 409–410. doi:10.1002/2013EO450001.

Williams, S. D. P., Bock, Y., Fang, P., Jamason, P., Nikolaidis, R. M., Prawirodirdjo, L., et al. (2004). Error analysis of continuous GPS position time series. *Journal of Geophysical Research, 109,* B03412. doi:10.1029/2003JB002741.

Williams, S. D. P., & Penna, N. T. (2011). Non-tidal ocean loading effects on geodetic GPS heights. *Geophysical Research Letters, 38,* L09314. doi:10.1029/2011GL046940.

Wu, C. L., & Chau, K. W. (2011). Rainfall–runoff modelling using artificial neural network coupled with singular spectrum analysis.

Journal of Hydrology, 399(3–4), 394–409. doi:10.1016/j.jhydrol.2011.01.017.

Xu, C., & Yue, D. (2015). Monte Carlo SSA to detect time variable seasonal oscillations from GPS-derived site position time series. *Tectonophysics, 665,* 118–126. doi:10.1016/j.tecto.2015.09.029.

Yan, H., Chen, W., Zhu, Y., Zhang, W., & Zhong, M. (2009). Contributions of thermal expansion of monuments and nearby bedrock to observed GPS height changes. *Geophysical Research Letters, 36,* L13301. doi:10.1029/2009GL038152.

Zerbini, S., Matonti, F., Raicich, F., Richter, B., & van Dam, T. (2004). Observing and assessing nontidal ocean loading using ocean, continuous GPS and gravity data in the Adriatic area. *Geophysical Research Letters, 31,* L23609. doi:10.1029/2004GL021185.

Zerbini, S., Raicich, F., Errico, M., & Cappello, G. (2013). An EOF and SVD analysis of interannual variability of GPS coordinates, environmental parameters and space gravity data. *Journal of Geodynamics, 67,* 111–124. doi:10.1016/j.jog.2012.04.006.

Zhang, J., Hassani, H., Xie, H., & Zhang, X. (2014). Estimating multi-country prosperity index: A two-dimensional singular spectrum analysis approach. *Journal of Systems Science and Complexity, 27*(1), 56–74. doi:10.1007/s11424-014-3314-3.

(Received December 16, 2016, revised February 2, 2017, accepted February 6, 2017, Published online February 15, 2017)

Pure Appl. Geophys. 175 (2018), 1841–1867
© 2018 The Author(s)
https://doi.org/10.1007/s00024-018-1856-3

Pure and Applied Geophysics

A Filtering of Incomplete GNSS Position Time Series with Probabilistic Principal Component Analysis

Maciej Gruszczynski,[1] Anna Klos,[1] and Janusz Bogusz[1]

Abstract—For the first time, we introduced the probabilistic principal component analysis (pPCA) regarding the spatio-temporal filtering of Global Navigation Satellite System (GNSS) position time series to estimate and remove Common Mode Error (CME) without the interpolation of missing values. We used data from the International GNSS Service (IGS) stations which contributed to the latest International Terrestrial Reference Frame (ITRF2014). The efficiency of the proposed algorithm was tested on the simulated incomplete time series, then CME was estimated for a set of 25 stations located in Central Europe. The newly applied pPCA was compared with previously used algorithms, which showed that this method is capable of resolving the problem of proper spatio-temporal filtering of GNSS time series characterized by different observation time span. We showed, that filtering can be carried out with pPCA method when there exist two time series in the dataset having less than 100 common epoch of observations. The 1st Principal Component (PC) explained more than 36% of the total variance represented by time series residuals' (series with deterministic model removed), what compared to the other PCs variances (less than 8%) means that common signals are significant in GNSS residuals. A clear improvement in the spectral indices of the power-law noise was noticed for the Up component, which is reflected by an average shift towards white noise from − 0.98 to − 0.67 (30%). We observed a significant average reduction in the accuracy of stations' velocity estimated for filtered residuals by 35, 28 and 69% for the North, East, and Up components, respectively. CME series were also subjected to analysis in the context of environmental mass loading influences of the filtering results. Subtraction of the environmental loading models from GNSS residuals provides to reduction of the estimated CME variance by 20 and 65% for horizontal and vertical components, respectively.

Key words: Probabilistic principal component analysis, common mode error, GNSS, time series analysis, missing data.

1. Motivation and introduction

The advantages of reliable coordinates provided by the globally distributed Global Navigation Satellite System (GNSS) stations have been appreciated by scientists since the early 90s. The position changes of antennae are expressed by coordinates that for decades have been continuously and regularly determined in global reference frame. The GNSS position time series provided by permanent observations have been used primarily to realize and maintain modern kinematic reference frames (Gross et al. 2009) as well as for geophysical studies as a measure of surface displacement or strain (Kreemer et al. 2014; Métivier et al. 2014). The linear trend due to plate tectonics and seasonal signals caused by i.e. environmental mass loading or thermoelastic deformation are very consistent for GNSS position time series recorded by nearby stations. Even the sophisticated time series modelling has not resulted in a total loss of spatial and temporal correlations (Wdowinski et al. 1997, Shen et al. 2013; Bogusz et al. 2015). The so-called Common Mode Error (CME) is the superposition of errors and geophysical phenomena, that similarly affect the coordinates of stations included in the regional networks. It is a spatially correlated type of error, which may be also temporally correlated depending on the temporal structure of the phenomena it absorbs. It is essential to remove, reduce or eliminate the impact of CME in the GNSS networks to improve the accuracy of the estimated velocities (He et al. 2017). The main theoretical contributors to the potential CME sources are (Wdowinski et al. 1997; King et al. 2010):

Electronic supplementary material The online version of this article (https://doi.org/10.1007/s00024-018-1856-3) contains supplementary material, which is available to authorized users.

[1] Faculty of Civil Engineering and Geodesy, Military University of Technology, Warsaw, Poland. E-mail: maciej.gruszczynski@wat.edu.pl

The original version of this chapter was revised. The correction to this chapter is available at https://doi.org/10.1007/978-3-319-96277-1_23

1. errors in the alignment to the reference frame;
2. errors related to satellites, which are usually observed in small networks as the mismodeling of the satellite: orbits, clocks, or antenna phase center variations;
3. signal emission media effects commonly influencing stations in regional network (troposphere and ionosphere modelling);
4. physical sources of station movements as the mismodeled (or unmodeled) large-scale atmospheric and hydrological effects, as well as small scale crust deformations;
5. errors caused by algorithms, software, and data processing strategies, including ambiguity resolution problem.

To reduce the effect of CME, a number of studies have been preceded by the filtering of the GNSS time series using different methods (Fig. 1). The Common Mode Error term was first introduced by Wdowinski et al. (1997), who described correlated errors in the regional networks of the GNSS stations. The authors used stacking approach assuming that CME is equal to an arithmetic mean of all available residuals at a specified epoch. Nikolaidis (2002) proposed to improve this method and called it a "weighted stacking", indicating that the GPS-derived coordinates with unequal formal errors cannot contribute equally to the final CME estimates. A group of methods known as "stacking" assumes the spatial uniformity of CME that has to be met over a regional network. This in turn imposes the condition that the estimates of CME at a single epoch are equal for the entire set of stations. In addition, a limit in the maximum size of the GNSS network which can be used to derive the CME estimates is set up. The stacking method can be used for networks with stations as far as 600 km from each other (Wdowinski et al. 1997; Márquez-Azúa and DeMets 2003). For larger networks and stations located up to 2 000 km from each other, various spatial filters can be introduced to differentiate the spatial response of any individual station. The stacking and spatial filtering methods should not be considered as similar approaches either in the sense of the character of the input and output data or mathematical formulas. Spatial filtering means that the CME is extracted with varying spatial responses and is individually and locally fitted to each individual station position time series. Spatial filters take into consideration the length of time series and the distance between stations (Márquez-Azúa and DeMets 2003) or the inter-station distances with correlations between residuals collected for any individual station (Tian and Shen 2011, 2016). It has been shown that CME can be also reduced using a 7-parameter (or 14-parameter) similarity transformation (Ji and Herring 2011; Blewitt et al. 2013). The limitations of aforementioned methods are related to: (1) a limitation of maximal area of network which can be subjected to filtering (mainly in case of stacking), (2) a partial inability to detect stations with strong local effects which will affect the CME estimates or (3) dependence on the conventional weighting procedure. Spatial filters and stacking methods significantly reduce the scattering of the position time series and, as a result, as soon as CME is removed, they improve the precision of velocity estimates. As an example, Wdowinski et al. (1997) reduced the time series root-mean-square (RMS) values for individual stations by more than 50%. Tian and Shen (2016) minimized the RMS values by 20.7, 13.2, and 14.4% on average for the North, East, and Up components, respectively. Despite the fact that the reduction in the scatter of the position time series is important and desirable, it should not be the only main indicator. The preferred filtering method should be able to accurately divide the GNSS residuals into two modes: the spatially

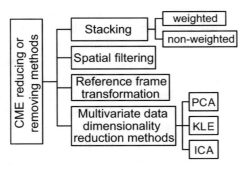

Figure 1
A diagram of various mathematical methods used so far to reduce or remove CME, including weighted and non-weighted stacking, spatial filtering, reference frame transformation and multivariate data dimensionality reduction methods such as Principal Component Analysis (PCA), Karhunen–Loeve Expansion (KLE) and Independent Component Analysis (ICA)

correlated part (CME) and spatially uncorrelated mode (noise). The unwanted effect of smoothing of the GNSS position time series appears when signals that do not correspond to real common effects are removed. It leads to undesired change in the long-term trend (station velocity) and also in the accuracy of this trend being misestimated.

Bearing in mind all of the aforementioned issues, many techniques which reduce the dimensionality of multivariate data have already been implemented to improve the CME filtering. Dong et al. 2006 and Serpelloni et al. 2013 proved, that the Empirical Orthogonal Function (EOF) decomposition provides a more solid numerical framework for the separation of modes than the stacking approach. In addition, this does not assume spatial uniformity of CME as stacking does, but instead employs a uniform temporal function which affects stations across regional network. Dong et al. (2006) were the first to apply the Principal Component Analysis (PCA) and Karhunen–Loeve Expansion (KLE) methods for CME extraction. They are based on different assumptions concerning the construction of the orthonormal vector basis. The former uses the covariance matrix of observations, while the latter applies the correlation matrix of observations. With regards to the fact that the traditional PCA can be applied only for complete data, Shen et al. (2013) proposed the use of a modified PCA (mPCA) to filter the position time series with missing data, which are reproduced from Principal Components (PCs). The PCA approach was further extended by Li et al. (2015), who introduced weighted spatio-temporal filtering. Similarly to weighted stacking, weighted PCA (wPCA) was proposed taking into consideration the individual errors of coordinates. This weighting procedure may cause an unwanted situation when time series with a weak CME response may significantly affect CME value. This may occur, when coordinates from stations affected by strong local effects e.g. local hydrology-induced or station-specific movements are determined by small standard errors. According to earlier publications, the weighting based on errors of observations does not refer to the nature of CME's. The advantages of EOF's for CME filtration have recently been confirmed by Gruszczynski et al. (2016), who showed significant improvement in the accuracy of stations velocities.

The main purpose of this research is to introduce for the first time a probabilistic PCA (pPCA) method for spatio-temporal filtering of GNSS position time series, and to employ it to filter the position time series whilst leaving the missing values without interpolation (Fig. 2a). Although pPCA has previously been employed in various areas of research, e.g. estimation of the EOF's for satellite-derived sea surface temperature (SST) data (Houseago-Stokes and Challenor 2004), a study on the precipitation and absorption squeeze (Andrei and Malandrino 2003), generation of the video textures (Fan and Bouguila 2009), detection of a small target (Cao et al. 2008), investigation of traffic flow volume (Qu et al. 2009), managing self-organizing maps (Lopez-Rubio et al. 2009), detection of outliers (Chen et al. 2009), tracking of the objects (Xiang et al. 2012), speaker recognition (Madikeri 2014), investigation of the nonlinear distributed parameter processes (Qi et al. 2012), nonlinear sensor fault diagnosis (Sharifi and Langari 2017), studying trends of mean temperatures and warm extremes (Moron et al. 2016), denoising of images (Mredhula and Dorairangaswamy 2016) or detection of the rolling element bearing fault (Xiang et al. 2015), according to the best of our knowledge, the pPCA filtering that is readily adapted to the position time series with missing data, has been presented for GNSS residuals (either position or ZTD) for the first time.

We present the pPCA as an alternative approach to spatio-temporal filtering PCA methods proposed by Dong et al. (2006) and by Shen et al. (2013), which will later be referred to as an iterative PCA (iPCA) and modified PCA (mPCA), respectively. Both methods are based on PCA algorithm and characterized by conventional approximates which modify standard PCA to deal with discontinuous time series. In mPCA approach it is assumed that the covariance matrix is initially constructed using all available time series. Gaps are then interpolated by minimizing the weighted quadratic norm of PC unknowns. In iPCA approach it is assumed that residuals can initially be spatially averaged, which means that any missing epoch may be completed using values from other stations that do not have gaps in this specified epoch. However, a problem occurs when there is a gap in the dataset which starts and

Real network stations coordinate availability

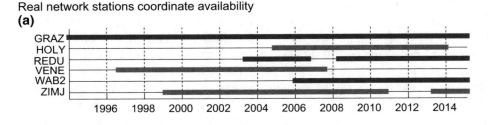

Incidentally occuring data time spans of GNSS time series

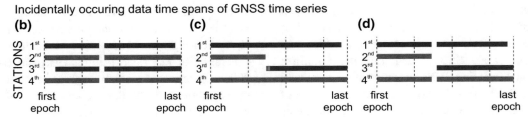

Figure 2

a Availability of observations for six selected stations distributed by the International GNSS Service (IGS); **b–d** graphs show simulated but frequently observed data gaps which may cause the iPCA method (**b**) or mPCA method (**c**) or both of them (**d**) unfunctional

ends at the same time almost for all stations in the network (Fig. 2b). In such a case, there are two options. First, the missing epochs from all series can be deleted, however, some amount of data containing important information for further analysis is removed. Second, during the first stage of interpolation real dependencies in GNSS residuals may be neglected due to the fact, that initial values adopted without a reliable probabilistic model can significantly influence further estimates. The mPCA method fails when any two time series of a network do not have any, or have only a few common epochs of observations (Fig. 2c). In this case, the covariance matrix cannot be set.

Figure 2d shows a theoretical time span of residuals subjected to filtering, where neither iPCA nor mPCA is able to perform orthogonal transformation since a certain gap is present in all data or two series do not have a single common observation. Unlike iPCA and mPCA, the pPCA method which we have introduced in this research, takes into account the probabilistic framework to determine the optimal model for the missing data. Since in pPCA the missing values are considered as latent variables, it is possible to filter even the series shown in Fig. 2d.

In this research, we applied the pPCA method to resolve the problem of a proper spatio-temporal filtering of GNSS position time series when gaps occur at the same time in the regional network and the series do not necessarily have the same observation time span. This method is presented as an alternative to the classic PCA approach and its modifications: mPCA and iPCA. The paper is organized as follows. We start with a set of complete data with a changing amount of simulated gaps to prove the effectiveness of the approach that we employed. Then, we continue with a set of 25 permanent GNSS stations which were included in the latest realization of the International Terrestrial Reference System (ITRS). At the end, we present hard numbers demonstrating the importance of spatio-temporal filtering before uncertainty of linear velocity being determined. It is worth mentioning, that the methodology presented in this research, although applied to the GNSS position time series, is universal and can be successfully adapted to data having spatial relationships gathered by GNSS, as e.g. ZTD (Zenith Total Delay), or any other geodetic instruments such as GRACE (Gravity Recovery and Climate Experiment) or altimetric satellites.

2. Probabilistic Principal Component Analysis

For the network formed by n GNSS stations with a time series spanning m days, before we attempt the spatio-temporal filtering, we are obliged to construct the observation matrix $\mathbf{R}(t_i, r_j)$ ($i = 1, 2, \ldots, m$ and $j = 1, 2, \ldots, n$) for each topocentric component (North, East or Up) separately. Residual time series $\mathbf{r}(t)$ constitute the matrix in such a way that each row corresponds to the epoch of observation t_i, while columns represent each subsequent GNSS coordinate time series r_j from the GNSS stations. To introduce the pPCA procedure, we firstly present the most common derivation of PCA of the matrix $\mathbf{R}$ through eigenvalue decomposition. At this stage, the time series are assumed to be complete. The 4-step basics are given as (Jolliffe 2002):

Step 1: computation of the mean-centered matrix $\mathbf{R}_c$ by subtracting the vector of means of all columns and from each row of $\mathbf{R}$,

Step 2: computation of the covariance matrix $\mathbf{C} \cdot v = \mathbf{R}_c' \cdot \mathbf{R}_c$ which is of n per n-dimension matrix,

Step 3: computation of the eigenvalue decomposition of $\mathbf{C} \cdot v$ given by $\mathbf{C} \cdot v = \mathbf{V} \cdot \mathbf{\Lambda} \cdot \mathbf{V}^{-1}$, where $\mathbf{\Lambda}$ is a matrix with k non-zero diagonal eigenvalues of the covariance matrix and $\mathbf{V}$ is the n per n-dimension matrix with the corresponding eigenvectors in individual columns. The number of eigenvalues may be less than or equal to the number of time series ($n \geq k$), but in most cases with real data, the matrix $\mathbf{C} \cdot v$ is usually of full rank and the number of eigenvectors is equal to the number of the time series ($n = k$),

Step 4: sorting of the eigenvectors and corresponding eigenvalues in a decreasing order. The eigenvalues represent the contribution of each Principal Component mode in the total variance of data. Those principal components are estimated as:

$$a_k(t_i) = \sum_{j=1}^{n} \mathbf{R}(t_i, r_j) \cdot \mathbf{v}_k(r_j) \qquad (1)$$

where $a_k(t_i)$ is the k-th PC of matrix $\mathbf{R}$ and $\mathbf{v}_k(r_j)$ is its corresponding eigenvector (a matching column adapted from $\mathbf{V}$).

A standard PCA approach is applicable only to the complete datasets and any attempt to use this method for data with missing values must be preceded by deleting the rows with missing data, interpolating or modifying PCA algorithm (Ilin and Raiko 2010; Zuccolotto 2012). Real geodetic data are susceptible to incompletion. Since coordinate time series are arranged in the observation matrix by time, any time series that starts later or ends earlier than other stations are also considered as missing. Furthermore, the hardware or software failure or replacement, physical disturbance, data loss or removal of outliers at the pre-analysis contributes to gaps in the data.

We employed a more complex procedure for eigenvalue decomposition in case the data matrix being incomplete. Probabilistic PCA presented here is based on the Expectation–Maximization (EM) algorithm (Roweis 1997; Tipping and Bishop 1999). The regularized EM algorithm has been recently used to interpolate missing values before traditional PCA and ICA were performed for the Chinese regional GNSS network (Ming et al. 2017). In contrast to an interpolation of incomplete time series, the EM algorithm employed in pPCA handles missing values by considering them as additional latent variables. Products of pPCA-based filtering can be interpreted in the same way as results from the traditional PCA, however, the pPCA method stands out by application of a flexible statistical model.

The probabilistic PCA is based on the following latent variable model:

$$\mathbf{t} = \mathbf{W} \cdot \mathbf{x} + \mathbf{\mu} + \mathbf{\varepsilon} \qquad (2)$$

where: $\mathbf{t}$ is a n-dimensional observation vector, $\mathbf{x}$ is a q-dimensional vector of latent variables, $\mathbf{W}$ is a n per q-dimensional transformation matrix, $\mathbf{\mu}$ is the vector mean of $\mathbf{t}$, $\mathbf{\varepsilon}$ is a noise model which compensates for the errors.

In case of filtering of GNSS-derived position residuals, $\mathbf{t}$ can be identified with time series of all available residuals at given epochs, while $\mathbf{x}$ are residuals that are not directly estimated in dataset, e.g. due to the lack of coordinates or as an effect of outliers removal. According to pPCA theorem missing values are rather inferred from other residuals that really exist in time series via the assumption of a spatially correlated CME. $\mathbf{W}$ is the matrix whose columns are composed of the scaled eigenvectors of sample covariance matrix of residuals, which are necessary to estimate CME.

There is no closed-form analytical solution for $\mathbf{W}$, which is the reason for the need to apply an iterative EM algorithm. In this research, we adopted the pPCA theorem proved by Tipping and Bishop (1999), who gave the analytic solution for the model showed in Eq. (2), EM algorithm description and a full characterization of its properties. The Maximum Likelihood (ML) solution for pPCA latent variable model is given by:

$$\mathbf{W}_{\mathrm{ML}} = \mathbf{V}_q\left(\mathbf{\Lambda}_q - \sigma^2 \cdot \mathbf{I}\right)^{1/2}\mathbf{B} \qquad (3)$$

where: $\mathbf{B}$ is an arbitrary q per q-dimension orthogonal rotation matrix, $\mathbf{I}$ is an identity matrix, σ^2 is an isotropic variance.

Each of the columns of matrix $\mathbf{V}_q$ (n per q-dimension) is the principal eigenvector of sample covariance matrix of the GNSS residuals, with corresponding eigenvalue in the q per q-dimension diagonal matrix $\mathbf{\Lambda}_q$. Since one of the most important steps of each PCA-based procedure is the decomposition of the covariance matrix into the matrix with eigenvalues and matrix with corresponding eigenvectors, this in case of pPCA the maximization of the likelihood function (Eq. 3) by using EM algorithm is a key issue to obtain the most probable elements of $\mathbf{V}_q$ and $\mathbf{\Lambda}_q$ matrixes. It results in the calculation of principal eigenvectors and eigenvalues necessarily for reliable CME estimation (Eq. 4).

The EM algorithm consists of two main steps: the E-expectation and M-maximization. The parameters of the model given in the Eq. (3) are resolved with the Maximum Likelihood Estimation (MLE) in an iterative manner (Tipping and Bishop 1999) by 3-step procedure:

Step 1 (E-step): calculation of the expected value of the log-likelihood function, given the considered data and the current estimates of the model parameters,

Step 2 (M-step): finding the new parameters by maximizing the log likelihood function using the expected parameters derived in the E-step,

Step 3: repeating Steps 1 and 2 until convergence. For our purposes the convergence criteria was set up as a relative change in the transformation matrix elements less than 10^{-4}.

Using the EM algorithm for finding the principal axes by iteratively maximizing the likelihood function (Eq. 3), the latent variable model defined by Eq. (2) affects mapping from the latent space into principal subspace of the observed data.

One of the most important features related to pPCA method is the fact that the q-number of EOFs to retain, can be specified at the very outset. The reason for limitation of this parameter is the fact that, in case of small value of q in relation to high value of n (number of dimensions—in our case number of GNSS stations) the dimension of $\mathbf{W}$ transformation matrix is much smaller than the covariance matrix for traditional EOF analysis. This makes pPCA method to be computationally much more efficient and less burdensome for computers. Many papers have focused on the issue of determining the optimum q number of retained EOFs prior to using EM algorithm (e.g. Jolliffe 1972; Houseago-Stokes and Challenor 2004). However, there is no satisfactory and versatile rule. In this research, at the pre-processing stage based on our dataset, we computed the maximum number of principal components which can be retained from pPCA. We found that only the first PC is significant when deterministic model was subtracted prior the pPCA analysis (please see data and methods described in section "GNSS time series"), which is consistent with the considerations of other authors (Dong et al. 2006; Shen et al. 2013). We adopted $q = 3$ value to allow for more variance to be retained. Furthermore, some aspects of computational as well as communication complexity of PCA-based methods were the subject of many analyses (e.g. Roweis 1997; Houseago-Stokes and Challenor 2004; Ilin and Raiko 2010) with leading conclusion that the probabilistic PCA is the most promising PCA approach, especially for large datasets.

Another important advantage of the pPCA method is the ability to interpret its products in the same manner as results of traditional PCA. This allows to adopt the definition and applications of CME estimates (Dong et al. 2006):

$$\mathbf{CME}_j(t_i) = \sum_{k=1}^{p} \mathbf{a}_k(t_i) \cdot \mathbf{v}_k(r_j) \qquad (4)$$

where p is a number of first significant PCs. Following Shen et al. (2013) and Tiampo et al. (2004), we used the Fisher-Snedecor test (Fisher, 1932) for the equality of two variances to decide on the number of significant PCs. CME is removed from the unfiltered residuals $\mathbf{r}(t)$ using arithmetic subtraction, thus obtaining so-called "filtered" residuals. The first few PCs (or just the first PC in some cases) reflect a common source function which affects the regional GNSS network, i.e. CME, and represents the highest contribution to the variance of the GNSS-derived position residuals (Dong et al. 2006).

3. GNSS Time Series

In this research, we employed the daily-sampled GNSS position time series that were produced by the International GNSS Service (Dow et al. 2009) by a network solution referred to as "repro2" to estimate coordinates (Rebischung et al. 2016). Each of the selected stations contributed to the newest realization of the International Terrestrial Reference System (namely ITRF2014, Altamimi et al. 2016). Since it is imperative to investigate the response of the newly adopted pPCA method to the number of missing data, we were obliged to find stations with almost complete time series for reference purposes. Since the distance between stations from the selected network is quite important for this research, a set of 25 stations located in Central Europe were chosen (Fig. 3). The distance between any two stations taken for analysis is shorter than 1870 km which is consistent with the overall assumptions related to applicability of PCA-based filtering methods. Márquez-Azúa and DeMets (2003) found that the spatial correlations of the residual time series are high within 1000 km distance and they gradually decrease to zero for c.a. 6000 km. Similar studies were performed for stations distant at the 10^3 km level, located in China (Li et al. 2015; Shen et al. 2013) or in Australia (Jiang and Zhou 2015).

For the purpose of this study, we used a time span of 2003.2–2015.0 (Fig. 4), when all selected stations were operating. The length of the GNSS coordinate time series is very important whilst the station velocity is expected to be determined with high

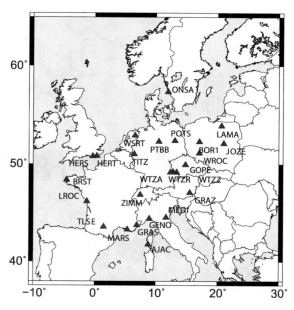

Figure 3
Geographical distribution of 25 stations employed in the analysis

reliability. The data time span commonly assumed by other authors (e.g. Blewitt and Lavallée 2002) as a minimum for reliable velocity estimation is 3 years. However, lately, Klos et al. (2018) argued that this span should be extended to 9 years. Time-dependent improvements in the consistency of GNSS-derived position residuals are explained by avoiding errors caused by mismodeling of seasonal signals and noises. Every incorrectly estimated seasonal signal for regional network of GNSS stations may increase spatially-correlated errors. As a result of studies performed both on GNSS time series and models of surface mass loading deformation, Santamaría-Gómez and Mémin (2015), found that at least 4 years of continuous data is necessary to meet requirements of the accurate modelling of the inter-annual deformations as a step towards reliable estimation of secular velocities. In case of time series included in this analysis, 12.2 years long data time span exceeds assumed minimal levels.

We used epochs of offsets compiled by IERS (International Earth Rotation and Reference System Service) ITRS Product Centre and available at http://itrf.ensg.ign.fr/ITRF_solutions/2014/doc/ITRF2014-soln-gnss.snx to eliminate the influence of discontinuities on final estimates. Any value was considered

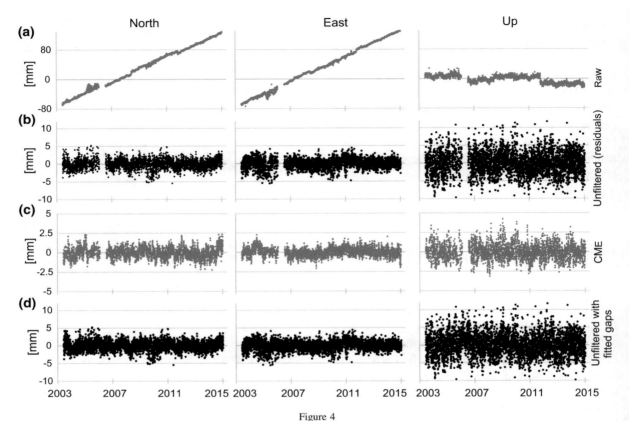

Figure 4
Exemplary GNSS coordinate time series for station BRST (Brest, France). **a** "raw" GNSS coordinate time series. Trend, offsets and seasonal signals can be noticed. **b** "unfiltered" time series subjected to filtering analysis, obtained when model from Eq. (5) was removed. **c** CME estimates based on the pPCA method performed for incomplete time series. **d** Residuals of BRST with gaps being filled. The different scale of vertical axes for (**a**–**d**) should be noticed

as an outlier, when it fell outside 3 times the interquartile range (IQR) below or above the median (Langbein and Bock 2004).

The time series were characterized by 3.8% of gaps on average. Time series form station ONSA (Onsala, Sweden) were the most complete with only 0.5% of missing data, while station BRST (Brest, France) had the greatest amount of missing data: 13.2%.

We modelled each of the topocentric (North, East or Up) position time series (Fig. 4a) $\mathbf{x}(t)$ with a mathematical function that takes the form of:

$$\mathbf{x}(t) = x_0 + v_x \cdot t + \sum_{i=1}^{j} [A_i \cdot \sin(\omega_i \cdot t + \phi_i) + \mathbf{r}(t)] \quad (5)$$

where: x_0 represents the initial coordinate at the reference epoch, v_x refers to the linear velocity, $\sum_{i=1}^{j} [A_i \cdot \sin(\omega_i \cdot t + \phi_i)]$ accounts for periodic signals with angular velocities ω_i, $\mathbf{r}(t) = \mathbf{CME}(t) + \varepsilon(t)$ are the residuals being a sum of spatially- and temporally-correlated CME and temporally-correlated noise.

In the following research, the parameters of the deterministic part were estimated with the Maximum Likelihood Estimation (MLE) method according to approach to the deterministic part given by Bogusz and Klos (2016). Unlike the vast majority of described modelling approaches found in literature (Dong et al. 2006; Shen et al. 2013), where only annual and semiannual signals were used, our deterministic model assumes different periodicities: fortnightly,

Chandlerian, tropical and draconitic (see Bogusz and Klos 2016).

CME contains some part of a flicker noise (Klos et al. 2016) with spectral index of -1, which was found to be mostly present in the GNSS position time series (Williams et al. 2004; Bos et al. 2008). The residual time series $\mathbf{r}(t)$ obtained after a deterministic model was subtracted (Fig. 4b), are subjected to further analysis and will be referred to later in this paper as the "unfiltered" time series.

Despite the fact that the time series were de-trended and seasonal signal were removed, we still notice the non-zero correlations between residuals $\mathbf{r}(t)$. For the purpose of proving correlation level, we performed two types of tests. Initially, to quantify the level of correlations, we used Lin's concordance correlation coefficient (Lin 1989), as it was recommended for GNSS residuals by Tian and Shen (2016). Lin's concordance coefficient, was introduced to provide a measure of reliability that is based on covariation and correspondence in contrast to commonly used Pearson correlation coefficient, which in turn is a measure of linear covariation between two sets of scores. Differences can be comprehended by geometric interpretation, where two time series are plotted in one scatterplot and best fitted line is imposed. Pearson correlation coefficient specifies how far from the line are data points, whilst Lin's concordance coefficient additionally taking into account the distance to the 45-degree line, which represents perfect agreement. Concordance correlation coefficient ranges from 0 to ± 1 and its interpretation is close to other correlation coefficients, which means, that values near 1 mean perfect concordance and 0 means no correlation. For unfiltered residuals used in this analysis we found that the average Lin's concordance correlation coefficient is equal to median values of 0.39 for horizontal and 0.54 for vertical components, respectively, when the distance between individual stations is less than 500 km (see the red dots in Fig. 11). Correlations are smaller for more distant stations, which is mostly evident for the Up component. Then, we employed the Kaiser–Meyer–Olkin (KMO) index to assess whether we are able to use multivariate analysis to efficiently extract the common signals from a set of stations we employed (Cerny and Kaiser 1977):

$$ \mathrm{KMO} = \frac{\sum_j \sum_{k \neq j} \rho_{jk}^2}{\sum_j \sum_{k \neq j} \rho_{jk}^2 + \sum_j \sum_{k \neq j} \hat{\rho}_{jk}^2} \quad (6) $$

where: $\hat{\rho}_{jk}$ represents a partial correlation, and ρ_{jk} is a correlation coefficient between variables j and k which mean time series from jth and kth stations. Partial correlation is the measure of association between two time series, while controlling or adjusting the effect of one or more additional time series.

The KMO index is a value that describes dataset applied to dimensionality reduction techniques (e.g. pPCA). This index measures the proportion of common variance among the all variables. By definition, the KMO index ranges between 0 and 1. Values close to 1 mean that common signals have a significant variance. For the observation matrix from the real unfiltered residuals of 25 GNSS stations we obtained KMO indices equal to 0.961, 0.966 and 0.988 for the North, East and Up components, respectively.

4. pPCA Filtering of Artificially Incomplete Time Series

In this part of the research, we analyzed and compared iPCA, mPCA and pPCA methods with traditional pre-interpolated PCA approach for the spatio-temporal filtering of the GNSS-derived position time series. Missing values were introduced to real GNSS position time series to simulate the number of gaps we might expect in the observations. In this way, we assessed the ability of each method to deal with incomplete time series and its sensitivity on the number of missing values.

The artificially incomplete residuals were produced in the following manner. First, we used the GNSS position residuals from a set of 25 stations presented in Fig. 3. We fully interpolated them, assuming adequate values of mean and standard deviation of inputted points in such a way that interpolation procedure did not change the variance of the time series. With these assumptions, we obtained time series that imitated 25 unfiltered, fully complete GNSS residuals. An example is shown in Fig. 4d. Then, we randomly chose epochs and removed observations to simulate incomplete data.

We introduced gaps with length from 5 to 40% of the total length of the series with 5% increment. We assumed that the gaps were missing at random (MAR, Little and Rubin 2002) and that the number of stations subjected to introduced gaps is approximately equal to the number of time series which remain complete. Therefore, data gaps were introduced to 13 randomly chosen stations from a set of 25. For the remaining 12 stations, no data were deleted simulating time series as being complete. We were tempted to accept this procedure by two issues. First, to investigate what is the impact of missing data on CME estimates. Second, how much the CME computed for complete time series is affected by values which are missing on other stations.

In case of traditional PCA with interpolated gaps we assumed a white noise model to simulate scatter of residuals to show, how the simple assumption of linear interpolation may bias the CME estimates. The response of each method was then analyzed according to the increased number of missing values. The relative errors of CME were computed based on the unfiltered residuals subjected to introduced gaps and compared to CME estimated for the complete unfiltered residuals. Complete time series were treated as a reference dataset for spatio-temporal filtering. It is worth noting that CME is identical for all methods where time series are complete. The relative error of CME determination was computed as (Shen et al. 2013):

$$\sigma_{CME} = \pm \sqrt{\frac{(\mathbf{CME}_i - \mathbf{CME}_0)^T (\mathbf{CME}_i - \mathbf{CME}_0)}{\mathbf{CME}_i^T - \mathbf{CME}_0)}} \cdot 100 \tag{7}$$

where $\mathbf{CME}_0$ and $\mathbf{CME}_i$ are the vectors of Common Mode Errors computed before data were deleted and after data gaps were interpolated, respectively. When gaps were introduced, we randomly chose stations and epochs to be deleted 100 times and we averaged results of simulations. Figure 5 presents the relative errors for stations where data were deleted, while Fig. 6 includes relative errors of CME for stations where no data were removed. When the interpolation of GNSS-based position residuals has been applied before PCA-based filtering, the CME values were biased reaching values of relative error equal to 10

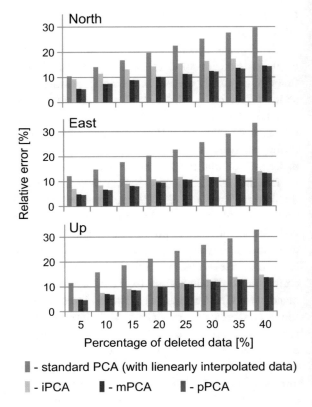

Figure 5
Relative errors of CME (σ_{CME}) where different percentages of data were randomly deleted. CME was computed with the use of pPCA (red bars), iPCA (green), mPCA (blue) and traditional PCA with gaps interpolated (grey). Relative errors relate to the subset of 13 stations that were affected by introduced gaps. The results were averaged for 100 different simulations in every step we performed

and 35%, when 5 and 40% of data were deleted, respectively. The relative errors of CME estimates are similar both for mPCA, iPCA and for pPCA which allows us to conclude that our method can constitute an alternative approach to both methods already mentioned. Only in a few cases the pPCA performance in CME estimation is slightly better than mPCA, but the difference between both of them is not significant and reaches the maximum of 0.1%.

The relative errors of CMEs estimated with pPCA ranged between 5 and 14% for the entire set of stations. The reconstruction of CME computed on the basis of incomplete data is slightly better for the vertical component for which the relative errors of CME are less by 1–3%. This fact may be explained by the higher correlation, which allows to determine

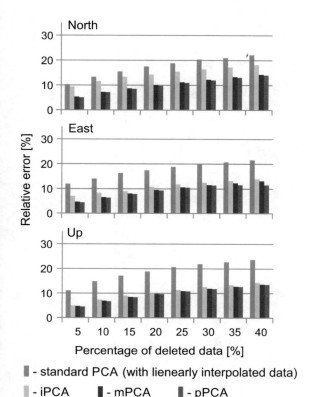

North

East

Up

Percentage of deleted data [%]

Relative error [%]

■ - standard PCA (with lienearly interpolated data)

■ - iPCA   ■ - mPCA   ■ - pPCA

Figure 6
The same as Fig. 5 but the relative errors are computed for the set of 12 stations with complete data for which no values were removed. However, although no data was deleted, the results were biased due to missing data in the remaining 13 time series shown in Fig. 5. The results were averaged for 100 different simulations in every step we performed

interpolation, the larger the number of missing values, the higher the relative error of CME, rising to 33%. In addition, the CME values were biased for 12 stations, where no data were removed. All relative errors of CME presented in Fig. 6 are non-zero. Despite the fact that each time series derived from 12 stations was not subjected to a deleting procedure, CME estimates were also incorrectly calculated due to the missing values in the remaining time series.

Then, we ran imitation of missing values to simulate a specific case when GNSS coordinate time series have just few common epochs [see stations VENE (Venezia, Italy) and WAB2 (Bern, Switzerland) in Fig. 2a]. As it was mentioned before, every inconsistency in the first and last observed epoch for GNSS time series has to be treated during spatio-temporal filtering as an incompleteness. We simulated a few time series which have relatively small number of common observations. First, we randomly selected 6 stations from a set of 25 complete residuals that we employed and made the time series to end at 2009.12, which means that we purposely deleted data after 2009.12. Later, other 6 stations were randomly selected and we artificially made them to start after 2009.12. Having those two datasets of 6 stations, we gradually added the previously deleted data to prolong the time of common observations in every stage. In each iteration, the remaining set of 13 stations was untouched. For this kind of dataset, spatio-temporal filtering of 25 GNSS residuals was carried out using 4 PCA-based methods. Relative errors of CME were estimated for all stations and the results were averaged. The selection of 12 stations was repeated 20 times independently at every stage to average results for various combinations of stations subjected to data deletion. We also estimated the time which is required to estimate CME for all methods. Table 1 presents relative errors of CME (σ_{CME}, Eq. 7) estimated with the procedure described above.

Based on the data presented in Table 1, we can conclude that pPCA method gives quite consistent results compared to other algorithms. The relative errors of CME averaged for 20 simulations do not exceed 19% in each direction. For analyzed set of stations, the mPCA method requires that the time series have a minimum of 400 common epochs of observations for horizontal components and a

more reliable parameters of latent variable model in pPCA. Another important feature, which can be seen in Figs. 5 and 6 is the fact, that compared to mPCA and pPCA, the iPCA method performs worse in the North component than in the East and Up components. Taking into consideration only North component, the differences in relative error of CME estimated with the use of iPCA and pPCA methods are about 4%. It may be due to inhomogeneous spatial response of individual stations to the CME source, which is presented in Fig. 7 and described in the next section.

Similarly to iPCA and mPCA, the relative error of CME reconstructed with the use of pPCA is always less than 14%, even in cases when 40% of residuals were deleted from the 13 stations selected out of 25 (Fig. 5). In the standard PCA approach with

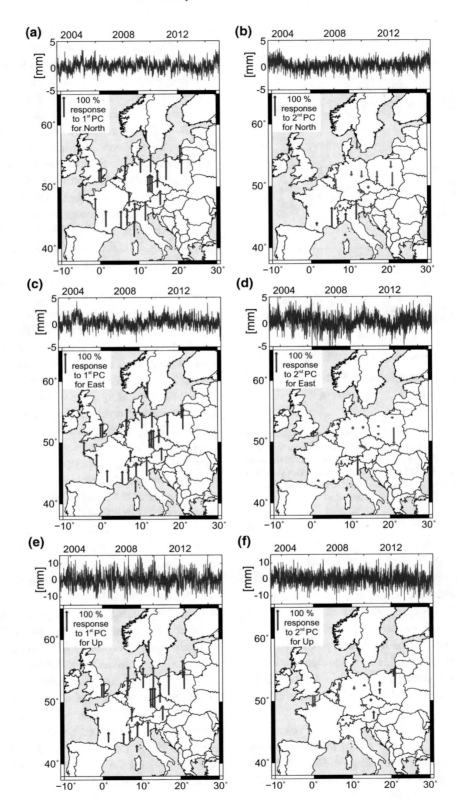

◄Figure 7

Maps of spatial distribution of GNSS residuals responses (normalized eigenvector elements) to the 1st (**a, c, e**) and 2nd (**b, d, f**) PC, for North, East and Up components, respectively, and corresponding scaled Principal Components (solid line at the top of each map). Normalized eigenvector elements can be identified with station contribution (positive-blue or negative-red) into the amount of variance in particular PC. Note different scales in horizontal and vertical directions

minimum of 1500 epochs for a vertical one. Otherwise, algorithm is unable to calculate Principal Components, because the covariance matrix estimated at the beginning of the algorithm is not positive semidefinite and some of its eigenvalues are negative. The differences between iPCA and pPCA methods can be seen for horizontal components, where time span of common epochs is shorter than 800 observations. In such a case, the differences in relative error of CME reach 14%. The iPCA, mPCA and pPCA algorithms performed similarly for Up component except in cases in which mPCA was unfunctional.

Standard PCA method is the fastest and also the least complex of those being considered, because residuals are fully interpolated a priori and eigen-decomposition is made only once, but results shown in Table 1 do not give grounds for including this

method for further analysis. Time needed to estimate CME depends on the computing power of the resources. In our case, we conducted each PCA-based filtering method simultaneously using the same HPC-class (High-Performance Computing) resource, therefore, we show relative values of calculation time referred to in pPCA method. This method calculates CME relatively faster than mPCA up to a maximum of 300%, when residuals are loaded with the largest number of missing values. Differences in processing time have decreased almost to zero for these two methods, where residuals have more than 2000 epochs of common observations. Computational time for iPCA method is very similar to pPCA method, i.e. iPCA is up to 20% slower. However, for more complete time series the iPCA seems to be faster than pPCA (up to 50%). Since in our case the number of stations and epochs for daily time series are relatively small, processing time does not seem to be a key factor for defining the superiority of filtering methods to be used for GNSS position time series. However, when long-term hourly (or even more frequent) GNSS time series from network formed by hundreds of stations would have to be employed for filtering procedure, the computational complexity can influence the choice of method.

Table 1

Relative errors of CME averaged for a set of 25 GNSS residuals, in which 12 randomly selected residuals have limited common time span of observations

Number of common epochs	North [%]				East [%]				Up [%]			
	Standard PCA	iPCA	mPCA	pPCA	Standard PCA	iPCA	mPCA	pPCA	Standard PCA	iPCA	mPCA	pPCA
50	99.4	32.6	–	18.6	85.9	22.5	–	18.5	99.9	13.9	–	13.6
100	98.2	27.2	–	18.6	81.6	22.0	–	18.1	91.4	12.9	–	13.2
200	96.1	24.6	–	18.3	77.1	21.6	–	17.3	90.6	12.5	–	12.7
400	69.8	20.0	20.1	18.1	75.5	21.7	30.7	16.7	89.6	12.2	–	13.0
600	56.4	19.7	17.4	17.4	75.2	19.1	20.5	16.5	76.9	12.0	–	12.2
800	55.5	17.9	16.1	17.2	68.8	18.8	17.0	16.5	71.7	11.4	–	12.4
1000	54.6	16.6	15.8	15.6	50.5	17.6	16.7	16.3	61.5	10.9	–	12.0
1500	47.5	15.0	15.6	15.0	45.2	16.6	16.6	16.2	50.3	10.7	10.8	10.9
2000	39.0	15.4	14.2	14.1	37.0	15.1	15.2	14.5	47.7	9.2	9.2	10.4
2500	35.1	14.7	12.5	12.1	26.7	14.4	13.4	13.4	46.7	9.5	9.9	10.2
3000	28.1	13.4	10.3	9.4	22.2	11.3	11.6	10.8	15.8	7.6	7.3	7.8
3500	15.3	12.9	9.1	8.6	15.1	7.9	7.3	6.9	12.5	5.6	5.5	5.5

Number of common epochs for those 12 stations is contained in the first column. It shows how much these 12 stations are overlapped. Number of epochs in the whole dataset is 4314. Symbol '−' means that a certain algorithm was unable to calculate Principal Components

Table 2

Eigenvalues shown as a percentage of variance of residuals $\mathbf{r}(t)$ *represented by first seven PCs*

Topocentric component principal Component	North [%]	East [%]	Up [%]
1st	36	36	49
2nd	7	8	7
3rd	6	7	5
4th	5	6	5
5th	4	4	4
6th	4	4	3
7th	4	3	3

On the basis of the results presented in this section, we may conclude that the pPCA method is able to be directly applied to the GNSS position time series with no need to interpolate the data before spatio-temporal filtering. In turn, GNSS time series do not have to start and end at the same epochs, they are not affected by the interpolation procedure. What is more, a gap present in all time series at one (or several) epoch, will facilitate the calculation of CME.

5. pPCA Filtering of Real Time Series

In the following section, we present the results of spatio-temporal filtering performed with pPCA for real dataset consisting of position time series from 25 IGS stations. Residuals are the result of standard pre-processing described previously and they are not subjected to intentional data deleting or interpolating procedure. We employed a set of 25 stations presented in Fig. 3 and used the "unfiltered" residuals $\mathbf{r}(t)$ of their position time series presented in Fig. 4b.

First, we analyzed the eigenvalues and eigenvectors related to each consecutive PCs. Table 2 presents the proportion of variance of all residuals, which is represented by first seven PCs. The eigenvalues can be interpreted as a fraction of the total variance of the residual time series corresponding to each eigenvector. Additionally, the analysis of eigenvalues allowed us to define significant PC (or PCs) which may be interpreted as CME.

The 1st PC explains 36 and 49% of the total variance for horizontal and vertical components,

respectively. Higher order PCs do not contribute to the total variance of residuals higher than 8%. These percentages support the hypothesis that regional phenomena affect the vertical component more than the horizontal components. The first PC is the only one which satisfies the criteria of CME consideration. As indicated by the Fisher-Snedecor test at the 95% confidence level, the variance of residuals which is explained by this mode significantly differs from the variances of the remaining PCs. Therefore, in the following part of the paper, the CME will be calculated using only the 1st PC.

Figure 7 shows scaled PCs obtained through pPCA procedure and their corresponding eigenvectors. Scaled Principal Components are obtained by multiplication of each PC by the normalization factor, which is equal to the maximum response of the network stations to this mode. A procedure to compute the normalized eigenvectors was adopted from Dong et al. (2006). The normalized eigenvector elements refer to the spatial response of individual stations to the CME source if the considered PC can be identified as CME. Those elements may be positive or negative with values between -100 and $+100\%$ (Fig. 7). The theoretical assumption of CME changeability within the considered GNSS network is supported by the results presented in the Fig. 7. The entire set of stations show a positive response to the 1st PC with values higher than 33% for all topocentric components. A minimum response was found for the AJAC (Ajaccio, France) station for Up component (Fig. 7e). The elements of the eigenvector related to 2nd PC are both positive and negative (Fig. 7b, d, f). Such result can be explained by the fact that signals extracted by the 2nd and also by subsequent PCs are due to an uncommon source for that network. They may result from local or regional effects and are unnoticeable for the entire set of stations. The consecutive PCs are characterized by the statistically negligible amount of variance explained by them. Both eigenvalues presented in Table 2 and spatial distribution of station responses shown in Fig. 7 analyzed together allow us to conclude, that 1st PC is the only one that fulfills the CME definition. This has been also confirmed previously by Fisher-Snedecor tests.

Results presented in Fig. 7 show a spatial pattern for the East and Up components found for network station responses to the 1st PC which is identified as CME. For these two components of position, the GNSS residuals responses to the CME are higher for stations situated in Northeastern Europe than for other selected stations. For North component, station responses are more homogeneous. The median value of normalized eigenvector corresponding to 1st PC is equal to 81, 73 and 74% for North, East and Up components, respectively. For the analyzed network, only 5 from 25 stations have relative responses less than 70% to 1st PC in North component. This result is different for Up and East components, where as many as 10 stations have relative responses less than 70% to 1st PC. From 10 stations characterized by the lowest response for CME in Up and East components, 8 of them are located in Southeastern Europe. These are: AJAC (Ajaccio, France), BRST (Brest, France), GRAS (Grasse, France), HERS (Herstmonceux, UK), HERT (Herstmonceux, UK), LROC (La Rochelle, France), MARS (Marseille, France), TLSE (Toulouse, France), see Fig. 3. Spatial pattern, which was found for 1st PC (Fig. 7) is similar to the distribution of power-law noise which was observed earlier by Klos and Bogusz (2017). They showed that vertical components from Central and Northern European stations may be characterized by smaller spectral indices of power-law noise than any other stations in Europe.

The small scale crustal deformations due to surface mass loading are considered to be a very important contributor to the spatially correlated errors in GNSS (Dong et al., 2006; Yuan et al., 2008). To support the explanations of spatial pattern shown in Fig. 7, we analyzed another dataset employing environmental loading models, which are freely available at the École et Observatoire des Sciences de la Terre (EOST) Loading Service (http://loading.u-strasbg.fr/displ_itrf.php). We used the environmental loadings calculated from three different models: ERA (ECMWF Reanalysis) Interim atmospheric model, Modern Era-Retrospective Analysis (MERRA) land hydrological loading and non-tidal ocean loading ECCO2 (Estimation of the Circulation and Climate of the Ocean version 2) model. We averaged the models to be sampled every 24 h to be consistent with GNSS position sampling rate. Then, we summed these models at corresponding epochs to obtain their superposition, which means a joint effect of environmental mass loading on the displacements of the selected ITRF2014 stations. We also limited their time span to be equal to the GNSS residuals (2003.2–2015.0). These time series will be referred to later in this paper as the "environmental loading time series". Subsequently, we submitted them to pPCA procedure to obtain their spatial responses to the 1st and 2nd PCs (Fig. 8) and corresponding scaled PCs.

Comparing spatial distribution of normalized eigenvectors computed for "environmental loading time series" (Fig. 8), to the eigenvectors computed for unfiltered GNSS residuals, a significant similarity can be noticed especially for 1st PC. The level of variance corresponding to the 1st PC reaches 66, 84 and 90% for North, East and Up components, respectively, and differs from each consecutive PC variance. As well as for GNSS residuals, environmental loading time series respond more to the 1st PC in Northeastern Europe with regards to East and Up components. It is worth emphasizing, that we may draw a North–South oriented line separating areas with different responses relating to 2nd PC. More extended investigations of this phenomena does not coincide with the scope of this paper, but we presume, that this effect is related to differences between the influence of the continental and oceanic climate.

As stated previously, environmental loadings, in particular, atmospheric, hydrological and non-tidal oceanic effects, are one of the potential sources of CME in the GNSS coordinate time series. According to the spatial pattern which was presented in Figs. 7 and 8, we estimated how large-scale environmental effects influence the character of CME. For this purpose, we first derived the CME of the "unfiltered" GNSS residual time series using pPCA for a real dataset as described above. Then, to assess the contribution of loading effects to this CME, we derived the CME of the "unfiltered" GNSS residuals adjusted for loading effects. Finally, to make a comparison we calculated CME variances and discussed the results of CME noise analysis together for two dataset.

GNSS-derived position residuals, as well as, other time series of measurements of wide variety of dynamic processes are usually characterized by

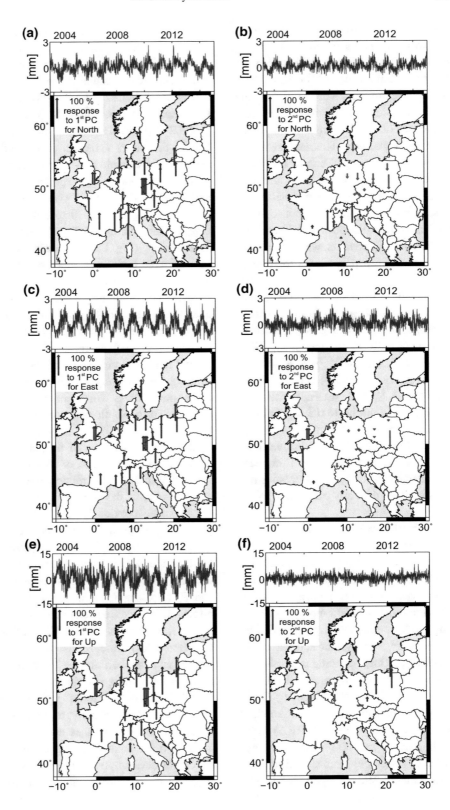

◄Figure 8
Maps of spatial distribution of environmental loading time series
responses (normalized eigenvector elements) to the 1st (**a, c, e**) and
2nd (**b, d, f**) PC, for North, East and Up component, respectively
and corresponding scaled Principal Components (solid line at the
top of each map). Loading correction time series are calculated for
the same locations as GNSS stations shown in Fig. 3

spectral indices equal to fractional numbers lower
than zero (e.g. Langbein and Johnson 1997). In this
research, noise analysis was performed with Maximum Likelihood Estimation, which was previously
applied in numerous studies describing noise character of GNSS position time series [i.e. Williams
et al. (2004), Teferle et al. (2008), Bos et al. (2010) or
Klos et al. (2016)]. These researches showed that the
noise of GNSS residuals has a character of power-law
process with spectral indices varying between -2
(random walk) and 0 (white noise), which are mainly
near to -1 (flicker noise). We assumed two different
noise models to describe the CME estimates from
GNSS "unfiltered" and "filtered" residuals, meaning
a combination of power-law and white noise model
and autoregressive process. The details of this analysis are described in next paragraphs.

Figure 9 presents the variance of CME, which is
identified with 1st PC. CME estimated for "unfiltered" GNSS residuals is characterized by the
variances between 0.26 and 1.04 mm^2 for horizontal
components and between 0.94 and 17.06 mm^2 for
vertical component. The variances of CME were
reduced to 0.19–0.87 mm^2 (20% of average reduction) in the horizontal directions and to
0.64–6.87 mm^2 (65%) for vertical direction. A
change in CME variances arises from the fact that the
environmental loading models remove much of CME
variance (Fig. 10), especially with a frequency band
between 9 and 12 cpy (cycles per year) mainly
affected, which was also noticed before by
Gruszczynska et al. (2018). The above described
results are consistent with the assertion that GNSS
residuals are highly affected by environmental mass
loading influences, mostly in the vertical direction.

Within this noise analysis we found that the character of CME is very close to a pure flicker noise for
horizontal components, however, it has a character of
autoregressive process of first order for vertical component (Fig. 10). Spectral indices we delivered using

MLE analysis computed for CMEs of unfiltered GNSS
residuals were equal to − 1.21 and − 1.16 for North
and East components, respectively. The contribution of
power-law noise was equal to 100.00%, meaning that
there is no white noise stored in CME series for horizontal components. Having removed the
environmental loadings, spectral indices were equal to
− 0.99 and − 0.93 for North and East components,
respectively. CME series estimated for Up component
are clearly affected by pure autoregressive process of
first order (AR(1)), which is flat for low frequencies
and stepped when moving to shorter periods. This may
indicate that CME is affected by large-scale atmospheric phenomena which also have a character of
autoregressive processes (Matyasovszky 2012).
Moreover, due to the fact that following stations:
BOR1 (Borowa Gora, Poland), GOPE (Pecny, Czech
Republic), GRAZ (Graz, Austria), JOZE (Jozefoslaw,
Poland), LAMA (Lamkowko, Poland), ONSA (Onsala, Sweden), POTS (Potsdam, Germany), PTBB
(Brunswick, Germany), WROC (Wroclaw, Poland),
WSRT (Westerbock, Netherlands), WTZA, WTZR,
WTZZ (all three in Bad Koetizng, Germany), situated
in Central Europe, contributed the most to CME estimates, which was described as the percentage response
to 1st PC, the CME we estimated reflects mainly the
character of residuals of these stations. A large cut off
between 3 and 14 cpy in a power of CME was noticed
for CME in Up direction when loading models were
removed from series, which causes the CME to
resemble a power-law noise. This may indicate that
CME in the vertical direction contains environmental
effects which affect stations located close to each
other.

6. Analysis of GNSS Position Residuals

The GNSS residuals with CME subtracted, which
we refer to as "filtered" residuals, were analyzed in
terms of the Lin's concordance coefficients between
individual pairs of stations in the network under
analysis. In addition, the parameters of noise derived
from MLE and the uncertainties of velocities estimated for the North, East and Up components using
Hector software (Bos et al. 2013) were considered.
The values presented in Fig. 11 confirm the benefits

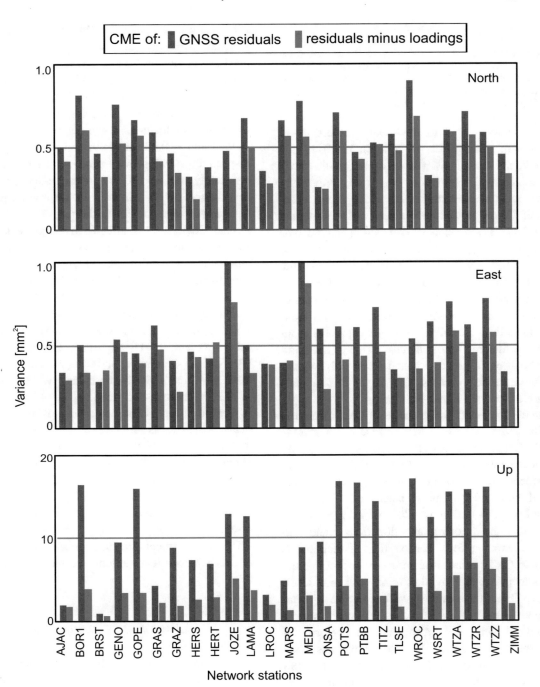

Figure 9
Variance of a CME derived by pPCA filtering method for residuals of 25 stations analyzed in this research. Red bars represent CME estimated for unfiltered GNSS residuals, whilst blue bars show CME of residuals when the superposition of environmental loading models is removed. Note different scales in particular charts

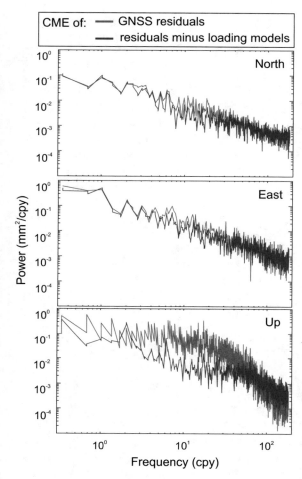

Figure 10
Stacked power spectral densities (PSDs) estimated for CME computed for 25 stations (in red) with a Welch periodogram (Welch 1967), as well as stacked PSDs of CME after environmental loading models were subtracted (in blue)

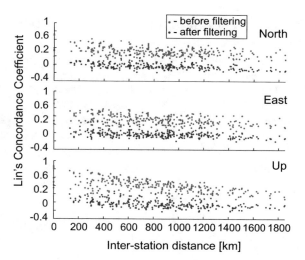

Figure 11
The Lin's concordance correlation coefficient between each pair of 25 stations computed for "unfiltered" (red) and "filtered" (blue) residuals of the position time series is shown as a function of the distance between stations

of spatio-temporal filtering. We showed that pPCA filtering reduces the absolute values of the concordance coefficient on average by 66, 67 and 67% for the North, East and Up components, respectively. The relative reduction is similar for stations in close proximity to each other as well as for stations situated far away, which means that the CME was efficiently reconstructed from 1st PC estimated by the pPCA method.

In the next stage we estimated the variance of residuals, the character of noise and the Bayesian Information Criterion (BIC, Schwarz 1978) which helped to assess the appropriateness of the employed noise model for all stations in the network (Fig. 12).

The changes in noise characteristics for "filtered" and "unfiltered" residuals allowed us to recognize the effect of spatio-temporal filtering on GNSS-derived position residuals.

First, we decided on the preferred noise model to be employed for any individual station. We examined the PSDs of "unfiltered" and "filtered" residuals and found that "unfiltered" residuals in Up direction for stations, BOR1 (Borowa Gora, Poland), GOPE (Pecny, Czech Republic), GRAZ (Graz, Austria), JOZE (Jozefoslaw, Poland), LAMA (Lamkowko, Poland), ONSA (Onsala, Sweden), POTS (Potsdam, Germany), PTBB (Brunswick, Germany), WROC (Wroclaw, Poland), WSRT (Westerbock, Netherlands), WTZA, WTZR, WTZZ (all three in Bad Koetizng, Germany) situated in Central Europe, are affected by pure autoregressive noise model (please see Figures in Supplementary Materials S1). However, when CME is removed from these vertical time series, "filtered" residuals are characterized by pure power-law noise, meaning that we remove the effect that, probably, the atmosphere has on vertical component. On the other hand, we need to be aware of the fact, that we also slightly change the character of other series, which were not affected by atmosphere as much as Central European stations. This makes that, what is noticed from PSDs, these "filtered"

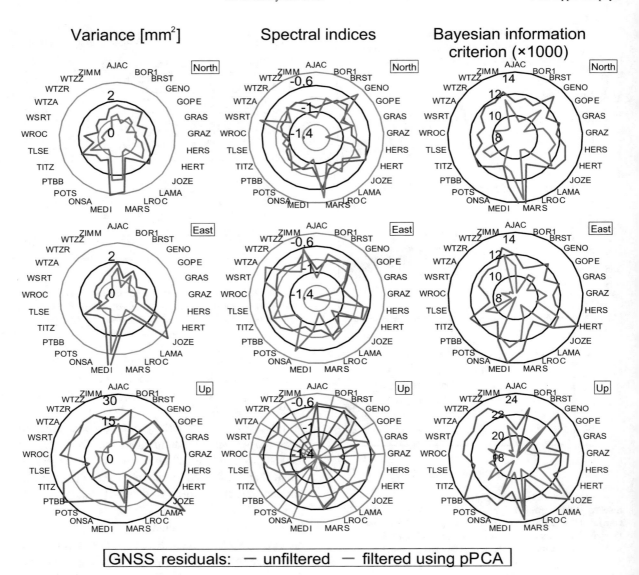

Figure 12
The GNSS residual variances and parameters of noise: spectral indices and BICs estimated for residuals computed before (red) and after (blue) pPCA filtering

residuals are much more affected by white noise than "unfiltered" residuals were, meaning that white noise contributes now more into white plus power-law noise combination.

The variance of "unfiltered" residuals ranged between 1 and 4 mm² for the North and East components, whilst it was significantly higher for the Up component and ranged between 10 and 38 mm². Spectral indices of power-law noise vary between

− 0.6 and − 1.0 for the North and East components and between − 0.6 and − 1.4 for Up component, keeping in mind that for Central European stations, the spectral indices are slightly underestimated because of the portion of AR(1) noise model in residuals. Having filtered the residuals by pPCA, we observed a significant reduction in the variances of between 10 and 74% for all stations with a median decrease estimated at 36, 37 and 46% for the North,

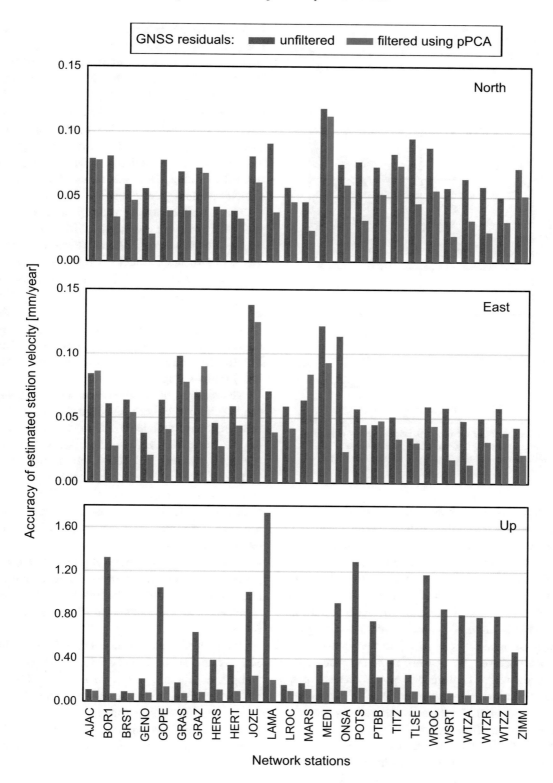

◀Figure 13
Uncertainties in GNSS station velocity determined before spatio-
temporal filtering (red bars) and after CME filtering using pPCA
(blue bars). Note the different scale on the Up component graph

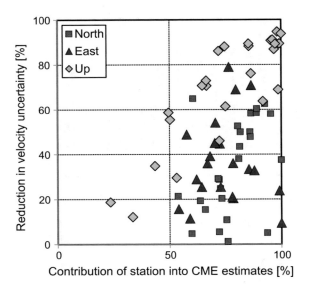

Figure 14
The contribution of each individual station to the CME estimates
(normalized eigenvector elements) presented versus reduction in
velocity uncertainty [%] when CME was removed. 100% of
reduction means that accuracy was reduced to zero

Figure 13 presents the uncertainties of velocity computed for the "unfiltered" incomplete time series from 25 stations and the uncertainties estimated for the "filtered" ones using the pPCA method. The filtering of the CME leads to a more reliable determination of the GNSS station velocity, especially in the case of the Up component. Prior to filtering, the velocity uncertainties were higher than 1 mm/yr for a few stations, while from Fig. 13 we noticed that after CME was removed by pPCA, the velocity is more precise than 0.2 mm/year.

The largest change of velocity uncertainty equal to 95 and 94% was estimated for the Up component of two Polish stations: BOR1 (Borowa Gora) and WROC (Wroclaw), which is caused by a change of noise model from AR(1) to pure power-law noise. The smallest changes in velocity uncertainty were estimated for the Up component for the AJAC (Ajaccio, France), BRST (Brest, France), GRAS (Grasse, France), LROC (La Rochelle, France), MARS (Marseille, France) and MEDI (Medicina, Italy) stations (Fig. 13).

The changes of velocity uncertainties are significantly different for each topocentric component. The contributions of individual stations to CME estimated for the North vary between 54% for station ONSA (Onsala, Sweden) to 100% for station WROC (Wroclaw, Poland). These contributions have led to a reduction in velocity uncertainty of 21 and 38%, respectively. However, stations with a reduction larger than 38% were also observed. Station WSRT (Westerbork, Netherlands) is one of them, with a maximum reduction of 65%. Both the parameters of noise and velocity accuracy computed before and after filtering, as well as, the reduction of variance in residuals show the importance of GNSS time series filtering.

We noticed a correlation between the contribution of individual stations to CME for the Up component and the reduction in velocity uncertainty. The higher the contribution, the greater the reduction in velocity uncertainty. Pearson correlation coefficients computed for these two variables amount to 46, 32 and 86% for North, East and Up components, respectively (Fig. 14).

East, and Up components, respectively. A clear improvement in the spectral indices (going towards 0) of the power-law noise was only noticeable for the Up component, but just for stations affected by AR(1) for "unfiltered" residuals, which is reflected by an average shift in the spectral indices towards white noise from − 0.98 to − 0.68 (improvement of almost 30%). This is mainly because a shift between preferred noise models from AR(1) to pure power-law noise was observed. We also estimated the changes in BIC values which confirm the appropriateness of a model to be fitted into certain residuals. We found an improvement in BIC values for all stations and all components after filtration.

Having filtered the CME values, we estimated the uncertainties of the station velocity of the GNSS position time series using the preferred noise model (PL + WN or AR(1) + WN) for each of them.

7. Discussion and Conclusions

The future of GNSS positioning augmented by continuous measurements provided by permanent stations, will lead to the installation of stations in many new places. Each inequality of operation time span in relation to spatio-temporal analysis should be considered as missing value. This results in the necessity of finding an appropriate method to perform spatio-temporal filtering with no need to limit the series for the same length or to interpolate the gaps. In this study, we proposed probabilistic PCA-based filtering method for the GNSS time series highly affected by missing values or for a situation where stations started and ended operation at different times. We compared the newly applied method with those widely used hitherto: iPCA and mPCA. Moreover, we proved that pPCA gives comparable results but due to its flexible probabilistic model it exceeds in performance both methods, especially in those cases where time series are not characterized by common observational epochs. We compared the traditional PCA filtering approach with the newly employed pPCA and found a few benefits. First, the observations do not have to be interpolated, since pPCA is able to retrieve CME from data with gaps treated in this approach as latent values. Second, the time series may start and end in any epoch, and what is more, they do not have to overlap. This benefit may introduce a fresh perspective of the CME values and may work in any type of network, where the stations do not operate at the same time.

Our analysis of the data from the selected ITRF2014 stations lead us to conclude that CME should not be considered as a uniform signal, homogeneous for all stations. We showed instead that the station spatial responses to the CME may deviate from each other in networks that span up to 1800 km. In case of the considered network, the GNSS stations located in central part of Europe (in Poland, Czech Republic and Germany), contributed the most to the common variability of CME with normalized responses of 87%. The remaining stations contributed 74% on average to CME. The explanation for this phenomenon may simply include the response of stations to environmental loading models, as similar patterns in both GNSS residuals and loading models were observed across the Europe.

It is well known that the vertical component of the GNSS position time series is not determined with the same precision as the horizontal (Wang et al. 2012; Ming et al. 2017). This is due to the principles of satellite navigation systems. The loading processes and spatially-correlated errors have a different effect on vertical component. With this in mind, we noticed a larger reduction in velocity uncertainties in the vertical direction, which is also strictly related to the improvement in the noise characteristics of height component. In addition, the correlation coefficient estimated for pairs of stations decreased much more in the vertical than in horizontal direction. This effect was also confirmed by eigenvalues obtained via the pPCA procedure. These can be interpreted as a percentage of residuals variance represented by each consecutive Principal Components. Since only the 1st PC is identified with CME and eigenvalues corresponding to this PC were equal to 36% of the total variance for horizontal and 49% for vertical direction, we may therefore conclude that CME variance is more significant in Up component. As a result of this, pPCA filtering performs very well especially in Up component. This is very important in the context of increasing expectations regarding to high accuracy of station velocities estimated from the GNSS position time series.

Our results considering environmental loading models are similar to those provided by Zhu et al. (2017) who showed that the RMS for CME estimated for the vertical component is reduced by up to 1.5 mm when loading models are removed. We showed that having removed the environmental loading models, the CME variances are reduced from 10.4 to 3.2 mm^2 on average for vertical components.

The spatial pattern that we noticed in the contribution of individual stations to CME estimations is similar to the spatial dependencies in the amplitudes of power-law noise shown by Klos and Bogusz (2017). The lower the spectral index, the higher the contribution of individual stations to CME. It agrees with the common effect of loadings, which was also investigated using the pPCA method on the basis of the superposition of the environmental models. The stations mostly affected by spatially homogeneous environmental effects also contributed the most to CME estimates. Following Jiang et al. (2013), stations situated in Central Europe are much more affected by loadings comparing to other parts of Europe. This causes that the vertical displacement we might expect from loading effects are few times higher for stations situated in Central Europe we employed. This dependence was noticed in a form of CME estimated for vertical component, as it resembled the autoregressive noise. When being compared with CMEs for horizontal components, which are of pure power-law character, we may conclude that this CME strictly reflects the atmospheric effect which Central European stations are affected the most. This behavior was also seen for individual PSDs estimated in this research. The autoregressive noise model is preferred over widely employed power-law character for all Central European stations, meaning, that if they contribute the most into CME estimates, this character will also transfer to CME itself. Having removed the CME from "unfiltered" residuals, a power-law noise model became a preferred one for stations affected by autoregressive noise model up till now. So, in other words, we removed the atmospheric effect, which appears in Central European stations and was enough powerful to be transferred to CME estimates. In its turn this brings us a question if the spatial extent of stations should not be limited to the joint environmental impact which loading effects have on position time series. So far, it was stated, that the networks can be as extent as 2 000 km, but then various spatial filters should be employed to differentiate the spatial response of individual stations. Our finding brings here a new light if the environmental loadings impact should not also be taken into consideration.

Hitherto, improvements in the GNSS position time series have resulted in reduction in the scatter of individual time series. Tian and Shen (2016) found an improvement in the scatter of residual time series of 20.7, 13.2, and 14.4% for the North, East and Up components, respectively. Ming et al. (2017) proved that the reduction in scatter when CME was removed was equal to 6.3% for all directions. We estimated the properties of CME using MLE analysis and demonstrated an improvement in colored noise parameters at almost all stations.

In conclusion, according to our analysis we can confidently state, that the newly applied probabilistic Principal Component Analysis is a powerful and efficient tool for the spatio-temporal filtering of any type of geodetic gapped data and not only for the GNSS observations investigated in this paper, being a good alternative for such algorithms as mPCA, iPCA and classical PCA.

Acknowledgements

This research was financed by the National Science Centre, Poland, grant no. UMO-2017/25/B/ST10/02818 under the leadership of Prof. Janusz Bogusz. GNSS time series were accessed from http://acc.igs.org/reprocess2.html on 2016-05-05. Maps were drawn using the Generic Mapping Tool (GMT) (Wessel et al. 2013). Loading models were accessed from EOST Loading Service, http://loading.u-strasbg.fr/ on 2016-05-05. Algorithms used in this work are partially modified from Matlab functions. Few of them have been recreated on the basis of their description given in the cited papers. This research was partially supported by the PL-Grid Infrastructure, grant name "plgmgruszcz2018a".

References

Altamimi, Z., Rebischung, P., Métivier, L., & Collilieux, X. (2016). ITRF2014: a new release of the International Terrestrial Reference Frame modeling nonlinear station motions. *Journal of Geophysical Research: Solid Earth*. https://doi.org/10.1002/2016JB013098.

Andrei, M., & Malandrino, A. (2003). Comparative coreflood studies for precipitation and adsorption squeeze with PPCA as the scales inhibitor. *Petroleum Science and Technology, 21*(7–8), 1295–1315. https://doi.org/10.1081/LFT-120018174.

Blewitt, G., Kreemer, C., Hammond, W. C., & Goldfarb, J. M. (2013). Terrestrial reference frame NA12 for crustal deformation studies in North America. *Journal of Geodynamics, 72,* 11–24. https://doi.org/10.1016/j.jog.2013.08.004.

Blewitt G., Lavallée D. (2002) Effect of annual signals on geodetic velocity, J. geophys. Res.: Solid Earth, vol. 107 (pg. ETG 9-1-ETG 9-11), https://doi.org/10.1029/2001jb000570.

Bogusz, J., Gruszczynski, M., Figurski, M., & Klos, A. (2015). Spatio-temporal filtering for determination of common mode error in regional GNSS networks. *Open Geosciences.* https://doi.org/10.1515/geo-2015-0021.

Bogusz, J., & Klos, A. (2016). On the significance of periodic signals in noise analysis of GPS station coordinates time series. *GPS Solutions, 20*(4), 655–664. https://doi.org/10.1007/s10291-015-0478-9.

Bos, M. S., Bastos, L., & Fernandes, R. M. S. (2010). The influence of seasonal signals on the estimation of the tectonic motion in short continuous GPS time-series. *Journal of Geodynamics, 49,* 205–209. https://doi.org/10.1016/j.jog.2009.10.005.

Bos, M. S., Fernandes, R. M. S., Williams, S. D. P., & Bastos, L. (2008). (2008): fast error analysis of continuous GPS observations. *J. Geod, 82,* 157–166. https://doi.org/10.1007/s00190-007-0165-x.

Bos, M. S., Fernandes, R. M. S., Williams, S. D. P., & Bastos, L. (2013). Fast error analysis of continuous GNSS observations with missing data. *Journal of Geodesy, 87*(4), 351–360. https://doi.org/10.1007/s00190-012-0605-0.

Cao, Y., Liu, R. M., & Yang, J. (2008). Infrared small target detection using PPCA. *International Journal of Infrared and Millimeter Waves, 29*(4), 385–395. https://doi.org/10.1007/s10762-008-9334-0.

Cerny, C. A., & Kaiser, H. F. (1977). A study of a measure of sampling adequacy for factor-analytic correlation matrices. *Multivariate Behavioral Research, 12*(1), 43–47.

Chen, T., Martin, E., & Montague, G. (2009). Robust probabilistic PCA with missing data and contribution analysis for outlier detection. *Computational Statistics & Data Analysis, 53*(10), 3706–3716. https://doi.org/10.1016/j.csda.2009.03.014.

Dong, D., Fang, P., Bock, Y., Webb, F., Prawirodirdjo, L., Kedar, S., et al. (2006). Spatio-temporal filtering using principal component analysis and Karhunen-Loeve expansion approaches for regional GPS network analysis. *Journal of Geophysical Research, 111,* B03405. https://doi.org/10.1029/2005JB003806.

Dow, J. M., Neilan, R. E., & Rizos, C. (2009). The international GNSS service (IGS) in a changing landscape of Global Navigation Satellite Systems. *Journal of Geodesy, 83*(3–4), 191–198. https://doi.org/10.1007/s00190-008-0300-3. **(IGS Special Issue)**.

Fan, W. T., Bouguila, N. (2009) Generating Video Textures by PPCA and Gaussian Process Dynamical Model. Progress in Pattern Recognition, Image Analysis, Computer Vision, and Applications, Proceedings, Book Series: Lecture Notes in Computer Science. Edited by: Bayro Corrochano, E. and Eklundh, JO, 5856:801–808. https://doi.org/10.1007/978-3-642-10268-4_94.

Fisher, R. A. (1932). Inverse probability and the use of likelihood. *Proceedings of the Cambridge Philosophical Society., 28*(3), 257–261. https://doi.org/10.1017/S0305004100010094.

Gross, R., Beutler, G., Plag, H-P. (2009) Integrated scientific and societal user requirements and functional specifications for the GGOS. In "Global Geodetic Observing System Meeting the Requirements of a Global Society on a Changing Planet in 2020" edited by Hans-Peter Plag and Michael Pearlman, ISBN 978-3-642-02686-7 e-ISBN 978-3-642-02687-4. https://doi.org/10.1007/978-3-642-02687-4, Springer Dordrecht Heidelberg London New York.

Gruszczynska, M., Rosat, S., Klos, A., Gruszczynski, M., & Bogusz, J. (2018). Multichannel singular spectrum analysis in the estimates of common environmental effects affecting GPS observations. *Pure and Applied Geophysics.* https://doi.org/10.1007/s00024-018-1814-0.

Gruszczynski, M., Klos, A., & Bogusz, J. (2016). Orthogonal transformation in extracting of common mode errors from continuous GPS networks. *Acta Geodynamics et Geomaterialia, 13*(3), 291–298. https://doi.org/10.13168/AGG.2016.0011.

He, X., Montillet, J.-P., Fernandes, R., Bos, M., Hua, X., Yu, K., et al. (2017). Review of current GPS methodologies for producing accurate time series and their error sources. *Journal of Geodynamics, 106,* 12–29. https://doi.org/10.1016/j.jog.2017.01.004.

Houseago-Stokes, R. E., & Challenor, P. G. (2004). Using PPCA to estimate EOFs in the presence of missing values. *Journal of Atmospheric and Oceanic Technology., 21*(9), 1471–1480. https://doi.org/10.1175/1520-0426(2004)021.

Ilin, A., & Raiko, T. (2010). Practical approaches to principal component analysis in the presence of missing values. *The Journal of Machine Learning Research, 11,* 1957–2000.

Ji, K. H., & Herring, T. A. (2011). Transient signal detection using GPS measurements: transient inflation at Akutan volcano, Alaska, during early 2008. *Geophysical Research Letters.* https://doi.org/10.1029/2011GL046904.

Jiang, W., Li, Z., van Dam, T., & Ding, W. (2013). Comparative analysis of different environmental loading methods and their impacts on the GPS height time series. *Journal of Geodesy, 87,* 687–703. https://doi.org/10.1007/s00190-013-0642-3.

Jiang, W. P., & Zhou, X. H. (2015). Effect of the span of Australian GPS coordinate time series in establishing an optimal noise model. *Science China: Earth Sciences, 58,* 523–539. https://doi.org/10.1007/s11430-014-4996-z.

Jolliffe, I. T. (1972). Discarding variables in a principal component analysis. I: artificial data. *Applied Statistics, 21,* 160–173. https://doi.org/10.2307/2346488.

Jolliffe, I. T. (2002). *Principal Component Analysis.* New York: Springer. https://doi.org/10.1007/b98835.

King, M., Altamimi, Z., Boehm, J., Bos, M., Dach, R., Elosegui, P., Fund, F., Hernndez-Pajares, M., Lavallèe, D., Mendes Cerveira, P., Penna, N., Riva, R., Steigenberger, P., van Dam, T., Vittuari, L., Williams, S., Willis, P. (2010) Improved constraints on models of glacial isostatic adjustment: a review of the contribution of ground-based geodetic observations. Surv Geophys 31:465–507. https://doi.org/10.1007/s10712-010-9100-4.

Klos, A., & Bogusz, J. (2017). An evaluation of velocity estimates with a correlated noise: case study of IGS ITRF2014 European stations. *Acta Geodynamics et Geomaterialia, 14*(3), 255–265. https://doi.org/10.13168/AGG.2017.0009.

Klos, A., Bogusz, J., Figurski, M., & Gruszczynski, M. (2016). Error analysis for European IGS stations. *Studia Geophysica et Geodaetica, 60*(1), 17–34. https://doi.org/10.1007/s11200-015-0828-7.

Klos, A., Olivares, G., Teferle, F. N., Hunegnaw, A., & Bogusz, J. (2018). On the combined effect of periodic signals and colored noise on velocity uncertainties. *GPS Solutions, 22,* 1. https://doi.org/10.1007/s10291-017-0674-x.

Kreemer, C., Blewitt, G., & Klein, E. C. (2014). A geodetic plate motion and global strain rate model. *Geochemistry, Geophysics, Geosystems, 15,* 3849–3889. https://doi.org/10.1002/2014GC005407.

Langbein, J., & Bock, Y. (2004). High-rate real-time GPS network at Parkfield: utility for detecting fault slip and seismic displacements. *Geophysical Research Letters, 31,* 15. https://doi.org/10.1029/2003GL019408.

Langbein, J., & Johnson, H. (1997). Correlated errors in geodetic time series: implications for time-dependent deformation. *Journal of Geophysical Research, 102*(B1), 591–603. https://doi.org/10.1029/96JB02945.

Li, W., Shen, Y., & Li, B. (2015). Weighted spatio-temporal filtering using principal component analysis for analyzing regional GNSS position time series. *Acta Geodaetica et Geophysica, 50*(4), 419–436. https://doi.org/10.1007/s40328-015-0100-1.

Lin, L. I.-K. (1989). A concordance correlation coefficient to evaluate reproducibility Biometrics. *International Biometric Society., 45*(1), 255–268. https://doi.org/10.2307/2532051.

Little, R. J. A., & Rubin, D. B. (2002). *Statistical analysis with missing data* (2nd ed.). Hoboken: Wiley. https://doi.org/10.1002/9781119013563.

Lopez-Rubio, E., Ortiz-de-Lazcano-Lobato, J. M., & Lopez-Rodriguez, D. (2009). Probabilistic PCA self-organizing maps. *IEEE Transactions on Neural Networks, 20*(9), 1474–1489. https://doi.org/10.1109/TNN.2009.2025888.

Madikeri, S. R. (2014). A fast and scalable hybrid FA/PPCA-based framework for speaker recognition. *Digital Signal Processing, 32,* 137–145. https://doi.org/10.1016/j.dsp.2014.05.012.

Márquez-Azúa, B., & DeMets, C. (2003). Crustal velocity field of Mexico from continuous GPS measurements. 1993 to June 2001: implications for the neotectonics of Mexico. *Journal of Geophysical Research.* https://doi.org/10.1029/2002JB002241.

Matyasovszky, I. (2012). Spectral analysis of unevenly spaced climatological time series. *Theor Appl Climatol, 111*(3–4), 371–378. https://doi.org/10.1007/s00704-012-0669-z.

Métivier, L., Collilieux, X., Lercier, D., Altamimi, Z., & Beauducel, F. (2014). Global coseismic deformations, GNSS time series analysis, and earthquake scaling laws. *J Geophys Res Solid Earth, 119,* 9095–9109. https://doi.org/10.1002/2014JB011280.

Ming, F., Yang, Y., Zeng, A., & Zhao, B. (2017). Spatiotemporal filtering for regional GPS network in China using independent component analysis. *J Geod, 91*(4), 419–440. https://doi.org/10.1007/s00190-016-0973-y.

Moron, V., Oueslati, B., Pohl, B., Rome, S., & Janicot, S. (2016). Trends of mean temperatures and warm extremes in northern tropical Africa (1961–2014) from observed and PPCA-reconstructed time series. *Journal of Geophysical Research-Atmospheres, 121*(10), 5298–5319. https://doi.org/10.1002/2015JD024303.

Mredhula, L., & Dorairangaswamy, M. (2016). An effective filtering technique for image denoising using probabilistic principal component analysis (PPCA). *Journal of Medical Imaging and Health Informatics, 6*(1), 194–203. https://doi.org/10.1166/jmihi.2016.1602.

Nikolaidis, R. (2002) Observation of geodetic and seismic deformation with the Global Positioning System. Ph.D. thesis. Univ. of Calif., San Diego.

Qi, C. K., Li, H. X., Li, S. Y., Zhao, X. C., & Gao, F. (2012). Probabilistic PCA-based spatiotemporal multimodeling for nonlinear distributed parameter processes. *Industrial and Engineering Chemistry Research, 51*(19), 6811–6822. https://doi.org/10.1021/ie202613t.

Qu, L., Li, L., Zhang, Y., & Hu, J. M. (2009). PPCA-based missing data imputation for traffic flow volume: a systematical approach. *IEEE Transactions on Intelligent Transportation Systems, 10*(3), 512–522. https://doi.org/10.1109/TITS.2009.2026312.

Rebischung, P., Altamimi, Z., Ray, J., & Garayt, B. (2016). The IGS contribution to ITRF2014. *J Geod, 90*(7), 611–630. https://doi.org/10.1007/s00190-016-0897-6.

Roweis, S. (1997). EM algorithms for PCA and SPCA. *Advances in Neutral Information Processing Systems, 10,* 626–632.

Santamaría-Gómez, A., & Mémin, A. (2015). Geodetic secular velocity errors due to interannual surface loading deformation. *Geophysical Journal International.* https://doi.org/10.1093/gji/ggv190.

Schwarz, G. E. (1978). Estimating the dimension of a model. *Annals of Statistics, 6*(2), 461–464. https://doi.org/10.1214/aos/1176344136.

Serpelloni, E., Faccenna, C., Spada, G., Dong, D., & Williams, S. D. P. (2013). Vertical GPS ground motion rates in the Euro-Mediterranean region: new evidence of velocity gradients at different spatial scales along the Nubia-Eurasia plate boundary. *Journal of Geophysical Research: Solid Earth, 118,* 6003–6024. https://doi.org/10.1002/2013JB010102.

Sharifi, R., & Langari, R. (2017). Nonlinear sensor fault diagnosis using mixture of probabilistic PCA models. *Mechanical Systems and Signal Processing, 85,* 638–650. https://doi.org/10.1016/j.ymssp.2016.08.028.

Shen, Y., Li, W., Xu, G., & Li, B. (2013). Spatio-temporal filtering of regional GNSS network's position time series with missing data using principle component analysis. *Journal of Geodesy, 88,* 351–360. https://doi.org/10.1007/s00190-013-0663-y.

Teferle, F. N., Williams, S. D. P., Kierulf, H., Bingley, R., & Plag, H.-P. (2008). A continuous GPS coordinate time series analysis strategy for high-accuracy vertical land movements. *Physics and Chemistry of the Earth, 33,* 205–216. https://doi.org/10.1016/j.pce.2006.11.002.

Tiampo, K. F., Rundle, J. B., Klein, W., Ben-Zion, Y., & McGinnis, S. (2004). Using eigenpattern analysis to constrain seasonal signals in southern California. *Pure and Applied Geophysics, 161,* 1991–2003. https://doi.org/10.1007/978-3-0348-7873-9_13.

Tian, Y., & Shen, Z. (2011). Correlation weighted stacking filtering of common-mode component in GPS observation network. *Acta Seismologica Sinica, 33*(2), 198–208. https://doi.org/10.3969/j.issn.0253-37822011.02.007.

Tian, Y., & Shen, Z.-K. (2016). Extracting the regional common-mode component of GPS station position time series from dense continuous network. *Journal of Geophysical Research: Solid Earth, 121*(2), 1080–1096. https://doi.org/10.1002/2015JB012253.

Tipping, M. E., & Bishop, C. M. (1999). Probabilistic principal component analysis. *Journal of the Royal Statistical Society, 61B,* 611–622.

Wang, W., Zhao, B., Wang, Q., & Yang, S. (2012). Noise analysis of continuous GPS coordinate time series for CMONOC. *Advances in Space Research, 49,* 943–956. https://doi.org/10.1016/j.asr.2011.11.032.

Wdowinski, S., Bock, Y., Zhang, J., Fang, P., & Genrich, J. (1997). Southern California permanent GPS geodetic array: spatial filtering of daily positions for estimating coseismic and postseismic displacements induced by the 1992 Landers earthquake. *Journal of Geophysical Research, 102*(B8), 18057–18070. https://doi.org/10.1029/97JB01378.

Welch, P. D. (1967). The use of fast fourier transform for the estimation of power spectra: a method based on time averaging over short, modified periodograms. *IEEE Transactions on Audio Electroacoustics, 15*(2), 70–73.

Wessel, P., Smith, W. H. F., Scharroo, R., Luis, J. F., & Wobbe, F. (2013). Generic mapping tools: improved version released. *Eos, Transactions AGU, 94,* 409–410.

Williams, S. D. P., Bock, Y., Fang, P., Jamason, P., Nikolaidis, R., Prawirodirdjo, L., et al. (2004). Error analysis of continuous GPS position time series. *Journal of Geophysical Research.* https://doi.org/10.1029/2003JB002741.

Xiang, Z.-Y., Cao, T.-Y., Zhang, P., Zhu, T., & Pan, J.-F. (2012). Object tracking using probabilistic principal component analysis based on particle filtering framework. *Advanced Materials Research, 341–342,* 790–797. https://doi.org/10.4028/www.scientific.net/AMR.341-342.790.

Xiang, J., Zhong, Y., & Gao, H. (2015). Rolling element bearing fault detection using PPCA and spectral kurtosis. *Measurement, 751,* 180–191. https://doi.org/10.1016/j.measurement.2015.07.045.

Yuan, L., Ding, X., Chen, W., Guo, Z., Chen, S., Hong, B., et al. (2008). Characteristics of daily position time series from the Hong Kong GPS fiducial network. *Chin J Geophys, 51*(5), 1372–1384. https://doi.org/10.1002/cjg2.1292.

Zhu, Z., Zhou, X., Deng, L., Wang, K., & Zhou, B. (2017). Quantitative analysis of geophysical sources of common mode component in CMONOC GPS coordinate time series. *Advances in Space Research.* https://doi.org/10.1016/j.asr.2017.05.002.

Zuccolotto, P. (2012). Principal component analysis with interval imputed missing values. *AStA Advances in Statistical Analysis, 96*(1), 1–23. https://doi.org/10.1007/s10182-011-0164-3.

(Received October 20, 2017, revised March 25, 2018, accepted March 28, 2018, Published online April 3, 2018)

Pure Appl. Geophys. 175 (2018), 1869–1888
© 2017 Springer International Publishing AG, part of Springer Nature
https://doi.org/10.1007/s00024-017-1712-x

Pure and Applied Geophysics

CrossMark

Overpressure and Fluid Diffusion Causing Non-hydrological Transient GNSS Displacements

GIULIANA ROSSI,[1] PAOLO FABRIS,[2] and DAVID ZULIANI[2]

Abstract—In this work, global navigation satellite system (GNSS) observations from the northern tip of the Adria microplate are analysed to differentiate non-periodic (transient) tectonic signals from other deviations from the linear trends primarily due to hydrological loading effects. We tested a recently proposed hypothesis that a porosity wave generated by fault-valve mechanisms in a seismogenic fault in the Bovec basin (western Slovenia) propagated throughout the surrounding region. After excluding potential spatially correlated common-mode errors in the considered time series, we investigated the relationship between the GNSS observations and periodic hydrological loading variations. The tests demonstrated that subtracting the hydrological term was effective at the global scale and that the frequency band of the transient signal ($1.5 < T < 3.5$ years) was not correlated with hydrological effects at the local scale (within a few kilometres of the station). Next, the results of previous works are used to calculate the permeability values and pore-pressure state at the source of the transient signal. The permeability values for the four main rock formations in the region are consistent with independent observations for similar lithotypes. The ratio between the effective stress and lithostatic load for different vertical profiles in the Bovec area indicated a state of overpressure, with pore-pressure close to the value of the lithostatic load. Thus, our results help define a scenario in which the porosity wave could have originated. Indeed, the formation of the domains of interconnected fractures, such as during the formation of a porosity wave, increases the permeability values, thereby relieving overpressure and restoring a state of equilibrium.

Key words: Transient signal in cGNSS, hydrological load, porosity wave, rock permeability, effective stress.

1. Introduction

The last 20 years have witnessed a rapid increase in the number of continuously operating global navigation satellite system (cGNSS) networks to monitor crustal deformations in tectonically active regions or provide a reference for surveying. Time series of more than decadal length enable us to detect non-periodic (transient) displacements resulting from volcanic, hydrological, or tectonic processes in various global regions (e.g., Feng and Newman 2009; Chamoli et al. 2014; Devoti et al. 2015; Silverii et al. 2016). Recently, Rossi et al. (2016, 2017) reported significant deviations from regional linear trends in cGNSS time series in a small area along the northern edge of the Adria microplate (across NE Italy and W Slovenia, see Fig. 1). In particular, these authors focused on a transient displacement with duration of roughly two and half years, as recorded by 13 cGNSS stations. The transient signal propagated through the region with an approximately circular/elliptical pattern, thereby spurring a movement that was initially oriented upward (with the exclusion of a site to the east of the area) and subsequently oriented downward. The horizontal component showed similar behaviour, with a smaller oscillation that was parallel to the main tectonic faults and fractures (Fig. 1b). A tomographic inversion of the arrival travel time of the transient signal supplied a laterally and vertically varying propagation velocity field and revealed the source location and origin time. Rossi et al. (2016, 2017) used the 3D tomographic algorithm described in detail by Böhm et al. (1999) and Vesnaver and Böhm (2000). The disturbance appeared to have originated approximately 3.5 months prior to the $M_w = 5.2$ event that occurred near Bovec (Slovenia) in 2004 and was located 6.5 km NW of the main shock epicentre along the continuation of the recognized seismogenic Ravne fault at a depth of 9.2 km. The authors hypothesized that this transient displacement process could have been attributed to deep fluids that were mobilized by valve behaviours of the Ravne fault and

[1] Istituto Nazionale di Oceanografia e di Geofisica Sperimentale - OGS, Centro di Ricerche Sismologiche, Borgo Grotta Gigante 42/c, Sgonico (Trieste), Italy. E-mail: grossi@inogs.it
[2] Istituto Nazionale di Oceanografia e di Geofisica Sperimentale - OGS, Centro di Ricerche Sismologiche, via Treviso 55, Udine, Italy.

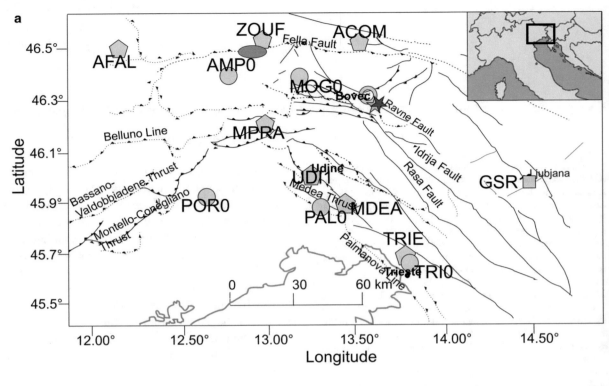

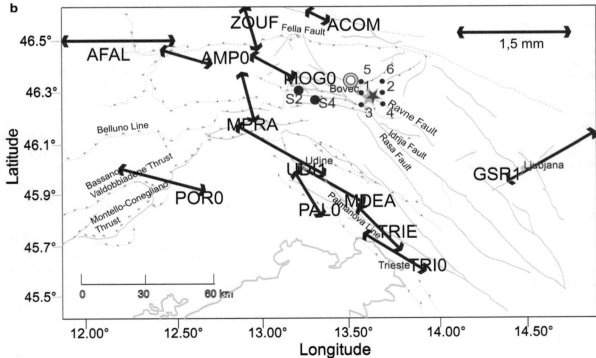

◀Figure 1

a Map of the region and its main tectonic lineaments. Solid lines: strike-slip faults; hachured lines: thrusts. Active features are shown in bold (redrawn from Moratto et al. 2012). The inset shows the location of the study region. Polygon: FReDNet cGNSS station (OGS); filled circle: Marussi cGNSS network station (Regional Council of the Autonomous Region Friuli Venezia Giulia); and square: GSR1 EUREF cGNSS station (Ljubljana, Slovenia). The dark grey ellipse marks the Zoncolan meteorological station (OSMER-FVG). The two stars show the locations of the main shocks from the 1998 (pale grey) and 2004 Bovec-Krn (lead grey) events, whereas the open circle marks the origin of the porosity wave (see Rossi et al. 2016). **b** Same map as in **a**; the arrows correspond to the cGNSS sites and indicate the horizontal directions along which the transient signal reached the maximum amplitude (see Rossi et al. 2016 for details on the procedure). The black circles refer to synthetic profiles for the porosity and density values and the lithostatic load (see text, Fig. 9). The two large circles (S2 and S4) denote the synthetic profiles of Faccenda et al. (2007), whereas the small circles (1–6) are profiles drawn from the model by Bressan et al. (2012)

that this transient displacement dispersed through the region in the form of porosity waves. Porosity waves are packets of fluid-filled interconnected cracks that self-propagate upward and in the direction of minimum horizontal stress following the pressure gradient and permeability variations to achieve more efficient transport (Wiggins and Spiegelman 1995; Connolly and Podladchikov 1998, 2013; Revil and Cathles 2002). Notably, porosity waves have recently been recognized as efficient media for the mobilization of fluids under different conditions. Skarbek and Rempel (2016) proposed that dehydration-induced porosity waves generated at subduction interfaces might act as a mechanism underlying the episodic tremor and slip (ETS) phenomenon. Joshi and Appold (2017) applied numerical modelling to verify the capacities for such waves to travel sufficiently fast to create the aseismic slip that was observed in the Nankai décollement, and their modelling revealed that the broad range of propagation velocities of such waves depends on the effective stress, hydraulic diffusivity, and permeability values. The magnitudes of the propagation velocities found by Rossi et al. (2016) are compatible with other observed porosity waves at both large and laboratory scales (Natale and Salusti 1996; Vardoulakis et al. 1998; Lupi et al. 2011). Moreover, Rossi et al. (2016, 2017) supported their model through a hydraulic tomographic inversion (Brauchler et al. 2013) of arrival travel times of the transient signal. The hydraulic diffusivity values they obtained

are compatible with rock formations. As a further validation of the hypothesis, the propagation velocity, and diffusivity values were used to calculate the initial effective stress (Revil and Cathles 2002), and similar values to those at same depths of brittle failure equilibrium were observed (Rossi et al. 2017).

The interpretation that tectonic processes cause transient signals follows from the removal of other potential effects that act at annual or multiyear scales, such as thermoelastic effects (e.g., Dong et al. 2002; Tsai 2011) or hydrological loading effects. The latter are predominant, and geodetic data are used to evaluate groundwater time variations (e.g., Rajner and Liwosz 2011; Argus et al. 2014; Fu et al. 2014; Devoti et al. 2015). Many authors have been confronted with the problem of removing hydrological loading effects that may occur in GNSS recordings over a broad frequency range (Blewitt and Lavallée 2002; Bennet 2008), and two approaches are usually adopted: parametric (Chamoli et al. 2014) and non-parametric method (Chanard et al. 2014). Zou et al. (2014) presented a comparison of these two approaches. Rossi et al. (2016) considered seasonal semi-annual and annual effects alongside possible multi-year fluctuations, which are essential to identify small transient signals, by applying the hybrid parametric approach by Chamoli et al. (2014). This method involves the subtraction of seasonal terms to identify hydrological loading variations at different time scales, and the scale factors of annual and semi-annual cyclic oscillations and other hydrological load signals are then estimated from GNSS data observations via the least-squares method. Both Chamoli et al. (2014) and Rossi et al. (2016) used the Global Land Data Assimilation System (GLDAS) model, which incorporates satellite- and ground-based observations of the land-surface temperature, humidity, precipitation, radiation, wind, and pressure into forcing parameters to constrain the modelled land surface states at global 0.25° and 1° grid squares (Rodell et al. 2004). A small limitation of this model is that it does not incorporate groundwater variations or oceanic loadings. In fact, changes in groundwater storage levels in aquifers and aquitards generate poroelastic deformations. Aquifer depletion induces subsidence, whereas recharging induces pore reopening and uplift (Wahr et al. 1998; Argus et al.

2014). Moreover, groundwater variations in complicated conduit systems can also produce recordable effects in regions dominated by carbonates at the surface and at depth, where karstic phenomena occur (e.g., Devoti et al. 2015; Silverii et al. 2016).

In contrast, the satellite gravimetry of the Gravity Recovery and Climate Experiment (GRACE) considers the effects of surface water and groundwater over continental areas, thereby constraining total water-storage levels (Tapley et al. 2004; Wahr et al. 2004; Argus et al. 2014). Therefore, Rossi et al. (2016) also used GRACE (equivalent water height) data to estimate the hydrological loading effects on the vertical component. As a drawback, GRACE data have a temporal resolution of 30 days in time and a spatial resolution of 1°; moreover, the effective spatial resolution is limited to a few hundred kilometres because of the truncation of spherical harmonics (Landerer and Swenson 2012). Rossi et al. (2016) used global models because the sensitivity of GNSS data observations to hydrological loading effects is reduced at local (1 km around the instrument) and regional (1–10 km) scales (Agnew 2001; Llubes et al. 2004; Longuevergne et al. 2009) relative to the sensitivity at global scales (Dill and Dobslaw 2013).

However, Argus et al. (2014) and Fu et al. (2014) showed that global positioning system (GPS) tools could integrate GRACE information with groundwater changes at smaller scales. Moreover, Devoti et al. (2015) found notable deformation caused by rainfall in a karstic plateau for the same region analysed by Rossi et al. (2016, 2017), recorded by GPS stations and mainly visible in the horizontal component recordings.

In this study, we analysed aspects of the research by Rossi et al. (2016) in more depth. First, we investigated the possibility that transient signals may be related to other effects, including spatially correlated common mode errors and hydrological impact at a more local scale (within a few kilometres of the station) (Sect. 3). Second, we expanded on results from the previous work to verify the reliability of the derived quantities. We calculated the permeability and state of the pore pressure at the source of the transient signal to further validate our hypothesis of fluid propagation from seismogenic depth (Sect. 4).

2. Regional Tectonics and Data

The northern tip of the Adria microplate is characterized by a complex tectonic structure, which was created by the superposition of different orogenic phases and two main tectonic systems. The first is the Alpine tectonic system, which is characterized by E–W-trending thrusts, and the second is the Dinaric tectonic system, which consists of NW–SE-trending transpressive structures. Figure 1 shows the main lineaments of the area studied in this work. A succession of carbonatic platforms alternates with terrigenous flysch formations, and this sequence is associated with repeated thrusting phases, thereby constituting the geologic frame of the study region. The complexity of the tectonic area is reflected by the spatial distribution of the seismicity trends, which follow major lithologic and rheological changes (Bressan et al. 2016). The local seismic activity is moderate, culminating in the last forty years in two events with magnitudes more than 6.0 in 1976 in Friuli (NE Italy), and in two earthquakes with magnitudes more than 5.0 in 1998 and 2004 in the Bovec-Krn area (W Slovenia). Before the availability of GPS measurements, a network of tiltmeters and strainmeters has been active for decades in the Friuli area. This network revealed the presence of two quasi-periodic signals with durations of 33 and 9 years that acted perpendicularly to the Alpine and Dinaric tectonic systems, respectively (Rossi and Zadro 1996; Braitenberg and Zadro 1999). Moreover, the horizontal pendulums of Grotta Gigante (Braitenberg and Zadro 1999; Braitenberg et al. 2006) recorded repeating tremors in the 3 years preceding the Friuli 1976 earthquake (Chiaruttini and Zadro 1976), and these events were interpreted a posteriori as slow earthquakes (Bonafede et al. 1983). Currently, the Istituto Nazionale di Oceanografia e Geofisica Sperimentale-OGS maintains the Friuli Regional Deformation Network (FReDNet), which consists of 16 GNSS permanent sites across NE Italy, to continuously monitor the region (Battaglia et al. 2003; D'Agostino et al. 2005; Zuliani et al. 2017). The Marussi cGNSS network of the Friuli Venezia Giulia Regional Council also maintains ten permanent monitoring sites. Records have revealed a strain rate of 3 mm/year in the region, thus confirming the

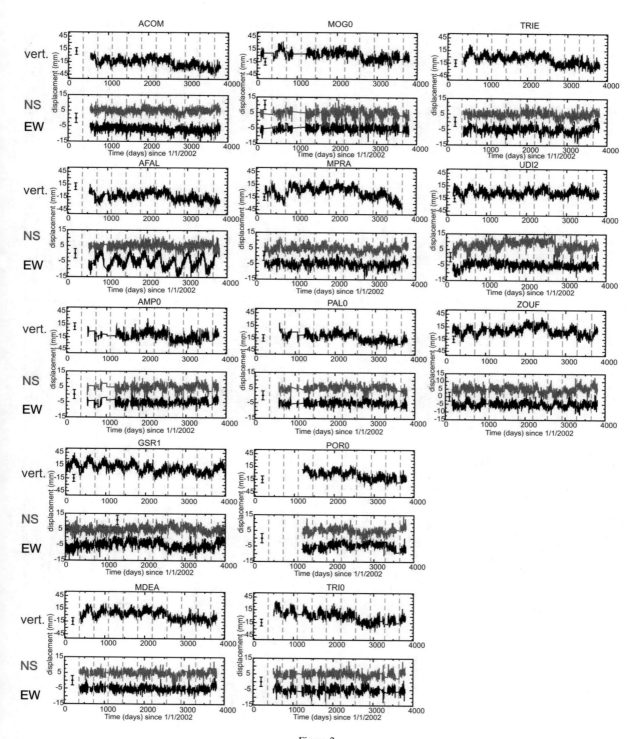

Figure 2

De-trended GNSS time series. The vertical component (upwards positive) and two horizontal components (EW in black, eastwards positive; NS in red, northwards positive) are marked for each site. The average error bar is shown in the plots. Dashed brown vertical lines mark the end of each year

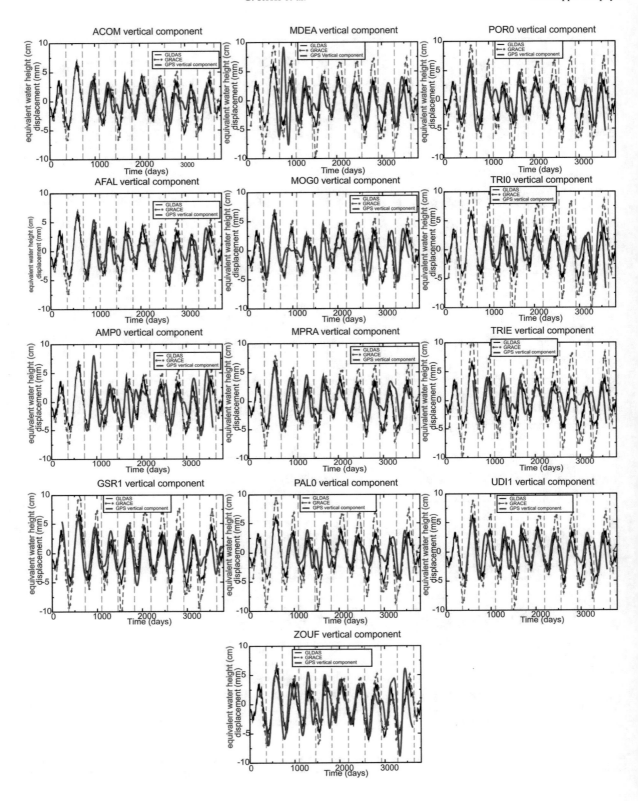

◀Figure 3
Filtered vertical GPS signal (blue line) for each site compared to the displacements from hydrological loading according to the GLDAS (black line) and GRACE equivalent water heights (red dashed line with circles). The sign of the GRACE signal is inverted. Dashed brown vertical lines mark the end of each year

occurrence of NS shortening (D'Agostino et al. 2005; Weber et al. 2010; Devoti et al. 2011).

High rainfall levels characterize this region from a hydrological perspective. The lowest levels are recorded along the coast (roughly 1000 mm/year), with the levels gradually increasing to a maximum in the Prealps region (2400–3100 mm/year) and decreasing across the Alps (1500–1700 mm/year). The primary period of precipitation runs from September to December, with a secondary peak occurring in March, while the driest month of the year is July. Groundwater forms a composite aquifer system that includes an unconfined aquifer in the upper plain and a multi-layered confined aquifer that reaches 500 m of thickness in the lower plain. The two aquifers are separated by a resurgence belt (Ceschia et al. 1991; Zini et al. 2013). Karstic phenomena, such as those in outcropping formations, are also present in the subsurface formations (Cimolino et al. 2010), implying the presence of complex groundwater flow in the rock matrix, fractures, and conduits (Devoti et al. 2015).

3. Data and Hydrological Load Analysis

The here analysed GNSS dataset included the longest time series (7–12 years) from the FReDNet and Marussi networks and data from the EUREF permanent GSR1 station (Ljubljana, Slovenia) (Fig. 1a). The data processing and test methods used to evaluate the stability of the solution were described in Rossi et al. (2016). In this study, we attempted to improve the signal-to-noise ratio in the GPS time series and verify whether the transient signal was common outside the study region. We filtered out possible spatially correlated common mode errors in the GPS time series (e.g., Serpelloni et al. 2013) using the "stacking" approach by Wdowinski et al.

(1997). We also considered eight additional EUREF sites outside the region: UPAD, WTZR, GRAZ, MEDI, PRAT, PFAN, ZIMM, and GENO. Figure 2 shows the time series of the three components for all the considered sites after this correction. The correction reduced the residual RMS by approximately 12% and more accurately determined both transient signals with a 2.5-years duration and the seasonal terms. The latter are evident from the vertical component and one or two of the horizontal components (see the ACOM, AFAL, and GSR1 EW components and the GSR1, MOG0, and UDI2 NS components). Figure 3 shows the filtered GPS vertical component observations ($0.5 < T < 1$ year) (blue curves). The main positive peak in the observations occurs during the summer for all the stations (between July and August), i.e., the driest period in the region. Therefore, this region elastically responds to loadings of snow and water (Argus et al. 2014). As previously noted by Chamoli et al. (2014), GPS signals also include a semi-annual component that is likely induced by secondary rainfall peak during spring, which modulates the annual term and changes the amplitude of the signal over time. The parametric approach of Rossi et al. (2016) takes into account this discrepancy. Rossi et al. (2016) estimated the hydrological loading effects by using both GLDAS- and GRACE-based approaches. Figure 3 also presents the vertical displacement from the modelled hydrological load via the GLDAS model and the GRACE equivalent water height for all the analysed GNSS sites (after inverting the sign of the GRACE signal). Notably, the two signals are consistent with the GPS observations and with each other regardless of the coarser sampling of the GRACE data. Moreover, the signals are fully consistent for sites in the mountainous area (AFAL, ZOUF, ACOM, AMP0, MOG0, and MPRA). A small discrepancy is observed for sites in the alluvial plain (UDI2, MDEA, PAL0, POR0, and GSR1) and the Trieste sites (TRI0 and TRIE). The GPS observations show slightly better agreement with the GLDAS curves. The seasonal term is smaller for the horizontal GNSS components and noisier than the vertical component. Two sites in the mountainous area and three in the alluvial plain show a more significant seasonal term in the horizontal plane. Along the vertical component, the sites in the alluvial plain show greater discrepancy and phase delay with respect to the modelled loading effect via the GLDAS model. Figure 4 shows the same time series as in Fig. 2 after subtracting any hydrological loading effects by using the GLDAS approach. The reduction in seasonal terms was effective, thus revealing the presence of a transient signal with a duration of approximately 2.5 years. Figure 5 shows a zoom of the same time series, limited to the interval June 2006–October 2011. The anomalous signal is evident in the vertical components of all the sites and is recognizable in most of the time series of the horizontal components, matching the trends in Fig. 1b. We performed an additional analysis based on a simplified hydrological balance (precipitation minus evapotranspiration) to exclude the possibility that the transient signal was caused by local groundwater variations (Zerbini et al. 2010). Unfortunately, time-series information on runoff across the entire region is not yet available. Zini et al. (2011, 2013) evaluated surface-runoff levels for a considerable portion of the study region based on the curve number (CN), modified for a long time series (Kannan et al. 2008). The values represented the mean over decadal periods and varied from 40 in karstic areas to 90 in areas where low-permeability rocks (e.g., flysch) crop out. The mean annual resulting runoff in the plain is 216 mm/year.

For our hydrological balance calculations, we used precipitation data to build a cumulative curve and subtracted the linear trend to identify any potential multiyear variations.

Retrieving the simplified hydrological balance requires evapotranspiration estimates from the Thornthwaite (1948) formula:

$$\text{PET} = 16\left(\frac{L}{12}\right)\left(\frac{N}{30}\right)\left(\frac{10T_a}{l}\right)^{\alpha} \quad (1)$$

where

$$l = \sum_{i=1}^{12}\left(\frac{T_{ai}}{5}\right)^{1.514} \quad (2)$$

and

$$\alpha = \left(6,75 \times 10^{-7}\right)l^3 - \left(7,71 \times 10^{-5}\right)l^2 + \left(1,792 \times 10^{-2}\right)l + 0,49239 \quad (3)$$

where PET is the estimated potential evapotranspiration level (mm/month), T_a is the average daily temperature, N is the number of days in a month, L is the average length of a day, and l is the heat index.

We used data drawn from meteorological stations in the regional council networks and selected the closest stations to each of the GNSS stations to evaluate the hydrological loading effects at a local scale. Monthly data from each station were used to build a cumulative curve, and the linear trend was subtracted to determine possible multiyear variations. Figure 6 shows the Zoncolan station, which is located 7 km from our GNSS site ZOUF (grey ellipse in Fig. 1a). Figure 6a shows the data and cumulative curve, whereas Fig. 6b presents the de-trended cumulative curve (black dashed line). The multiyear variation results indicate that an oscillation occurred from 2005 to 2009. In Fig. 6b, the grey dashed curve represents the simplified hydrological balance after subtracting the PET from the de-trended cumulative precipitation curve of the Zoncolan station. The PET values range from 100 to 200 and are consistent with the estimated values from Zini et al. (2013) for the same area. The resulting hydrological balance is approximately 1000 mm/year.

To verify whether the transient signal analysed by Rossi et al. (2016) was linked to the water-balance variations at high frequencies and multiyear scales, we correlated the hydrological balance in Fig. 6b with the vertical component of the ZOUF time series before and after applying a bandpass filter for the transient frequency band $(1.5 < T < 3.5 \text{ years})$ (Table 1). The original data and the data from which the hydrological load is subtracted with the GLDAS model show a negative correlation before filtering, whereas the data when using the GRACE model for the hydrological load subtraction are not correlated with the hydrological balance. The correlation with all three datasets at the $1.5 < T < 3.5$ years frequency band is null, which supports the interpretation that the transient signal was not related to hydrological loading effects. Figure 7 shows the results of the cross-correlation of the hydrological balance with the ZOUF vertical original signal and with the same, after the subtraction of the hydrological load with the GRACE model. The correlation, in fact, is lower than

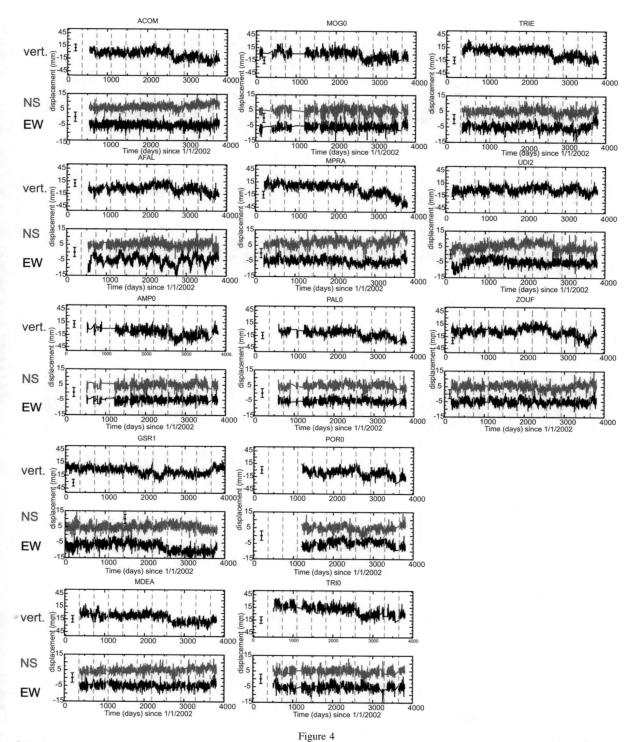

Figure 4

Same time series as in Fig. 2 after subtracting the hydrological terms from the GLDAS model. The vertical component (upwards positive) and two horizontal components (EW in black, eastwards positive; NS in red, northwards positive) are marked for each site. The average error bar is shown in the plots. Dashed brown vertical lines mark the end of each year

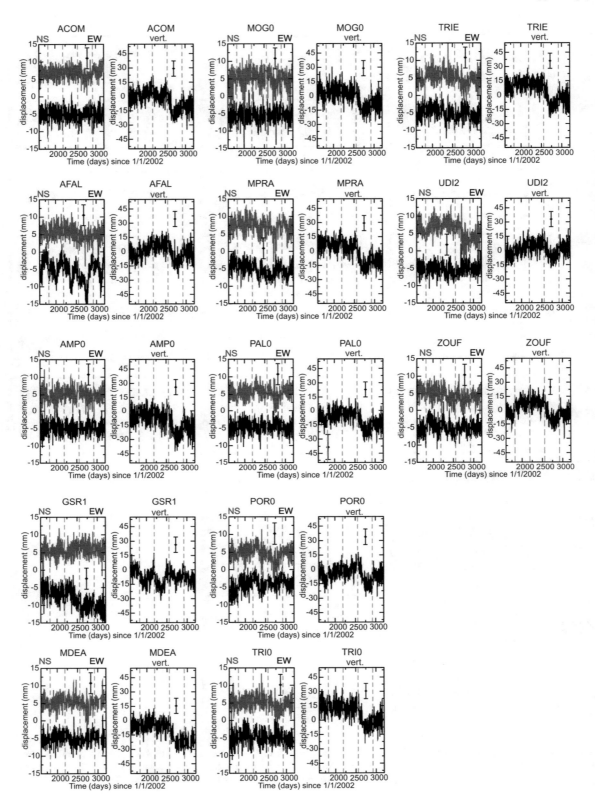

◄Figure 5
Zoom on the time series of Fig. 4 from June 2006 to October 2011, when the transient signal is observed. The diagram shows the horizontal components (EW in black, eastwards positive; NS in red, northwards positive) and vertical component for each site. The average error bar is shown in the plots. Dashed brown vertical lines mark the end of each year

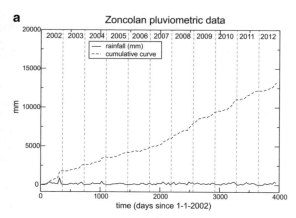

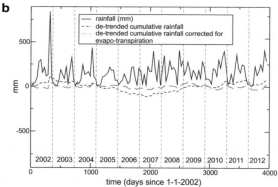

Figure 6
Rainfall and hydrological balance values recorded at the Zoncolan station (dark grey ellipse in Fig. 1a). **a** Rainfall data (solid line) and cumulative curves (dashed line). **b** Rainfall data (solid line) and detrended cumulative rainfall data before (dashed black line) and after correcting for evapotranspiration (dashed grey line). Dashed pale grey vertical lines mark the end of each year

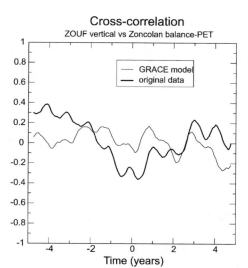

Figure 7
Cross-correlation of the de-trended cumulative rainfall data at Zoncolan station after correcting for evapotranspiration and the ZOUF vertical signal: original data (thick line), after subtracting the hydrological load effects according to GRACE modelling (thin line)

0.4 in the former case and smaller than 0.25 in the second case (Fig. 7), showing that the procedure also eliminated possible delayed effects.

4. Physical Properties and Pressure State

Figure 8 presents a hodogram for the plane vertical/horizontal direction of the maximum transient's amplitude for each site on the map from Fig. 1b. The horizontal component is slightly exaggerated in the graph, to enable the evaluation of the changes from site to site. The similarities among most of the signals are striking. In particular, the horizontal component increases for GSR1, UDIN, PAL0, POR0, and AFAL,

Table 1

Pearson correlation coefficient between simplified hydrological balance (Zoncolan pluviometric station) and the GNSS data (ZOUF vertical GPS data) without (original data) and with hydrological correction (GLDAS model based and GRACE model based) and at different frequency bands (raw data and $1.5 < T < 3.5$ years)

	Original data	GLDAS corrected data	GRACE corrected data
Raw data	− 0.45	− 0.41	− 0.34
$1.5 < T < 3.5$ years	− 0.26	− 0.26	− 0.13

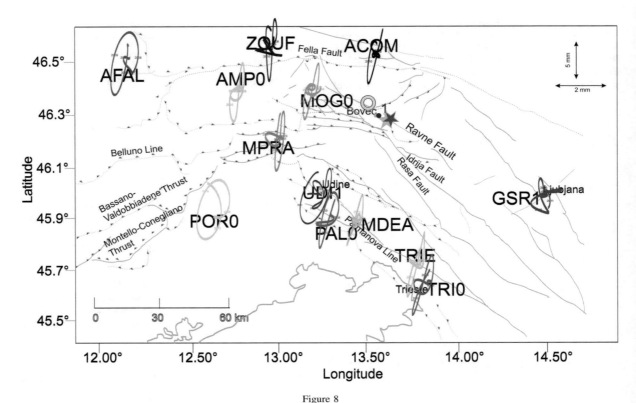

Figure 8
Hodograms in the plane vertical/direction of the maximum transient's amplitude for every station on the same map as shown in Fig. 1b. The vertical and horizontal scale factors are shown in the top right corner of the figure

i.e., at the largest distances from the apparent origin of the transient, whereas the smallest values are observed for ACOM, MOG0, MDEA and TRIE, which are closest to the transient's origin. The pattern in Fig. 8 suggests that the disturbance has originated in an elliptical area centred at the origin of the transient signal and likely had a NW–SE-oriented major axis.

Rossi et al. (2016, 2017) used the transient signal propagation velocity and the hydraulic diffusivity to calculate the effective stress at the origin to support their hypothesis of a porosity wave generated by fault-valve behaviours of the Ravne fault (Sibson 1992). The effective stress was calculated from the laboratory measurement outcomes for the four most representative rock formations in the region.

In this work, we attempted to confirm the proposed model further.

First, we used the hydraulic diffusivity from Rossi et al. (2017) and calculated the permeability k via Eq. (4) (Talwani et al. 1999):

$$k = D_{\mathrm{h}}\eta_f \left[\varphi\beta_f + (1 - \varphi)\beta_{\mathrm{r}} \right] \qquad (4)$$

where η_{f}, ϕ, β_{f}, and β_{r} are the fluid dynamic-viscosity, rock porosity, and fluid and rock compressibility, respectively, whereas D_{h} is the tomographic hydraulic diffusivity. We considered the four principal lithologies in the region for which the density and porosity were available from borehole and laboratory measurements (Venturini 2002; Faccenda et al. 2007). As in Rossi et al. (2016, 2017), we used values from the hydraulic tomographic inversion for locations wherein each lithotype could be assumed to be dominant at the depths that were crossed by the rays. The permeability values are comprised between 5.2×10^{-17} and 2.87×10^{-15} m^2. In particular, terrigenous flysch formations are characterized by a permeability value on the order of 10^{-17} m^2, whereas limestone and dolomitic limestone show much higher permeability ($\sim 7 \times 10^{-16}$ to 3×10^{-15} m^2) (Table 2).

Table 2

Transient-causing effective stress (σ) and permeability (k) calculated for the four main regional rock formations

Lithology	Φ^a	ρ_g^a (kg/m³)	C^b (km/year)	D_h^b(m²/s)	σ_0^b(MPa)	k^c (m²)
Triassic "Dolomia Principale" (dolomitic limestone)	0.051	2860	34.80 ± 0.14	$7.64 \times 10^{-02} \pm 1.22 \times 10^{-03}$	23.4 ± 0.3	$2.87 \times 10^{-15} \pm 4.61 \times 10^{-17}$
Cenozoic flysch	0.032	2750	16.00 ± 0.25	$2.194 \times 10^{-02} \pm 0.5 \times 10^{-03}$	22.41 ± 0.1	$5.22 \times 10^{-17} \pm 1.16 \times 10^{-19}$
Jurassic limestone	0.004	2630	15.20 ± 0.25	$2.70 \times 10^{-03} \pm 0.006 \times 10^{-03}$	22.3 ± 0.1	$6.65 \times 10^{-16} \pm 4.61 \times 10^{-17}$
Paleozoic sandstone	0.011	2700	7.01 ± 0.07	$3.44 \times 10^{-03} \pm 0.005 \times 10^{-03}$	23.2 ± 0.1	$7.60 \times 10^{-17} \pm 1.11 \times 10^{-19}$

Principal regional rock formation lithology; we assumed $\eta_f = 8.90 \times 10^{-04}$ Pa s, $\beta_f = 4.6 \times 10^{-10}$ Pa⁻¹, and $\beta_r = 2 \times 10^{-11}$ Pa⁻¹ (Talwani et al. 1999)

φ porosity at room pressure and temperature, ρ_g bulk density, C transient propagation velocity, D_h hydraulic diffusivity, σ_0 calculated initial effective stress, assuming water density, k permeability

[a]Data from Faccenda et al. (2007)

[b]Data from Rossi et al. (2016; 2017)

[c]From our calculations based on Eq. 4

Rossi et al. (2016, 2017) also determined the original effective stress values by hypothesizing that the saturating fluid was water, obtaining 22–23 MPa (Rossi et al. 2017). According to Terzaghi (1923), the effective vertical stress as a function of the depth z is as follows:

$$\sigma_v = \rho_b gz - P_p = (1 - \lambda)\rho_b gz \quad (5)$$

where g is the gravitational acceleration, P_p is the pore pressure, and ρ_b is the bulk density of sediment, which can be calculated using the density of the grains in the rock ρ_g:

$$\rho_b = \varphi\rho_f + (1 - \varphi)\rho_g \quad (6)$$

The term λ in Eq. (5) is the pore fluid ratio (Terzaghi 1923; Hubbert and Rubey 1959), which is defined as follows:

$$\lambda = \left(P_p - \rho_f gH\right)/\rho_b gz \quad (7)$$

where ρ_f denotes the water density, and H denotes the water-table depth. The hydrostatic pore-pressure conditions are represented by $\lambda = 0.4$, whereas conditions in which the pore-fluid overpressure reaches the value of the lithostatic load P_L ($P_L = \rho_b gz$) are represented by $\lambda = 1$.

To verify the plausibility of the values and subsequent effects on the pore-pressure conditions, we calculated the lithostatic load from the surface to a depth of 9.2 km in the area of the Bovec basin, i.e.,

consistent with the source location of the transient signal (Rossi et al. 2016). We used the density and porosity of the most representative lithological formations of the region (Faccenda et al. 2007) and the density from vertical profiles traced through the 3D model and generated by a seismo-gravimetric inversion (Bressan et al. 2012). Then, we calculated λ from Eq. (7) to evaluate the pore-pressure conditions from the effective stress according to Rossi et al. (2016, 2017).

We calculated the lithostatic load P_L from the surface to the depth of the transient's source (9.2 km) based on six vertical profiles through the nodes of the 3D velocity/density model that was developed by Bressan et al. (2012) for the Bovec area and the 2004 seismic sequence (see Fig. 1b for the location and Fig. 9a for the profiles). Following Bressan et al.'s (2012) assumptions on rock formations, we reconstructed corresponding porosity profiles from the laboratory measurements of Faccenda et al. (2007) (Fig. 9b). We also considered the vertical synthetic profiles S2 and S4 developed by Faccenda et al. (2007), which represent the stratigraphy of the area and were translated into density and porosity profiles from laboratory values for comparison (see Fig. 1b for the location and Fig. 9a and b for the profiles). However, the two synthetic profiles were located west of the Bovec basin in the Prealps belt of Friuli. Both profiles show density and porosity differences

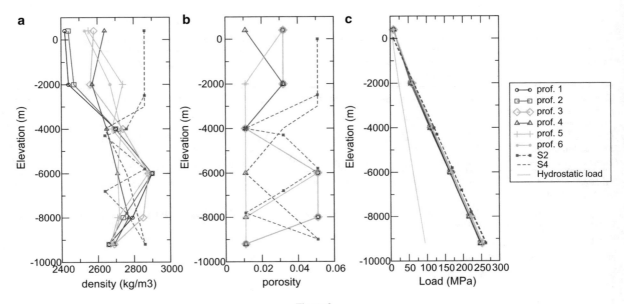

Figure 9
Vertical profiles of the density (**a**), porosity (**b**), and lithostatic load (**c**). The solid lines refer to the six vertical profiles traced through the nodes of the 3D velocity/density model by Bressan et al. (2012) for the Bovec area and the 2004 seismic sequence (small circles in Fig. 1b). The dashed lines refer to the profiles S2 and S4 by Faccenda et al. (2007) and represent the stratigraphy of the area (large circles in Fig. 1b). The light blue curve in **c** denotes the hydrostatic load

relative to the six profiles from the 3D seismo-gravimetric inversion (Bressan et al. 2012), the greatest being observed for the S4 profile (Fig. 9a, b). The main difference between the tomographic and the synthetic profiles occurs in the shallowest section (from the topographic surface to a depth of approximately 2 km), where the synthetic profiles assumed the same lithology while the inversion provided smaller values for both the density and porosity at the same depths. Notwithstanding these differences, the resulting lithostatic load does not differ by more than 4% between the two datasets (Fig. 9c). The lithostatic load at a depth of 9.2 km for the vertical profiles from the 3D model (P_{L1}) is approximately 252 MPa with a standard deviation of 2.7. The load determined at the same depth based on the synthetic profiles from Faccenda et al. (2007) (P_{L2}) is 263 MPa with a standard deviation of 1 (Fig. 9c).

The average effective stress from the data in Table 2 is 22.8 MPa with a standard deviation of 0.55. Hence, the quantity $(1 - \lambda)$ is approximately 0.1 for P_{L1} and P_{L2}, which implies a value of $\lambda \sim 0.9$, i.e., a pore pressure that is close the value of the lithostatic load.

5. Discussion

Our analysis of the longest time series of cGNSS observations of the northern edge of the Adria microplate confirmed the presence of a transient non-periodic signal that had an apparent duration of 2.5 years and slowly propagated through the region. Our investigation of potentially spatially correlated common-mode errors enabled us to reduce noise levels while confirming the presence of an anomalous signal (Fig. 5).

According to our first comparison of the predicted hydrological loading effects for each site when using the GLDAS global model and GRACE information, the sites located in the alluvial plain of the studied region showed the most significant differences between the two approaches (Fig. 3). Such a discrepancy could be caused by the presence of a multi-layered aquifer system in the Friuli plain (Martelli and Granati 2006) or loading effects from the Adriatic Sea (Zerbini et al. 2004), which are not considered in the GLDAS model. However, the main positive peak of the GNSS observations is observed during the summer (between July and August) for all the stations, which is the driest period for the region (blue curves in Fig. 3). According to Argus et al. (2014), this characteristic reflects the elastic response of the solid earth to loadings of snow and water. In contrast, the top of an aquifer should be dominated by poroelastic effects from aquifer recharging and depletion. All the studied sites exhibit the former behaviour, indicating that they elastically respond to hydrological loads. Even sites within the alluvial plain, such as POR0, do not exhibit a porous response to groundwater regardless of the multi-layered aquifer system in the region. Hence, we interpret that loading effects from the Adriatic Sea most likely cause the discrepancy between the GLDAS and GRACE models (Fig. 3).

We correlated original data and data from which hydrological loading effects were subtracted for different frequency bands with the curve of the hydrological balance from the pluviometric data of nearby meteorological stations (Figs. 6, 7). The Thornthwaite hydrological balance used here is simplified and does not include detailed information on time variations in surface-runoff patterns. The runoff in the mountain areas and plains is 70–90 mm/year and 60–70 mm/year, respectively. A lack of information on the time variations in the runoff levels and withdrawals prevented us from conducting more detailed analyses. Although the Thornthwaite water-balance model was developed for North American conditions, this model also applies to monthly stream flows in different areas with marked topography, such as the study region (Calvo 1986). Moreover, the PET values range between 100 and 200, which is consistent with the estimated values from Zini et al. (2013) for the same area based on crop-evapotranspiration data. The resulting hydrological balance is approximately 1000 mm/year, which is similar to the estimates by Zini et al. (2011).

The data that were corrected using GRACE information showed a low correlation even without filtering (Fig. 7), whereas the correlation in the frequency band of the transient signal was null also for the original time series, as when using the GLDAS model (Table 1).

Rossi et al. (2016, 2017) concluded that the GNSS transient signal was caused by a porosity wave that originated at 9.2-km depth in the Bovec basin and propagated through the region. This porosity wave was generated via a fault-valve behaviour of the Ravne Fault, as in the model by Sibson (1992). The propagation velocity values from Rossi et al. (2016; 2017) ranged from one to tens of kilometres per year, which is consistent with the values by Revil and Cathles (2002) and Vardoulakis et al. (1998) and those by Joshi and Appold's (2017) models. The hydraulic diffusivity values reported in the literature cover a broad range of magnitudes for the crust in an interval from 1×10^{-2} to 10 m^2/s (e.g., Simpson et al. 1988). The results of the tomographic inversion were slightly smaller than these values and ranged from 3×10^{-4} to 8×10^{-2} m^2/s, which is consistent with the observations by Shapiro et al. (2002) and Doan et al. (2006). In this work, we used hydraulic diffusivity values to calculate the permeability of the four dominant rock formations in the region (Table 2). The values obtained vary from 5.22×10^{-17} to 2.87×10^{-15}, which is consistent with the permeability values that characterize similar lithologies in the Vienna Basin (from 9.8×10^{-18} to 2.46×10^{-13}) (Hawle et al. 1967) and other observations (e.g., Sibson and Rowland 2003). In particular, 'Dolomia principale' dolomitic limestone is more permeable than micrite-rich limestones, such as the Jurassic limestones in our region, and the limestones in the Southern Apennines as observed by Giorgioni et al. (2016). Following these observations, we can assimilate the Dolomia principale into Dolomite B, a coarse form of crystalline dolomite characterized by similar values to ours.

The hodograms in Fig. 8 provide new insights into the effects of the propagation of the porosity wave in the region. The amplitude of the vertical signal is similar at all the sites. Conversely, the horizontal displacement values vary between the stations closest to and farthest from the hypothesized transient source (ACOM, MOG0, MDEA, and TRIE and GSR1, UDIN, PAL0, POR0, and AFAL, respectively), with the highest values found for the latter group of stations. According to Wahr et al. (2013), the horizontal displacement for a circular mass change is zero in the centre and increases closer to the boundary of the circle, while the displacement starts to decrease after crossing the perimeter of the circle. In our case, the pattern in Fig. 8 suggests that the mass change occurred in an elliptical area around the origin of the transient signal. The major axis appears to be oriented NW–SE, i.e., oriented along the station's tilting direction and coinciding with the orientation of the Dinaric transpressive faults (similar to the Ravne fault; Fig. 1b).

Our calculation of the lithostatic load (Fig. 9) allowed us to relate the effective stress to the state of the pore pressure through the λ value. We found a state of high overpressure close to the lithostatic load ($\lambda \sim 0.9$). Moreover, the range of permeability from our analyses is within the range that characterizes the transition between forced overpressure and advective/convective flow (Sibson and Rowland 2003) (Table 2). Overpressure conditions manifest when less-permeable layers with $k < 1 \times 10^{-17}$ m^2 impede on fluid drainage (e.g., Neuzil 1995; Sibson and Rowland 2003), whereas so-called seismogenic permeability occurs in a range of $10^{-16} < k < 10^{-14}$ m^2 (Talwani and Acree 1984). However, the observations of Townend and Zoback (2000) suggest that earthquakes are precluded under conditions of fluid overpressure, whereas seismic faulting occurs under hydrostatic fluid pressure conditions (Sibson and Rowland 2003). When the pore-pressure levels exceed the lithostatic load, dilatation and natural hydrofracturing can occur to accommodate the fluid (Neuzil 1995). Neuzil's (1995) description of the formation of interconnected fractures in such cases to relieve high-pressure levels by increasing permeability levels is not considerably different from the principles of the proposed model of porosity waves (Connolly and Podladchikov 2013). Complex tectonics from repeated thrusting and folding phases in the region can favour the formation of local overpressure in neighbouring faults and thrusts, where less-permeable terrigenous lithotypes contact carbonate rocks with much higher levels of permeability. Such conditions may have occurred at the NW tip of the Ravne fault. Such fault is hypothesized to continue to the NW and remain hidden under Quaternary sediment in the Bovec basin (Kastelic et al. 2008). Kastelic et al. (2008) found that the Bovec basin acted as a barrier to the segment that activated during the 1998 sequence. We identified at the transient's origin, a few months before the $M_w = 5.2$ earthquake in 2004, an initial overpressure phase close to the lithostatic load, which would have prevented earthquake faulting and subsequently induced dilation to accommodate fluids. This process does not differ from the process described by Sibson's (1992) model of the fault-valve behaviour of a fault that crosses supra-hydrostatic gradients. The damaged volume around a fault creates a favourable environment for the formation of interconnected fractures, which help release pore pressure and disperse fluids from the seismogenic zone towards areas of lower intermediate stress (Connolly and Podladchikov 2000; Revil and Cathles 2002).

6. Conclusions

Our results support the interpretation of transient signals reported by Rossi et al. (2016, 2017) and show that hydrological loading effects did not cause these signals. First, we excluded potential spatially correlated common-mode errors in the time series and concluded that the region reflects the elastic response of the solid earth to the loading of snow and water and shows minimal poroelastic effects. We demonstrated how subtracting hydrological loading at a global scale is effective for reducing seasonal effects. Moreover, the frequency band of interest ($1.5 < T < 3.5$ years) is not correlated with local (one to a few kilometres from the station) hydrological effects.

The patterns of motion in the various sites is compatible with a mass change that occurred in the transient signal's origin volume and was then distributed over an elliptical area along a NW–SE-oriented major axis coinciding with the orientation of the Ravne fault. The permeability values for the four most representative rock formations in the region (two flysch formations of Palaeozoic and Cenozoic age, Dolomitic limestone, and micritic limestone) are consistent with the permeability measurements for similar lithotypes. The calculated ratio between the effective stress and lithostatic load (P_L) for the different vertical profiles in the Bovec area revealed a state of overpressure with pore-pressure levels that are close to the value of the lithostatic load at the transient signal's original depth ($P_p = 0.9\ P_L$).

These results help elucidate the source of the transient signal that was reported and analysed by Rossi et al. (2016, 2017). Together with the tectonic stress, the complex juxtaposition of layers with low and high levels of permeability can favour the local surge of overpressure conditions. However, less-permeable flysch formations impede the drainage of fluids from dehydration, thereby resulting in disequilibrium. Equilibrium conditions are restored through dilatation and the formation of domains of interconnected fractures, which increase the permeability levels and relieve overpressure conditions (Neuzil 1995). Future modelling may confirm and better define the valve behaviour of the Ravne fault during the spring–summer period of 2004, which caused the transient signal recorded by the cGNSS stations. Confirming such behaviours may provide insights for similar phenomena in other contexts.

Acknowledgements

We dedicate this paper to the memory of our colleague and friend Marco Mucciarelli, who encouraged us in this research. We are indebted to Manuele Faccenda for providing information on the laboratory methods that were used in the rock experiments and Chiara Calligaris for engaging in fruitful discussions and providing data for the hydrological parameters of the studied region. We thank the four anonymous reviewers for their punctual and constructive comments, which helped us improve this manuscript. We thank Giorgio Durì, Michele Bertoni, Elvio Del Negro, and Sandro Urban for their support with the maintenance of the FReDNet GNSS network. FReDNet is managed by the Istituto Nazionale di Oceanografia e di Geofisica Sperimentale-OGS with the support from the Friuli Venezia Giulia Civil Protection and Provincia Autonoma di Trento. We thank OSMER FVG for providing the rainfall and climate data (https://www.osmer.fvg.it/clima/clima_fvg).

REFERENCES

Agnew, D. C. (2001). Map projections to show the possible effects of surface loading. *Journal of the Geodetic Society of Japan, 47,* 255–260. https://doi.org/10.11366/sokuchi1954.47.255.

Argus, D. F., Fu, Y., & Landerer, F. W. (2014). Seasonal variation in total water storage in California inferred from GPS observations on vertical land motion. *Geophysical Research Letters, 41,* 1971–1980. https://doi.org/10.1002/2014GL059570.

Battaglia, M., Zuliani, D., Pascutti, D., Michelini, A., Marson, I., Murray, M. H., et al. (2003). Network assesses earthquake potential in Italy's Southern Alps. *Eos Transactions AGU, 84,* 262–264. https://doi.org/10.1029/2003EO2800032003.

Bennet, R. A. (2008). Instantaneous deformation from continuous GPS: contributions from quasi-periodic loads. *Geophysical Journal International, 174,* 1052–1064. https://doi.org/10.1111/j.1365-246X.2008.03846.x.

Blewitt, G., & Lavallée, D. (2002). Effect of annual signals on geodetic velocity. *Journal of Geophysical Research, 107*(B7), ETG 9-1–ETG 9-11. https://doi.org/10.1029/2001JB000570.

Böhm, G., Rossi, G., & Vesnaver, A. (1999). Minimum-time ray-tracing for 3-D irregular grids. *Journal of Seismic Exploration, 8,* 117–131.

Bonafede, M., Boschi, E., & Dragoni, M. (1983). Viscoelastic stress relaxation on deep fault sections as possible source of very long period elastic waves. *Journal of Geophysical Research, 88,* 2251–2260. https://doi.org/10.1029/JB088iB03p02251.

Braitenberg, C., Romeo, G., Taccetti, Q., & Nagy, I. (2006). The very broad-band long-base tiltmeters of Grotta Gigante (Trieste, Italy): secular term tilting and the great Sumatra–Andaman Islands earthquake of December 26, 2004. *Journal of Geodynamics, 41,* 164–174. https://doi.org/10.1016/j.jog.2005.08.015.

Braitenberg, C., & Zadro, M. (1999). The Grotta Gigante horizontal pendulums—instrumentation and observations. *Bollettino di Geofisica Teorica e Applicata, 40*(3/4), 577–582.

Brauchler, R., Böhm, G., Leven, C., Dietrich, P., & Sauter, M. (2013). A laboratory study of tracer tomography. *Hydrogeology Journal, 21*(6), 1265–1274. https://doi.org/10.1007/s10040-013-1006-z.

Bressan, G., Gentile, G. F., Tondi, R., De Franco, R., & Urban, S. (2012). Sequential Integrated Inversion of tomographic images and gravity data: an application to the Friuli area (north-eastern Italy). *Bollettino di Geofisica Teorica ed Applicata, 53,* 191–212. https://doi.org/10.4430/bgta0059.

Bressan, G., Ponton, M., Rossi, G., & Urban, S. (2016). Spatial organization of seismicity and fracture pattern in NE-Italy and W-Slovenia. *Journal of Seismology, 20*(I2), 511–534. https://doi.org/10.1007/s10950-015-9541-9.

Calvo, J. C. (1986). An evaluation of Thornthwaite water-balance technique in predicting stream runoff in Costa-Rica. *Hydrological Sciences Journal-Journal Des Sciences Hydrologiques, 31*(1), 51–60. https://doi.org/10.1080/02626668609491027.

Ceschia, M., Micheletti, S., & Carniel, R. (1991). Rainfall over Friuli-Venezia Giulia: high amounts and strong geographical gradients. *Theoretical and Applied Climatology, 43,* 175–180. https://doi.org/10.1007/BF00867452.

Chamoli, A., Lowry, A. R., & Jeppson, T. N. (2014). Implications of transient deformation in the northern Basin and Range, western United States. *Journal of Geophysical Research Solid Earth, 119,* 4393–4413. https://doi.org/10.1002/2013JB010605.

Chanard, K., Avouac, J. P., Ramillien, G., & Genrich, J. (2014). Modeling deformation induced by seasonal variations of continental water in the Himalaya region: sensitivity to Earth elastic structure. *Journal of Geophysical Research, Solid Earth, 119*(6), 5097–5113. https://doi.org/10.1002/2013JB010451.

Chiaruttini, C., & Zadro, M. (1976). Horizontal pendulum observations at Trieste. *Bollettino di Geofisica Teorica e Applicata, 19,* 441–455.

Cimolino, A., Della Vedova, B., Nicolich, R., Barison, E., & Brancatelli, G. (2010). New evidence of the outer Dinaric deformation front in the Grado area (NE-Italy). *Rendiconti Lincei, 21*(1), 167–179. https://doi.org/10.1007/s12210-010-0096-y.

Connolly, J. A. D., & Podladchikov, Y. Y. (1998). Compaction driven fluid flow in viscoelastic rock. *Geodinamica Acta, 11,* 55–84. https://doi.org/10.1016/S0985-3111(98)80006-5.

Connolly, J. A. D., & Podladchikov, Y. Y. (2000). Temperature-dependent viscoelastic compaction and compartmentalization in sedimentary basins. *Tectonophysics, 324,* 137–168. https://doi.org/10.1016/S0040-1951(00)00084-6.

Connolly, J. A. D., & Podladchikov, Y. Y. (2013). Metasomatism and the chemical transformation of rock (599–658), Lecture Notes in Earth System Sciences. In D. E. Harlov & H. Austrheim (Eds.), *A Hydromechanical model for lower crustal fluid flow.* Berlin: Springer. https://doi.org/10.1007/978-3-642-28394-9_14.

D'Agostino, N., Cheloni, D., Mantenuto, S., Selvaggi, G., Michelini, A., & Zuliani, D. (2005). Strain accumulation in the southern Alps (NE Italy) and deformation at the northeastern boundary of Adria observed by CGPS measurements. *Geophysical Research Letters, 32,* L19306. https://doi.org/10.1029/2005GL024266.

Devoti, R., Esposito, A., Pietrantonio, G., Pisani, A. R., & Riguzzi, F. (2011). Evidence of large scale deformation patterns from GPS data in the Italian subduction boundary. *Earth and Planetary Science Letters, 311*(3–4), 230–241. https://doi.org/10.1016/j.epsl.2011.09.034.

Devoti, R., Zuliani, D., Braitenberg, C., Fabris, P., & Grillo, B. (2015). Hydrologically induced slope deformations detected by GPS and clinometric surveys in the Cansiglio Plateau, southern Alps. *Earth and Planetary Science Letters, 419,* 134–142. https://doi.org/10.1016/j.epsl.2015.03.023.

Dill, R., & Dobslaw, H. (2013). Numerical simulations of global-scale high-resolution hydrological crustal deformations. *Journal of Geophysical Research, Solid Earth, 118,* 5008–5017. https://doi.org/10.1002/jgrb.50353.

Doan, M. L., Brodsky, E. E., Kano, Y., & Ma, K. F. (2006). In situ measurement of the hydraulic diffusivity of the active Chelungpu Fault, Taiwan. *Geophysical Research Letters, 33,* L16317. https://doi.org/10.1029/2006GL026889.

Dong, D., Fang, P., Bock, Y., Cheng, M. K., & Miyazaki, S. (2002). Anatomy of apparent seasonal variations from GPS-derived site position time series. *Journal of Geophysical Research, 107,* ETG 9-1–ETG 9-16. https://doi.org/10.1029/2001JB000573.

Faccenda, M., Bressan, G., & Burlini, L. (2007). Seismic properties of the upper crust in the central Friuli area (northeastern Italy) based on petrophysical data. *Tectonophysics, 445*(3–4), 210–226. https://doi.org/10.1016/j.tecto.2007.08.004.

Feng, L., & Newman, A. V. (2009). Constraints on continued episodic inflation at Long Valley Caldera, based on seismic and geodetic observations. *Journal of Geophysical Research, 114,* B06403. https://doi.org/10.1029/2008JB006240.

Fu, Y., Argus, D. F., & Landerer, F. W. (2014). GPS as an independent measurement to estimate terrestrial water storage variations in Washington and Oregon. *Journal of Geophysical Research, 120,* 552–566. https://doi.org/10.1002/2014JB011415.

Giorgioni, M., Iannace, A., D'Amore, M., Dati, F., Galluccio, L., Guerriero, V., et al. (2016). Impact of early dolomitization on multi-scale petrophysical heterogeneities and fracture intensity of low-porosity platform carbonates (Albian-Cenomanian, southern Apennines, Italy). *Marine and Petroleum Geology, 73,* 462–478. https://doi.org/10.1016/j.marpetgeo.2016.03.011.

Hawle, H., Kratochvil, H., Schmied, H., & Wieseneder, H. (1967). Reservoir geology of the carbonate oil and gas reservoir of the Vienna basin. *Proceedings of Seventh World Petroleum Congress, 2–9 April, Mexico City, Mexico, 2,* 371–395.

Hubbert, M. K., & Rubey, W. W. (1959). Mechanics of fluid filled porous solids and its application to overthrust faulting. Role of fluid pressure in mechanics of overthrust faulting. *Bulletin of the Geological Society of America, 70,* 115–166. https://doi.org/10.1130/0016-7606(1959)70[115:ROFPIM]2.0.CO;2.

Joshi, A., & Appold, M. S. (2017). Numerical modelling of porosity waves in the Nankai accretionary wedge décollement, Japan: implications for aseismic slip. *Hydrogeologic Journal, 25,* 249–264. https://doi.org/10.1007/s10040-016-1479-7.

Kannan, N., Santhi, C., Williams, J. R., & Arnold, J. G. (2008). Development of a continuous soil moisture accounting procedure for curve number methodology and its behaviour with different evapotranspiration methods. *Hydrological Processes, 22,* 2114–2121. https://doi.org/10.1002/hyp.6811.

Kastelic, V., Vrabec, M., Cunningham, D., & Gosar, A. (2008). Neo-Alpine structural evolution and present-day tectonic activity of the eastern Southern Alps: the case of the Ravne Fault, NW Slovenia. *Journal of Structural Geology, 30,* 963–975. https://doi.org/10.1016/j.jsg.2008.03.009.

Landerer, F. W., & Swenson, S. C. (2012). Accuracy of scaled GRACE terrestrial water storage estimates. *Water Resources Research, 48,* W04531. https://doi.org/10.1029/2011WR011453.

Llubes, M., Florsch, N., Hinderer, J., Longuevergne, L., & Amalvict, M. (2004). Local hydrology, the Global Geodynamics Project and CHAMP/GRACE perspective: some cases studies. *Journal of Geodynamics, 38,* 355–374. https://doi.org/10.1016/j.jog.2004.07.015.

Longuevergne, L., Boudin, F., Boy, J.P., Oudin, L., Florsch, N., Vincent, T. & Kammenthaler, M. (2009). Physical modelling to remove hydrological effects at local and regional scale: application to the 100-m hydrostatic inclinometer in Sainte-Croix-aux-Mines (France). IUGG—Observing our Changing Earth, 2007, Perugia, Italy, 133 (3), 533–539, doi:https://doi.org/10.1007/978-3-540-85426-563.

Lupi, M., Geiger, S., & Graham, C. M. (2011). Numerical simulations of seismicity-induced fluid flow in the Tjörnes Fracture Zone, Iceland. *Journal of Geophysical Research, 116,* B07101. https://doi.org/10.1029/2010JB007732.

Martelli, G., & Granati, G. (2006). The confined aquifer system of Friuli Plain (North Eastern Italy): analysis of sustainable groundwater use. *Giornale di Geologia Applicata, 3,* 59–67. https://doi.org/10.1474/GGA.2006-03.0-08.0101.

Moratto, L., Suhadolc, P., & Costa, G. (2012). Finite-fault parameters of the September 1976 M > 5 aftershocks in Friuli (NE Italy). *Tectonophysics, 536–537,* 44–60. https://doi.org/10.1016/j.tecto.2012.02.002.

Natale, G., & Salusti, E. (1996). Transient solutions for temperature and pressure waves in fluid-saturated porous rocks. *Geophysical Journal International, 124,* 649–656. https://doi.org/10.1111/j.1365-246X.1996.tb05630.

Neuzil, C. E. (1995). Abnormal pressures as hydrodynamic phenomena. *American Journal of Science, 295,* 742–786.

Rajner, M., & Liwosz, T. (2011). Studies of crustal deformation due to hydrological loading on GPS height estimates. *Geodesy and Cartography, 60*(2), 135–144. https://doi.org/10.2478/v10277-012-0012-y.

Revil, A., & Cathles, L. M., III. (2002). Fluid transport by solitary waves along growing faults. A field example from the South Eugene Island basin, Gulf of Mexico. *Earth and Planetary Sciences Letters, 202,* 321–335. https://doi.org/10.1016/S0012-821X(02)00784-7.

Rodell, M., Houser, P. R., Jambor, U., Gottschalck, J., Mitchell, K., Meng, C.-J., et al. (2004). The Global Land Data Assimilation System. *Bulletin of the American Meteorological Society, 85*(3), 381–394. https://doi.org/10.1175/BAMS-85-3-381.

Rossi, G., & Zadro, M. (1996). Long-term crustal deformations in NE-Italy revealed by tilt-strain gauges. *Physics of the Earth and Planetary Interiors, 97,* 55–70. https://doi.org/10.1016/0031-9201(96)03166-4.

Rossi, G., Zuliani, D., & Fabris, P. (2016). Long-term GNSS measurements through Northern Adria microplate reveal fault-induced fluid mobilization. *Tectonophysics, 690,* 142–159. https://doi.org/10.1016/j.tecto.2016.04.031.

Rossi, G., Zuliani, D., & Fabris, P. (2017). Corrigendum to: 'Long-term GNSS measurements through Northern Adria microplate reveal fault-induced fluid mobilization'. *Tectonophysics, 694,* 486–487. https://doi.org/10.1016/j.tecto.2016.10.035.

Serpelloni, E., Faccenna, C., Spada, G., Dong, D., & Williams, S. D. P. (2013). Vertical GPS ground motion rates in the Euro-Mediterranean region: new evidence of velocity gradients at different spatial scales along the Nubia-Eurasia plate boundary. *Journal of Geophysical Research, Solid Earth, 118,* 6003–6024. https://doi.org/10.1002/2013JB010102.

Shapiro, S. A., Rothert, E., Rath, V. & Rindschwentner, J. (2002). Characterization of fluid transport properties of reservoirs using induced microseismicity. *Geophysics,* 67, Special section—Seismic signatures of fluid transport, 212–220, doi:https://doi.org/10.1190/1.1451597.

Sibson, R. H. (1992). Implications of fault-valve behaviour for rupture nucleation and recurrence. *Tectonophysics, 211,* 283–293. https://doi.org/10.1016/0040-1951(92)90065-E.

Sibson, R. H., & Rowland, J. V. (2003). Stress, fluid pressure and structural permeability in seismogenic crust, North Island, New Zealand. *Geophysical Journal International, 154,* 584–594. https://doi.org/10.1046/j.1365-246X.2003.01965.x.

Silverii, F., D'Agostino, N., Métois, M., Fiorillo, F. & Ventafridda, G. (2016). Transient deformation of karst aquifers due to seasonal and multi-year groundwater variations observed by GPS in Southern Apennines (Italy). *Journal of Geophysical Research, Solid Earth,* 121, 8315–8337, doi. https://doi.org/10.1002/2016JB013361.

Simpson, D. W., Leith, W. S., & Scholz, C. H. (1988). Two types of reservoir-induced seismicity. *Bulletin of the Seismological Society of America, 78,* 2025–2040.

Skarbek, R. M., & Rempel, A. W. (2016). Dehydration-induced porosity waves and episodic tremor and slip. *Geochemistry, Geophysics, Geosystems, 17,* 442–469. https://doi.org/10.1002/2015GC006155.

Talwani, P., & Acree, S. (1984). Pore pressure diffusion and the mechanism of reservoir-induced seismicity. *Pure and Applied Geophysics, 122,* 947–965. https://doi.org/10.1007/BF00876395.

Talwani, P., Cobb, J. S., & Schaeffer, M. F. (1999). In situ measurements of hydraulic properties of a shear zone in northwestern South Carolina. *Journal of Geophysical Research, 104*(B7), 14993–15003. https://doi.org/10.1029/1999JB900059.

Tapley, B. D., Bettadpur, S., Ries, J. C., Thompson, P. F., & Watkins, M. M. (2004). GRACE Measurements of Mass Variability in the Earth System. *Science, 305*(5683), 503–505. https://doi.org/10.1126/science.1099192.

Terzaghi, K. (1923). Die Berechnung der Durchlässigkeitsziffer des Tones aus dem Verlauf der hydrodynamischen Spannungserscheinungen. Sitzungber. Akademie der Wissenschaften in Wien. *Mathematish-Naturwissenschaftiliche Klasse, 132,* 125–138.

Thornthwaite, C. W. (1948). An approach toward a rational classification of climate. *Geographical Review, 38*(1), 55–94.

Townend, J., & Zoback, M. D. (2000). How faulting keeps the crust strong. *Geology,* 28, 399–402. https://doi.org/10.1130/0091-7613(2000)28<399:HFKTCS>2.0.CO;2.

Tsai, V. C. (2011). A model for seasonal changes in GPS positions and seismic wave speeds due to thermoelastic and hydrologic variations. *Journal of Geophysical Research, 116,* B04404. https://doi.org/10.1029/2010JB008156.

Vardoulakis, I., Stavropoulou, M., & Skjaerstein, A. (1998). Porosity waves in a fluidized sand-column test. *Philosophical Transactions of the Royal Society of London A, 356,* 2591–2608. https://doi.org/10.1098/rsta.1998.0288.

Venturini, S. (2002). Il pozzo Cargnacco 1: un punto di taratura stratigrafica nella pianura friulana. *Memorie della Società Geologica Italiana, 57,* 11–18.

Vesnaver, A., & Böhm, G. (2000). Staggered or adapted grids for seismic tomography? *The Leading Edge, 19*(9), 944–950. https://doi.org/10.1190/1.1438762.

Wahr, J., Khan, S. A., van Dam, T., Liu, L., van Angelen, J. H., van den Broeke, M. R., et al. (2013). The use of GPS horizontals for loading studies, with applications to northern California and southeast Greenland. *Journal of Geophysical Research, Solid Earth, 118,* 1795–1806. https://doi.org/10.1002/jgrb.50104.

Wahr, J., Molenaar, M., & Bryan, F. (1998). Time variability of the Earth's gravity field: hydrological and oceanic effects and their possible detection using GRACE. *Journal of Geophysical Research, 103*(B12), 30205–30229.

Wahr, J., Swenson, S., Zlotnicki, V., & Velicogna, I. (2004). Time-variable gravity from GRACE: first results. *Geophysical Research Letters, 31,* L11501. https://doi.org/10.1029/2004GL019779.

Wdowinski, S., Bock, Y., Zhang, J., Fang, P., & Genrich, J. (1997). Southern California Permanent GPS Geodetic Array: spatial filtering of daily positions for estimating coseismic and postseismic displacements induced by the 1992 Landers earthquake. *Journal of Geophysical Research, 102*(B8), 18057–18070.

Weber, J., Vrabec, M., Pavlovčič-Prešeren, P., Dixon, T., Jiang, Y., & Stopar, B. (2010). GPS-derived motion of the Adriatic microplate from Istria Peninsula and Po-Plain sites, and geodynamic implications. *Tectonophysics, 483,* 214–222. https://doi.org/10.1016/j.tecto.2009.09.001.

Wiggins, C., & Spiegelman, M. (1995). Magma migration and magmatic solitary waves in 3-D. *Geophysical Research Letters, 22,* 1289–1292. https://doi.org/10.1029/95GL00269.

Zerbini, S., Matonti, F., Raicich, F., Richter, B., & van Dam, T. (2004). Observing and assessing nontidal ocean loading using ocean, continuous GPS and gravity data in the Adriatic area.

Geophysical Research Letters, 31, L23609. https://doi.org/10.1029/2004GL021185.

Zerbini, S., Raicich, F., Richter, B., Gorini, V., & Errico, M. (2010). Hydrological signals in height and gravity in Northeastern Italy inferred from principal components analysis. *Journal of Geodynamics, 49*(3–4), 190–204. https://doi.org/10.1016/j.jog.2009.11.001.

Zini, L., Calligaris, C., Treu, F., Iervolino, D. & Lippi, F. (2011). *Risorse idriche sotterranee del Friuli Venezia Giulia: sostenibilità dell'attuale utilizzo.* EUT edizioni Università di Trieste, ISBN: 9-788883-033148.

Zini, L., Calligaris, C., Treu, F., Zavagno, E., Iervolino, D., & Lippi, F. (2013). Groundwater sustainability in the Friuli Plain. *AQUA mundi: Am 07058,* 041–054. https://doi.org/10.4409/Am-058-13-0051.

Zou, R., Freymueller, J. T., Ding, K., Yang, S., & Wang, Q. (2014). Evaluating seasonal loading models and their impact on global and regional reference frame alignment. *Journal of Geophysical Research- Solid Earth, 119*(2), 1337–1358. https://doi.org/10.1002/2013JB010186.

Zuliani, D., Fabris, P. & Rossi, G. (2017). FReDNet: evolution of permanent GNSS receiver system. In: New Advanced GNSS and 3D Spatial Techniques Applications to Civil and Environmental Engineering, Geophysics, Architecture, Archeology and Cultural Heritage, *Lecture notes in Geoinformation and Cartography,* Cefalo R., Zieliński J., Barbarella, M (Eds) (pp.123–137) Springer, Switzerland.

(Received December 16, 2016, revised October 17, 2017, accepted October 26, 2017, Published online November 4, 2017)

Pure Appl. Geophys. 175 (2018), 1889–1907
© 2017 Springer International Publishing AG
https://doi.org/10.1007/s00024-017-1595-x

Contemporary State of the Elbrus Volcanic Center (The Northern Caucasus)

Vadim Milyukov,[1] Eugeny Rogozhin,[2] Andrey Gorbatikov,[2] Alexey Mironov,[1] Andrey Myasnikov,[1] and
Marina Stepanova[2]

Abstract—The Elbrus volcanic center is located in southern Russia on the northern slope of the main ridge of the Greater Caucasus. Current classifications define Elbrus as a dormant volcano that could become active even after millennia of quiescence. In this study, we use two new geophysical methods to assess the contemporary state of the Elbrus volcano. The first method is based on an evaluation of parameters of resonant modes "reemitted" by the resonant structure (i.e., volcanic chamber) in response to the excitation of a seismic impact and recorded by a precise laser interferometer–strainmeter. The second method is based on low-frequency microseismic sounding and allows determination of the deep structure of complicated geological objects. Our study locates the magma chamber at depths of 1–8 km and extended magma source at depths of 15–40 km beneath the Elbrus eastern summit. An unknown magmatic structure, comparable to the Elbrus magmatic structure but currently much colder, was also identified 50 km from Mt. Elbrus. Based on our analysis, we assess the Elbrus volcano to be currently in a quasi-stable state of thermodynamic equilibrium.

Key words: Elbrus volcano magmatic structures, resonant modes, low-frequency microseismic sounding, laser interferometer–strainmeter.

1. Introduction

Mount Elbrus is the highest mountain in Europe, which forms a part of the Caucasus Mountains in southern Russia, and has two summits. The slightly taller western summit stands at 5642 m and the eastern summit at 5621 m. The Elbrus volcanic center is confined to a sub-meridional system of faults including Mount Elbrus, a double-peaked, polygenetic, stratovolcano, as well as a number of minor volcanic centers concentrated on its western slope. The western summit has a well-preserved volcanic crater of about 250 m in diameter.

Mount Elbrus was formed more than 2.5 million years ago. The Elbrus volcano was active the Pleistocene and the Holocene. Modern geochronological studies of Elbrus volcanic eruptions using different methods, such as K-Ar and Sr-Nd-Pb (Lebedev et al. 2005, 2010) and electron paramagnetic resonance dating (Bogatikov et al. 2002a), show that such eruptions have occurred repeatedly and for a long time (i.e., 1 million years) (Chernyshev et al. 2001). The catastrophic eruption was characterised by intense explosive emissions and ended with an effusion of andesitic–dacitic lavas distributed considerable distances from the volcano. The last eruption took place probably less than one thousand years ago (Gurbanov et al. 2013). Evidence of recent volcanism includes several lava flows on the mountain, which look fresh, and roughly 260 square kilometers of volcanic debris. Current classifications define Elbrus as a dormant volcano that could become active even after millennia of quiescence (Bogatikov et al. 2003). According to some data (Gurbanov et al. 2013), the dormant volcano Elbrus has recently manifested a number of signs indicative of its transition from a passive to more active phase, including the intensive melting of glaciers on near-surface magmatic chambers and increased fumarole activity.

Current ideas about the internal structure of the Elbrus volcano are based on modern technologies for deciphering space images of the Earth's surface and electromagnetic tomography of the volcanic edifice using magnetotelluric sounding methods. In the first case, two areas with anomalously low values of the

---

[1] Sternberg Astronomical Institute, Lomonosov Moscow State University, Universitetsky pr., 13, Moscow 119234, Russia. E-mail: milyukov@sai.msu.ru

[2] Schmidt Institute of Physics of the Earth of the Russian Academy of Sciences, Bolshaya Gruzinskaya str., 10-1, Moscow 123242, Russia.

tectonic fragmentation field of the lithosphere are located near the Elbrus volcanic center (Bogatikov et al. 2002b). In the second case, two conductive objects, separated by depth with low resistance have been identified. A joint correlation was used to process results obtained using the two applied methods to construct a three-dimensional model of the Elbrus volcanic internal structure. This model indicates that the magmatic chamber is strictly situated under the volcanic edifice and is characterised by its considerable size with the lower edge confined to depths of about 8 km. At depths of about 5 km, the width of the camera reaches 9 km. A sharp reduction in the size of the camera starts with a depth of about 2 km, and at a depth of 1 km, the chamber dimensions are 2–2.5 km. The magmatic source is located in the depth interval from 15 to 45 km and its dimensions are 35 and 15 km in latitude and longitude, respectively (Spichak et al. 2007). The goal of this work is to assess the contemporary state of the Elbrus volcano using new geophysical methods.

2. Resonance Parameters of Elbrus Magmatic Structures

2.1. The Resonance Method Outline

Volcanic activity is often accompanied by seismic activity, for which there are two types of seismic signals (Kumagai and Chouet 2000). The first type is volcano-tectonic (VT) earthquakes and the second is long-period (LP) events and tremor. LP events and tremor represent the volumetric sources and are associated with the movement of magmatic and hydrothermal fluids. Pressure fluctuations are associated with mass transport and can create a trigger mechanism of acoustic vibrations of the magmatic resonance systems that generate tremor and/or LP events. In particular, the observation of LP events can play an important role in the quantitative assessment of volcanic processes, because the resonant properties of magmatic structures (i.e., source of such events) can be determined from the complex frequencies of the damped harmonic oscillations corresponding to these events. Damped oscillations are characterised by two parameters: the frequency of the dominant

oscillation mode and Q-factor of the oscillatory system (Chouet 2003).

A model of various shapes can be used to represent the magmatic structure as a resonator, such as a sphere (Crosson and Bame 1985; Fujita et al. 1995), a cylinder (Chouet 1985), or a fluid-filled crack (Kumagai and Chouet 2000) embedded in rock matrix. The resonant properties of such a resonator depend on the model parameters including the: P-wave velocity of the rock matrix (α_s); speed of sound in the fluid (α_f); fluid density (ρ_f); and rock density (ρ_s). The state of the system is characterised by the impedance contrast:

$$Z = \left(\frac{\rho_s}{\rho_f}\right)\left(\frac{\alpha_s}{\alpha_f}\right). \qquad (1)$$

The solution of the equation of elasticity under the assumption of spherically symmetric deformation defines the dimensionless complex frequencies of eigen oscillations of the system, $\Omega_i = (r_0/\alpha_f)\,\omega_i$, scaled by the characteristic time $\tau_0 = r_0/\alpha_f$ (i.e., time necessary for the elastic wave to travel a distance r_0 from the center to the surface of the sphere). $\mathrm{Re}\,\Omega_i$ and $\mathrm{Im}\,\Omega_i$ determine the frequency and the damping factor of relevant oscillation modes. The quality factors for a spherical resonator take almost the same values for all regular modes and are determined by the empirical relation:

$$Q^{-1} = \mathrm{Im}\,\Omega \approx 0.7\,\frac{\mathrm{Re}\,\Omega_1}{2\pi}\,\ln\left(\frac{Z+1}{Z-1}\right), \qquad (2)$$

where Ω_1 is the frequency of the fundamental mode of the oscillating sphere.

The eigen frequencies are found from another empirical equation which is also a function of the impedance contrast:

$$f_i = \frac{\mathrm{Re}\,\Omega_1}{2\pi(r_0/\alpha_f)} \approx \frac{1}{2\pi(r_0/\alpha_f)}\left(a_i + b_i \cdot \tanh\left(c_i\left(Z-1\right)\right)\right). \qquad (3)$$

The coefficients a_i, b_i, and c_i are determined by the least-squares method. Thus, Eqs. (2) and (3) are used to determine approximate values of the frequencies and Q-factors of the eigen oscillations of a spherical magma resonator for any set of parameters α and ρ (Fujita et al. 1995).

The mechanism of resonance excitation may be different. It can be various processes occurring within the magma chamber (e.g., excitation in the form of a pressure step inside of a chamber; Kumagai and Chouet 2000). According to the Geophysical Service of the Russian Academy of Sciences (GSRAS), seismic activity of the Elbrus volcano is currently very low. Apparently, the state of the volcano is sufficiently far away from its pre-eruptive stage when the magmatic structures and movement of magmatic fluids are capable of generating seismic events. Therefore, we are considering another excitation mechanism, an external source, which is a broadband and powerful seismic signal. Upon the incidence of such a signal, magmatic structures (i.e., resonators) generate secondary seismic waves, which have a set of resonant modes and contain information about the physical and mechanical properties of structural inhomogeneities. These resonant modes are determined using the geometrical parameters and elastic properties of the magma chamber, as well as magmatic properties. By observing the response of a resonant magmatic structure excited by powerful teleseismic earthquakes, one can determine the geometrical size and physical properties of this structure including acoustic properties of the magmatic fluid that fills it.

According to Lay and Wallace (1995), the energy (in joules) released in the earthquake of magnitude M_s is given as

$$E_{EQ} = 10^{\frac{3}{2}M_s+4} .$$

(4)

In the case of shallow-focus earthquakes, about 60% of the earthquake energy is realised in the form of Rayleigh and Love waves (Puzyrev 1997). The energy incident on a magma chamber (ΔE) is determined by the angular size of the chamber relative to the earthquake epicentre, i.e., ratio of the transverse chamber size (r_0) to the epicentral distance, L:

$$\beta = 2 \tan^{-1}\left(\frac{r_0}{L}\right).$$

Thus, the seismic energy incident on a magma chamber is defined by the expression:

$$\Delta E = 0.6 \, E_{EQ} \, \frac{\beta}{2\pi}.$$

(5)

The portion of the energy that falls on the camera is stored by the magmatic chamber (ΔE_{in}) and the other portion passes through the chamber (ΔE_{out}). The relationship between these two parts is determined by the impedance contrast (Eq. 1). The greater the impedance contrast, the greater the energy stored by the camera. The sound speed of the magmatic fluid is low (200–500 m/s) compared with the P-wave velocity of the rock matrix. Therefore, the impedance contrast for the magmatic chamber embedded in rock matrix is high (e.g., 15–40). In this case, the majority of the energy incident on the camera is stored.

A surface wave with energy ΔE_{in} reaching the magma chamber excites a resonator, which is filled very quickly with a certain amount of energy. A portion of energy stored by the magma chamber is intrinsic losses (ΔE_{in}^i) and the rest is radiated at resonant frequencies, ΔE_{in}^r. These two parts are represented by quality factors Q_r^{-1} (radiation losses) and Q_i^{-1} (intrinsic losses). The ratio of these two components can vary widely depending on the composition of the magmatic fluids. For magmas consisting of a liquid–gas mixture with gas-volume fractions less than 10% (i.e., "bubbly liquids"), the model values of Q_i factors strongly depend on the bubble radius and are rather low ($10^1 \div 10^3$). On the contrary, for magma with gas-volume fractions more than 90% (i.e., gas-gas mixture), the calculated Q_i factors range between 10^{10} and 10^{11}, indicating that the attenuation due to these intrinsic loss mechanisms is trivial.

Magmas with a gas-volume fraction between 10 and 90% are classified as "foam". There are no reliable estimates of Q_i for foams. This issue remains unresolved (Kumagai and Chouet 2000). According to our estimations (see Sect. 2.3), the gas-volume fractions of the Elbrus magmatic fluids are between 0.3 and 0.7, such that this magma is classified as foam. As there is no model estimate for intrinsic losses in the Elbrus magmatic chamber, but there is a tendency that such losses decrease with increasing gas-volume fractions, it can be expected that the energy of intrinsic losses will be less than the energy radiated by the magmatic chamber. Thus, $\Delta E = \Delta E_{out} + \Delta E_{in}^i + \Delta E_{in}^r$. Under the assumptions outlined above, $\Delta E_{in}^r > \Delta E_{in}^i$ and $\Delta E_{in}^r > \Delta E_{out}$. In other words, the majority of the energy incident on a magma chamber is radiated in the form of secondary seismic waves.

The distance from the Elbrus volcano to the Baksan interferometer ($L \approx 20$ km) is comparable to the size of the magma chamber ($2r_0 \approx 10$ km). Due to the proximity of the detector to the radiating object, the divergence of seismic waves can be neglected. Following this assumption, the total radiated energy is proportional to the size (i.e., diameter) of the magmatic chamber. Half of the energy is emitted in the direction of the interferometer. Part of the energy that reaches the interferometer is proportional to the ratio $l/2r_0$, where l is the arm length of the Baksan laser interferometer. In this study, $l \approx 75$ m. Consequently, the energy incoming to the interferometer, ΔE_{BLI}, can be estimated as

$$\Delta E_{BLI} \approx \frac{1}{2} \Delta E_{in}^r \frac{l}{2r_0} \approx 2 \times 10^{-3} \Delta E. \qquad (6)$$

Thus, approximately one thousand of the energy incident on the Elbrus magma chamber reaches BLI. This estimation was calculated based on $l \approx 75$ m, $2r_0 \approx 10^4$ m, and $\Delta E_{in}^r \approx \Delta E$. Nevertheless, in the present study, we proceed from the requirement that ΔE_{BLI} should be much larger than the energy that can be confidently detected by this instrument (i.e., $\Delta E_{BLI} \gg E_{min}$). This method of studying the resonant properties of magmatic structures excited by a broadband teleseismic signal is called the "resonance method", and was developed to study the Elbrus volcano (Milyukov 2006).

2.2. Baksan Laser Interferometer–Strainmeter

Long-base laser interferometers are instruments used for high precision measurement of relative changes in the distance between two points. The frequency of laser interferometers ranges from practically zero up to tens of MHz and is limited only by the performance of the employed electronics. The dynamic range is also practically unlimited and the threshold for sensitivity exceeds other devices by several orders of magnitude. An important advantage of the laser interferometer is its own intrinsic length reference, the wavelength of the laser radiation, which is very stable and known within high accuracy.

A monitor of the local earth's crustal deformation, including possible vibrations associated with the Elbrus volcano, is carried out by the Baksan long-base laser interferometer–strainmeter. The Baksan station of Moscow University is located in the Baksan gorge (Kabardino-Balkaria Republic), 30 km south-west of the town of Tyrnyauz and 18 km from Mount Elbrus. The interferometer is mounted at a level of 650 m and depth of 400 m inside the main tunnel of the Baksan Neutrino Observatory within the Institute of Nuclear Research of the Russian Academy of Sciences. The coordinates of the interferometer are 43°12′N and 42°43′E and its azimuth is 150°37′. The general scheme of the Baksan laser interferometer–strainmeter is presented in Fig. 1.

The Baksan laser interferometer (BLI) is a Michelson two-beam unequal-arm interferometer operating in the regime of separated beams. The length of its larger (measuring) arm is 75 m with an optical length of 150 m, and the length of the minor (reference) arm is 0.3 m. The optical elements of the interferometer are mounted in two cylindrical vacuum chambers with lower parts that are built into pedestals. The supporting pedestals of the interferometer are concrete pillars, deepened to 1.5 m, and rigidly connected to bedrock. The chambers are

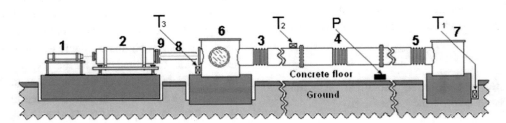

Figure 1
General scheme of the Baksan laser interferometer and the position of monitoring sensors: (1, 2, 8, 9) laser; telescopic system and light guide; (3, 4, 5) bellows; (6, 7) vacuum chambers; (T1, T2, T3) temperature sensors; (P) pressure gauge (Milyukov et al. 2010)

connected by three bellows and vacuum tubes (30 cm in diameter) that form a light guide. The light emitted by a He–Ne laser 1 is incident on the telescopic system 2, which is composed of two objective lenses. The beam formed by the telescopic system, which has a large curvature radius, is incident to the beam-splitting cube that divides the beam into two parts, according to their amplitudes, and guides each towards measurement and reference corner cube prisms. The beam-splitting cube and reference corner cube prism are located in vacuum chamber 6 and the measuring corner cube prism is in vacuum chamber 7. The reflected beams recombine on the beam-splitting cube. An interference pattern produced by a recombination of the reference and measuring beams consists of straight fringes. The optical signal is converted into an electrical signal by means of the electronic readout system, which ensures operation of the interferometer over a wide range of frequencies from ultra-low (limited only by the time interval of continuous observations) to thousands of Hz. The instrumental resolution of the interferometer is 3×10^{-13} for crustal strain measurements. The design and technical layout of the Baksan interferometer are described in detail in Milyukov and Myasnikov (2005) and Milyukov et al. (2005).

2.3. Observation Data and Results

Since 2005, the resonance method has been used for monitoring the state and dynamic processes of the Elbrus volcano and has also been used to estimate the parameters of the shallow magmatic chamber of Mt. Etna (Iafolla et al. 2008). Spectral analysis of strain data containing records of a large number of earthquakes on both global and regional scales revealed a stable set of resonant modes of high repeatability, which was associated with the resonant modes of the magmatic chamber of the Elbrus volcano (Milyukov 2006; Milyukov et al. 2010).

An estimation of the resonant parameters of modes excited by moderate-power earthquakes (i.e., magnitude usually below 6) in the so-called "near zone" of the Elbrus volcano (≤ 600 km) is of particular interest. The choice of such events owes to the relative proximity of these earthquakes to the volcanic edifice, which creates the possibility for excitation of resonant structures (i.e., magmatic chambers) of the Elbrus volcano, while at the same time, the energy of these moderate earthquakes is not sufficient to excite free oscillations of the Earth. Thus, identification of the stable resonance modes should clearly indicate excitation of regional structures only. Such an analysis was performed for ten events within the "near zone" observed between 2005 and 2014. All ten earthquakes are characterised by moderate magnitude with M_s values between 4.8 and 6.0. Five earthquakes occurred in the Northern Caucasus, another four in Turkey and one in the eastern part of the Black Sea. Parameters of earthquakes and their locations are presented in Table 1 and Fig. 2. Values of ΔE (Eq. 5) and ΔE_{BLI} (Eq. 6) from each of the considered earthquakes are listed in Table 1.

Table 1

Parameters of moderate-power earthquakes occurred in the "near zone" of the Elbrus volcano (~ 600 km) during 10-year observation

No.	Date	Latitude	Longitude	Depth, km	M_s	E_{EQ}, J	L, km	ΔE, J	ΔE_{BLI}, J
1	2005-01-25	37.622	43.703	41	5.6	1.6×10^{13}	645	2.1×10^{10}	4.2×10^{7}
2	2005-03-14	39.354	40.890	5	5.7	2.2×10^{13}	462	4.2×10^{10}	8.3×10^{7}
3	2008-10-11	43.372	46.254	16	5.5	1.1×10^{13}	307	3.1×10^{10}	6.3×10^{7}
4	2009-09-07	42.660	43.443	15	5.8	3.2×10^{13}	111	2.4×10^{11}	4.9×10^{8}
5	2010-03-08	38.864	39.986	12	6.0	6.3×10^{13}	539	1.0×10^{11}	2.0×10^{8}
6	2011-11-14	38.658	43.170	10	4.8	1.0×10^{12}	524	1.6×10^{9}	3.3×10^{6}
7	2012-12-23	42.420	41.075	15	5.6	1.6×10^{13}	152	9.0×10^{10}	1.8×10^{8}
8	2013-09-17	42.146	45.812	5	5.3	5.6×10^{12}	305	1.6×10^{10}	3.2×10^{7}
9	2014-06-29	41.603	46.665	34	5.0	2.0×10^{12}	396	4.3×10^{9}	8.7×10^{6}
10	2014-09-29	41.197	48.100	13	5.4	7.9×10^{12}	522	1.3×10^{10}	2.6×10^{7}

E_{EQ} is the energy released in the earthquake of magnitude M_s, ΔE is the seismic energy incident on the Elbrus magma chamber, and ΔE_{BLI} is the energy of the secondary seismic waves incoming to BLI. Coordinates, time, and magnitude of earthquakes from catalog (International Seismological Centre 2014)

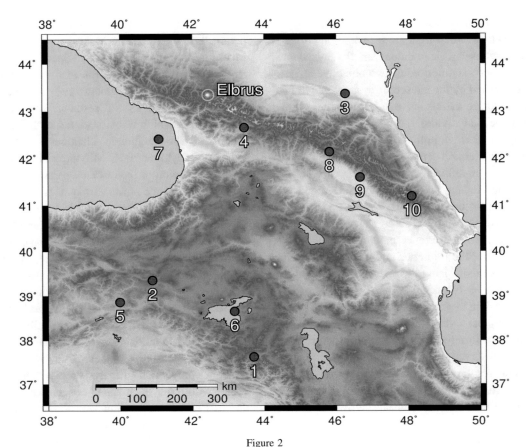

Figure 2
Map of the Caucasus and neighbouring territories of Turkey and Iran: the epicenters of moderate-power earthquakes occurred in the "near zone" of the Elbrus volcano. The Baksan laser interferometer is located 20 km from the foot of Mt. Elbrus. The *numbers* indicate the seismic events (EQ) in accordance with Table 1

The assessment of the regional resonant mode parameters was performed on the basis of the spectral least-squares algorithm (Kozyreva and Milyukov 2003), described below. The analysed signal is the sum of the useful geophysical signal (GS) and additive noise. We suppose that the GS parameters are unknown and must be estimated, while the shape of the GS is known. The reaction of a distributed system to an external broadband excitation is a linear superposition of quasi-harmonic oscillations related to single modes. In the single-frequency approximation, the spectral density of the excited GS single mode is

$$N_S(\omega) = N_0 \frac{\delta^2}{\delta^2 + (\omega - \Omega)^2}, \quad \omega > 0, \quad (7)$$

where $N_0 = \max_\omega N_S(\omega)$ is the maximum value of the spectral density of GS single mode; ω and Ω are the angular and resonant angular frequencies, respectively, and δ is the half-width of the spectral line.

The theoretical evaluation related to calculation of the parameters for typical magmatic structures (Sobisevich et al. 2002) shows that for a chamber with a characteristic size of approximately 5–50 km, the dominant mode frequency should be in the range of 0.001–0.1 Hz. According to modern studies, the characteristic size of the Elbrus magma chamber is ~9 km (Spichak et al. 2007). For a spherical model of the magma chamber, the frequency of the fundamental resonant mode, estimated according to Eq. 3, is ~0.015 Hz (Milyukov 2006). To obtain a

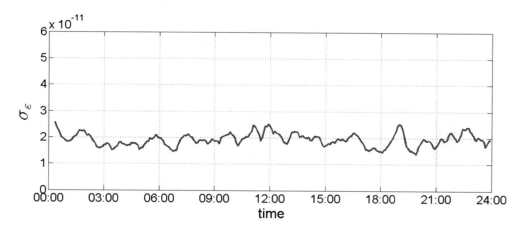

Figure 3

Baksan Laser interferometer–strainmeter. Daily variations of the seismic noise standard deviation in the frequency range of 0.006–0.03 Hz averaged over a 5-min time window over 6 days

confident registration of secondary waves emitted by the magmatic chamber (with signal-to-noise ratio ~ 10), the amplitude A of the wave incoming to the interferometer should cause a deformation of $\sim 10^{-10}$, while the seismic noise level measured by BLI is $\sim 10^{-11}$ in the frequency range of $\sim 10^{-2}$ Hz (Fig. 3). The least confident detectable amplitude $A_{\min}$ can be estimated from the relation:

$$\frac{A_{\min}}{\lambda/4} = \frac{\Delta l}{l} = 10^{-10},$$

where λ is the wavelength of the resonant frequency and l and Δl are the arm length of the interferometer and its variation, respectively. For $\lambda = T \times v \approx 2.8 \times 10^5$ m ($T \approx 70$ s is the period and $v \approx 4 \times 10^3$ m/s is the P-wave velocity of the rock matrix), the minimal detectable amplitude is $A_{\min} \approx 7$ μm. Accordingly, the energy of such seismic wave emitted by the magma chamber is equal to

$$E_{\min} = \frac{1}{2} m \omega^2 A_{\min}^2 \approx 1.5 \times 10^2 J, \qquad (8)$$

where $m = \rho_f V = \frac{4}{3} \pi \rho_f r_0^3 \approx 7.6 \times 10^{14}$ kg is the mass of the magma chamber, and $\omega \approx 9 \times 10^{-2}$ s^{-1} and $A_{\min} \approx 7$ μm are the frequency and amplitude of the eigen oscillations, respectively. Thus, the amount of energy emitted by the magmatic chamber and incoming to the Baksan laser interferometer–strainmeter should be not less than $E_{\min} \approx 1.5 \times 10^2$ J. According to Table 1, the condition $\Delta E_{BLI} \gg E_{\min}$ is valid for all earthquakes considered.

Within the statistical processing of the data, stable groups of resonant modes have been defined in most recorded earthquakes taking into consideration features of the spectrum section related to the defined mode. The regional modes defined in the spectral analysis are grouped into families that have a strongly pronounced linear character, which means that the regional structures responsible for excitation of these modes possess resonant properties. The analysis revealed ten families of spectral modes in the period range of 50–150 s (Table 2). The statistics of the selected families, including the mean mode period value, $\overline{T}_i$ (where i is the family number) and the period's standard deviation (STD), are also characterised by the number of mode observation in this family and its repeatability. In this set of resonant modes, there are three modes with periods of 62.1, 64.2, and 67.9 s, which are observed in the vast majority of seismic events and the latter two modes are excited by all events. The intensity of these modes in the spectra is at its maximum or close to its maximum. Scattering of the periods of the spectral modes that form each of these families does not exceed a relative value of 10^{-3}.

To compare spectra, the spectral amplitude was normalised to the maximum mode amplitude for each spectrum in the frequency range between 0.005 and 0.03 Hz. As such, the spectral amplitude may vary between 0 and 1 with the most intense mode represented by an amplitude of 1. Two such spectra, excited by the most powerful earthquakes with magnitudes of 5.8 and

Table 2

Evaluated period values (in s) of resonant modes of regional structures excited by the "near zone" earthquakes (2005–2014 years) in the period range of 50–150 s

Events	Period, s									
	1	2	3	4	5	6	7	8	9	10
1. 25.01.2005	55.64	–	62.09	64.21	66.49	67.93	86.01	101.4	129.3	132.5
2. 14.03.2005	55.56	–	62.09	64.21	66.37	68.0	86.66	–	130.96	133.1
3. 11.10.2008	55.56	59.57	62.09	64.32	66.49	67.93	87.01	101.31	129.11	133.0
4. 07.09.2009	55.56	59.62	62.09	64.32	66.49	67.93	–	–	128.93	–
5. 08.03.2010	–	59.57	62.09	64.21	66.49	67.93	85.81	101.90	–	132.0
6. 14.11.2011	–	–	61.86	64.15	–	67.98	85.80	101.64	–	132.5
7. 23.12.2012	55.72	59.91	62.09	64.31	–	67.96	85.82	101.89	–	–
8. 17.09.2013	–	59.64	62.11	64.23	66.44	67.94	85.99	–	130.6	132.98
9. 29.06.2014	55.50	59.62	62.21	64.35	66.32	67.92	85.26	–	130.1	–
10. 29.09.2014	55.81	59.54	–	63.94	66.54	67.93	–	101.75	131.34	133.7
Mean value, s	55.61	59.63	62.08	64.22	66.44	67.94	86.04	101.62	130.03	132.85
STD, s	±0.10	±0.11	±0.08	±0.1	±0.08	±0.03	±0.51	±0.21	±0.87	±0.51
Number of observ.	7	7	9	10	8	10	8	6	7	7
Repeatability, %	70	70	90	100	80	100	80	60	70	70

The table also shows the mean period value of the mode family, its standard deviation, number of observations of given mode, and the repeatability (i.e., number of mode observations in relation to number of earthquakes)

6.0 (event numbers 4 and 5, Table 1), are shown in Fig. 4. The peak numbers 3, 4, and 6 indicate the periods of 62.1, 64.2, and 67.9 s, respectively (Table 2).

Further analysis consisted of constructing the averaged spectrum using an ensemble of the normalised spectra. Theoretically, the maximum conventional mode amplitude in the averaged spectrum can be equal to 10 if this mode is maximised in all initial spectra. The width of the spectral peak depends on how close the resonant modes are to one another, such that smaller distances between modes yield a narrower peak width and vice versa. Thus, the peak width is proportional to the standard deviation of the period determination (Table 2). The averaged spectrum is shown in Fig. 5. In the averaged spectrum, the most intense modes are numbers 3 ($T = 62.1$ s), 4 ($T = 64.2$ s), and 6 ($T = 67.9$ s). The low-frequency cumulative modes are also revealed; however, their intensity is significantly lower. Thus, the spectral analysis highlights the three most intense regional resonant modes that can be associated with magmatic structures of the Elbrus volcano.

Evaluation of the quality factor for each selected mode was based on the algorithm (Eq. 7), according to estimation of the mode line width in the corresponding spectrum. Refinement of the Q parameter is

significantly worse than the resonant frequency. Therefore, Table 3 shows the values of Q-factors that have been determined with sufficient confidence. Substantial scattering of the Q-factor values, which range from 215 to 315, is caused by both the imperfection of the assessment method and seismic noise.

Acoustic properties of magmatic fluids determine the Q-factor of the resonant modes and do not largely affect the resonance frequencies themselves. The Q-factors of the rock matrix (e.g., basalts, granites) are characterised by values of 100–150 (Lay and Wallace 1995). However, the resonant properties of magmatic structures of volcanoes can vary over a very wide range depending on magmatic fractions (Kumagai and Chouet 2000; Chouet 1996, 2003). Thus, the Q-factor can vary from a few units, for water or liquid basalt, to several hundreds for magmas consisting of a gas and liquid mixture.

Geological data on the composition of the Elbrus volcano indicate a significant role of mantle components in the melts that form the Elbrus volcanic center. Most likely, the magma formed from a mixing of basalt or basaltic andesitic melts with an acidic rhyodacitic crustal melt. It is also noted that the vast majority of Elbrus lava flows consist of

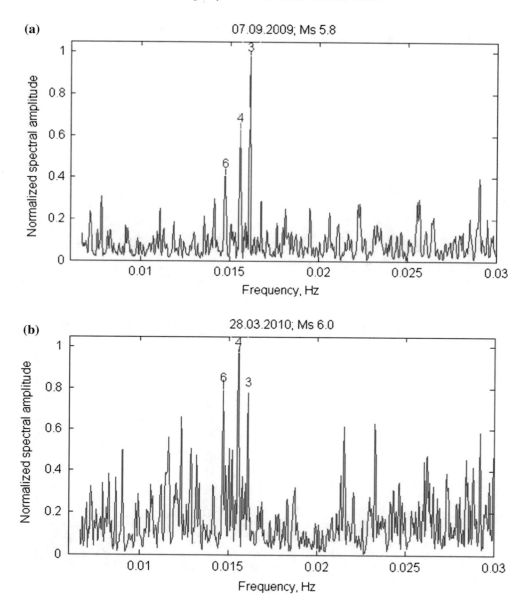

Figure 4
Normalised amplitude spectra of the regional resonant modes excited by the most powerful earthquakes from the "near zone" of the Elbrus volcano: **a** M_s 5.8 (event 4, Table 1); **b** M_s 6.0 (event 5, Table 1). The *numbers* indicate the periods in accordance with Table 1

dacites (wt. 70% SiO_2) and less prominent rhyolites of calc-alkaline series (Naumov et al. 2002). Thus, modeling of the acoustic properties of Elbrus volcanic fluids has been carried out with the assumption that the magmatic chamber is filled with magma consisting of dacite and dissolved carbon dioxide (CO_2). The magma chamber is embedded within a granitic matrix at temperatures of the order of

1200 °C. Within the framework of this model, the dependence on magma density, sound speed, and the Q-factor on a gas component of magmatic fluid for pressure values in the range of 0.5–4 kbar have been obtained (Milyukov et al. 2010). We compare the model Q-values with the real Q-factors estimated in this work, as shown in Fig. 6. The estimated values of the Q-factors for revealed modes are 250–270. These

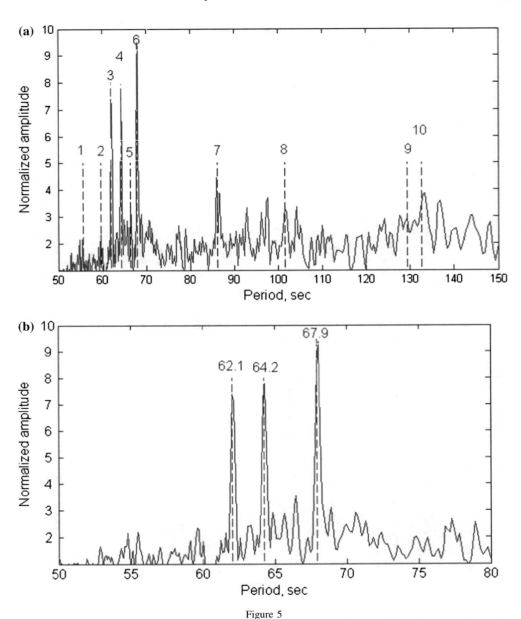

Figure 5
Spectrum of the regional resonance modes obtained by averaging over the ensemble of normalised spectra of ten earthquakes in the "near zone" of the Elbrus volcano: **a** in the period range of 50–150 s and **b** three most intensive modes. The periods of the modes are specified in seconds. *Numbers* of the spectral modes correspond to the notations in Table 2

values of the Q-factor satisfy the gas-volume fraction of 0.3–0.7 and pressure values of 0.5–2 kbar, corresponding to depths of 1–8 km.

Variation of the Q-factor over time has been investigated over an observation interval of 10 years. Due to the fact that Q-values are estimated with large uncertainties, an improvement of the estimation was achieved by increasing statistical analysis of events. Such analysis was based on a large number of events (~ 300) recorded by the strainmeter on both global and regional scales. For each particular earthquake, the Q-value was calculated as the average $\langle Q \rangle_i$ (i is the serial number of the earthquake) for all modes identified in a particular event and coincident with

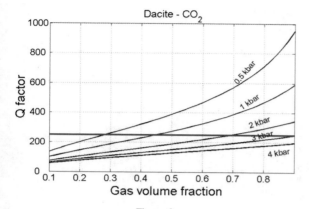

Figure 6
Model Q-factor values versus gas-volume fraction for pressures of
0.5–4 kbar and the estimated Q-factor values (*solid thick line*)

modes, as listed in Table 2. Results of the analysis are presented in Fig. 7.

3. Deep Structure and Geological Activity of the Volcano Elbrus

3.1. Method Outline

To estimate the deep structure beneath the Elbrus edifice and the entire region, we applied a method of passive seismo-prospecting known in the literature as the "microseismic sounding method" or MSM. MSM employs the assumption that the vertical component of the natural microseismic field consists predominantly of the fundamental Rayleigh modes, such that contributions from the higher modes are insignificant, and as a consequence, their effects can be treated as noise. MSM was initially formulated as a phenomenological method (Gorbatikov et al. 2008) based on the observation that the inclusion of seismic waves with velocities lower than those in the hosting medium causes an increase of amplitudes in background microseismic field above it, whereas the inclusion of higher velocities manifests itself by a decrease in the amplitude. This phenomenon was more thoroughly studied in a series of numerical simulations (Gorbatikov and Tsukanov 2011).

The MSM belongs to the group of passive methods of seismic prospecting and has demonstrated its ability to solve geological, geophysical, and structural problems for different classes of geological objects in various geographic and climatic conditions. To date, extensive insight has been gained through application of the MSM in scientific and industrial projects (Gorbatikov et al. 2008, 2009; Gorbatikov and Tsukanov 2011; Gorbatikov et al. 2013).

The MSM has a number of well-known analogs, including: (1) modifications of surface wave tomography that are based on estimating the phase part of the Green function from the cross correlation

Table 3

Evaluated Q-factor values of resonant modes of regional structures excited by the "near zone" earthquakes (2005–2014 years)

Events	1	2	3	4	5	6	7	8	9	10
	Period, s (mean value)									
	55.61	59.63	62.08	64.22	66.44	67.94	86.04	101.62	130.03	132.85
	Q-factor									
1. 25.01.2005	271	—	287	215	287	264	125	202	315	117
2. 14.03.2005	315	—	276	239	292	252	207	—	299	109
3. 11.10.2008	292	315	273	314	252	241	275	192	111	134
4. 07.09.2009	279	311	261	274	270	256	—	—	301	—
5. 08.03.2010	—	279	253	266	269	276	289	290	—	176
6. 14.11.2011	—	—	—	297	—	197	227	312	—	173
7. 23.12.2012	296	306	289	289	—	211	279	199	—	—
8. 17.09.2013	—	279	253	266	269	276	189	—	297	204
9. 29.06.2014	227	226	221	213	216	203	254	—	198	—
10. 29.09.2014	239	215	—	218	211	183	—	242	237	139
Mean value	269	272	264	259	257	265	230	239	251	151
STD	±35	±34	±20	±34	±29	±21	±52	±46	±69	±33

The table also shows the mean period value and mean Q-factor value with standard deviations for each mode family revealed in the analysis

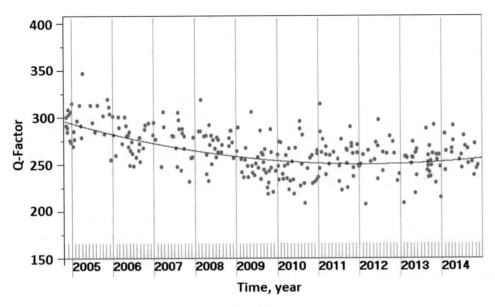

Figure 7
Q-factor dynamics of the resonant modes of the volcano Elbrus magma chamber over the period of 2005–2014

function; (2) modifications of the method of spatial correlation (SPAC); and (3) modifications of techniques analysing the ratios between the horizontal to vertical surface wave components (i.e., H/V method). Despite the wide variety of technology employed in its implementation (e.g., seismic arrays or single point measurements), all such approaches share the common fact that their underlying model of microseismic field formation requires a consistent locally layered medium. Therefore, the horizontal resolution of these methods can be estimated as $(3 \div 5)\,\lambda$, where λ is the wavelength of the fundamental Rayleigh mode that interacts with the heterogeneity.

In contrast to these approaches, the model of microseismic field formation assumed within the MSM does not require the medium to be layered. In this respect, it is believed that the main contribution to the microseismic field is provided by the fundamental Rayleigh modes, while the presence of the higher modes is minimal. The role of the informative parameter in the MSM is played by the distortion of the amplitude field caused by interaction with velocity inhomogeneities. The phase information is not used. The shape and depth of the inhomogeneity are estimated from the surface distribution of the distortions at the given frequency that such a disturbance

manifests itself. As demonstrated by (Gorbatikov and Tsukanov 2011), above the high-velocity heterogeneities (i.e., on the Earth's surface), velocities of the elastic waves in the heterogeneity are higher than in the hosting medium and the spectral amplitudes in a certain frequency band decrease. Above the low-velocity structures, on the other hand, where the velocities are lower than in the hosting medium, the spectral amplitudes increase.

There is a critical frequency f of the Rayleigh wave for which distortions from a inhomogeneity located at a depth H are maximal compared to the similar inhomogeneities located at other depths. This frequency is linked to H and the corresponding velocity of the fundamental Rayleigh mode, $V_R(f)$, by the formula $H \approx 0.4\, V_R(f)/f$. This relationship has been confirmed by both field studies of geological objects of different scales and origins, as well as by numerical simulations, and is used for the inverse procedure of estimating the depth of an unknown inhomogeneity that forms amplitude distortions observable on the surface at frequency f.

According to the numerical simulations (Gorbatikov and Tsukanov 2011), resolution of the method in reconstructing images along the horizontal is $(0.25 \div 0.3)\,\lambda$, where λ is the effective wavelength of the sounding wave. The vertical resolution is estimated as $(0.3 \div 0.5)\,\lambda$, where λ is the effective wavelength for the average depth between the inhomogeneities. It was also shown that the presence of an isolated small inclusion can be detected even if it is as small as one tenth of the wavelengths, or even smaller.

The fieldwork technology is reduced to the accumulation of the power spectrum of the microseismic signal during a certain time successively at each subsequent point along the profile (or network) with one or a few portable sensors. Simultaneously, the microseismic signal should be recorded at the reference point located around the studied area to apply a correction to eliminate the non-stationary sounding of the microseismic signal. Both profile and aerial surveys can be carried out depending on the stated problem.

The direct problem of scattering of the Rayleigh fundamental modes by individual separate bodies of a simple shape was considered in detail in (Gorbatikov and Tsukanov 2011). The effects of the buried inclusion on the spatial distribution of the power spectrum of Rayleigh waves on the surface were calculated using a software package developed for this purpose. Bodies of different shapes, sizes, depths, and velocity contrasts relative to the hosting medium were analysed. However, the study was limited only to the cases in which the inclusion and hosting medium had equal Poisson ratios. It has been shown that the amplitude response is insensitive to the shape of the inclusion with contrasting velocity if the size of the inclusion does not exceed $\lambda/4$, where λ is the wavelength of the fundamental Rayleigh mode corresponding to the depth of the inclusion. Nevertheless, it is possible to detect the presence of the inclusion itself and to determine the sign of the velocity contrast. In terms of depth, we may state that images of two inclusions with velocity contrasts of the same sign should merge if the distance between them is half of their occurrence depth or less.

However, some of the obtained images found in the field observations disagree with results of the numerical simulations. This particularly concerns cases when two sub-vertical bodies manifested separately at depths where they had been predicted to merge (i.e., the effect of super-resolution). This inconsistency between simulations and observations required an additional study, which showed that super-resolution can exist when the Poisson ratio of the inclusion material approaches zero (Gorbatikov et al. 2013). For natural objects, this can indicate the presence of microfracturing or porosity. To date, numerical modeling has shed light on the problems of the influence of the Poisson ratio of the inclusion on its image in the microseismic field, as well as the estimation of the nonlinear effect in the amplitude response using the MSM for two closely located buried bodies.

The solution to the inverse problem in the MSM is not unique, as is the case in the most of geophysical methods. It is impossible to reconstruct a complex combination of the medium parameters from simply the amplitude response distribution in a microseismic field. Nevertheless, the MSM sections provide sufficiently useful information for geological and tectonic interpretations.

It is interesting to mention that sub-vertical geological structures and velocity boundaries are preferable for the MSM, while sub-horizontal boundaries are more favorably investigated using seismic reflection methods. This can be explained by the mutual spatial position of the wave fronts and velocity boundaries relative to each other. For example, an ideal horizontal (within the profile) velocity boundary will not be visible at all in a cross-sections using the MSM, similar to the way the presence and positions of ideal sub-vertical faults can be hardly found using the seismic reflection method.

3.2. Observations with MSM and Results

During the fieldwork period of 2014–2015, the MSM profile was measured to establish the deep structure of the Elbrus volcano, as well as the structure of the region. The profile (Fig. 8) extended along the Baksan River valley from the foot of the eastern summit of Mount Elbrus (5100 m above sea level) to the southern edge of the town of Tyrnyauz. The average distance between measurement stations was maintained at about 500 m. The profile broke naturally into three fragments of different directions: a sub-meridional section from ~ 0 to 7 km of the profile; a sub-latitudinal section from ~ 7 to 34 km; and a northeastern section from ~ 34 to 48 km. The measurement process consisted of: (1) installing (i.e., burying the small hole) broadband high-resolution

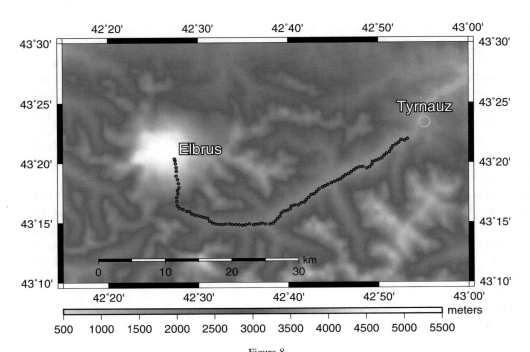

Figure 8
Map of the MSM observation area. The MSM profile stations are indicated by *red points* and altitudes are scaled by *colour (horizontal bar)*

seismometer (Guralp CMG-6TD) at each subsequent point along the profile; (2) recording the microseismic background signal in the frequency band of 0.03–50 Hz over a period of at least 2 h; and (3) subsequent relocation of the instrument to next point of the profile. At the same time as recording the signal at the profile points, the microseismic signal was also recorded at the basic station placed near the center of the profile. Simultaneous recordings are required to reduce the natural trend deviations and avoid confusion with spatial variations.

The data processing included: (1) estimation of the accumulated power spectrum of microseismic signal for every point of the profile based on the 2-h record (Fig. 9); (2) calculation of the logarithmic difference between the power spectrum at the profile point and the power spectrum at the basic station; and (3) estimation of spatial amplitude variations along the profile for every frequency of the spectrum by means of subtracting the spatial average amplitude from the amplitudes at every point of profile. Finally, every spatial variation curve corresponding to each frequency f_i in the spectrum was linked to the depth according to $H \approx 0.4\,V_R(f_i)/f_i$, as mentioned

previously. Figure 10 shows examples of the velocity amplitude distribution along the profile for several chosen periods of the microseismic signal related to the corresponding amplitudes on the basic station. A more detailed description of the measuring and processing stages with examples of records can be found in (Gorbatikov et al. 2013).

The constructed cross section along the profile is shown in Fig. 11. Deviations of shear seismic waves velocities from the averaged regional seismic velocity model are observed over a depth range of 0–40 km. One can distinguish three low-velocity areas located one above the other and deeper tracing into the crust in the section under the central part of Mount Elbrus. Another low-velocity layer associated with deposits of volcanic products can be identified in the altitude range of 2.5–3.8 km inside the edifice of the Elbrus volcano. A relatively distinct low-velocity body (indicated by the number 1 in Fig. 11) can be seen over a depth range of 7–12 km. It is logical to assume that this body formed due to the presence of magmatic melt. This assumption can also be confirmed by the occurrence of fumaroles near the eastern summit of Mount Elbrus that were studied during expeditions

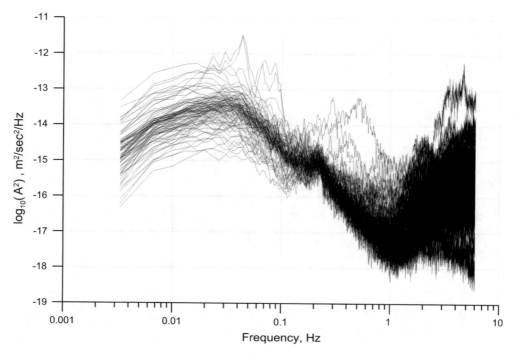

Figure 9
Power spectrum densities for all points of the MSM profile

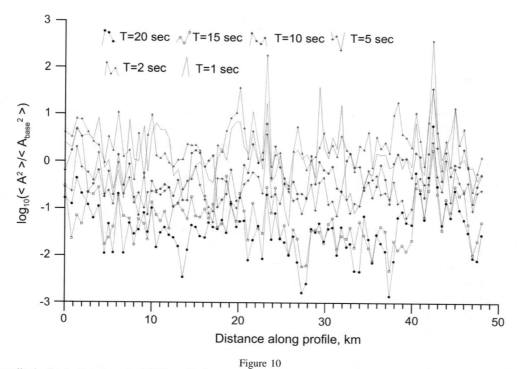

Figure 10
Velocity amplitude distribution along the MSM profile for several chosen periods of the microseismic signal related to the corresponding
amplitudes recorded by the basic seismometer at the same time as recordings made at the profile points

309

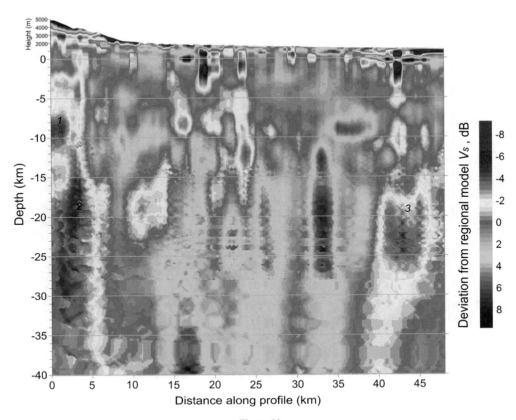

Figure 11
Cross section along the Elbrus–Tyrnyauz profile based on the microseismic sounding data. The degree of deviation (in decibels) from the seismic shear wave smoothed regional model is scaled by *colour* (*vertical bar*)

under the leadership of I.I. Serdyukov, a mountaineer and scientist. Finally, the deepest and largest low-velocity body can be observed in the middle crust over a depth range of 15–30 km (indicated by the number 2 in Fig. 11). The biggest body has no distinct lower boundary and continuously extends to a depth of 40 km and deeper. No other sites of the section show such a distinct extent deep into the lithosphere.

In the section of the sub-aerial edifice of the Elbrus volcano itself, one can see two distinct layers, one with higher and another with lower seismic velocities. The upper high-velocity layer has a thickness of about 1 km near the eastern summit and lenses out near measuring point 3 (end of 5th km of profile, Fig. 11). The underlying low-velocity layer is located in the altitude range of 2800–3800 m and is exposed at the surface in the area of between measuring points 5–7 (~6th km of profile, Fig. 11). A high-velocity layer, composed of young

glassy viscous dacite lavas of moderately acidic composition from the last eruption phase (less than 35,000 years old), covers the ledges and most gentle slope areas within the profile. The low-velocity layer is composed of lavas from earlier phases of volcanic activity, which overlie Hercynian basement rocks (Fig. 11). Horizons made of lavas from early phases of volcanic activity of the Elbrus volcano crop out in a large erosive window (points 4–8, ~6 to 8 km of profile), washed by Malyi Azau Creek that originates from the glacier.

In the area of point 20 (~14th km of profile), an SSW-trending fault zone can be inferred. This fault zone was identified on the surface owing to the reorientation of rock foliation and strong plicative structures. On the section, the fault looks like an inclined beam, characterised by low-velocity seismic waves, which plunges to a depth of about 10 km from the surface.

The vertical structure ($\sim$42th km of the profile) that looks like a cooled upwelling magma channel is the apparent center of smooth subsidence of subhorizontal boundaries to depths of 1–2 km. One can assume that this is related to tectonic subsidence above the depleted magma chamber. On the surface, this submerged structure is manifested as a negative gravity anomaly. A similar but even more intensive negative gravity anomaly accompanies the Elbrus volcano.

4. Discussion

In the Elbrus volcanic edifice, directly beneath its eastern summit in the depth range of 7–12 km, a low-velocity volume was identified by the MSM that is clearly associated with the assumed Elbrus magma chamber. According to the resonant method, obtained values of the Q-factor suggest that the identified modes are generated by structures containing magmatic fluids, with the three most intensive modes possessing periods of 62.1, 64.2, and 67.9 s, which delineates the low-frequency part of the spectrum of the Elbrus magma chamber. According to the spherical model (Fujita et al. 1995; Milyukov 2006), the size of the magma chamber, which has a main resonance at these periods, is equal to $\sim$9 km. According to our data, the Elbrus magma can be classified as gas–liquid foam with gas-volume fraction values of 0.3–0.7. This result is consistent with the fact that products of the Elbrus volcanic eruptions have a porous structure, which may indicate a high gas component (Bogatikov et al. 2003). The depth of magma chamber is 1–7 km.

The acoustic properties of the chamber as a mechanical resonator are defined by the state of the geological medium. The Q-factor of the resonator is an indicator of this state. The change of the Q-factor of resonant modes over time can be due to two processes: (1) changes in the gas component of the magma (i.e., solution or dissolution of volatile components within the magmatic fluids) and (2) changes in the pressure conditions within a magma chamber. Both of these processes can be caused by the following thermodynamic states of the magma chamber:

1. Adiabatic process (i.e., no heat flow) at low pressures near the surface. In a magma chamber, volatile components are dissolved or exsolved due to a change in pressure. The Q-factor value (indicator of the state) may both decrease and increase.

2. Exothermic process (with changing the energy conditions) at moderately high pressures. The pressure increases due to the arrival of hot lava from a deeper source. The volatiles can remain unchanged or even decrease. The Q-factor decreases.

Estimation of the Q-factor over a 10-year range shows different dynamics (Fig. 7). From 2004 to 2009, the Q-factor decreased due to increasing chamber pressure that could have been caused by both the process of outgassing and/or arrival of hot lava. Since 2009, there is a tendency for a possible stabilisation of the Q-factor values that may indicate a transition of the magmatic structures into a quasi-stable state of thermodynamic equilibrium.

In regard to the magma source of the Elbrus volcano, the MSM indicates a major low-velocity volume, as well as beneath the eastern summit of Mt. Elbrus, which begins at a depth of 15 km and extends continuously down to depths of 40 km. Apparently, this deep structure represents the magmatic source of the volcano.

The low-velocity structure detected by the MSM in the eastern part of the Elbrus–Tyrnyauz cross section at depths of 15–30 km suggest the presence of a hidden major magma source, comparable in magnitude with the Elbrus source, but currently much colder (area 3 in Fig. 11). The depths and comparable spatial dimensions of the volcanic chambers (areas 2 and 3 in Fig. 11) beneath the volcanic structures are noteworthy.

5. Conclusion

The results obtained in the present work confirm previous reports (Bogatikov et al. 2002b; Spichak et al. 2007) on the presence of the magma chamber within the Elbrus volcanic edifice at depths of 1–8 km with a characteristic size of about 9 km and an extended magma source at depths of 15–40 km. In addition, a new magmatic structure, located 50 km to the east of

Mt. Elbrus, has been discovered. This unknown magmatic structure is comparable to the Elbrus magmatic structure, but is currently much colder.

The aim of our study is to report the contemporary state of the Elbrus volcanic structures. The data on the dynamic state of fluids of the magma chamber over the past decade indicate possible pressure variations in the chamber, which may be associated with outgassing processes. However, we can now conclude with greater certainty regarding the likelihood of a quasi-stable state of thermodynamic equilibrium. Therefore, previous assumptions about the transition of the Elbrus volcano from a passive to a more active stage (Gurbanov et al. 2013) are perhaps somewhat exaggerated.

The experimental data, obtained by the resonance method and MSM, and interpretation of the results indicate that Elbrus is in a long seismic lull phase yet continues to retain the risk of volcanic activation. There is the well-known case of Mt. St. Helens (California, USA) when the previously observed weak seismic activity increased dramatically over a 2-week period prior to eruption in 1980. In this regard, regular monitoring of various volcanic processes and phenomena is the most effective method to solve the problem of the forecast probability of volcano activation. The development of new methods and comprehensive monitoring of the Elbrus volcano is highly relevant and can play an important role in obtaining reliable forward-looking information.

Acknowledgements

This work is partly supported by the Grant RFBR No. 16-05-00122.

REFERENCES

Bogatikov, O. A., Gurbanov, A. G., Koshchug, D. G., Gazeev, V. M., & Shabalin, R. V. (2002a). The EPR dating of the rock-forming quartz from volcanic rocks of the Elbrus volcano, Northern Caucasus, Russia. *Doklady Earth Sciences, 385*(5), 570–573.

Bogatikov, O. A., Nechaev, Yu V., & Sobisevich, A. L. (2002b). Application of space technologies for the monitoring of geological structures of the Elbrus volcano. *Doklady Earth Sciences, 387*(9), 993–998.

Bogatikov, O. A., Gurbanov, A. G., Koshchug, D. G., Gazeev, V. M., Shabalin, R. V., Dokuchaev, A. Y., et al. (2003). Main evolutionary phases of Elbrus volcano (Northern Caucasus, Russia) based on EPR dating of quartz. *Journal of Volcanology and Seismology, 3*, 3–14. **(in Russian)**.

Chernyshev, I. V., Lebedev, V. A., Bubnov, S. N., Arakelyants, M. M., & Goltsman, Y. V. (2001). Stages of magmatic activity in the Elbrus volcanic center (Greater Caucasus): Evidence from isotope-geochronological data. *Doklady Earth Sciences, 380*(7), 848–852.

Chouet, B. A. (1985). Exitation of a buried magmatic pipe: A seismic source model for volcanic tremor. *Journal of Geophysical Research: Solid Earth, 90*(B2), 1881–1983. doi:10.1029/JB090iB02p01881.

Chouet, B. A. (1996). Long-period volcano seismicity: Its source and use in eruption forecasting. *Nature, 380*, 309–316. doi:10.1038/380309a0.

Chouet, B. A. (2003). Volcano seismology. *Pure and Applied Geophysics, 160*(3—-4), 739–788. doi:10.1007/PL00012556.

Crosson, R. S., & Bame, D. A. (1985). A spherical source model for low frequency volcanic earthquakes. *Journal of Geophysical Research: Solid Earth, 90*(B12), 10237–10247. doi:10.1029/JB090iB12p10237.

Fujita, E., Ida, Y., & Oikawa, J. (1995). Eigen oscillation of a fluid sphere and source mechanism of harmonic volcanic tremor. *Journal of Volcanology and Geothermal Research, 69*(3–4), 365–378. doi:10.1016/0377-0273(95)00027-5.

Gorbatikov, A. V., & Tsukanov, A. A. (2011). Simulation of the Rayleigh waves in the proximity of the scattering velocity heterogeneities. Exploring the capabilities of the microseismic sounding method. *Izvestiya, Physics of the Solid Earth, 47*(4), 354–369. doi:10.1134/S1069351311030013.

Gorbatikov, A. V., Stepanova, M Yu., & Korablev, G. E. (2008). Microseismic field affected by local geological heterogeneities and microseismic sounding of the medium. *Izvestiya, Physics of the Solid Earth, 44*(7), 577–592. doi:10.1134/S1069351308070082.

Gorbatikov, A. V., Larin, N. V., Moiseev, E. I., & Belyashov, A. V. (2009). The microseismic sounding method: Application for the study of the buried diatreme structure. *Doklady Earth Sciences, 428*(1), 1222–1226. doi:10.1134/S1028334X0907040X.

Gorbatikov, A. V., Montesinos, F. G., Arnoso, J., Stepanova, M. Y., Benavent, M., & Tsukanov, A. A. (2013). New features in the subsurface structure model of El Hierro Island (Canaries) from low-frequency microseismic sounding: An insight into the 2011 seismo-volcanic crisis. *Surveys in Geophysics, 34*(4), 463–489. doi:10.1007/s10712-013-9240-4.

Gurbanov, A. G., Bogatikov, O. A., Karamurzov, B. S., Lexin, A. B., Gazeev, V. M., Tsukanov, L. E., et al. (2013). Results of the evaluation of the current status of "sleeping" Elbrus. *Journal of Vladikavkaz Scientific Center, 13*(4), 36–50. **(in Russian)**.

Iafolla, V., Milyukov, V., & Nozzoli, S. J. (2008). Observations of mount Etna seismicity during the 2002–2003 eruption based on deep-sea gravimeter data. *Journal of Volcanology and Seismology, 2*(4), 289–299. doi:10.1134/S0742046308040064.

International Seismological Centre. (2014). *On-line Bulletin*. Internatl. Seismol. Cent., Thatcham, United Kingdom. http://www.isc.ac.uk

Kozyreva, A. V., & Milyukov, V. K. (2003). Assessment of resonance characteristics of the Elbrus volcano magma source by strain observations. *Moscow University Physics Bulletin, 58*(2), 66–72.

Kumagai, H., & Chouet, B. A. (2000). Acoustic properties of a crack containing magmatic or hydrothermal fluids. *Journal of Geophysical Research: Solid Earth*, *105*(B11), 25493–25512.

Lay, T., Wallace, T. (1995). *Modern Global Seismology, International Geophysics Series* (vol. 58, 1st ed.). San Diego: Academic.

Lebedev, V. A., Chernyshev, I. V., Bubnov, S. N., & Medvedeva, E. S. (2005). Chronology of magmatic activity of the Elbrus volcano (Greater Caucasus): Evidence from K-Ar isotope dating of lavas. *Doklady Earth Sciences*, *405*(9), 1321–1326.

Lebedev, V. A., Chernyshev, I. V., Chugaev, A. V., Goltsman, Y. V., & Bairova, E. D. (2010). Geochronology of eruptions and parental magma sources of Elbrus volcano, the Greater Caucasus: K-Ar and Sr-Nd-Pb isotope data. *Geochemistry International*, *48*(1), 41–67. doi:10.1134/S0016702910010039.

Milyukov, V., Kopaev, A., Zharov, V., Mironov, A., Myasnikov, A., Kaufman, M., et al. (2010). Monitoring crustal deformations in the Northern Caucasus using a high precision long base laser strainmeter and the GPS/GLONASS network. *Journal of Geodynamics*, *49*(3–4), 216–223. doi:10.1016/j.jog.2009.10.003.

Milyukov, V. K. (2006). Monitoring of the state of Elbrus magmatic structures using lithospheric strain observations. *Journal of Volcanology and Seismology*, *1*, 3–15. **(in Russian)**.

Milyukov, V. K., & Myasnikov, A. V. (2005). Metrological characteristics of the Baksan laser interferometer. *Measurement Techniques*, *48*(12), 1183–1190. doi:10.1007/s11018-006-0042-7.

Milyukov, V. K., Klyachko, B. S., Myasnikov, A. V., Striganov, P. S., Yanin, A. F., & Vlasov, A. N. (2005). A laser interferometer-deformograph for monitoring the crust movement. *Instruments and Experimental Techniques*, *48*(6), 780–795. doi:10.1007/s10786-005-0140-9.

Naumov, V. B., Tolstoy, M. L., Gurbanov, A. G., Gazeev, V. M. (2002) The composition of magmatic melt inclusions in phenocrysts of quartz, plagioclase from the successive evolutionary series of lava and pyroclastic flows of the Elbrus volcanic center. In M. Laverov (ed) *Catastrophic processes and their impact on the environment, Volcanism* (vol. 1, pp. 311–320). Moscow: Publishing House of Institute of Physics of the Earth, Russian Academy of Sciences **(in Russian)**.

Puzyrev, N. N. (1997) *Methods and objects of seismic research. Introduction to general seismology.* Novosibirsk: Publishing House of Siberian Branch of the Russian Academy of Sciences **(in Russian)**.

Sobisevich, A. L., Rudenko, O. V., Milyukov, V. K., Nechaev, Y. V. (2002) Monitoring induced geophysical processes in heterogeneous structures of the geological environment of central volcanoes. In M. Laverov (ed) *Catastrophic processes and their impact on the environment, Volcanism* (Vol. 1, pp. 365–397). Moscow: Publishing House of Institute of Physics of the Earth, Russian Academy of Sciences **(in Russian)**.

Spichak, V. V., Borisova, V. P., Fainberg, E. B., Khalezov, A. A., & Goidina, A. G. (2007). Electromagnetic 3D tomography of the Elbrus volcanic center according to magnetotelluric and satellite data. *Journal of Volcanology and Seismology*, *1*(1), 53–66. doi:10.1134/S0742046307010046.

(Received December 16, 2016, revised June 13, 2017, accepted June 16, 2017, Published online June 26, 2017)

Pure Appl. Geophys. 175 (2018), 1909–1923
© 2017 Springer International Publishing AG
https://doi.org/10.1007/s00024-017-1639-2

Pure and Applied Geophysics

Strain Accumulation and Release of the Gorkha, Nepal, Earthquake (M_w 7.8, 25 April 2015)

Federico Morsut,[1] Tommaso Pivetta,[1] Carla Braitenberg,[1] and Giorgio Poretti[1]

Abstract—The near-fault GNSS records of strong-ground movement are the most sensitive for defining the fault rupture. Here, two unpublished GNSS records are studied, a near-fault-strong-motion station (NAGA) and a distant station in a poorly covered area (PYRA). The station NAGA, located above the Gorkha fault, sensed a southward displacement of almost 1.7 m. The PYRA station that is positioned at a distance of about 150 km from the fault, near the Pyramid station in the Everest, showed static displacements in the order of some millimeters. The observed displacements were compared with the calculated displacements of a finite fault model in an elastic halfspace. We evaluated two slips on fault models derived from seismological and geodetic studies: the comparison of the observed and modelled fields reveals that our displacements are in better accordance with the geodetic derived fault model than the seismologic one. Finally, we evaluate the yearly strain rate of four GNSS stations in the area that were recording continuously the deformation field for at least 5 years. The strain rate is then compared with the strain released by the Gorkha earthquake, leading to an interval of 235 years to store a comparable amount of elastic energy. The three near-fault GNSS stations require a slightly wider fault than published, in the case of an equivalent homogeneous rupture, with an average uniform slip of 3.5 m occurring on an area of 150 km × 60 km.

1. Introduction

The Gorkha district (Nepal) was struck on the 25th of April 2015 by a 7.8 M_w earthquake that caused over 8000 victims, over 20,000 injured and destroyed several villages and cities in the Kathmandu area. The earthquake occurred in one of the most tectonically active areas of the Earth, where the collision between the Indian and Eurasia plates generates several seismic sources, capable of catastrophic earthquakes up to 8 M_w (Z12 zone of Chaulagain et al. 2015; Rajendran and Rajendran 2011). The superficial effect of this earthquake was

recorded by different geodetic and seismological networks spread over the whole Nepal area: in particular, several continuous GNSS stations were active in the surrounding of the epicenter during the earthquake. For this reason and obviously for the terrible impact on the population and human activities, the Gorkha earthquake was widely studied (Avouac et al. 2015; Galetzka et al. 2015; Grandin et al. 2015; Sreejith et al. 2016; Arora et al. 2017).

Galetzka et al. (2015) performed an analysis on GNSS (Global Navigation Satellite System) and InSar (Interferometric Synthetic Aperture Radar) interferometry data which were jointly inverted to retrieve the spatial distribution of the slip on the fault. Similarly, Wang and Fialko (2015) performed an inversion of geodetic data, investigating different assumptions of the fault geometries. Galetzka et al. (2015) additionally derived from the continuous GNSS time series and the teleseismic records a model for the slip rate release on the earthquake, providing a kinematic model of the rupture. Grandin et al. (2015) included strong-motion data, teleseismic, high rate GNSS, and InSAR in a joint inversion scheme to gain insights into the whole rupture process. The inverted slip pattern is in accordance with Galtezka et al. (2015), but seems a little greater in certain zones. Avouac et al. (2015) analyze teleseismic records and SAR images to study the earthquake in terms of nucleation and rupture process. Sreejith et al. (2016) analyzed coseismic and post-seismic deformations from a combination of GNSS and InSAR observations. They consider the fault to be constituted by a combination of flat and a ramp structure; the authors inverted for both coseismic and post-seismic deformations. The retrieved fault slip model agrees with Galetzka et al. (2015) in terms of both magnitude and direction. The difference between observation and calculation reported in the contribution evidences a

---

[1] Department of Mathematics and Geosciences, University of Trieste, Via Weiss 1, Trieste, Italy. E-mail: tpivetta@units.it

slight underestimation of the GNSS data. For this model, the coseismic slip occurred prevalently on the ramp, while the post-seismic slip involved the northern sector along the flat structure. Denolle et al. (2015) studied the dynamics of the event by computing P wave spectra. Fan and Shearer (2015), from the analysis of teleseismic P waves, revealed that the earthquake was an almost unilateral rupture that propagated in the east–west direction with a rupture length of 165 km. Supplementary GNSS stations were also added immediately after the main shock, to monitor the post-seismic fault movements. Gualandi et al. (2016) and Mencin et al. (2016) testified the presence of a post-seismic afterslip that released an aseismic moment equivalent of 7.1 M_w (Mencin et al. 2016).

Arora et al. (2017) analyze the time space pattern of aftershock activity and find evidence for high pore pressure fluid fluxes into the crust.

Regarding the coseismic slip, the contributions, although diverging regarding the assumptions on the physical model, revealed that the slip occurred on a 150 km $\times$ 50-km portion of the fault that gently dips towards North–North-East (azimuth 290°E from N, dip 8–10°) and constitutes an important fault of the Main Himalayan Thrust. The extent of the fault area is in accordance also with the distribution and location of the aftershocks. According to the focal mechanism solution from seismology, the earthquake is almost a pure compressional event (USGS).

The scientific group EverestK2 (EvK2), in collaboration with the CNR (Italian National Council of Research), placed a continuous GNSS station (PYRA) in 2009 next to the Pyramid Laboratory in the Everest Mountain. Another one was placed in Nagarkot (NAGA) near Kathmandu, in Nepal. Both stations were programmed to sample the position every 30 s; from 2014, both the time series present gaps in the data, due to logistic problems in the data management which occurred during this period.

Fortunately, the stations were active in the area in the period from March 2015 to October 2015; hence, they were able to record the geodetic effect of the Gorkha earthquake and also the relevant 7.3 M_w aftershock which occurred in May 2015. The NAGA station is above the fault, increasing the near-fault observation from two to three stations. Our station

confirms that the models underestimate the dislocation in the near field.

In this contribution, we present the elaboration of the GNSS time series from the 23rd to the 26th April 2015 and a comparison with other geodetic data available for the area. We also tried to verify the compatibility of our coseismic superficial deformation with the predicted deformation calculated from a fault model, using the Okada dislocation model in an elastic halfspace (Okada 1985).

Finally, we present an analysis on the recurrence time of the earthquake in the area based on the comparison of the strain released by the Gorkha fault and the interseismic strain rate estimated from other long period GNSS time series. The work contributes in demonstrating the usefulness of the geodetic measurements in defining the strain accumulation and release during the seismic cycle, which is an important aspect for the characterization of the seismic risk of an area.

2. GNSS Elaboration

To process our GNSS data from PYRA and NAGA stations, we used 30 s sample RINEX files for both of them, as well as other files for the Bernese Software 5.2 like various Center for Orbit Determination in Europe (CODE) products for raw corrections (satellite ephemerides, Earth parameters, clock bias corrections, ionosphere parameters, and infra-code biases). For a very precise position, we need also the Vienna Mapping Function 1 parameters (VMF1, Böhm et al. 2006), used to mitigate tropospheric errors.

The first step consisted in calculating station positions precisely, in fact using the precise point position technique (PPP-PPP_VMF1.PCF script), by which it is possible to determine the absolute position of our stations in the ITRF 2008 system. As another useful result, we estimated a first plate velocity (Eurasian and Indian for both stations). During this step, we also calculated other station positions to realize a second more precise estimation of positions: the Lhasa (LHAZ) and Kitab (KIT3) stations belonging to the IGS (International GNSS Service) Network.

The second step consisted in using the double-difference technique (DD-RNX2SNX_VMF1.PCF) with a kinematic approach, where we have a position estimation for every sample (every 30 s). This approach is more precise than PPP; in fact, we can have a millimeter precision.

The time series of displacements of our stations were estimated from double-difference solutions using a kinematic approach. The locations of the two stations are presented in Table 1. In Fig. 2, we present the three components (N, E, and U) in UTM 45 R from original Bernese Earth Centered, Earth Fixed (ECEF) coordinates. We analyzed 3 days from 24th to 26th of April 2015. The two stations have been analyzed thus with the same procedure, that is first with the PPP procedure, and then with the double difference. The millimetric noise level is similar to the two stations, but the different scales used in Figs. 1 and 2 for illustrating the time series are very different, due to the fact that the dislocation is over 1 m in NAGA, and at centimetric level in PYRA.

3. GNSS Data Analysis: Comparison Between Modelled and Observed Coseismic Data

The NAGA station clearly detected on all the three components of the coseismic surface displacement, as is evident from Fig. 1. The station is close to

other two GNSS stations used by Galetzka et al. (2015) to retrieve the slip on the fault: magnitude and direction of the displacement are in agreement with those stations. With our station, the near-fault coseismic movement has been detected by three stations, increasing the number of near-fault observations. The observation of the new station with a movement in the southern direction (1.76 m) and an uplift of almost 1.2 m is coherent with the existing stations, allowing a robust estimate of the coseismic displacement. In the next 36 h, other three greater 6 M_w events occurred according to USGS catalogue (6.1 M_w, 25/04/2015 at 06:15:22; 6.6 M_w, 25/04/2015 at 06:45:21; and 6.7 M_w, 26/04/2015 at 07:09:10); however, no coseismic displacement has been produced.

The PYRA station (Fig. 2) shows instead a slighter displacement: the effect of the earthquake is more visible on the North (2 cm) component with respect to the East (1 cm). The up/down movement seems to be absent, but could be masked by the higher noise on the height component.

As already stated, GNSS time series acquired at high sampling rates requires precise removing of the atmospheric effects to not alter the tectonic signal. In particular, in these mountainous areas, snow precipitation that accumulates on the top of the GNSS antenna could significantly introduce errors in the positioning system.

To exclude the influence of rapid changes in the atmospheric conditions, not modelled by the VMF1 corrections, we plot temperature, pressure, and snow level for the PYRA station, next to the observed time series. The data were obtained from a CNR meteorological station placed near the GNSS station. All the data have been down sampled from the original 1 s to a 30 s sampling after an anti-alias filter was applied. As is evident from Fig. 2, no clear correlation exists between the deformation time series and the atmospheric conditions. Calculating the Pearson coefficient between each component of displacement and pressure and temperature time series, we obtain values ranging between 0.08 and 0.12 testifying scarce influence of the atmospheric variables and displacement. We calculated the Pearson coefficient also for the GNSS-filtered time series, with a 51 point running average, which minimizes the high-

Table 1

Location of the GNSS stations analyzed in this study. The first ten stations were considered also in Galetzka et al. (2015)

#	Name	Latitude (°)	Longitude (°)
1	DNGD	28.754	80.582
2	DNSG	28.345	83.763
3	JMSM	28.805	83.743
4	KKN4	27.801	85.279
5	NAST	27.657	85.328
6	NPGJ	28.117	81.595
7	PYUT	28.101	82.987
8	RMTE	26.991	86.597
9	SMKT	29.969	81.807
10	TPLJ	27.352	87.710
11	PYRA	27.957	86.815
12	NAGA	27.693	85.521

The last two are the new stations, which are presented in detail in the following

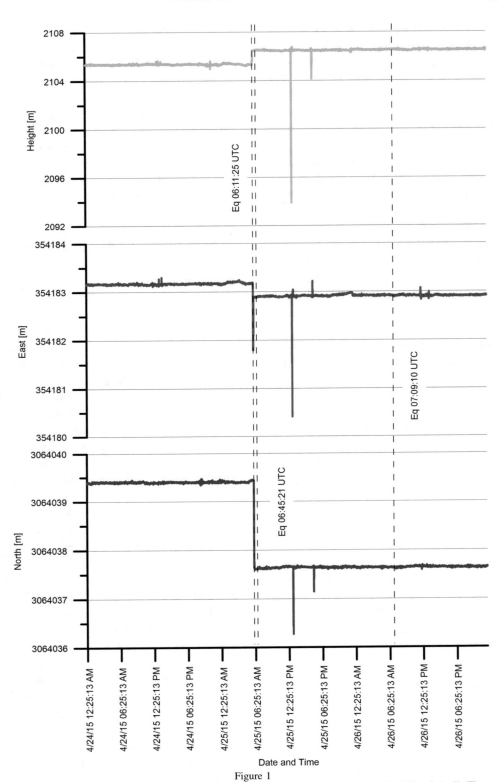

Figure 1
Displacement time series of the three components observed at NAGA station from the 24th to the 26th of April. The earthquake and aftershock ($M_w > 6$) occurrence times are marked as *bold vertical dashed lines*

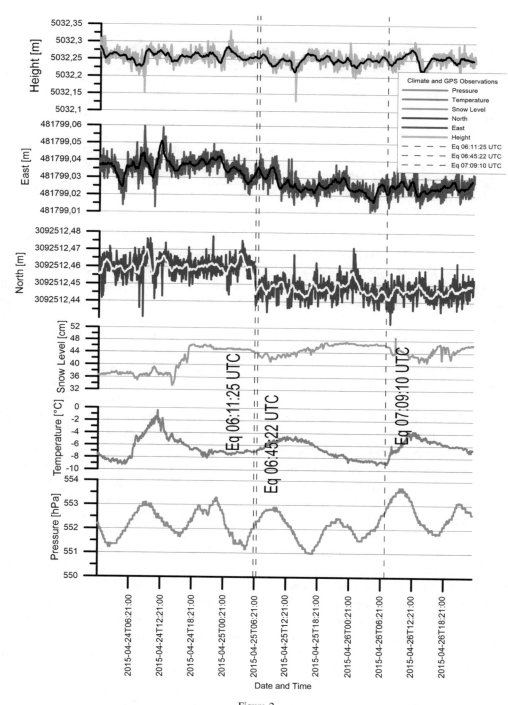

Figure 2

Displacement time series of the three components observed at the PYRA station from the 24th to the 26th of April 2015 reported together with the filtered time series, with a 51 points running average (*black* and *white lines*). The earthquake time is marked as *vertical dashed line*. Temperature, snow level, and pressure are also plotted. The meteorological data are courtesy of EvK2-CNR Pyramid Laboratory. As is evident, no correlation exists between the time series and the atmospheric parameters

Table 2

Summary of the position statistics before and after the Gorkha earthquake for NAGA and PYRA stations

PYRA	North (m)	RMS (m)	East (m)	RMS (m)	Height (m)	RMS (m)
24/04/2015	3092512.46151	0.00049	481799.03685	0.00056	5032.25657	0.00220
26/04/2015	3092512.44653	0.00044	481799.02426	0.00051	5032.25448	0.00198
Delta	−0.01498	0.00066	−0.01259	0.00076	−0.00209	0.00296
NAGA	North (m)	RMS (m)	East (m)	RMS (m)	Height (m)	RMS (m)
24/04/2015	3064039.39881	0.00052	354183.16418	0.00063	2105.35817	0.00236
26/04/2015	3064037.63777	0.00047	354182.91115	0.00054	2106.49761	0.00204
Delta	−1.76104	0.0007	−0.25303	0.00083	1.13944	0.00312

The uncertainty of each component, expressed in terms of root mean square (RMS) error, is also included. The RMS Delta is calculated with the error propagation, assuming that the covariance terms equal zero

frequency noise. In addition, in this case, the Pearson coefficient is low with values between 0.09 and 0.16. As a consequence, the E–W component, that showed a smoother displacement with respect to the step-like deformation which occurred in the N–S component, could not be due to delays in the GNSS time arrival due, for example, to local changes in the atmospheric conditions.

A summary of the observed coseismic displacements derived from the GNSS data is reported in Table 2.

A solution to the problem of finding the superficial coseismic displacements given a finite rectangular fault in a homogeneous elastic half space was given in 1985 by Okada. This solution is of great interest to seismologists and geodesists, because exploiting the superposition principle even a very complicated rupture process could be simulated. The Okada model (Okada 1985) has been successfully applied to various earthquakes, leading to a better comprehension of the rupture mechanics (e.g., Toda et al. 1998; Caporali et al. 2005). It was also implemented in several inversion (Battaglia et al. 2013; Cheloni et al. 2010) schemes to retrieve the slip which occurred on fault given the observed static displacements.

In our contribution, we used the software Coloumb3 (Toda et al. 2011), distributed by the USGS to construct the fault model, discretized into patches, and to calculate the surface displacements according to the Okada (1985) equations. We test two fault models: one from the USGS, which was derived from only seismological data and the other from Galetzka et al. (2015) which included continuous

GPS and InSAR data in a joint inversion approach. In both cases, the fault model considers an area of about 200 km × 150 km; however, most of the slip occurred in a narrower area of about 150 km × 50 km located in the deeper section of the Main Himalayan Thrust. According to both the inverted models, the slip has not involved the most superficial section of the fault which is locked, as also pointed out by other authors (Mencin et al. 2016).

Both the models reveal a similar strike and dip of the fault: the most evident differences are in the slip magnitude distribution in each patch in which the fault has been discretized.

Figure 3 presents the comparison between the observed data and the modelled values using the fault model from Galetzka et al. (2015). In addition to our stations NAGA and PYRA, we include also the GNSS stations of the UNAVCO project that were used in the study of Galetzka. The stations NAGA and PYRA had not been used by Galetzka et al. (2015). The slip on fault according to Galetzka et al. (2015) is illustrated with the contour representation. In general, the fitting is quite good, especially in terms of direction of movement. The calculated magnitude of the displacement in the NAGA station seems to be a little underestimated, confirming the underestimation of the other two near-fault stations. Changing the fault parameters, such as varying the values of the halfspace rigidity or slightly changing the dip and strike angles, does not increase the fit. For instance, varying the dip angle, from 9 to 7 degrees, increases the fit in the horizontal plane, but at the same time, the vertical component fit is worsened. In the following chapter, an equivalent uniform slip

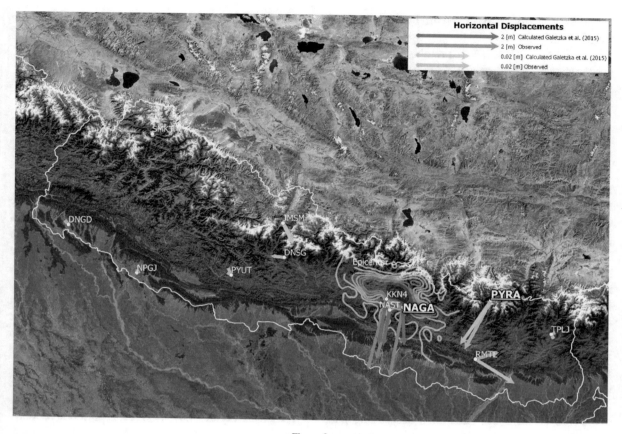

Figure 3
Observed (*orange/light orange*) and modelled (*blue/light blue*) horizontal coseismic displacements in our stations PYRA and NAGA and in the other stations used by Galetzka et al. (2015). The fault model, used for the calculations, is reported with the contours of equal slip: the maximum slip occurred is 5.77 m. The location of the epicenter of 7.8 M_w is plotted with a *red star*

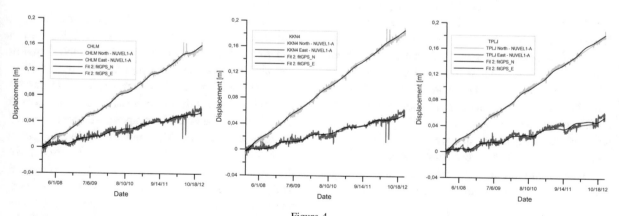

Figure 4
5-year time series of the stations CHLM, KKN4, and TPLJ: cyan lines are the North component, while red the East. From all models, the velocity model NUVEL1-A (DeMets et al. 1990, 1994) has been subtracted. The *black lines* are the fitted linear and periodic components according to Eq. 2

321

model is used to improve the fit by changing the width of the fault.

The USGS solution (2016) on the other hand fits the data worse with respect to the Galetzka et al. (2015) model; the predicted surface displacements are too small in terms of magnitude.

Regarding NAGA station, the USGS solution (2016) is able to explain less than 50% of the observed magnitude, while the Galetzka et al. (2015) model reproduces the magnitudes to almost 60% of the observed signal. The PYRA station shows a better fit with Galetzka et al. (2015) model, with a perfect match between observed and modelled displacements.

An explanation of the discrepancy in the NAGA station could be that Galzetka et al. (2015) invert for the slip occurred on fault using GPS and InSAR interferometry data; interferometry data usually are a smoothed representation of the observed displacements, so locally, discrepancies with GPS data could be present. In the supplementary materials of the article, Galetzka et al. (2015) reported the processing of the InSAR images: the displacement was obtained by repeated observations of the SAR scenes at May 3 and February 22 2015, both on a descending path with 56 m baseline difference. The processing included a 500 m Gaussian low-pass filter to improve the coherence. The final superficial deformation according to InSAR interferometry data shows maximum displacements along Line of Sight of only 1.1–1.2 m against 1.7 m that we have found for NAGA, the almost 1.8 m in NAST, and 1.3 m in KKN4. The model underestimates the InSAR observations in the area, where the stations NAGA and KKN4 are located, as shown in the supplementary materials (Galetzka et al. 2015). The PYRA station on the other hand showed an almost perfect fit in terms of both magnitude and direction of the movement with the modelled coseismic deformation.

From the slip on the fault, we evaluate an average strain released by the earthquake from the formula:

$$\varepsilon = \Delta u / L. \tag{1}$$

where ε is the shear strain along slip direction, Δu is the average slip, and L corresponds in our case to the width of the fault along dip (50 km), since the slip occurred prevalently as a thrust movement. The average slip is 3.5 m. Therefore, the released strain by the Gorkha earthquake amounts to about 7×10^{-5} strain.

4. Long Period GNSS Time Series: Estimation of the Interseismic Deformation

The time series derived from GNSS data allows us to measure the deformation to which an area is subjected and evaluate its strain rate.

According to the UNAVCO database (2016), the area includes over 20 continuous GPS stations in a 500-km radius around the epicenter and all of them present different time coverages, with some of them recently placed (in the last 2 years). Some stations present also temporal gaps in the acquired time series, in some cases up to 2–3 months. Taking into account these limitations, we selected three stations that guarantee at least 5 years of continuous observation with temporal gaps limited to some days/year. The stations selected were: CHLM, KKN4, and TPLJ.

The elaboration process for these stations was almost the same as for the coseismic time series, presented before, but we used a different double-difference technique, with a daily position for stations. Using a parallel approach with the Bernese software, we calculated stations positions over 5 years, from 2008 to 2012 in a fast high precision procedure.

The time series of these four stations are plotted in Fig. 4. The stations CHLM, KKN4, and TPLJ show a coherent North-East secular trend in which different periodic signals are superposed. The periodicities were estimated by spectral analysis. We find that all the stations are subject to yearly and semi-annual periodicities connected to temperature variations and hydrological effects. Multiannual oscillations, centered at the frequency 0.6 1/year, are also evident in the east/west and up/down components. Detailed analysis on the hydrological effects in the Himalaya and their implications in the recovery of the tectonic motions are discussed in Fu and Freymueller (2012). Interestingly, in the Up and East components, pluri-annual oscillations appear. To correctly estimate the secular trend at each station, it is important to isolate the yearly and semi-annual oscillations from the

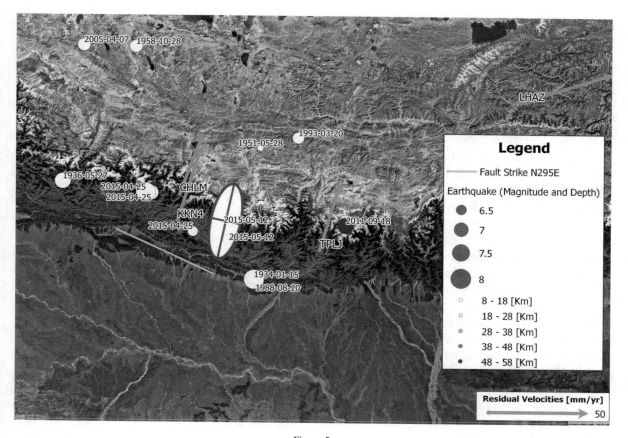

Figure 5
Vectors show the velocity residuals, with respect to EURA plate (NUVEL1-A model). The strain rate ellipse is reported together with the principal axes. Earthquakes from the USGS catalogue in the past 100 years are shown with circles: size is proportional to magnitude, and colors report the hypocenter depth (*white* is shallow earthquake)

GNSS data series. We adopt a widely used approach (Ader et al. 2012) that fits the GNSS time series through the following relation and efficiently recovers the linear trend:

$$x(t) = k + mt + A\cos\left(\frac{\pi}{Ta}t\right) + B\cos\left(\frac{\pi}{Ta}t\right)$$
$$+ C\cos\left(\frac{2\pi}{Ta}t\right) + D\cos\left(\frac{2\pi}{Ta}t\right) \quad (2)$$

where m is the secular trend, A and B are the annual amplitude oscillations, and C and D are the semi-annual components. The parameters are easily calculated through a least squares regression. We did not take advantage of GRACE data for removing the seasonal deformation transients from our GNSS time series, as proposed by other authors (Fu et al. 2012), because the resolution of GRACE data is not adequate to recover possible local variations in the response to hydrologic loads in our GNSS stations configuration. As shown in Devoti et al. (2015), the hydrologic effects are local, at much smaller scale than the 300-km resolution of GRACE, especially in mountain areas.

The three stations show an impressive 35 mm/year movement in the N15°E direction, as evident from Fig. 5. In this figure, we include also the LHAZ station that shows a more pronounced eastward movement, indicating that the station is located on the Eurasia plate and subjected to the eastern extrusion of crustal material (Shin et al. 2015). Obviously, the relative movement between this station and the other 3 is accommodated by the thrusts at the feet of the Himalayan chain and is the cause of the high seismic activity of the region, as demonstrated also by the Gorkha earthquake.

From the velocity field, we could estimate the interseismic strain rate for the area. Using the approach of Shen et al. (1996), we calculated an average strain rate of the area. The method relies on a least squares approach to interpolate the strain variations in a regular grid, given the velocity observations of the GNSS network. In our case, spatial variations of the strain rate could not be estimated, due to the scarcity of GNSS long-term observations of the Nepalese network.

The strain rate is reported in Fig. 5 with the ellipse that shows the maximum axis of deformation: the maximum axis of compression is oriented N15°E with magnitude 132 nanostrain/year, while the minor axis is a dilatation with modulus 2 nanostrain/year. The maximum axis is almost perfectly perpendicular to the strike of Gorkha's fault. The spatial configuration of the GNSS network indicates that the calculated strain rate is representative for an area located 90-km northward of the fault trace in the direction perpendicular to the strike.

To validate our estimate of the strain rate, that is important for recurrence time calculation, we implement a simple fault model (Savage 1983) apt to describe subduction events. In this model, the subducting plate on average moves into the asthenosphere relative to the overriding plate with the same speed of the relative convergence rate of the two plates. The subduction process is divided into a locked upper zone, from the surface to about 40-km depth, and a steadily sliding zone from below and reaching greater depths. The locked zone moves in a stick slip movement, with times of non-sliding interseismic phase, and a sudden rupture at the earthquake occurrence. The earthquake down-dip slip according to this model equals to the horizontal convergence rate multiplied by the time between successive ruptures. The strain at surface assimilates a sawtooth, as it starts at zero after the rupture, increasing linearly to a maximal value at rupture, with constant strain rate during an interseismic cycle. The measured strain and strain rate are modulated by a nonlinear dependence of the position relative to the fault, with maximum horizontal compression just above the mid-line of the fault, and the smallest values in correspondence with the projection of top and bottom of the locked fault. It is of interest to investigate the consistency of the modelled fault, the near-fault strain rate, the amount of down-dip fault slip, the convergence rate, and the rupture recurrence times in the frame of this model. Based on the Savage (1983) model, the following relations should hold.

With u = down-dip slip with rupture (m), T = recurrence time of earthquake on fault (year), B = plate convergence rate (m/year), it should hold:

$$u = BT. \tag{3}$$

The strain rate ε_{xx} as a function of the position perpendicular to the strike (x) is modelled in Savage(1983) by the equation:

$$\varepsilon_{xx} = \frac{2BT}{\pi} s \sin\alpha (s - x\cos\alpha)(x - s\cos\alpha)/D^4 \tag{4}$$

where s is the surface width, α is the dip angle, and $D^2 = x^2 + s^2 - 2xs\cos\alpha$.

From Eq. 4, the strain rate is easily calculated:

$$\dot{\varepsilon}_{xx} = \frac{2B}{\pi} s \sin\alpha (s - x\cos\alpha)(x - s\cos\alpha)/D^4. \tag{5}$$

Setting the parameters of the Gorkha earthquake into Eq. 4, $\alpha = 11°$, $s = 50$ km and assuming $B = 20$ mm/year, as reported in Mencin et al. (2016), we could compare the strain rate calculated from the three GNSS time series and the outcome from the model. Figure 6 shows the strain rate variations according to Savage (1983) along a profile perpendicular to the fault trace (0 m corresponds to fault plane intersecting topographic surface); the red line reports the 132 nanostrain/year estimated for our stations. The intersection of this line with the modelled strain rate occurs at different positions along the profile; an intersection is found at around 70 km, slightly in a southern position with respect to our GNSS estimates (90 km). This means that the measured strain rate is too big for a position of 90-km distance, and it would be correct for a position at 70-km distance. This is coherent to the previous observation that the dislocation model underestimates the observed deformation.

Recovering spatial variations in the strain rate would be interesting in the light of the validation of the above-mentioned model. However, up to now, an

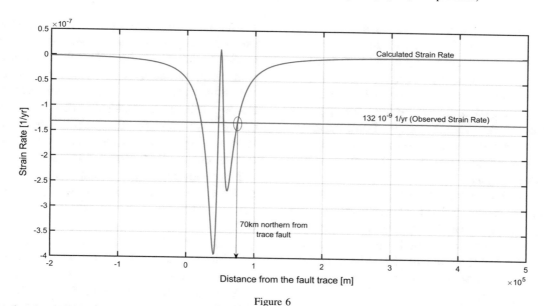

Figure 6

Calculated strain rate profile perpendicular to the fault trace (*blue line*), according to Savage (1983). The observed strain rate is reported with the *red horizontal line*. The intersection occurs at 70 km, while the observed strain rate is located 90-km northward from the fault trace

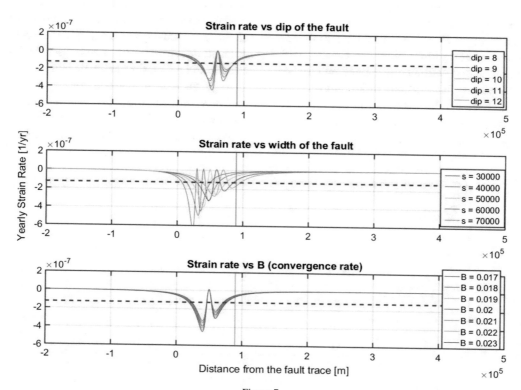

Figure 7

Sensitivity analysis of strain rate in function of dip (°), width (s, in m) of the fault, and plate convergence rate (B, in m/year)

adequate GNSS data set, in terms of spatial and temporal resolutions, is lacking for the Gorkha area. Some new stations have been placed in the Northern and Southern portions of the fault for monitoring the subsequent post-seismic events (Mencin et al. 2016); however, the data could not be used for long-term estimates of the tectonic trends. Considering the GNSS configuration used in the study, we could only make some rough estimates of the errors, with a sensitivity analysis.

Figure 7 shows various strain rate profiles as a function of the dip of the fault, the width, and the convergence rate; the red dashed line is the 132 nanostrain/year level estimated in our GNSS network and the vertical line reports the 90-km distance between the estimated strain rate location and the trace of the fault.

It appears that the width of the fault seems to be a critical parameter, while the other parameters show less influence. At the end considering a fault that is 60-km wide instead of 50 km, a better fit of the observed strain rate is obtained.

The sensitivity analysis proves that our strain rate measurement is compatible with the Gorkha fault model presented and with a convergence rate for the India-Himalaya (B) that amounts to 20 mm/year.

From the average slip occurred on the fault, we could estimate a recurrence time according to the Savage model of 235 year.

In Fig. 5, we plot together with the strain ellipse from our analysis, the fault trace, and the historical seismicity ($M_w > 6.5$) from the USGS database. The fault trace extends over the length marked by the 2015 aftershock sequence. The area was not active in the past 100 years: earthquakes occurred in neighboring segments of the subducting plate. In 1934–1936 both segments, west and east of the 2015 rupture broke with M_w 8 (east, 1934) and M_w 7 (west, 1936) earthquakes. The eastward fault broke again in 1988 with a smaller and deeper event (M_w 6.5). The 2015 sequence broke the segment joining these two faults.

Thapa and Guoxin (2013) compiled a catalogue of the historical seismicity of the whole Nepal: concerning the Gorkha area, they found that it was struck by four >7.5 M_s events in 1255, 1408, 1681, and 1810. Such data appear to be consistent with our estimates from strain rate analysis. Ader et al. (2012) also estimate the recurrence time in the whole Himalaya chain. Taking advantage of the Savage model (1983), they calculate the slip deficit occurring along the Himalayan thrust and consequently the moment deficiency. The authors used Molnar (1979) relation to predict, for the whole Himalayan chain, an average 50-year recurrence for $M_w = 8$. The discrepancy with our estimate could be explained by the different areas considered in the study: our analysis focuses on a specific seismogenic structure which

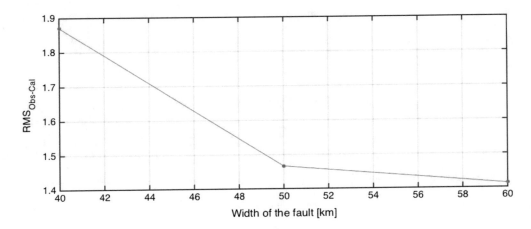

Figure 8

Root mean square residual (m) between observed and calculated displacements summed for all stations as a function of the width of the fault. Increasing the width from 50 to 60 km leads to a slightly better accordance with observed displacement field

caused the Gorkha earthquake, while Ader et al. (2012) conducted a large scale study, giving repeat times for the entire Himalayan arch.

However, we remark that an improvement of the strain rate estimate will be possible in the future by considering the GNSS stations located around the fault that were placed just immediately after the earthquake (Gualandi et al. 2016; Mencin et al. 2016).

We test whether a wider fault gives a better fit also to the GNSS coseismic movement. The above finding regarding the strain rate would suggest that the near-fault observations are better explained by a fault that is 10 km wider than the present model. We compute displacement for an equivalent model, with a homogeneous fault slip, keeping the other fault parameters, such as strike, dip, and rake, the same as the previously used model. In Fig. 8, the root mean square (RMS) error between observed and calculated displacements for different widths of the fault is reported. A decrease in error is seen as the fault width slightly increases from 40 to 60 km.

5. Conclusions

In this contribution, we analyzed the GNSS coseismic signal due to the catastrophic 7.8 M_w Gorkha earthquake which occurred on 25th April 2015. Two GNSS stations were added to the geodetic network present in the area that is available from the UNAVCO database. The NAGA station increases the amount of strong-motion near-fault stations, allowing a robust estimate of the coseismic displacement. The NAGA station confirms that the present earthquake rupture model underestimates the strong-motion dislocation, calling for a coseismic slip on a slightly wider fault. Both these two stations PYRA and NAGA displayed a coseismic offset due to the fault movement. The PYRA station, located next to the Pyramid observatory on Everest, showed signals close to the noise level of GNSS observations. However, a first comparison with meteorological data excluded the effect of a change in the weather conditions in the observed displacements. Hence, we verified the compatibility of the observed

deformation field with the modelled displacement field calculated using the Okada approach. Two fault solutions have been tested: a major concordance is found for the Galetzka et al. 2015 model that exploits also geodetic data, with respect to the only seismological solution of the USGS. Some differences in terms of magnitude between observed and modelled data are present too: it seems that the predicted movement is a little underestimated for the strong-motion near-fault stations. This effect could be explained by the fact that the model had used the InSAR interferometry data in the inversion scheme that could smooth the retrieved slip on the fault.

From the slip on the fault, we derived the shear strain released by the earthquake. The processing of long period data from other GNSS stations located in the area allowed us to estimate the strain rate to which the area is subjected and compare it with the predictions of a model that describes the yearly accumulation of strain due to a locked fault (Savage 1983). Since the GNSS network for the Gorkha area includes only few stations, spatial variations of the strain rate could not be retrieved. However, the estimates of the model are in accordance with the Gorkha fault parameters and the yearly convergence rate, except that for the coseismic observations, the width of the fault should be a bit greater. From the convergence rate, we established a recurrence time of about 235 years.

The comparison with the recurrence time of similar earthquakes in the area, derived from seismological catalogues and historical seismicity, let us infers that the recurrence time of the earthquake from geodetic analysis is realistic. Better estimates of the strain accumulation and release, and retrieving the local spatial variation of the strain rate could only be possible if a dense and continuous geodetic network is established and maintained in the Gorkha area and extended also to the Indian region.

Acknowledgements

Niraji Manandhar is gratefully thanked for providing GNSS data for the Nagarkot station. Gianpietro Verza from EvK2-CNR is acknowledged for giving meteorological data. We acknowledge the UNAVCO consortium for the GNSS data of the stations KIT3, KKN4, LAHZ, TPLJ, and CHLM. The reviewers are kindly acknowledged for their fruitful comments that contributed in improving the quality of the paper.

REFERENCES

Ader, T., Avouac, J.-P., Liu-Zeng, J., Lyon-Caen, H., Bollinger, L., Galetzka, J., et al. (2012). Convergence rate across the Nepal Himalaya and interseismic coupling on the Main Himalayan Thrust: implications for seismic hazard. *Journal of Geophysical Research, 117,* B04403. doi:10.1029/2011JB009071.

Arora, B. R., Bansal, B. K., Prajapati, S. K., Sutur, A. K., & Nayak, S. (2017). Seismotectonics and seismogenesis of M_w 7.8 Gorkha Earthquake and its Aftershocks. *Journal of Asian Earth Sciences, 133,* 2–11. doi:10.1016/j.jseaes.2016.07.018.

Avouac, J.-P., Meng, L., Wei, S., Wang, T., & Ampuero, J.-P. (2015). Lower edge of locked main Himalayan thrust unzipped by the 2015 Gorkha earthquake. *Nature Geoscience, 8,* 708–711. doi:10.1038/ngeo2518.Battaglia.

Battaglia, M., Cervelli, P. F., & Murray, J. R. (2013). dMODELS: a MATLAB software package for modeling crustal deformation near active faults and volcanic centers. *Journal of Volcanology and Geothermal Research, 254,* 1–4.

Böhm, J., Werl, B., & Schuh, H. (2006). Troposphere mapping functions for GPS and very long baseline interferometry from European Centre for Medium-Range Weather Forecasts operational analysis data. *Journal of Geophysical Research: Solid Earth, 111,* B02406. doi:10.1029/2005JB003629.

Caporali, A., Braitenberg, C., & Massironi, M. (2005). Geodetic and hydrological aspects of the Merano earthquake of 17 July 2001. *Journal of Geodynamics, 39,* 317–336.

Chaulagain, H., Rodrigues, H., Silva, V., Spacone, E., & Varum, H. (2015). Seismic risk assessment and hazard mapping in Nepal. *Natural Hazards, 78*(1), 583. doi:10.1007/s11069-015-1734-6.

Cheloni, D., D'Agostino, N., D'Anastasio, E., Avallone, A., Mantenuto, S., Giuliani, R., et al. (2010). Coseismic and initial post-seismic slip of the 2009 M_w 6.3 L'Aquila earthquake, Italy, from GPS measurements. *Geophysical Journal International, 181*(3), 1539–1546.

DeMets, C., Gordon, R. G., Argus, D. F., & Stein, S. (1990). Current plate motions. *Geophysical Journal International, 101,* 425–478.

DeMets, C., Gordon, R. G., Argus, D. F., & Stein, S. (1994). Effect of the recent revisions to the geomagnetic reversal timescale. *Geophysical Research Letters, 21,* 2191–2194.

Denolle, M. A., Fan, W., & Shearer, P. M. (2015). Dynamics of the 2015 *M* 7.8 Nepal earthquake. *Geophysical Research Letters, 42,* 7467–7475. doi:10.1002/2015GL065336.

Devoti, R., Zuliani, D., Braitenberg, C., Fabris, P., & Grillo, B. (2015). Hydrologically induced slope deformations detected by GPS and clinometric surveys in the Cansiglio Plateau, Southern Alps. *Earth and Planetary Science Letters, 419,* 134–142.

Fan, W., & Shearer, P. M. (2015). Detailed rupture imaging of the 25 April 2015 Nepal earthquake using teleseismic P waves. *Geophysical Research Letters, 42,* 5744–5752.

Fu, Y., & Freymueller, J. T. (2012). Seasonal and long-term vertical deformation in the Nepal Himalaya constrained by GPS and GRACE measurements. *Journal of Geophysical Research, 117,* B03407. doi:10.1029/2011JB008925.

Fu, Y., Freymueller, J. T., & Jensen, T. (2012). Seasonal hydrological loading in southern Alaska observed by GPS and GRACE. *Geophysical Research Letters, 39,* L15310. doi:10.1029/2012GL052453.

Galetzka, J., Melgar, D., Genrich, J. F., Geng, J., Owen, S., Lindsey, E. O., et al. (2015). Slip pulse and resonance of the Kathmandu Basin during the 2015 Gorkha earthquake, Nepal. *Science, 349,* 1091–1095.

Grandin, R., Vallée, M., Satriano, C., Lacassin, R., Klinger, Y., Simoes, M., et al. (2015). Rupture process of the M_w = 7.9 2015 Gorkha earthquake (Nepal): insights into Himalayan megathrust segmentation. *Geophysical Research Letters, 42,* 8373–8382.

Gualandi, A., Avouac, J-P., Galetzka, J., Genrich, J.F., Blewitt, G., Adhikari,L.B., Koirala, B.P., Gupta, R., Upreti, B.N., Pratt-Sitaula, B., Liu-Zeng, J. (2016). Pre- and post- seismic deformation related to the 2015, M_w7.8 Gorkha earthquake, Nepal, *Tectonophysics,* in press, doi:10.1016/j.tecto.2016.06.014.

Mencin, D., Bendick, R., Upreti, B. N., Adhikari, D. P., Gajurel, A. P., Bhattarai, R.R., Shrestha, H.R., Bhattarai, T. N., Manandhar, N., Galetzka, J., Knappe, E., Pratt-Sitaula, B., Aoudia, A., Bilham, R. (2016). Himalayan strain reservoir inferred from limited afterslip following the Gorkha earthquake, Nature Geoscience, 9, 533-537, doi:10.1038/ngeo2734.

Molnar, P. (1979). Earthquake recurrence intervals and plate tectonics. *Bulletin of the Seismological Society of America, 29,* 211–229.

Okada, Y. (1985). Surface deformations due to shear and tensile faults in a halfspace. *Bull. Seism. Soc. Am., 75*(4), 1135–1154.

Rajendran, K., & Rajendran, C. P. (2011). Revisiting the earthquake sources in the Himalaya: perspectives on past seismicity. *Tectonophysics, 504*(1–4), 75–88.

Savage, J. C. (1983). A dislocation model of strain accumulation and release at a subduction zone. *Journal of Geophysical Research, 88,* 4984–4996.

Shen, Z.-K., Jackson, D. D., & Ge, B. X. (1996). Crustal deformation across and beyond the Los Angeles basin from geodetic measurements. *Journal of Geophysical Research, 101*(B12), 27957–27980.

Shin, Y. H., Shum, C. K., Braitenberg, C., Lee, S. M., Na, S.-H., Choi, K. S., et al. (2015). Moho topography, ranges and folds of Tibet by analysis of global gravity models and GOCE data. *Scientific Reports, 5,* 1–7. doi:10.1038/srep11681.

Sreejith, K. M., Sunil, P. S., Agrawal, R., Saji, A. P., Ramesh, D. S., & Rajawat, A. S. (2016). Coseismic and early postseismic deformation due to the 25 April 2015, M_w 7.8 Gorkha, Nepal, earthquake form InSAR and GPS measurements. *Geophysical Research Letters, 43,* 3160–3168.

Thapa, D. R., & Guoxin, W. (2013). Probabilistic seismic hazard analysis in Nepal. *Earthquake Engineering and Engineering Vibration, 12,* 577–586.

Toda, S., Stein, R.S., Reasenberg, P.A., Dieterich, J.H. (1998). Stress transferred by the M_w = 65 Kobe, Japan, shock: Effect on aftershocks and future earthquake probabilities. J Geophys Res 103(24), 543–24,565.

Toda, S., Stein, R.S., Sevilgen, V., Lin, J. (2011). Coulomb 3.3 Graphic-rich deformation and stress-change software for earthquake, tectonic, and volcano research and teaching—user guide. U.S. Geological Survey Open-File Report. http://pubs.usgs.gov/of/2011/1060/. Accessed 10 Dec 2016.

UNAVCO (2016) ftp://data-out.unavco.org/pub/rinex, http://www.unavco.org/data/gps-gnss/data-access-methods/dai2/app/dai2.html#. Accessed 10 Dec 2016.

USGS. (2016). Event page M7.8–36 km E of Khudi, Nepal. https://earthquake.usgs.gov/earthquakes/eventpage/us20002926#executive. Accessed 10 Dec 2016.

Wang, K., & Fialko, Y. (2015). Slip model of the 2015 M_w 7.8 Gorkha (Nepal) earthquake from inversions of ALOS-2 and GPS data. *Geophysical Research Letters, 42,* 7452–7458.

(Received December 30, 2016, revised July 28, 2017, accepted July 31, 2017, Published online August 4, 2017)

Correction to: Geodynamics and Earth Tides Observations from Global to Micro Scale

Carla Braitenberg and Giuliana Rossi

Correction to:
C. Braitenberg, G. Rossi (eds.), *Geodynamics and Earth Tides Observations from Global to Micro Scale,* **Pageoph Topical Volumes, https://doi.org/10.1007/978-3-319-96277-1**

The below mentioned chapters were published as Open access with incorrect Copyright holder name as "© Springer Nature Switzerland AG". This has now been updated as "© The Author(s)" in these chapters. Accordingly, the Cover and the FrontMatter have been updated to reflect these changes.

1. Interferometric Water Level Tilt Meter Development in Finland and Comparison with Combined Earth Tide and Ocean Loading Models
2. Multichannel Singular Spectrum Analysis in the Estimates of Common Environmental Effects Affecting GPS Observations
3. A Filtering of Incomplete GNSS Position Time Series with Probabilistic Principal Component Analysis

---

The updated online versions of the chapters can be found at
https://doi.org/10.1007/978-3-319-96277-1
https://doi.org/10.1007/978-3-319-96277-1_7
https://doi.org/10.1007/978-3-319-96277-1_17
https://doi.org/10.1007/978-3-319-96277-1_19